de Gruyter Studies in Mathematics 6

Editors: Heinz Bauer · Peter Gabriel

Ulrich Krengel

# Ergodic Theorems

With a Supplement by Antoine Brunel

Walter de Gruyter
Berlin · New York 1985

*Author*
Dr. Ulrich Krengel
Professor of Mathematics
Universität Göttingen



*Library of Congress Cataloging in Publication Data*

88-2653

Krengel, Ulrich, 1937–
Ergodic theorems.
(De Gruyter studies in mathematics ; 6)
Bibliography: p.
Includes index.
1. Ergodic theory. I. Brunel, Antoine. II. Title.
III. Series.
QA313.K74 1985 515.4′2 85-4457
ISBN 0-89925-024-6 (U.S.)

*CIP-Kurztitelaufnahme der Deutschen Bibliothek*

**Krengel, Ulrich:**
Ergodic theorems / Ulrich Krengel. With a suppl. by Antoine Brunel.
– Berlin ; New York : de Gruyter 1985.
(De Gruyter studies in mathematics ; 6)
ISBN 3-11-008478-3

NE: GT


Cover design: Rudolf Hübler, Berlin. Typesetting and Printing: Tutte Druckerei GmbH, Salzweg-Passau. Binding: Lüderitz & Bauer, Berlin.

Lesen heißt borgen,
daraus erfinden abtragen.

*G. C. Lichtenberg,*
*Sudelbuch F*

# Preface

The study of ergodic theorems is the oldest branch of ergodic theory. It was started in 1931 by von Neumann and Birkhoff, having its origins in statistical mechanics. While new applications to mathematical physics continued to come in, the theory soon earned its own rights as an important chapter in functional analysis and probability.

So far, a comprehensive treatment has been neglected, and this book tries to provide it. Most of its material has not appeared in any other book. This applies even to older results, but the main body of the results is less than twenty years old and several interesting topics have just been added in the last decade.

Roughly speaking we ask: When do averages of quantities generated in a stationary manner converge? In the classical situation the stationarity is described by a measure preserving transformation $\tau$, and one considers averages taken along a sequence $f, f \circ \tau, f \circ \tau^2, \ldots$ for integrable $f$. This corresponds to the probabilistic concept of stationarity. More generally, $\tau$ can be replaced by an operator $T$ in a function space and $f \circ \tau^i$ by $T^i f$. As $T^i$ is the result of the iterated action of the *same* operator we again have some kind of stationarity. Generalizing further, we study semigroups $\{T_g, g \in G\}$ of operators and limits of averages of $T_g f$ over subsets $I_n \subset G$. The term "ergodic theorem" has been used by some authors for quite distinct limit theorems, but we reserve it for convergence theorems dealing with such averages and for their close relatives. This meaning seems most widely accepted. Among the relatives we count subadditive ergodic theorems, local ergodic theorems (generalizing the differentiation of integrals), ratio ergodic theorems and ergodic theorems for information.

The modes of convergence under consideration mostly are norm convergence for "mean" ergodic theorems, and convergence almost everywhere for "individual" (or "pointwise") ergodic theorems. Recently, weak convergence has gained importance for nonlinear ergodic theorems and almost uniform convergence for ergodic theorems in von Neumann algebras. Convergence in distribution will not be considered. Typically, it applies to renormed averages rather than to averages, and it requires different tools.

I have tried to make the various parts of the book independently readable. The reader should start with any section which is of interest to him. He will then notice which previous results enter and find that often just a few will suffice. Of course, this implies some redundancy. On the other hand, I hope that this way large portions of the book can serve as a textbook, and that this approach will render this monograph useful and readily accessible for non-specialists.

I have not always given the shortest proof. Sometimes a longer proof seemed more transparent. Another aspect has been the wish to introduce a variety of methods. In some "additive" ergodic theorems the proof of convergence could have been simplified by the use of subadditive theorems. However, the longer additive arguments give access to an evaluation of the limit.

I presuppose knowledge of basic measure theory, and, for many sections, some functional analysis. But I tried to help the non-experts with references even for standard theorems.

Surely this book is biased towards my personal interests and even more so since I have included a number of new results and proofs. But I also tried hard not to miss any important result and to give it fair coverage. If a good presentation existed I sometimes may have just quoted it. I apologize in advance to anyone whose contributions were overlooked.

Concerning convergence almost everywhere the book of Stout [1974] covers many of the themes in the complement of this book.

Most sections end with Notes containing additional information. But I did include credits in the main text when it seemed possible without much delay.

Theorems, lemmas, definitions etc. are numbered consecutively in each section. A quotation "theorem 2.3.4" refers to chapter 2, section 3, theorem 3.4. A quotation "theorem 3.4" refers to the current chapter.

My indebtedness extends to many. First, I would like to thank my teacher, Konrad Jacobs, for generating my interest in ergodic theory, for giving me a sound introduction, and for suggesting a fertile area of research. I also owe much to Louis Sucheston and his infectious enthusiasm. I am very grateful to M. Lin, A. del Junco, Y. Derriennic, G. Keller, M. Akcoglu, R. Nagel, M. Mathieu, H.-J. Borchers, A. Bellow, R. Jajte, W. Takahashi, J. Fritz, M. Keane, M. Denker and many others for useful hints and encouragement. Martina Hochhaus helped with the bibliography. Marrie Powell contributed her unusual skill at mathematical typing. Special thanks go to Heinz Bauer for inviting the book into this series.

The idea for this book arose in 1976 at the end of a pleasant and interesting sabbatical I spent at the University of Paris VI. Antoine Brunel and I agreed that a book covering the whole spectrum of ergodic theorems was badly needed and planned to write it jointly. Unfortunately, grave personal reasons prevented this. I am the more grateful to Antoine Brunel for writing the supplement on Harris processes, a topic to which he contributed so much.

I devote this book to my wife Beate, and to Jeannette Brunel who died of cancer in 1981. They provided the environment for us in which devotion to mathematical work was possible.

Göttingen, April 1985 Ulrich Krengel

# Contents

# Chapter 1: Measure preserving and null preserving point mappings

We begin with the classical ergodic theorems for measure preserving transformations and with their role in the theory of stationary processes. Then the subadditive ergodic theorem is proved. We use it to derive the multiplicative ergodic theorem of Oseledeč, a powerful tool in the study of dynamical systems. Recurrence is discussed for the wider class of null preserving transformations. The dominated ergodic theorem provides important estimates of the supremum of averages. Some topics on measure preserving transformations like weak mixing, multiparameter semigroups, vector valued ergodic theorems, and the ergodic theorem for information are postponed although they could be read right away.

## § 1.1 Von Neumann's mean ergodic theorem, ergodicity

**1. Definitions and examples.** A *measurable space* $(\Omega, \mathscr{A})$ consists of a non empty set $\Omega$ and a $\sigma$-algebra $\mathscr{A}$, i.e., a non empty class of subsets of $\Omega$, closed under the formation of complements and countable unions. A *measure* $\mu$ on $(\Omega, \mathscr{A})$ is a non negative set function $\mu$: $\mathscr{A} \to \mathbb{R}^+ \cup \{\infty\}$ with $\mu(\emptyset) = 0$ which is $\sigma$-additive. The triple $(\Omega, \mathscr{A}, \mu)$ is called a *measure space*. In the case $\mu(\Omega) = 1$, $\mu$ is called a *probability measure*, and $(\Omega, \mathscr{A}, \mu)$ a *probability space*.

Let $(\Omega, \mathscr{A})$ and $(\Omega', \mathscr{A}')$ be measurable spaces. A mapping $\tau$: $\Omega \to \Omega'$ is called *measurable* (or $\mathscr{A} - \mathscr{A}'$-measurable) if $\tau^{-1}\mathscr{A}' = \{\tau^{-1}A' \colon A' \in \mathscr{A}'\} \subset \mathscr{A}$. $\tau$ is called a *homomorphism* of $(\Omega, \mathscr{A}, \mu)$ into $(\Omega', \mathscr{A}', \mu')$ if $\tau$ is measurable and the measure $\mu \circ \tau^{-1}$ defined on $\mathscr{A}'$ by $(\mu \circ \tau^{-1})\,(A') = \mu(\tau^{-1}A')$ agrees with $\mu'$.

$\tau$: $\Omega \to \Omega$ is called *measure preserving* if it is measurable and satisfies $\mu \circ \tau^{-1} = \mu$. In this case $\mu$ is called *invariant* or $\tau$-invariant. A measure preserving transformation will also be called *endomorphism* (of $(\Omega, \mathscr{A}, \mu)$). If $\tau$ is an invertible endomorphism of $\Omega$ onto $\Omega$ for which $\tau^{-1}$ is an endomorphism, then $\tau$ is called an *automorphism*.

Examples of endomorphisms of measure spaces turn up in many branches of mathematics. Perhaps the simplest examples are translations in $\mathbb{R}^n$ and rotations $x \to x + \alpha$ (mod 1) in $[0, 1[$ with Lebesgue measure and, more generally, translations in locally compact groups $\Omega$ with left Haar measure.

Another class of measure preserving transformations in the same space is given by the continuous grouptheoretic automorphisms of $\Omega$.

A simple number theoretic example can be defined via the expansion of $\omega \in \Omega = [0,1] \setminus \mathbb{Q}$ as a continued fraction: Identify $\omega$ with $(\omega_1, \omega_2, \ldots) \in \mathbb{N}^{\mathbb{N}}$, where

$$\omega = \frac{1|}{|\omega_1} + \frac{1|}{|\omega_2} + \frac{1|}{|\omega_3} + \ldots .$$

Now $\omega \to \tau\omega = (\omega_2, \omega_3, \ldots)$ defines an endomorphism in $(\Omega, \mathscr{A}, \mu)$ when $\mu$ is the measure with density $(1+x)^{-1}$ with respect to Lebesgue measure; see e.g. Billingsley [1965].

If $\Omega$ is a compact Hausdorff space and $\tau: \Omega \to \Omega$ continuous, there always exists an invariant measure on the $\sigma$-algebra $\mathscr{A}$ of Baire sets. As we want to return to this example later and a proof is simple if we make use of Banach limits, we take this liberty.

A *Banach limit* $L$ is a linear functional defined on the space $\ell_\infty$ of bounded sequences $x = (x_0, x_1, \ldots)$ of real numbers such that

(i) $L(x) \geqq 0$ holds for all $x$ with $x_i \geqq 0 \quad (i = 0, 1, \ldots)$,

(ii) $L((x_1, x_2, x_3, \ldots)) = L((x_0, x_1, x_2, \ldots)) \quad (x \in \ell_\infty)$ and

(iii) $L((1, 1, 1, \ldots)) = 1$.

Banach limits exist; see theorem 3.4.1. Using a fixed Banach limit $L$ and a fixed $\omega \in \Omega$, we can define a positive linear functional $\mu_\omega$ on the space $C(\Omega)$ of continuous functions on $\Omega$ by $\mu_\omega(f) = L((f(\tau^n\omega))_{n=0}^\infty)$.

By the Riesz representation theorem (see Bauer [1981]) this linear functional is of the form $\mu_\omega(f) = \int f(\eta)\,\mu_\omega(d\eta)$ for some measure $\mu_\omega$ on $\mathscr{A}$. The properties of $L$ imply that $\mu_\omega$ is an invariant probability measure.

The classical examples of endomorphisms which have originally motivated the search for ergodic theorems arise in statistical mechanics. A theorem of Liouville asserts the invariance of the $6r$-dimensional Lebesgue measure under the Hamiltonian flow in phase space, see Khintchine [1949].

Still other examples can be found in § 1.4. A rich collection of examples is given in the book of Cornfeld, Fomin, Sinai [1982].

If $(\Omega, \mathscr{A}, \mu)$ is a measure space, $\mathscr{L}_p = \mathscr{L}_p\,(\Omega, \mathscr{A}, \mu)$ denotes the space of real or complex valued measurable functions $f$ with $\|f\|_p := (\int |f|^p\,d\mu)^{1/p} < \infty$, $(1 \leqq p < \infty)$. $\mathscr{L}_\infty$ denotes the space of measurable functions for which $\|f\|_\infty := \inf\{\alpha > 0: \mu(\{|f| > \alpha\}) = 0\}$ is finite. We shall also write $\mathscr{L}_p(\mu)$ or $\mathscr{L}_p(\mathscr{A})$ if we want to mention the underlying measure or $\sigma$-algebra. $\{|f| > \alpha\}$ is a shorthand for $\{\omega \in \Omega: |f(\omega)| > \alpha\}$. We shall use such a shorthand notation also for other sets defined by properties of functions. Frequently we abreviate even further and write $\mu(|f| > \alpha)$ for $\mu(\{|f| > \alpha\})$.

Recall that $f = g$ (mod $\mu$) means $\mu(f \neq g) = 0$, and that equality mod $\mu$ is an equivalence relation in the space of measurable functions and in each space $\mathscr{L}_p$. The space $L_p = L_p(\Omega, \mathscr{A}, \mu)$ of equivalence classes in $\mathscr{L}_p$ is a Banach space with norm $\|\cdot\|_p$.

Most of the time we shall not distinguish between elements $f \in L_p$ and their representatives. In all statements involving only a sequence of elements of $\mathscr{L}_p$ or $L_p$ and holding only mod $\mu$, the difference is irrelevant.

The function $1_A$ which is 1 on $A \subset \Omega$ and 0 on the complement $A^c$ is called *indicator function* of $A$. $A$ and $B$ are equal mod $\mu$ if the measure of their *symmetric difference* $A \Delta B = (A \cap B^c) \cup (A^c \cap B)$ is zero. Again, sets which are equal mod $\mu$ will usually not be distinguished. $\mathbb{1}$ denotes the function $\equiv 1$.

A measure $\nu$ on $\mathscr{A}$ is called *$\mu$-continuous* if $\nu(A) = 0$ holds for all $A \in \mathscr{A}$ with $\mu(A) = 0$. We then write $\nu \ll \mu$. Call $\nu$ *equivalent* to $\mu (\nu \sim \mu)$ if $\nu \ll \mu \ll \nu$.

If $(\Omega, \mathscr{A})$ is a measurable space and $\tau: \Omega \to \Omega$ a measurable mapping, the operator $f \to Tf := f \circ \tau$ is called *composition with* $\tau$. It is a linear operator in the space of measurable functions on $\Omega$.
A measurable map $\tau: \Omega \to \Omega$ with $\mu \circ \tau^{-1} \ll \mu$ is called *null preserving*. An invertible null preserving map $\tau$ of $\Omega$ onto $\Omega$ for which also $\tau^{-1}$ is null preserving is called *nonsingular*. For null preserving $\tau$ the composition operator is well defined in the spaces of equivalence classes because then $f_1 = f_2$ mod $\mu$ implies $f_1 \circ \tau = f_2 \circ \tau$ mod $\mu$.

We shall make frequent use of the following notions from functional analysis:

**Definition 1.1.** If $\mathfrak{X}$, $\mathfrak{Y}$ are normed vector spaces and $T$ a linear operator mapping $\mathfrak{X}$ into $\mathfrak{Y}$, the *norm* $\|T\|$ of $T$ is given by

$$\|T\| = \sup_{\|f\| \leqq 1} \|Tf\|.$$

$T$ is called *bounded* if $\|T\|$ is finite, and $T$ is called a *contraction* if $\|T\| \leqq 1$. $T$ is called an *isometry* if $\|Tf\| = \|f\|$ holds for all $f$.

If a partial order is defined in $\mathfrak{X}$, the *positive cone* $\{f \in \mathfrak{X}: f \geqq 0\}$ of $\mathfrak{X}$ is denoted by $\mathfrak{X}^+$. $T$ is called *positive* if $T\mathfrak{X}^+ \subset \mathfrak{Y}^+$. In the case $\mathfrak{X} = \mathfrak{Y}$ we speak of a linear operator (contraction, ...) in $\mathfrak{X}$. $\mathscr{L}(\mathfrak{X})$ denotes the set of bounded linear operators in $\mathfrak{X}$. I denotes the identity operator.

If $\tau$ is an endomorphism of $(\Omega, \mathscr{A}, \mu)$ the composition with $\tau$ is a positive isometry in each $L_p$. If $\tau$ is only null preserving or nonsingular, it still is a positive contraction in $L_\infty$, but it need not map $L_p$ into $L_p$ for $1 \leqq p < \infty$.

**2. Contractions in Hilbert space.** We denote the scalar product of two elements $f$, $h$ of a Hilbert space $\mathfrak{H}$ by $\langle f, h \rangle$. The dual $T^*$ of a bounded linear operator $T$ in $\mathfrak{H}$ is characterized by $\langle Tf, h \rangle = \langle f, T^* h \rangle$, valid for all $f$ and $h$.

**Lemma 1.2.** *If $T: \mathfrak{H} \to \mathfrak{H}$ is a contraction in a real or complex Hilbert space and $g \in \mathfrak{H}$, then $g = Tg$ holds if and only if $g = T^* g$.*

*Proof.* If, for some $g$, $\langle g, Tg \rangle = \|g\|^2$, then $\langle g, Tg \rangle$ is real and $\langle g, Tg \rangle = \langle Tg, g \rangle$. We then get

$$\|Tg-g\|^2 = \langle Tg-g, Tg-g\rangle = \|Tg\|^2 + \|g\|^2 - 2\langle g, Tg\rangle$$
$$\leqq 2\|g\|^2 - 2\|g\|^2 = 0.$$

Thus, $g = Tg$ is equivalent to $\|g\|^2 = \langle g, Tg\rangle = \langle T^*g, g\rangle$. Applying this equivalence to $T^*$ the identity $g = T^*g$ follows. □

**Lemma 1.3.** *Let* $\mathcal{T}$ be a family of contractions $T$ *in a Hilbert space* $\mathfrak{H}$. *Then the orthogonal complement* $F^\perp$ *of* $F = \{g \in \mathfrak{H}: Tg = g \forall T \in \mathcal{T}\}$ *is the closure of the subspace* $N$ *spanned by* $\{h - Th: h \in \mathfrak{H}, T \in \mathcal{T}\}$.

*Proof.* Write $g \perp \mathfrak{H}_0$ if $g$ is orthogonal to a subspace $\mathfrak{H}_0$. Now $g \perp N \Leftrightarrow \langle g, (T-I)h\rangle = 0 \forall h \in \mathfrak{H}, T \in \mathcal{T} \Leftrightarrow \langle T^*g - g, h\rangle = 0 \forall h, T \Leftrightarrow T^*g = g \forall T \Leftrightarrow Tg = g \forall T \Leftrightarrow g \in F$.

Thus $N$, and hence also its closure $cl\, N$, is orthogonal to the closed subspace $F$. As any vector orthogonal to $cl\, N$ belongs to $F$ we have $F^\perp = cl\, N$. □

The following notation will be used frequently for linear operators $T$:

$$S_n f := S_n(T)f := \sum_{i=0}^{n-1} T^i f, \quad A_n f := A_n(T)f := n^{-1} S_n(T)f.$$

If the operator under consideration is $S$ we may also write $A_n f$ for $A_n(S)f$. If $T$ is the composition with $\tau$ we sometimes write $A_n(\tau)$ for $A_n(T)$.

**Theorem 1.4** (Mean ergodic theorem of von Neumann). *If $T$ is a contraction in a Hilbert space $\mathfrak{H}$, and $P$ the projection on $F = \{g \in \mathfrak{H}: Tg = g\}$, then $A_n f$ converges in norm to $Pf$ for $f \in \mathfrak{H}$, $(n \to \infty)$.*

*Proof.* If $f = (T-I)h$ for some $h$ then

$$\|A_n f\| = n^{-1}\|Th - h + T^2h - Th + \ldots + T^n h - T^{n-1}h\|$$
$$= n^{-1}\|T^n h - h\| \leqq 2n^{-1}\|h\| \to 0.$$

By approximation this yields $\|A_n f\| \to 0$ for all $f$ in the closure of $(T-I)\mathfrak{H}$, and now the assertion follows from Lemma 1.3. □

If $\tau$ is an endomorphism in $(\Omega, \mathcal{A}, \mu)$ and $p = 2$, it follows that for any $f \in L_p$ there is an $\bar{f} \in L_p$ with $\bar{f} = \bar{f} \circ \tau$ and $\|\bar{f}\|_p \leqq \|f\|_p$ such that $\|A_n(\tau)f - \bar{f}\|_p \to 0$ $(n \to \infty)$. Approximation arguments (or theorem 2.1.1) show that this remains true for $1 < p < \infty$. If $\mu(\Omega) = \infty$ the analogous statement does not always hold for $p = 1$; e.g., take $\Omega = \mathbb{R}^1$, $\mu =$ Lebesgue measure, $\tau\omega = \omega + 1$, and $f = 1_{[0,1[}$. For $f \in L_\infty$, $\|A_n(\tau)f - \bar{f}\|_\infty \to 0$ need not even hold in the case $\mu(\Omega) < \infty$. One can take $\Omega = [0, 1[$, $\mu =$ Lebesgue measure, $\tau\omega = \omega + \alpha \pmod 1$ with an irrational $\alpha$, and a suitable (highly discontinuous) $f \in L_\infty$. We leave this as an exercise to the reader.

**3. Absorbing and invariant sets.** In many cases the property $\bar{f} = \bar{f} \circ \tau$ of the limit function can be used to find the explicit form of $\bar{f}$. For future use we discuss the relevant notions in some more generality than needed here.

**Definition 1.5.** Let $\tau$ be null preserving in $(\Omega, \mathscr{A}, \mu)$. A set $A \in \mathscr{A}$ is called $\tau$-*absorbing* if $A \subset \tau^{-1} A$, $\tau$-absorbing mod $\mu$ if $\mu(A \setminus \tau^{-1} A) = 0$, $\tau$-*invariant* if $\tau^{-1} A = A$, and $\tau$-*invariant* mod $\mu$ if $\mu(A \Delta \tau^{-1} A) = 0$. An $\mathscr{A}$-measurable function $f$ is called $\tau$-*invariant* if $f = f \circ \tau$ and $\tau$-invariant mod $\mu$ if $\mu(f \neq f \circ \tau) = 0$.

Thus $A$ is $\tau$-absorbing if no orbit $\omega, \tau\omega, \tau^2\omega, \ldots$ starting in $A$ leaves $A$. If $A$ is $\tau$-absorbing mod $\mu$ the set $A^1 = A \setminus \bigcup_{k=0}^{\infty} \tau^{-k}(A \setminus \tau^{-1} A)$ differs from $A$ only by a set of measure 0 and is $\tau$-absorbing. Similarly, if $A$ is $\tau$-invariant mod $\mu$ the set $A^2 = \bigcup_{k=0}^{\infty} \tau^{-k} A^1$ is $\tau$-invariant and equal to $A$ mod $\mu$. Thus, in all considerations where sets of measure 0 do not matter, we need not distinguish the notions $\tau$-invariant and $\tau$-invariant mod $\mu$. A real valued $f$ is $\tau$-invariant if and only if $\{f > \alpha\}$ is $\tau$-invariant for all $\alpha \in \mathbb{R}$. A complex valued $f$ is $\tau$-invariant if both the real and the complex part are $\tau$-invariant. These observations can be used to show that for functions $f$ which are $\tau$-invariant mod $\mu$ there exists a $\tau$-invariant $f'$ with $\mu(f \neq f') = 0$, and we need not distinguish the notions $\tau$-invariant and $\tau$-invariant mod $\mu$ for functions either.

For any null preserving $\tau$ the family $\mathscr{I}$ of invariant sets clearly is a $\sigma$-algebra. By the above remarks a function $f$ is $\tau$-invariant if and only if it is $\mathscr{I}$-measurable. If $\tau$ is an endomorphism and $\mu$ finite, the $\sigma$-algebra $\mathscr{I}$ can be used to to express the limit $\bar{f}$ of $A_n(\tau)f$ as a *conditional expectation*:

Let $\mathscr{F}$ be a sub-$\sigma$-algebra of $\mathscr{A}$, such that the restriction of $\mu$ to $\mathscr{F}$ is $\sigma$-finite. Recall that for any $f \in L_1(\Omega, \mathscr{A}, \mu)$ there exists (by the Radon-Nikodym theorem) an $f_0 \in L_1$ which is $\mathscr{F}$-measurable and satisfies

$$\int_A f_0 \, d\mu = \int_A f \, d\mu \quad \forall A \in \mathscr{F},$$

and that $f_0$ is uniquely determined mod $\mu$. $f_0$ is called the conditional expectation of $f$ with respect to $\mathscr{F}$, and denoted by $E(f \mid \mathscr{F})$ or by $E_\mu(f \mid \mathscr{F})$ if we emphasize the fact that the basic measure is $\mu$. It follows from the Jensen inequality that the map $f \to E(f \mid \mathscr{F})$ is a contraction $E_{\mathscr{F}}$ in each space $L_p(\Omega, \mathscr{A}, \mu)$.

**Proposition 1.6.** *If $\tau$ is an endomorphism in $(\Omega, \mathscr{A}, \mu)$, and $\mu$ $\sigma$-finite on the $\sigma$-algebra $\mathscr{I}$ of $\tau$-invariant sets, then, for any $f \in L_2(\Omega, \mathscr{A}, \mu)$, the norm-limit $\bar{f}$ of $A_n(\tau)f$ is given by $\bar{f} = E(f \mid \mathscr{I})$.*

*Proof.* We may assume $\mu(\Omega) < \infty$. $\bar{f}$ is $\mathscr{I}$-measurable. For any $A \in \mathscr{I}$ we have

$\int_A f \circ \tau^k d\mu = \int_A f d\mu$ because $A$ is $\tau$-invariant and $\tau$ an endomorphism. Now $\int_A f d\mu = \langle A_n(\tau) f, 1_A \rangle \to \langle \bar{f}, 1_A \rangle = \int_A \bar{f} d\mu$ because strong convergence implies weak convergence. □

**Definition 1.7.** A null preserving transformation $\tau$ in $(\Omega, \mathscr{A}, \mu)$ is called *ergodic* if all $\tau$-invariant sets $A$ have the property that $\mu(A) = 0$ or $\mu(A^c) = 0$.

Thus, $\tau$ is ergodic if the space cannot be decomposed into two non trivial $\tau$-invariant subsets. If an endomorphism is ergodic and $0 < \mu(\Omega) < \infty$, the limit $\bar{f}$ is simply given by $\bar{f} = \mu(\Omega)^{-1} \int f d\mu$. This means that for large $n$ the space average $\mu(\Omega)^{-1} \int f d\mu$ is very close to the time average $A_n(\tau) f$. The important *ergodic hypothesis* in statistical mechanics is the assumption that these two averages are asymptotically equal for the endomorphisms arising in the Hamiltonian flow in phase space. Von Neumann's theorem made it clear that the limit of the time averages does exist in the sense of $L_2$-convergence and that it is equal to the space average in the ergodic case. The question of ergodicity of the transformations constituting the Hamiltonian flow was left open. For many endomorphisms the proof of their ergodicity is fairly simple, but for some, including the endomorphisms coming from the Hamiltonian flow, the question whether they are ergodic is very deep; see the Notes.

**4. Criteria of ergodicity.** We end this section by describing some general necessary and sufficient conditions for ergodicity. They do not go very far beyond reformulations of the definition. The proof of the ergodicity for specific examples usually requires arguments which exploit the specific nature of the examples.

(a) For general null preserving $\tau$ we can just say that $\tau$ is ergodic iff each measurable $\tau$-invariant function $f$ is constant $\mu$-a.e. This follows from our above remarks on $\tau$-invariant sets and functions. (As usual "iff" means "if and only if").

(b) If $\tau$ is nonsingular, $\tau$ is ergodic iff $\mu(A) > 0$ implies that the complement of $A^* = \bigcup_{k=-\infty}^{+\infty} \tau^{-k} A$ is a nullset. $A^*$ is the smallest $\tau$-invariant set containing $A$. An obvious equivalent condition is that $\mu(A) > 0$, $\mu(B) > 0$ imply the existence of an integer $k$ with $\mu(\tau^k A \cap B) > 0$.

**Proposition 1.8.** *For an endomorphism $\tau$ of a finite measure space $(\Omega, \mathscr{A}, \mu)$ each of the following conditions is equivalent to the ergodicity of $\tau$:*

*(e 1)* $\mu(A) > 0 \Rightarrow \mu((A^-)^c) = 0$, *where* $A^- = \bigcup_{k=0}^{\infty} \tau^{-k} A$;

*(e 2)* $\mu(A) > 0$, $\mu(B) > 0 \Rightarrow \exists k \geqq 0$ *with* $\mu(\tau^{-k} A \cap B) > 0$;

*(e 3) For all* $A, B \in \mathscr{A}$, $\lim_{n \to \infty} n^{-1} \sum_{k=0}^{n-1} \mu(\tau^{-k} A \cap B) = \mu(A) \mu(B) \mu(\Omega)^{-1}$;

*(e 4) There exists a family* $\mathcal{M} \subset \mathcal{A}$ *such that the linear combinations of the functions* $1_E$ $(E \in \mathcal{M})$ *lie dense in* $L_2$ *and which has the property that* $A_n(\tau) 1_E$ *converges weakly in* $L_2$ *to a constant;*

*(e 5) For all* $f, g \in L_2$, $\lim_{n\to\infty} n^{-1} \sum_{k=0}^{n-1} \langle f \circ \tau^k, g \rangle = \langle f, \mathbb{1} \rangle \langle \mathbb{1}, g \rangle \mu(\Omega)^{-1}$.

*Proof.* The equivalence of (e 1) and (e 2) to ergodicity follows because $\tau^{-1} A^- \subset A^-$ and $\mu(\tau^{-1} A^-) = \mu(A^-)$ imply $\mu(A^- \setminus \tau^{-1} A^-) = 0$. Von Neumann's theorem yields the existence of the limit in (e 5) (and hence in (e 3) and (e 4)) and the limit is $\langle \bar{f}, g \rangle$. If $\tau$ is ergodic $\bar{f}$ is constant so that the identity $\langle \bar{f}, \mathbb{1} \rangle = \langle f, \mathbb{1} \rangle$ implies $\bar{f} = \mu(\Omega)^{-1} \langle f, \mathbb{1} \rangle$. But $\langle \bar{f}, g \rangle = \langle f, \mathbb{1} \rangle \langle \mathbb{1}, g \rangle \mu(\Omega)^{-1}$. Taking $f = 1_A$, $g = 1_B$ we get (e 3). If $\tau$ is not ergodic, (e 3) and hence (e 5) cannot hold because (e 2) doesn't. The equivalence of (e 4) is obtained by an approximation argument. □

**Notes**

Koopman [1931] observed that, using Liouville's theorem and the measure preserving property, the Hamiltonian flow could be studied via the induced group of unitary operators in Hilbert space. This idea led von Neumann [1931] to a proof of his ergodic theorem via spectral theory.

The measure theoretic concept of ergodicity seems to go back to Birkhoff and Smith [1928], who used the term *metrically transitive*, still favoured by some authors.

Sinai [1963] announced a theorem which – roughly speaking – says that a system of $n$ balls of equal diameter following the laws of elastic reflection in a cubic box is ergodic. No published proof of this result seems to exist. Sinai [1970] considered the movement of only *one* ball in a 2-dimensional domain with smooth strictly positive curvature (dispersing billiards). This was simplified and generalized by Bunimovich-Sinai [1973]. Also Kubo [1976], Kubo-Murata [1981], Gallavotti [1975], and Keller [1977] contributed to this subject.

Assuming, in addition, "finite horizon", Gallavotti and Ornstein [1974] proved that dispersing billiards are isomorphic to Bernoulli shifts.

We refer to Bunimovich [1982] for a survey on recent developments in this area.

## § 1.2 Birkhoff's ergodic theorem

**1. Discrete time.** Our next aim is to give a proof of George D. Birkhoff's famous pointwise ergodic theorem. For later use we formulate some of the arguments in the more general operator theoretic setting. We write

$$(2.1) \qquad M_n^S f = \operatorname{Max}(S_1 f, \ldots, S_n f), \quad M_n f = \operatorname{Max}(A_1 f, \ldots, A_n f),$$

and

$$M_\infty f = \sup_{n \geqq 1} M_n f.$$

When this notation is applied to an endomophism $\tau$, $T$ will be the composition with $\tau$.

The key step in the proof of the pointwise ergodic theorem is the following *maximal ergodic theorem*, which was discovered for endomorphisms by Yosida-Kakutani [1939] and for general positive contractions in $L_1$ by Hopf [1954]:

**Theorem 2.1.** *Let $T$ be a positive contraction in $L_1(\Omega, \mathscr{A}, \mu)$. For real valued $f \in L_1$ put $E_n = \{M_n f \geqq 0\}$ $(= \{M_n^S f \geqq 0\})$ and $E_\infty = \bigcup_{n=1}^{\infty} E_n$. Then*

$$\int_{E_n} f d\mu \geqq 0 \quad \text{and} \quad \int_{E_\infty} f d\mu \geqq 0 .$$

*Proof.* We follow the elegant argument of A.M. Garsia [1965]: For $k = 1, \ldots, n$, $(M_n^S f)^+ \geqq S_k f$ and hence $f + T(M_n^S f)^+ \geqq f + TS_k f = S_{k+1} f$. Thus $f \geqq S_k f - T(M_n^S f)^+$ holds for $k = 1, \ldots, n$ because it is trivial for $k = 1$. Passing to maxima one obtains $f \geqq M_n^S f - T(M_n^S f)^+$. Now integrate over $E_n$:

$$\begin{aligned}
\int_{E_n} f d\mu &\geqq \int_{E_n} (M_n^S f - T(M_n^S f)^+) d\mu \\
&= \int_{E_n} ((M_n^S f)^+ - T(M_n^S f)^+) d\mu \\
&= \int_{\Omega} (M_n^S f)^+ d\mu - \int_{E_n} T(M_n^S f)^+ d\mu \\
&\geqq \int_{\Omega} (M_n^S f)^+ d\mu - \int_{\Omega} T(M_n^S f)^+ d\mu \geqq 0
\end{aligned}$$

because $\int h d\mu \geqq \int T h d\mu$ holds for $h \in L_1^+$. A passage to the limit yields the second inequality. □

This proof is a bit miraculous. In section 3.2 a discussion of the filling scheme will provide a longer but more intuitive proof.

The following *maximal ergodic inequality* was already known to Wiener [1939]. It admits also a simple direct proof (see theorem 5.2), and suffices for the proof of Birkhoff's theorem given in theorem 7.3.

**Corollary 2.2.** *If $\tau$ is an endomorphism in a measure space $(\Omega, \mathscr{A}, \mu)$, then*

$$\mu(M_n f \geqq \alpha) \leqq \alpha^{-1} \|f\|_1$$

*holds for any real valued $f \in L_1$ and $\alpha > 0$.*

*Proof.* It is enough to prove $\|f\|_1 \geqq \alpha \mu(A)$ for arbitrary sets $A$ of finite measure contained in $\{M_n f \geqq \alpha\}$. Put $E_{n,A} = \{M_n(f - \alpha 1_A) \geqq 0\}$. By theorem 2.1,

$$\int_{E_{n,A}} (f - \alpha 1_A)\, d\mu \geqq 0.$$

For any $\omega \in A \subset \{M_n f \geqq \alpha\}$ there exists a $k \leqq n$ with $A_k f(\omega) \geqq \alpha$. This implies $S_k(f - \alpha \mathbb{1})(\omega) \geqq 0$ and $S_k(f - \alpha 1_A)(\omega) \geqq 0$. Thus $A$ is contained in $E_{n,A}$, and

$$\|f\|_1 \geqq \int_{E_{n,A}} f\, d\mu \geqq \alpha \int_{E_{n,A}} 1_A\, d\mu = \alpha\mu(A). \quad \square$$

**Theorem 2.3** (Birkhoff's ergodic theorem). *If $\tau$ is an endomorphism in a measure space $(\Omega, \mathscr{A}, \mu)$ and $f \in L_1$ (real or complex), then the averages $A_n f$ converge $\mu$ — a.e. to some $\tau$-invariant $\bar{f}$ with $\|\bar{f}\|_1 \leqq \|f\|_1$. For each $\tau$-invariant $A \in \mathscr{A}$ with $\mu(A) < \infty$*

(2.1) $$\int_A \bar{f}\, d\mu = \int_A f\, d\mu.$$

*Proof.* First consider a real valued $f$. Because of

$$A_{n+1} f = (n+1)^{-1} S_{n+1} f = (n+1)^{-1} f + \frac{n}{n+1}\left(\frac{1}{n} S_n f\right) \circ \tau$$

the functions $f^u = \limsup_{n\to\infty} A_n f$ and $f^l = \liminf_{n\to\infty} A_n f$ are $\tau$-invariant. We show that $f^u$ and $f^l$ assume the values $\pm\infty$ only on a set of measure 0:

For any $\beta > 0$ the $\tau$-invariant set $D_\beta = \{f^u > \beta\}$ is contained in the union of the increasing sequence $\{M_n f \geqq \beta\}$. By the maximal inequality $\mu(M_n f \geqq \beta) \leqq \beta^{-1}\|f\|_1$. Hence $\mu(D_\beta) \leqq \beta^{-1}\|f\|_1$. Passing with $\beta$ to infinity $f^u < \infty$ a.e. follows. By symmetry $f^l > -\infty$ a.e., and $\mu(f^l < \alpha) = \mu(-\limsup A_n(-f) < \alpha) \leqq |\alpha|^{-1}\|f\|_1$ for $\alpha < 0$.

If $A_n f$ does not converge a.e. there exist rational numbers $\alpha < \beta$ such that the set $B = \{f^l < \alpha < \beta < f^u\}$ has positive measure. $\mu(B)$ cannot be infinite because $\alpha < 0$ or $\beta > 0$. By the $\tau$-invariance of $B$ the function $f' = (f - \beta) 1_B$ has the property that $f' \circ \tau^k$ vanishes outside $B$ for all $k \geqq 0$ and $B = \{\omega\colon \exists n \geqq 1 \text{ with } S_n f'(\omega) > 0\}$.

The maximal ergodic theorem implies $\beta\mu(B) \leqq \int_B f\, d\mu$.

A symmetric argument with $f'' = (\alpha - f) 1_B$ gives us $\int_B f\, d\mu \leqq \alpha\mu(B)$. Together this contradicts $\alpha < \beta$.

We have shown that $f^u = f^l$ a.e., and that these functions $\mu$ — a.e. assume only finite values. By the decomposition $f = f^+ - f^-$ we may assume $f \in L_1^+$ for the proof of $\|\bar{f}\|_1 \leqq \|f\|_1$. The lemma of Fatou then yields

$$\int \bar{f}\, d\mu = \int \liminf A_n f\, d\mu \leqq \liminf \int A_n f\, d\mu = \int f\, d\mu$$

where the last equality follows from $\int f \circ \tau^k\, d\mu = \int f\, d\mu$.

To prove the last statement in the theorem we may assume $\Omega = A$, $\mu(\Omega) < \infty$, and $f \geqq 0$. For any $\varepsilon > 0$ there exists a $K_\varepsilon \geqq 0$ such that $g_\varepsilon = f - (f \wedge K_\varepsilon)$ has norm $\|g_\varepsilon\|_1 < \varepsilon$. Now

$$\int (A_n f - K_\varepsilon)^+ \, d\mu \leqq \int A_n g_\varepsilon \, d\mu < \varepsilon$$

shows that the sequence $A_n f$ is uniformly integrable. As any uniformy.integrable sequence converging $\mu$-a.e. converges in $L_1$-norm, the assertion $\int \bar{f} d\mu = \int f d\mu$ follows from $\int A_n f = \int f d\mu$.

For complex valued $f$ one can use the decomposition into the real and the imaginary parts to prove the existence of the limit, and observe $|\bar{f}| \leqq \lim A_n |f|$. □

The condition (2.1) determines $\bar{f}$ uniquely: We can assume that $\mu$ is $\sigma$-finite. There exists a set $\Omega_0 \in \mathscr{I}$ such that every $A \in \mathscr{I}$ with $\mu(A) < \infty$ is contained mod $\mu$ in $\Omega_0$, and $\mu$ is $\sigma$-finite on $\Omega_0 \cap \mathscr{I}$. $\bar{f}$ vanishes in $\Omega_0^c$ and is given by $E(f|\mathscr{I})$ in $\Omega_0$. In particular, if $\tau$ is an ergodic endomorphism in a $\sigma$-finite infinite measure space, then $\bar{f}$ must vanish for all $f$.

**2. Continuous time.** We have stated the ergodic theorem for a single endomorphism $\tau$. Sometimes one is interested in a continuous time motion of the points $\omega$, and in a corresponding continuous time theorem. There is no difficulty to derive such a result from theorem 2.3. By a *flow* $\{\tau_t, t \in \mathbb{R}\}$ we mean a group of measurable transformations $\tau_t: \Omega \to \Omega$ with $\tau_0 =$ identity, $\tau_{t+s} = \tau_t \circ \tau_s$, $(t, s \in \mathbb{R})$. The flow will be called measure preserving if the $\tau_t$ are measure preserving. The flow is called *measurable*, if the map $(\omega, t) \to \tau_t \omega$ from $\Omega \times \mathbb{R}^1$ into $\Omega$ is $\tilde{\mathscr{A}} - \mathscr{A}$-measurable, where $\tilde{\mathscr{A}}$ is the completion of the product-$\sigma$-algebra $\mathscr{A} \otimes \mathscr{B}$ of $\mathscr{A}$ with the Borel sets, and the completion is taken with respect to the product $\tilde{\mu}$ of $\mu$ on $\mathscr{A}$ and the Lebesgue measure $\lambda$ on $\mathscr{B}$. The completely analoguous definitions can be given for *semiflows* $\{\tau_t, t \geqq 0\}$, (in which the $\tau_t$ need not be invertible). If $\{\tau_t, t \geqq 0\}$ is a measurable measure preserving semiflow in a $\sigma$-finite measure space, and if $f: \Omega \to \mathbb{R}$ is integrable, the function $\tilde{f}$ defined by $\tilde{f}(\omega, t) = f(\tau_t \omega)$ is, for all $T > 0$, integrable in $\Omega \times [0, T]$ (Fubini), and, hence, for $\mu$-a.e. $\omega \in \Omega$ the integrals $\int_0^T f(\tau_t \omega)\, dt$ are well-defined. Note that $\int_0^n f(\tau_t \omega)\, dt = \sum_{i=0}^{n-1} F(\tau_1^i \omega)$ with $F(\omega) = \int_0^1 f(\tau_t \omega)\, dt$ and that $F$ is integrable. Therefore the ergodic theorem implies that $n^{-1} \int_0^n f(\tau_t \omega)\, dt$ converges a.e. when the integers $n$ tend to infinity. The ergodic theorem also implies $n^{-1} F_0 \circ \tau_1^{n-1} \to 0$ a.e., where $F_0 = \int_0^1 |f \circ \tau_t|\, dt$. For $n \leqq T < n+1$, $|\int_0^T f \circ \tau_t dt - \int_0^n f \circ \tau_t dt| \leqq F_0 \circ \tau_1^{n-1}$. Thus the convergence a.e. of $T^{-1} \int_0^T f \circ \tau_t dt$ when $T$ tends to infinity along the reals is a rather trivial consequence of the discrete time theorem. Many of the discrete parameter theorems in this book will have such continuous parameter analogues and we shall usually not care to state the latter.

There is, however, another class of continuous parameter theorems which will be of more interest to us. These are the *local ergodic theorems*. They assert the convergence of continuous time averges over time intervals $[0, \varepsilon]$ when $\varepsilon \to 0 + 0$. They were introduced by N. Wiener [1939] for measurable measure preserving flows. In this case they are a consequence of a form of the fundamental theorem of calculus, which says that for Lebesgue integrable $f$ on $[0, \infty[$ one has

$$\lim_{\varepsilon \to 0} \frac{1}{\varepsilon} \int_s^{s+\varepsilon} f(t)\,dt = f(s)$$

for $\lambda$-almost all $s$; see Royden [1968: Ch. 5].

**Theorem 2.4** (Wiener's local ergodic theorem). *If $(\Omega, \mathscr{A}, \mu)$ is a $\sigma$-finite measure space and $\{\tau_t, t \geqq 0\}$ a measure preserving measurable semiflow, $f \in \mathscr{L}_1(\mu)$, then* $\lim_{\varepsilon \to 0+0} \varepsilon^{-1} \int_0^\varepsilon f(T_\alpha \omega)\,d\alpha = f(\omega)$ *holds* $\mu$-a.e..

*Proof.* Let $\tilde{N}$ denote the complement of the set of points $(\omega, t)$ in $\Omega \times [0, \infty[$ with $\lim_{\varepsilon \to 0+0} \varepsilon^{-1} \int_0^\varepsilon f(\tau_{t+\alpha}\omega)\,d\alpha = f(\tau_t \omega)$ and let $N_\omega = \{t \in [0, \infty[ : (\omega, t) \in \tilde{N}\}$ and $N^t = \{\omega \in \Omega : (\omega, t) \in \tilde{N}\}$. For $\mu$-almost all $\omega$, $\int_0^t f(\tau_\alpha \omega)\,d\alpha$ is well defined for all $t$ and $\tilde{f}(\omega, \alpha)$ is integrable in each $[0, T]$. The fundamental theorem of calculus implies $\lambda(N_\omega) = 0$ for these $\omega$.

$\tilde{\mu}(\tilde{N}) = 0$ follows, and, by Fubini, $\lambda$-almost all $t$ have the property that $\mu(N^t) = 0$. But $N^t = \tau_t^{-1} N^0$. As $\tau_t$ is measure preserving $\mu(N^0) = 0$ follows. □

The integrability of $\tilde{f}(\omega, \cdot)$ with respect to $\lambda$ is also clear for bounded $f$, when the $\tau_t$ are not necessarily measure preserving. If $\mu(N^t) = 0$ for almost all $t > 0$ and the $\tau_t$ are null preserving $\mu(N^t) = 0$ follows for all $t > 0$. Therefore the above argument also proves:

**Theorem 2.5.** *If $(\Omega, \mathscr{A}, \mu)$ is a $\sigma$-finite measure space and $\{\tau_t, t \geqq 0\}$ a null preserving measurable semiflow with the property that*

$$\mu(\tau_t^{-1} A) = 0 \quad \textit{for all } t > 0 \quad \textit{implies } \mu(A) = 0$$

*then*

$$\lim_{\varepsilon \to 0+0} \varepsilon^{-1} \int_0^\varepsilon f(\tau_t \omega)\,dt = f(\omega) \quad \mu\text{-a.e.}$$

*holds for all bounded measurable $f$.*

**3. Uniform convergence.** Simple examples like the Bernoulli shifts (§ 1.4) show that $A_n f$ need not converge *everywhere* even when $\Omega$ is a compact topological space, and $\tau$ and $f$ are continuous. However, for some interesting examples one obtains even uniform convergence by the following special case of the mean ergodic theorem in Banach spaces:

**Theorem 2.6.** *Let $\Omega$ be a compact metric space with metric $\varrho$. If $\tau\colon \Omega \to \Omega$ is continuous, and the functions $A_n f$, $(n \geqq 1)$, are equicontinuous, then $A_n f$ converges uniformly in $\Omega$.*

(Recall that the functions $f_n$ are called *equicontinuous* if for any $\varepsilon > 0$ there exists $\delta > 0$ such that $\varrho(\omega, \omega') < \delta$ implies $|f_n(\omega) - f_n(\omega')| < \varepsilon$ for all $n$).

*Proof.* Let $\bar{\mathcal{O}}(\omega)$ denote the closure of the orbit $\{\omega, \tau\omega, \ldots\}$ of a point $\omega$. If some $g \in C(\Omega)$ is 0 on $\bar{\mathcal{O}}(\omega)$, then $\mu_\omega(g) = 0$, where $\mu_\omega$ is the linear functional constructed in subsection 1 of § 1.1. Thus $\mu_\omega(\bar{\mathcal{O}}(\omega)) = 1$.

By the Birkhoff theorem there exists an $\omega^* \in \bar{\mathcal{O}}(\omega)$ for which $A_n f(\omega^*)$ converges to some finite $\bar{f}(\omega^*)$. Given $\varepsilon > 0$ find $\delta > 0$ by the equicontinuity of the $A_n f$ and then find $m$ with $\varrho(\tau^m \omega, \omega^*) < \delta$. $|A_n f(\omega) - A_n f(\tau^m \omega)|$ tends to 0 for $n \to \infty$, since all but $2m$ terms in the sums cancel and $f$ is bounded. $|A_n f(\tau^m \omega) - A_n f(\omega^*)| < \varepsilon$ implies $|A_n f(\omega) - \bar{f}(\omega^*)| < 2\varepsilon$ for large n. As $\varepsilon$ was arbitrary $\lim A_n f(\omega)$ exists.

By the compactness of $\Omega$ there exist $\omega_1, \ldots, \omega_K \in \Omega$ so that the open $\delta$ – balls $B_k$ with these centers cover $\Omega$. If $N$ is large enough, then $|A_n f(\omega_k) - \bar{f}(\omega_k)| < \varepsilon$, for $n \geqq N$ and $k = 1, \ldots, K$. If $\omega$ is arbitrary there exists $k$ with $\varrho(\omega, \omega_k) < \delta$. Now $|A_n f(\omega) - A_n f(\omega_k)| < \varepsilon$ shows that the convergence is uniform. □

**Example.** Consider the $d$-dimensional torus $\Omega = [0, 1[^d$, which is a compact group with coordinatewise addition mod 1. Take some $\alpha = (\alpha_1, \ldots, \alpha_d) \in \Omega$, for which $\alpha_1, \alpha_2, \ldots, \alpha_d, 1$ are *integrally independent*, i.e., $m_1 = m_2 = \ldots = m_d = 0$ shall be the only integers for which $\sum_{k=1}^{d} \alpha_k m_k$ is an integer.

The translation $\tau\colon \omega \to \omega + \alpha$ clearly preserves the $d$-dimensional Lebesgue measure $\mu = \lambda^d$. $\tau$ is ergodic. This can be proved using Fourier analysis; see e.g. Petersen [1983: p. 51]. We sketch a proof using only measure theory. First show by induction on $d$ that the sequence $\ldots, -2\alpha, -\alpha, 0, \alpha, 2\alpha, 3\alpha, \ldots$ is dense in $\Omega$: As all these points are different there is an accumulation point, and, hence, for any $\varepsilon > 0$, a $k \in \mathbb{Z}$ with $0 < |\alpha_i'| < \varepsilon/2d$, where $\alpha_i' = k\alpha_i \bmod 1$. Because of the induction hypothesis the line $\{x\alpha'\colon x \in \mathbb{R}\}$ through $\alpha' = (\alpha_1', \ldots, \alpha_d')$ and $(0, 0, \ldots, 0)$ lies densely in $\Omega$. (It intersects $\{0\} \times [0,1[^{d-1}$ in a dense subset.) For any point $z$ on this line there is some $m\alpha'$ $(m \in \mathbb{Z})$ in an $\varepsilon/2$-neighbourhood of $z$. Thus, for any $y \in \Omega$, there is some $mk\alpha$ in the $\varepsilon$-neighbourhood of $y$. (A similar argument shows the density of the "forward orbit" $\alpha, 2\alpha, \ldots$, which was already known to Kronecker [1884]).

Now let $A$ be a $\tau$-invariant set of positive measure, and let $0 < \eta < 1$ be given. For $B, C \in \mathscr{A}$ we say that $B$ fills out $(1-\eta)$ of $C$ when $\mu(B \cap C) \geqq (1-\eta)\mu(C)$. If $\xi > 0$ is sufficiently small one can fill out $(1-\eta)$ of $\Omega$ with say $k(\xi)$ disjoint cubes of side-length $\xi$ which lie at a strictly positive distance of each other. There are arbibrarily small $\xi$ such that $A$ fills out $(1-\eta)$ of *some* cube $C_\xi$ of side length $\xi$. By the density statement above, we now can find $k(\xi)$ disjoint translates $\tau^i C_\xi$ of $C_\xi$. As $\tau^i A = A$, $A$ fills out $(1-\eta)$ of $\tau^i C_\xi$, and, hence, $(1-\eta)^2$ of $\Omega$. As $\eta > 0$ was arbitrarily small $\mu(A) = 1$, and the ergodicity of $\tau$ follows.

An application of theorem 2.6 to this example now proves the theorem of Weyl [1916] on uniform distribution mod 1:

**Theorem 2.7.** *If $\tau$ in $\Omega = [0,1[^d$ is the translation* mod 1 *by $\alpha = (\alpha_1, \ldots, \alpha_d)$ and the numbers $\alpha_1, \ldots, \alpha_d, 1$ are integrally independent, then, for each continuous $f$, $A_n f$ converges uniformly to $\int f d\mu$.*

*Proof.* It is simple to check that the $A_n f$ are equicontinuous. By the ergodicity the limit must be the constant $\int f d\mu$. □

Sometimes Weyl's theorem is spelled out for Riemann integrable functions, but this is really an equivalent formulation which can be obtained by a simple approximation. (To prove convergence a.e. for $f \in \mathscr{L}_1$ one needs a maximal inequality even in this special case).

## Notes

Birkhoff [1931] has based his proof on the following (weaker) maximal inequality: If $F_\alpha$ is the $\tau$-invariant set $\{\limsup A_n f \geqq \alpha\}$, then $\int_{F_\alpha} f d\mu \geqq \alpha\mu(F_\alpha)$.

He actually has formulated his a result only for indicator functions $f = 1_B$ (in the setting of a closed analytic manifold having a finite invariant measure). Khintchine [1933] then showed that Birkhoffs result remained true for integrable $f$ on an abstract finite measure space. (Therefore theorem 2.3 is called Birkhoff-Khintchine-Theorem in a few countries. However, Khintchine himself emphasized that the idea of his proof was precisely that of Birkhoff).

Birkhoff's theorem (for finite $\mu$) easily implies norm convergence of $A_n f$ in $L_p$ for $f \in L_p$ $(1 \leqq p < \infty)$, and therefore contains the special case of von Neumann's theorem which motivated his work. As Birkhoff's paper appeared earlier, it is of interest that von Neumann's theorem was proved first and was known to Birkhoff.

The original proof of Yosida-Kakutani's maximal ergodic theorem, which used ideas of Kolmogorov [1937], remains of interest; see Petersen [1983]. Simplified and alternative arguments (sometimes for continuous time or special cases) were given by Riesz [1931], [1932], [1942], [1945], Pitt [1942], Hopf [1947], Dowker [1950] and others. Kamae [1982] gave a proof using nonstandard analysis. Using some of his ideas, Katznelson and Weiss [1982] gave a short proof without explicit use of a maximal ergodic theorem.

E. Bishop [1966], [1967], [1968] gave a proof of Birkhoff's theorem using *upcrossing inequalities* similar to those in martingale theory; see § 4.1. These inequalities are constructively valid, but Birkhoff's theorem is not. Nuber [1972] has sought conditions which constructively imply the conclusion of Birkhoff's theorem.

The first *ratio ergodic theorem* was proved by Stepanov [1936] (in the ergodic case) and Hopf [1937]: If $\tau$ is an endomorphism of a $\sigma$-finite $(\Omega, \mathscr{A}, \mu)$, $f \in \mathscr{L}_1$ and $g \in \mathscr{L}_1^+$, then $S_n f / S_n g$ converges a.e. on each $\{S_k g > 0\}$. This is now a special case of the Chacon-Ornstein theorem.

It is easy to see that Birkhoff's theorem implies the convergence a.e. of $B_n^\lambda f := (2n+1)^{-1} \sum_{k=-n}^{+n} e^{2\pi i k \lambda} f \circ \tau^k$ for fixed $\lambda \in \mathbb{R}$. This was strengthened by Wiener and Wintner [1941] who proved:

**Theorem 2.8.** *If $\tau$ is an automorphism of a finite measure space, then there exists, for each $f \in \mathscr{L}_1$, a nullset $N_f$ such that, for $\omega \in N_f^c$, $B_n^\lambda f(\omega)$ converges for all $\lambda \in \mathbb{R}$.*

Wiener and Wintner [1941a] also investigated when $F(t) = f(\tau_t \omega)$ is almost periodic and studied $\lim (2t)^{-1} \int_{-t}^{+1} F(s+u) \bar{F}(s)\, ds$, where $\bar{F}$ is the complex conjugate function.

**Speed of convergence.** When a convergence statement has been proved, one of the questions of interest is whether one can assert something about the speed of convergence. The famous law of the iterated logarithm is an example where a positive result is possible. In the form proved by Hartman and Wintner [1941] it can be stated in our language as follows: If $\mu(\Omega) = 1$, $\tau$ is an endomorphism, $Tf = f \circ \tau$, $f$ has integral $\int f d\mu = 0$ and $L_2$-norm $\|f\|_2 = 1$, and if $f, f \circ \tau, f \circ \tau^2, \ldots$ are independent, then

$$\limsup_{n \to \infty} A_n f / \sqrt{2 \log \log n / n} = 1 \quad \mu\text{-a.e.}.$$

By symmetry $\liminf = -1$, so that $A_n f = 0\,((n^{-1} \log \log n)^{1/2})$ a.e. Independence is crucial! No general positive ergodic theoretic result of this type is possible even for slower speeds. Indeed, Krengel [1978a] has shown: If $\tau$ is an ergodic endomorphism of the torus $[0,1[$ with Lebesgue measure, and $(\alpha_n)$ any null sequence of positive numbers, there exists a *continuous* $f$ with integral 0 and

$$\text{(2.2)} \qquad \limsup \alpha_n^{-1} |A_n f| = \infty \quad \text{a.e.}.$$

On the other hand, Halász [1976] proved: For any non decreasing sequence $(c_n)$ of positive numbers with $c_1 \geqq 2$ and tending to $\infty$, and for any ergodic automorphism of $[0,1[$ there exists $A$ with $\lambda(A) = \frac{1}{2}$ and $|S_n 1_A - n/2| \leqq c_n$ for all $n$. Thus, the convergence *can* be arbitrarily fast.

The following deep theorem of O'Brien [1983] contains a limit version of Halász' result and (2.2) for measurable $f$:

**Theorem 2.9.** *If $(b_n)$ is a sequence of positive numbers tending to $\infty$ and satisfying $\liminf n^{-1} b_n = 0$, there exists $\tau$ and a $\{+1, -1\}$-valued $f$ with $\limsup b_n^{-1} S_n f = 1$, and this $f$ can be constructed in such a way that the sequences $f, f \circ \tau, f \circ \tau^2, \ldots$ and $-f, -f \circ \tau, -f \circ \tau^2, \ldots$ have the same joint distribution.*

Kakutani and Petersen [1981] proved another strengthening of (2.2) for measurable $f$: They construct, for any sequence $(b_n)$ of positive numbers with divergent sum, a bounded measurable $f$ with integral 0 and $\sup_k |\sum_{i=1}^{k} b_i A_i f| = \infty$ a.e. Dowker and Erdös [1959] showed the existence of a bounded $f$ with $\int f d\lambda = 0$ for which $\sum_{i=1}^{k} b_i f \circ \tau^i$ fails to converge in measure on any subset of $[0,1[$ having positive measure, and with $\sup |\sum_{i=1}^{k} b_i f \circ \tau^i| = \infty$ a.e., see also Halmos [1948]. Baum and Katz [1965] showed: $\sum \mu(|A_n f| > \varepsilon)/n$ converges for $f$ with $\int f d\mu = 0$ if $f, f \circ \tau \ldots$ are independent.

The uniform boundedness principle implies that the existence of a sequence $\alpha_n$ tending to $\infty$, with $\limsup \alpha_n \| A_n f - Pf \| < \infty$ for all $f$, is equivalent to $\| A_n - P \| \to 0$. It is therefore easy to see that there is no speed of convergence in von Neumann's mean ergodic theorem.

There are a few positive results on speed of convergence for specific transformations and functions; see e.g. Kuipers and Niederreiter [1974], or Kowada [1973].

**Non-integrable functions.** If $\tau$ is an ergodic endomorphism in a probability space, and a non negative $f$ has a infinite integral, then Birkhoff's theorem implies $\lim_{n\to\infty} A_n f = \infty$, because

$$\liminf_{n\to\infty} A_n f \geqq \lim_{n\to\infty} A_n(f \wedge k) = \int (f \wedge k) d\mu \quad \text{for all } k.$$

J. Aaronson [1977] has proved that also other norming factors of $S_n f$ than $n^{-1}$ cannot produce convergence a.e. to a non zero finite function:

**Theorem 2.10.** *If $\tau$ is an ergodic endomorphism of a probability space $(\Omega, \mathscr{A}, \mu)$, $f \geqq 0$, and $\int f d\mu = \infty$, then for any sequence $(b_n)$ of positive reals, one has either*

$$\limsup_{n\to\infty} b_n^{-1} S_n f = \infty \quad \text{a.e.} \quad \textit{or} \tag{2.3}$$

$$\liminf_{n\to\infty} b_n^{-1} S_n f = 0 \quad \text{a.e.}. \tag{2.4}$$

There is a similar theorem for $\sigma$-finite measure spaces; see also Aaronson [1979].

Aaronson [1981] also considered this problem for functions $f$ which need not be non negative: If $b_n/n$ is non decreasing and tends to $\infty$, then (2.3) or (2.4) hold with $S_n f$ replaced by $|S_n f|$. In the case where the $f \circ \tau^i$ are independent Feller [1946] proved this assertion for arbitrary non decreasing sequences. But in the general ergodic case the condition $b_n/n \to \infty$ cannot be deleted: Aaronson [1977] gave an example of an $f$ with $\int |f| d\mu = \infty$ and $S_n f/n \to 1$ a.e. A related result is that of Kesten [1975], who showed that $\liminf A_n f > 0$ a.e. if $S_n f \to \infty$ a.e..

Let $(a_n)$ be a non decreasing sequence of positive numbers. Tanny [1974] showed that the condition $\liminf a_{k \cdot n}/a_n > 1$ for some $k > 1$ implies that for any ergodic $\tau$ in a probability space and all $f$, $\limsup f \circ \tau^n / a_n = \infty$ a.e. or $\limsup f \circ \tau^n / a_n = 0$ a.e. O'Brien [1982] proved that the condition is also necessary.

Dowker and Erdös [1959] have given several more examples: E.g. they showed that $S_n(\tau)g/S_n(\tau^{-1})g$ may diverge for ergodic automorphisms of a $\sigma$-finite measure space.

Del Junco and Steele [1977] have shown that for any ergodic endomorphism $\tau$ in $[0,1]$ (which Lebesgue measure) and for any increasing sequence $0 < b_1 \leqq b_2 \leqq$ of integers $b_n$ with $n^{-1} b_n \to 0$ there exists an indicator function $f = 1_A$ such that $\limsup b_n^{-1} \sum_{i=n}^{n+b_n} f \circ \tau^i = 1$ $\mu$-a.e. and the corresponding lim inf is 0 $\mu$-a.e. (Wiener and Wintner [1941a] have made a similar observation; see also Pfaffelhuber [1975]).

By an example of Burkholder [1962] the averages considered in Birkhoff's theorem may diverge a.e. after an application of a single conditional expectation operator. Isaac [1973] has studied similar questions for ratios.

We refer to Kuipers and Niederreiter [1974] for the theory of uniform distribution mod 1.

For abstract ergodic theorems in a Boolean algebra and in a logic, see Bunjakov [1973], Dvurecenskij and Riecan [1980], and Pulmannová [1982].

## § 1.3 Recurrence

**1. The conservative and dissipative part.** We call a null preserving $\tau$ *recurrent* if, for all $A \in \mathscr{A}$, $\mu$-almost all $\omega \in A$ belong to the set $A_{ret} = A \cap \bigcup_{k=1}^{\infty} \tau^{-k} A$ of points returning at least once, and *infinitely recurrent* if, for all $A \in \mathscr{A}$, $\mu$-almost all $\omega \in A$ belong to the set

$$A_{inf} = \{\omega \in A\colon \tau^k \omega \in A \text{ for infinitely many } k \geqq 1\} = A \cap \bigcap_{n=1}^{\infty} \bigcup_{k=n}^{\infty} \tau^{-k} A.$$

A set $W \in \mathscr{A}$ is called *wandering* if the sets $\tau^{-k} W$ $(k \geqq 0)$ are disjoint, or, equivalently, if no point in $W$ returns to $W$. $\tau$ is called *conservative* if there exists no wandering set of positive measure. Finally, $\tau$ is called *incompressible*, if there exists no $A \in \mathscr{A}$ with $A \subset \tau^{-1} A$ and $\mu(\tau^{-1} A \setminus A) > 0$. Passing to complements one sees that $\tau$ is incompressible if and only if $\tau^{-1} B \subset B$ implies $\mu(B \setminus \tau^{-1} B) = 0$.

**Theorem 3.1** (Recurrence theorem). *Let $\tau$ be null preserving in a measure space $(\Omega, \mathscr{A}, \mu)$. The following conditions are equivalent:*

*(i) $\tau$ is conservative,*

*(ii) $\tau$ is recurrent,*

*(iii) $\tau$ is infinitely recurrent,*

*(iv) $\tau$ is incompressible.*

*Proof.* (i) $\Rightarrow$ (ii): For any $A$ the set $A_0 = A \setminus A_{ret}$ of points in $A$ which never return is wandering, and, hence, a nullset.

(ii) $\Rightarrow$ (iii): For any $A \in \mathscr{A}$, $\mu(A_0) = 0$. The set $\tau^{-k} A_0$ is the set of points which visit $A$ at time $k$ for the last time. If $\omega \in A$ does not return to $A$ infinitely often there must be some $k \geqq 0$ with $\omega \in \tau^{-k} A_0$. Hence $A_{inf} = A \setminus \bigcup_{k=0}^{\infty} \tau^{-k} A_0$ differs from $A$ by a nullset.

(iii) $\Rightarrow$ (ii) is trivial.

(ii) $\Rightarrow$ (iv): If $\tau^{-1} B \subset B$, then the sequence $\tau^{-k} B$, $k = 1, 2, \ldots$ is decreasing, and $B \setminus \tau^{-1} B = B \setminus \bigcup_{k=1}^{\infty} \tau^{-k} B$ is the set of points in $B$ which never return. As $\tau$ is recurrent $\mu(B \setminus \tau^{-1} B) = 0$.

(iv) $\Rightarrow$ (i): If $W$ is wandering put $B = \bigcup_{k=0}^{\infty} \tau^{-k} W$. Clearly $\tau^{-1} B \subset B$. By the disjointness of the sets $\tau^{-k} W$ we have $W = B \setminus \tau^{-1} B$. Hence $\mu(W) = 0$. □

Clearly, if $\tau$ is an endomophism of a finite measure space, there cannot exist a wandering set of positive measure, and, therefore, $\tau$ must be infinitely recurrent. This is the recurrence theorem of Poincaré [1899], perhaps the oldest result in ergodic theory. Actually, Poincaré was most interested in a topological type of recurrence, which can be deduced from the result stated above: If $\Omega$ is a topological space with a countable basis $\mathscr{B} \subset \mathscr{A}$, and $\tau$ is infinitely recurrent, then almost every $\omega \in \Omega$ returns infinitely often to any neighborhood of itself. To see this, observe that the union $E$ of all sets $A \setminus A_{inf}$ with $A \in \mathscr{B}$ has measure 0. If $U$ is a neighborhood of some $\omega \in E^c$, there is an $A \in \mathscr{B}$ with $\omega \in A \subset U$. As $\omega \notin A \setminus A_{inf}$, $\omega$ returns infinitely often to $A$, and, hence, to $U$.

In general, if $\mu$ is $\sigma$-finite one can find a maximal subset of $\Omega$ on which $\tau$ is conservative.

**Theorem 3.2** (Hopf decomposition). *If $\tau$ is null preserving in the $\sigma$-finite measure space $(\Omega, \mathscr{A}, \mu)$, there exists a decomposition of $\Omega$ into two disjoint measurable sets $C$ and $D$, the conservative and the dissipative part, such that*

*(i) $C$ is $\tau$-absorbing,*

*(ii) the restriction of $\tau$ to $C$ is conservative, and*

*(iii) $D = \Omega \setminus C$ is an at most countable union of wandering sets.*

*If $\tau$ is even nonsingular, $C$ is $\tau$-invariant and there exists a wandering $W_0$ with $D = \bigcup_{k=-\infty}^{+\infty} \tau^k W_0$.* ($\tau$ is called *dissipative* if $\Omega = D$.)

The proof is based on an *exhaustion argument*. As this type of argument is used frequently, let us explain this simple technique: Let **P** be a certain property of measurable sets in the $\sigma$-finite measure space $(\Omega, \mathscr{A}, \mu)$ which is such that any subset of a set with this property again has property **P**. As $\mu$ is $\sigma$-finite, there exists a finite measure $\nu$ equivalent to $\mu$. Put $\alpha_1 = \sup\{\nu(A)\colon A \text{ has property } \mathbf{P}\}$. Pick some $A_1 \in \mathscr{A}$ with property **P** and $\nu(A_1) \geqq 2^{-1}\alpha_1$. If $A_1, \ldots, A_n$ have been determined, put $\alpha_{n+1} = \sup\{\nu(A)\colon A \text{ has property } \mathbf{P} \text{ and } A \subset \Omega \setminus \bigcup_{k=1}^{n} A_k\}$. Then find $A_{n+1} \subset \Omega \setminus \bigcup_{k=1}^{n} A_k$ with property **P** such that $\nu(A_{n+1}) \geqq 2^{-1}\alpha_{n+1}$. Let $\Omega_1 = \bigcup_{i=1}^{\infty} A_i$. The complement $\Omega_2$ of $\Omega_1$ is such that no subset of $\Omega_2$ with positive measure has property **P**. Moreover, if $\Omega$ is the disjoint union of $\Omega_1'$ and $\Omega_2'$ such that $\Omega_1'$ is a countable union of sets with property **P**, and no subset of $\Omega_2'$ has positive measure and property **P**, then $\Omega_1$ coincides up to nullsets with $\Omega_1'$.

*Proof of theorem 3.2.* Take the property of being wandering and put $\Omega_1 = D$, $\Omega_2$

$= C$. As $\tau^{-1} W$ is wandering if $W$ is wandering we see that $\tau^{-1} D \subset D$ up to nullsets. Hence $C$ is $\tau$-absorbing mod $\mu$. Changing $C$ on a nullset we can assume that it is $\tau$-absorbing.

If $\tau$ is nonsingular, also $\tau W$ is wandering if $W$ is, and we can infer that $D$ and $C$ are $\tau$-invariant. $W_0$ is constructed as follows: Take the $A_i$ from the construction of $D$, put $A_i^* = \bigcup_{k=-\infty}^{+\infty} \tau^k A_i$, and

$$W_0 = \bigcup_{i=1}^{\infty} (A_i \setminus \bigcup_{j=1}^{i-1} A_j^*). \quad \square$$

Also for general null preserving $\tau$ the description of $D$ can be made somewhat more precise: Put $W_1 = A_1$, and for $n \geqq 1$ put

$$W_{n+1} = A_{n+1} \cup (W_n \cap (\bigcup_{i=0}^{\infty} \tau^{-i} A_{n+1})^c).$$

Then the sequence $D_n = \bigcup_{i=0}^{\infty} \tau^{-i} W_n$ is increasing with union $D$, and each $W_n$ is wandering.

The following complement to the Hopf decomposition has been observed by J. Feldman [1962] in a more general situation. We leave it as an exercise.

**Proposition 3.3.** *Also in the case where $\tau$ is an endomorphism, $C$ is $\tau$-invariant.*

Halmos [1947] proved that the powers of a conservative $\tau$ are conservative. In fact we have:

**Theorem 3.4.** *If $\tau$ is null preserving in a $\sigma$-finite measure space $(\Omega, \mathscr{A}, \mu)$, the dissipative parts $D^n$ of $\tau^n$ are the same for all $n \geqq 1$.*

*Proof.* If $W$ is wandering for $\tau$ it is wandering for $\tau^n$, so that $D^1 \subset D^n$. Now let $W_1, W_2, \ldots$ be wandering sets for $\tau^n$ with union $D^n$. Put $h = \sum_{k=1}^{\infty} 2^{-k} 1_{W_k}$. Then $h$ is strictly positive on $D^n$ and

$$\sum_{i=0}^{\infty} h \circ \tau^{in} = \sum_{k=1}^{\infty} 2^{-k} \sum_{i=0}^{\infty} 1_{W_k} \circ \tau^{in} \leqq \sum_{k=1}^{\infty} 2^{-k} \leqq 1 \text{ on } \Omega.$$

If $D^n$ is not contained in $D^1 = D$, there exists an $\varepsilon > 0$ and $A \subset C$ with $\mu(A) > 0$ and $h \geqq \varepsilon 1_A$. Now

$$\sum_{i=0}^{\infty} 1_A \circ \tau^i \leqq \varepsilon^{-1} \sum_{i=0}^{\infty} h \circ \tau^i = \varepsilon^{-1} \sum_{j=0}^{n-1} (\sum_{i=0}^{\infty} h \circ \tau^{in}) \circ \tau^j \leqq \varepsilon^{-1} n \text{ on } \Omega.$$

But then no point in $A$ can return to $A$ more than $\varepsilon^{-1} n$ times, a contradiction to $A \subset C$. $\square$

The transformation $k \to k+1$ on $\Omega = \mathbb{Z}$ with the counting measure is an example of an ergodic measure preserving invertible transformation which is dissipative. This is essentially the only invertible example because of the last assertion in theorem 3.2. In particular a nonsingular ergodic $\tau$ in a non atomic measure space must be conservative. There exist endomorphisms in non atomic $\sigma$-finite measure spaces which are dissipative and yet ergodic; see § 3.1 (Notes).

**Proposition 3.5.** *If a null preserving transformation $\tau$ is conservative and $A$ $\tau$-absorbing* mod $\mu$, *then $A$ is $\tau$-invariant* mod $\mu$.

*Proof.* We may assume $\tau^{-1}A \supset A$ by modifying $A$ on a nullset. Almost every point of $E = \tau^{-1}A \setminus A$ returns to $E$. But $\tau^{-1}A \supset A$ implies that, for all $\omega \in E$, the orbit $\tau\omega, \tau^2\omega, \ldots$ is contained in $A \subset E^c$. Thus $E$ must have measure 0. □

It follows that a conservative null preserving transformation is ergodic if and only if $\mu(A) > 0$ implies $\mu((\bigcup_{k=0}^{\infty} \tau^{-k}A)^c) = 0$.

**2. Induced transformations.** If $\tau$ is recurrent, the *return time*

$$r_A(\omega) := \inf\{k \geqq 1 : \tau^k\omega \in A\}$$

to $A \in \mathscr{A}$ is finite a.e. in $A$. In the measure preserving case we can evaluate the integral of $r_A$:

**Theorem 3.6** (Recurrence theorem of Kac [1947]). *If $\tau$ is a conservative endomorphism of $(\Omega, \mathscr{A}, \mu)$ and $A \in \mathscr{A}$, then*

$$\int_A r_A(\omega)\,\mu(d\omega) = \mu(\bigcup_{n=0}^{\infty} \tau^{-n}A).$$

*In particular, if $\tau$ is ergodic and $\mu(A) > 0$, then $\int_A r_A\,d\mu = \mu(\Omega)$.*

*Proof.* We may assume $0 < \mu(A) < \infty$. Put $A_0 = A$ and

$$A_k = \tau^{-1}A_{k-1} \cap A^c, \quad R_k = \tau^{-1}A_{k-1} \cap A \quad (k \geqq 1).$$

$R_k$ is the set of points in $A$ with $r_A(\omega) = k$. For $k \geqq 1$, $A_k$ is the set of points in $A^c$ which visit $A$ for the first time at time $k$.

We have $\mu(A_k) = \mu(A_{k+1}) + \mu(R_{k+1})$, $(k \geqq 0)$, and, hence,

$$\mu(A_n) = \sum_{k=n+1}^{\infty} \mu(R_k) + \lim_{k\to\infty} \mu(A_k).$$

By the recurrence of $\tau$, $\mu(A_0) = \sum_{k=1}^{\infty} \mu(R_k)$, which implies $\lim_{k\to\infty} \mu(A_k) = 0$. Now the assertion of the theorem follows from

$$\mu(\bigcup_{n=0}^{\infty} \tau^{-n} A) = \sum_{n=0}^{\infty} \mu(A_n) = \sum_{n=0}^{\infty} \sum_{k=n+1}^{\infty} \mu(R_k)$$

$$= \sum_{k=1}^{\infty} k\,\mu(R_k) = \int_A r_A \, d\mu. \quad \square$$

Higher moments of $r_A$ have been studied by Blum and Rosenblatt [1967] and by Wolfowitz [1967].

For conservative null preserving $\tau$, we can define a map $\tau_A$ of $A_{ret}$ into $A$ by $\omega \to \tau^{r_A(\omega)}\omega$. $\omega$ is mapped to the place where it first returns to $A$. Because of $\tau_A^{-1} B = \bigcup_{k=1}^{\infty} R_k \cap \tau^{-k} B$, this map is measurable and null preserving. It maps $A_{inf}$ into $A_{inf}$ and one has $\tau_{A_{inf}} = \tau_A$ on $A_{inf}$. Neglecting a nullset, we assume $A = A_{inf}$. $\tau_A$ is called the transformation *induced* by $\tau$ on $A$. Obviously, $\tau_A$ is conservative.

**Theorem 3.7** (Kakutani [1943]). *If $\tau$ is a conservative endomorphism of $(\Omega, \mathscr{A}, \mu)$, the induced transformation is an endomorphism of $(A, A \cap \mathscr{A}, \mu_A)$, where $\mu_A$ is the restriction of $\mu$ to $A$.*

*Proof.* Take some measurable $B \subset A$, and put $B_0 = B$ and

$$B_{k+1} = \tau^{-1} B_k \cap A^c, \quad B'_{k+1} = \tau^{-1} B_k \cap A \quad (k \geqq 0).$$

We have

$$\mu(B) = \mu(B'_1) + \mu(B_1) = \mu(B'_1) + \mu(B'_2) + \mu(B_2) = \dots$$

$$= \sum_{k=1}^{\infty} \mu(B'_k) + \lim_{k\to\infty} \mu(B_k) \geqq \sum_{k=1}^{\infty} \mu(B'_k).$$

But $B'_k$ is the set $R_k \cap \tau^{-k} B = R_k \cap \tau_A^{-1} B$. Thus $\mu(B) \geqq \mu(\tau_A^{-1} B)$. The theorem now follows by an application of the following lemma to $\tau_A$. $\square$

**Lemma 3.8.** *If $\tau$ is conservative in $(\Omega, \mathscr{A}, \mu)$ and $\mu(\tau^{-1} A) \leqq \mu(A)$ holds for all $A \in \mathscr{A}$, then $\mu$ is $\tau$-invariant.*

*Proof.* Otherwise there exists an $A$ with $\mu(A) > \mu(\tau^{-1} A)$. Construct $A_0, A_1, \dots$ $R_1, R_2, \dots$ as in the proof of the Kac recurrence theorem. Then $\mu(A) > \mu(\tau^{-1} A) = \mu(A_1) + \mu(R_1) \geqq \mu(R_1) + \mu(R_2) + \mu(A_2) \; \dots \; \geqq \sum_{k=1}^{\infty} \mu(R_k) = \mu(A)$ yields a contradiction. $\square$

**Proposition 3.9.** *If a null preserving transformation $\tau$ in $(\Omega, \mathscr{A}, \mu)$ is conservative, and $\mathscr{I}$ is the $\sigma$-algebra of $\tau$-invariant sets, then $\mathscr{I} \cap A$ is the $\sigma$-algebra of $\tau_A$-invariant sets. In particular the ergodicity of $\tau$ implies that of $\tau_A$.*

*Proof.* $I \in \mathscr{I}$ means that $\omega \in I$ is equivalent to $\tau\omega \in I$, and, hence, to $\tau^2\omega \in I$, etc. Therefore $\omega \in A \cap I$ is equivalent to $\tau_A\omega \in A \cap I$, and $A \cap I$ is $\tau_A$-invariant. If, for some $I'$, $A \cap I'$ is not $\tau_A$-invariant, there exist an $\omega \in A$ such that $\omega \in I'$ and $\tau_A\omega \notin I'$, or $\omega \notin I'$ and $\tau_A\omega \in I'$. So the equivalence of the assertions $\tau^n\omega \in I'$ must fail for some $n$ and $I'$ is not $\tau$-invariant. □

Sometimes it is useful to reverse the process of inducing a transformation on a subset: Let $\tau$ be an endomorphism of $(\Omega, \mathscr{A}, \mu)$ and let $R_1, R_2, \ldots$ be a *partition* of $\Omega$, i.e., the $R_i$ form a sequence of disjoint measurable sets with union $\Omega$. Let $\Omega_0 = \Omega$, $\Omega_i = \bigcup_{k=i+1}^{\infty} R_k$, and $\Omega^* = \{(k, \omega) : k \geqq 0, \omega \in \Omega_k\}$, and for measurable $A \subset \Omega_k$ put $\mu^*(\{k\} \times A) = \mu(A)$. We have defined a measure space consisting of disjoint copies of the $\Omega_i$. Define an endomorphism $\tau^*$ in $\Omega^*$ by

$$\tau^*(k, \omega) = \begin{cases} (k+1, \omega) & \text{if} \quad \omega \in \Omega_{k+1} \\ (0, \tau\omega) & \text{if} \quad \omega \notin \Omega_{k+1}. \end{cases}$$

If we identify $\Omega$ with $\{(0, \omega) : \omega \in \Omega_0\}$, then $\tau^*_\Omega = \tau$. This is Kakutani's *skyscraper* construction. Clearly one can do the same thing with a null preserving $\tau$ and come out with a null preserving $\tau^*$. If $\tau$ is an automorphism of $(\Omega, \mathscr{A}, \mu)$, $\tau^*$ is again an automorphism. If $\tau$ is nonsingular, $\tau^*$ is. Clearly $\tau^*$ is ergodic if and only if $\tau$ is ergodic. Therefore the skyscraper construction is a very simple way of constructing ergodic automorphisms in non atomic infinite measure spaces. Simply start with an ergodic automorphism, say, in the unit interval, and choose the $R_k$ with $\sum_{k=1}^{\infty} k \cdot \mu(R_k) = \infty$.

## Notes

In his famous book, Hopf [1937] described the decomposition $\Omega = C \cup D$ for invertible transformations, and, in 1954, for the much more general situation of a positive contraction in $L_1$; see section 3.1. But concepts like wandering set or induced transformation have no equally simple and intuitive extensions to the operator case. Sucheston [1957], Helmberg [1965], [1965a], [1966], Simons [1965], [1971], Wright [1961], Roos [1964], Tsurumi [1958], Choksi [1961], Helmberg and Simons [1969], and others have therefore studied the case of non invertible null preserving transformations. One can split $D$ further into the set of points in $D$ whose orbit enter $C$ and the rest, see Kopf [1982], [1978].

Call two measure spaces *isomorphic* mod 0 if – after deleting nullsets $N$ and $N'$ from $\Omega$ and $\Omega'$-there exists a bijective map $\varphi$: $\Omega \setminus N \to \Omega' \setminus N'$ such that $\varphi$ and $\varphi^{-1}$ are measurable and $\mu \circ \varphi^{-1} = \mu'$. Endomorphisms $\tau$ and $\tau'$ in $\Omega$ and $\Omega'$ are called isomorphic mod 0 if such a $\varphi$ can be found with $\tau' \circ \varphi = \varphi \circ \tau$; the sets $N, N'$ shall be such that $\Omega \setminus N$ and $\Omega' \setminus N'$ are invariant under $\tau$ and $\tau'$ respectively. Two dissipative automorphisms $\tau$ and $\tau'$ in $\sigma$-finite spaces $(\Omega, \mathscr{A}, \mu)$ and $(\Omega', \mathscr{A}', \mu')$ are isomorphic mod 0 if and only if the measure spaces $(W_0, W_0 \cap \mathscr{A}, \mu_{W_0})$ and $(W_0', W_0' \cap \mathscr{A}', \mu'_{W_0'})$ are isomorphic mod 0, where $W_0$, $W_0'$ are the sets with $\Omega = \bigcup_{k=-\infty}^{+\infty} \tau^k W_0$, $\Omega' = \bigcup_{k=-\infty}^{+\infty} \tau'^k W_0'$ constructed in theorem 3.2.

In particular the set $W_0$ is determined uniquely up to measure theoretic isomorphism mod 0; see Krengel [1968/69: part I], where also dissipative flows are classified by showing that they are isomorphic to a flow $\tau_s$: $(\omega_0, t) \to (\omega_0, t+s)$ in a space $W_0 \times \mathbb{R}$, with a $W_0$ which is uniquely determined mod 0. The recurrence properties of nonsingular flows and null preserving semiflows were discussed in part II of this paper. Helmberg [1969] has derived a number of results on the mean recurrence time of measure preserving flows and semiflows, thereby giving a subtle continuous time version of the theorem of Kac.

If $\tau$ is an automorphism of a probability space $(\Omega, \mathscr{A}, \mu)$, and we put, for some $A \in \mathscr{A}$, $\nu_A(\omega) = r_A(\omega)$ $(\omega \in A)$, $\nu_A(\omega) = 0$ $(\omega \in A^c)$, then $\tau^{\nu_A}$ is again an automorphism. Neveu [1969] has studied the question, when, for a random variable $\nu$: $\Omega \to \mathbb{Z}^+$, $\tau^\nu$ is again an automorphism. It turns out that $\nu$ must be such that $\tau^\nu$ is obtained by iterated compositions of induced transformations on a decreasing sequence of subsets of $\Omega$.

Related results for more general groups of automorphisms were given by Geman and Horowitz [1975].

Jacobs [1967] has shown that the assertion of the Poincaré recurrence theorem remains valid if one replaces the stationarity assumption $\mu = \mu \circ \tau^{-1}$ by a recurrence property for the measure: Let $\mu$ be a probability measure on a complete, separable, metric space and let $\tau$: $\Omega \to \Omega$ be continuous. If the orbit $\mu \circ \tau^{-1}$, $\mu \circ \tau^{-2}, \dots$ of $\mu$ returns into each neighborhood of $\mu$ (in the weak topology given by $\mu_n \to \mu$ if $\int f d\mu_n \to \int f d\mu$ for all bounded continuous $f$), then the orbit $\tau\omega$, $\tau^2\omega, \dots$ of $\mu$-a.e. $\omega$ returns into each neighborhood of $\omega$ infinitely often.

Kurth [1975] considered homeomorphisms $\tau$ of a topological space $\Omega$ with countable basis and with a finite invariant measure, and gave a decomposition related to Poincarés theorem into departing points, asymptotic points and recurrent points.

Barone and Bhaskara Rao [1981] studied the recurrence theorem for *finitely* additive measures on a $\sigma$-algebra. Then a.e. point is $k$ times recurrent for fixed $k$, but not necessarily infinitely recurrent.

The following result of Khintchine [1934] is proved in many books (and never applied): If $\tau$ is an automorphism of a probability space, then for $\varepsilon > 0$ and $B \in \mathscr{A}$ there exists $L$ such that any interval of length $L$ contains at least one $k$ with $\mu(\tau^k B \cap B) \geqq \mu(B)^2 - \varepsilon$.

D. Maharam [1964] has "extended" nonsingular transformations $\tau$ of $\Omega$ to endomorphisms $\tilde{\tau}$ of $\Omega \times \mathbb{R}^+$ mapping the fiber of $\omega$ onto the fiber of $\tau\omega$. $\tilde{\tau}$ is conservative iff $\tau$ is conservative.

## § 1.4 Shift transformations and stationary processes

**1. Canonical representation of processes.** We now discuss a class of examples which is of great importance in probability theory.

Let $(E, \mathscr{F})$ be a measurable space and $J \neq \emptyset$ an arbitrary index set. The product space $E^J$ is the space of all functions $\omega$: $J \to E$. If $\omega_j$ is the value the function $\omega$ takes in $j \in J$, the map $X_j$: $\omega \to \omega_j$ is called the *j-th coordinate map* and $X_j(\omega) = \omega_j$ the $j$-th coordinate of $\omega$. In the case $J = \mathbb{Z}^+$ we can identify $\omega$ with the *unilateral* sequence $(\omega_0, \omega_1, \omega_2, \dots)$, and in the case $J = \mathbb{Z}$ with the *bilateral* sequence $(\dots, \omega_{-1}, \omega_0, \omega_1, \omega_2, \dots)$.

An $E$-valued random variable $Z$ defined on an abstract probability space $(\Omega', \mathscr{A}', P)$ is nothing but a measurable map $Z$: $\Omega' \to E$. A family $Y = \{Y_j, j \in J\}$

of $E$-valued random variables is called an $E$-valued *stochastic process* with *parameter space* $J$. We may write $Y_j = X_j \circ Y$, when $Y$ is considered as a map $\Omega' \to E^J$. $E$ is called *state space* of $Y$. The *product $\sigma$-algebra* $\mathscr{F}^J$ is the smallest $\sigma$-algebra in $E^J$ containing all $\sigma$-algebras $X_j^{-1}\mathscr{F}$ $(j \in J)$. $Y$ can be considered as a random variable $(\Omega', \mathscr{A}') \to (E^J, \mathscr{F}^J)$.

The *distribution* of $Z$ is the measure $P \circ Z^{-1}$ on $\mathscr{F}$, and the distribution of $Y$ is the probability measure $P \circ Y^{-1}$ on $\mathscr{F}^J$. For distinct $j_1, j_2, \dots, j_n \in J$ and $J_0 = \{j_1, \dots, j_n\}$, $E^{J_0}$ can be identified with $E^n$. The map $Y_{J_0}$: $\omega \to (Y_{j_1}(\omega), \dots, Y_{j_n}(\omega))$ is an $E^n$-valued random variable (for the $\sigma$-algebra $\mathscr{F}^n = \mathscr{F}^{J_0}$). The distribution of $Y_{J_0}$ is called the *n-dimensional marginal distribution* corresponding to $\{j_1, \dots, j_n\}$.

The distribution of $Y$ is uniquely determined by the finitedimensional marginal distributions: $P \circ Y^{-1}$ is the unique measure on $\mathscr{F}^J$ such that

$$(P \circ Y^{-1})\Big(\bigcap_{j \in J_0} X_j^{-1} A_j\Big) = P\Big(\bigcap_{j \in J_0} Y_j^{-1} A_j\Big)$$

holds for all finite $J_0 \subset J$ and all choices of $A_j \in \mathscr{F}$.

If $\{\mu_j, j \in J\}$ is a family of probability measures in $(E, \mathscr{F})$ the *product measure* $\mu = \prod_{j \in J} \mu_j$ is the unique probability measure in $E^J$ such that

$$\mu\Big(\bigcap_{j \in J_0} X_j^{-1} A_j\Big) = \prod_{j \in J_0} \mu_j(A_j)$$

holds for all $J_0$ and all choices of $A_j \in \mathscr{F}$. The random variables $Y_j$ $(j \in J)$ are called *independent* if

$$P\Big(\bigcap_{j \in J_0} Y_j^{-1} A_j\Big) = \prod_{j \in J_0} P(Y_j^{-1} A_j)$$

holds for all $J_0$, and all $A_j$. Thus, the random variables $Y_j$ $(j \in J)$ are independent iff $P \circ Y^{-1}$ is the product of the distributions $\mu_j = P \circ Y_j^{-1}$.

Now assume $J = \mathbb{Z}$ or $J = \mathbb{Z}^+$. The shift $\theta$: $E^J \to E^J$ is the transformation defined by $X_k(\theta\omega) = X_{k+1}(\omega)$. The shift in $\Omega := E^{\mathbb{Z}}$ is often called *bilateral shift*. It is bijective and both $\theta$ and $\theta^{-1}$ are measurable with respect to $\mathscr{A} = \mathscr{F}^{\mathbb{Z}}$. The shift in $\Omega^+ := E^{\mathbb{Z}^+}$ is measurable with respect to $\mathscr{A} = \mathscr{F}^{\mathbb{Z}^+}$ and surjective, but not invertible. It is called *unilateral shift*. (It will be convenient to use some notation like $\theta$, $\mathscr{A}$, $X_k$ both in $\Omega$ and $\Omega^+$. It should always be clear from the context, which interpretation applies).

If $E$ is a topological space, the unilateral shift is continuous in the product topology, and the bilateral shift is even a homeomorphism.

The simplest way to define a $\theta$-invariant measure in $\Omega$ is to take a product measure. It is easy to see that $\mu = \prod_{j \in \mathbb{Z}} \mu_j$ is $\theta$-invariant iff all $\mu_j$ are identical. The automorphism $\theta$ in $(\Omega, \mathscr{A}, \mu)$ is then called a (bilateral) *Bernoulli shift*. Unilateral Bernoulli shifts are defined in the same way in $\Omega^+$ with $\mu = \prod_{j \in \mathbb{Z}^+} \mu_j$ and identical $\mu_j$.

Random variables $Y_j (j \in J)$ are called *identically distributed* if their distributions $\mu_j$ agree. Consequently, if $Y = (Y_0, Y_1, \dots)$ is a sequence of independent identically distributed random variables, the distribution $\mu = P \circ Y^{-1}$ of $Y$ is the the shift invariant measure used for the definition of a Bernoulli shift.

Kolmogorov's *strong law of large numbers* asserts that $n^{-1} \sum_{k=0}^{n-1} Y_k$ converges *almost surely* (i.e., $P$-a.e.) to $E(Y_0) = \int Y_0 \, dP$, when the $Y_i$ are real valued, independent, identically distributed, integrable random variables. Let us see how this follows from Birkhoff's theorem. We need the ergodicity of Bernoulli shifts and show a bit more:

An endomorphism $\tau$ of a finite measure space $(\Sigma, \mathscr{B}, \nu)$ is called *mixing* if

$$(4.1) \qquad \lim_{n \to \infty} \nu(A \cap \tau^{-n} B) = \nu(A)\, \nu(B) / \nu(\Sigma)$$

holds for all $A, B \in \mathscr{B}$. Mixing implies ergodicity because (4.1) cannot hold for $A = B$ when $A$ is a $\tau$-invariant set with $0 < \nu(A) < \nu(\Sigma)$. Let $\mathscr{A}(m, n)$ denote the $\sigma$-algebra which is generated by the $\sigma$-algebras $X_k^{-1} \mathscr{F}$ $(m \leqq k \leqq n, k \neq \pm\infty)$. If $\theta$ is a Bernoulli shift, $\mu(A \cap \theta^{-n} B) = \mu(A)\, \mu(B)$ holds for all $A \in \mathscr{A}(k_1, k_2)$, $B \in \mathscr{A}(j_1, j_2)$ with $k_1 \leqq k_2 < \infty$ and $j_1 \leqq j_2 < \infty$ as soon as $j_1 + n > k_2$, because $\mu$ is a product measure and $\theta^{-n} B \in \mathscr{A}(j_1 + n, j_2 + n)$. Thus

$$(4.2) \qquad \lim_{n \to \infty} \mu(A \cap \theta^{-n} B) = \mu(A)\, \mu(B)$$

holds for all $A, B$ in the union of the $\sigma$-algebras $\mathscr{A}(k, l)$ with $k, l \neq \pm\infty$. This union is dense in $\mathscr{A}$ in the metric $d(A, A') = \mu(A \Delta A')$. Therefore (4.2) holds for all $A, B \in \mathscr{A}$ and $\theta$ is mixing.

Now put $K_x = \{\omega \in \Omega^+ : \lim_{n \to \infty} n^{-1} \sum_{k=0}^{n-1} X_0 \circ \theta^k(\omega) = E(Y_0)\}$ and $K_y = \{\omega' \in \Omega' : \lim_{n \to \infty} n^{-1} \sum_{k=0}^{n-1} Y_k(\omega') = E(Y_0)\}$. Birkhoff's theorem implies $\mu(K_x) = 1$. $X_k = X_0 \circ \theta^k$ and $Y_k = X_k \circ Y$ yield $K_y = Y^{-1}(K_x)$, so that $P(K_y) = (P \circ Y^{-1})(K_x) = \mu(K_x) = 1$. But this is the assertion of Kolmogorov's strong law.

Here we have applied the ergodic theorem only to a function on $\Omega^+$ depending on only one coordinate. But we could take any integrable function $f \colon \Omega^+ \to \mathbb{R}$, and then the sequence $f \circ \theta^k$ would in general not be a sequence of independent random variables. Thus, even in the special case of a Bernoulli shift the ergodic theorem is strictly more informative than the strong law. The main usefulness of the ergodic theorem, however, is due to the fact that there are many processes $Y = (Y_0, Y_1, \dots)$ for which $P \circ Y^{-1}$ is $\theta$-invariant although the $Y_0, Y_1, \dots$ are not independent.

If $Y = (Y_0, Y_1, \dots)$ is an $E$-valued process, the distribution $\mu = P \circ Y^{-1}$ is $\theta$-invariant iff $\mu = P \circ Y^{-1} \circ \theta^{-1} = P \circ (\theta \circ Y)^{-1}$. Now $\theta \circ Y$ is the process $(Y_1, Y_2, \dots)$ for which the time scale is shifted by one. It is therefore natural to introduce the following definition: An $E$-valued stochastic process $Y = (Y_j, j \in J)$ with $J = \mathbb{Z}$ or $J = \mathbb{Z}^+$ is called *stationary* if $Y$ and $\theta \circ Y$ have the same distribution. As the distribution is determined by the probabilities of events of the type $\{Y_{t_1} \in A_1, \ldots, Y_{t_n} \in A_n\}$, we have the equivalent definition: $Y$ is called stationary if

$$(4.3) \qquad P(Y_{t_1} \in A_1, \ldots, Y_{t_n} \in A_n) = P(Y_{t_1+s} \in A_1, \ldots, Y_{t_n+s} \in A_n)$$

holds for all $n$, all choices of $A_1, \ldots, A_n \in \mathscr{F}$, all $t_1, \ldots, t_n \in \mathbb{Z}^+$ (or $\mathbb{Z}$), and for $s = 1$, (and hence for all $s$). Intuitively, a process is stationary if the random mechanism generating the process is invariant under a translation of the time scale.

For a process $Y = (Y_t, t \in J)$ with $J = \mathbb{R}$ or $J = \mathbb{R}^+$, observed continuously, one can proceed very similarly: For an additive semigroup $J$ we define a family of shifts $\theta_t$ in the space $E^J$ by $X_s(\theta_t \omega) = X_{t+s}(\omega)$, where $X_s$ is the $s$-th coordinate. One can then call $Y$ stationary if the distribution is $\theta_s$-invariant for all $s \in J$. This is the case when (4.3) holds for all $t_1, \ldots, t_n \in J$ and all $s \in J$. (This time it is not sufficient to ask for the validity of (4.3) for a single $s$, because $\mathbb{R}$ is not a group generated by a single element).

Let us return to the case of discrete time. By the definition of stationarity a process is stationary if it generates an endomorphism. Conversely, an endomorphism generates many stationary processes:

**Proposition 4.1.** *If $\tau$ is an endomorphism of a probability space $(\Omega, \mathscr{A}, \mu)$, $(\tilde{E}, \tilde{\mathscr{F}})$ a measurable space, and $f\colon \Omega \to \tilde{E}$ measurable, then the sequence $(Z_i = f \circ \tau^i$, $(i \geq 0))$ is a stationary process. Similarly, if $\tau$ is an automorphism, $(Z_i, (i \in \mathbb{Z}))$ is stationary.*

*Proof.* If $n \in \mathbb{N}$, $A_1, \ldots, A_n \in \tilde{\mathscr{F}}$ and $t_1, \ldots, t_n \in \mathbb{Z}^+$ (or $\mathbb{Z}$) are given, we have

$$\begin{aligned} \mu(Z_{t_1} \in A_1, \ldots, Z_{t_n} \in A_n) &= \mu(f \circ \tau^{t_1} \in A_1, \ldots, f \circ \tau^{t_n} \in A_n) \\ &= \mu(\tau^{-1}\{f \circ \tau^{t_1} \in A_1, \ldots, f \circ \tau^{t_n} \in A_n\}) \\ &= \mu(f \circ \tau^{t_1+1} \in A_1, \ldots, f \circ \tau^{t_n+1} \in A_n) \\ &= \mu(Z_{t_1+1} \in A_1, \ldots, Z_{t_n+1} \in A_n). \quad \square \end{aligned}$$

**Corollary 4.2.** *If $Y = (Y_i, i \in J)$ is an E-valued stationary process on $(\Omega', \mathscr{A}', P)$, $(J = \mathbb{Z}$ or $\mathbb{Z}^+)$, and $f\colon E^J \to \tilde{E}$ measurable, then $\tilde{Y}_i = f \circ \theta^i \circ Y$ $(i \in J)$ defines an $\tilde{E}$-*valued stationary process. *(For $J = \mathbb{Z}^+$ this means that $\tilde{Y}_i = f(Y_i, Y_{i+1}, Y_{i+2}, \ldots)$ is stationary).*

*Proof.* Assume $J = \mathbb{Z}^+$. Apply proposition 4.1 with $\Omega^+$ instead of $\Omega$, and take $\tau = \theta$ and $\mu = P \circ Y^{-1}$. The stationarity of $Z = (Z_0, Z_1, \ldots)$ implies that $\mu \circ Z^{-1}$ is shift-invariant in $\tilde{\Omega}^+ = \tilde{E}^{\mathbb{Z}^+}$. We have $\mu \circ Z^{-1} = P \circ Y^{-1} \circ Z^{-1} = P \circ (Z \circ Y)^{-1}$. But $Z \circ Y$ is the process $\tilde{Y} = (\tilde{Y}_0, \tilde{Y}_1, \ldots)$. The same proof works for $J = \mathbb{Z}$. $\square$

For example, if the process $Y_0, Y_1, \ldots$ is real valued and stationary, the process

$\tilde{Y}_i$ defined by the *moving averages* $\tilde{Y}_i = M^{-1} \sum_{k=i}^{i+M-1} Y_k$, $M > 1$ fixed, is stationary. Clearly, the $\tilde{Y}_i$ will in general be dependent even if the $Y_i$ are independent. For the probabilist, the object of primary interest is the stationary process, and not the measure preserving transformation. Therefore, it is of interest to express the ergodicity of $\theta$ in terms of the process:

If $Y = (Y_j, j \in \mathbb{Z}^+)$ is defined on $(\Omega', \mathscr{A}', P)$, call $B \in \mathscr{A}'$ *invariant* if there exists some $A \in \mathscr{A}$ (in $\Omega^+$) such that

$$(4.4) \qquad B = \{(Y_n, Y_{n+1}, Y_{n+2}, \ldots) \in A\}$$

is true for all $n \geqq 0$. This is equivalent to the existence of an $A^* \in \mathscr{A}$ with $B = Y^{-1} A^*$ and $\theta^{-1} A^* = A^*$. One direction of this equivalence ist obvious: If such an $A^*$ exists, we have $\theta^{-n} A^* = A^*$ and $B = Y^{-1} \theta^{-n} A^* = \{(Y_n, Y_{n+1}, \ldots) \in A^*\}$ for all $n \geqq 0$. If $A$ with (4.4) exists $A$ must be $\theta$-invariant mod $\mu$, because $Y^{-1}(A \Delta \theta^{-1} A) = \{(Y_0, Y_1, \ldots) \in A$ and $(Y_1, Y_2, \ldots) \notin A\} \cup \{(Y_0, Y_1, \ldots) \notin A$ and $(Y_1, Y_2, \ldots) \notin A\}$ is empty. ($A = \theta^{-1} A$ need not hold!) In section 1.1 we have sketched how to find for $A$ a $\theta$-invariant $A^2$ with $\mu(A \Delta A^2) = 0$. It is an exercise to show that (4.4) implies $B = Y^{-1} A^2$, so that we can use $A^2$ for $A^*$. The invariant sets in $\Omega'$ for the process $Y$ form a $\sigma$-algebra $\mathscr{I}'$, and we have $\mathscr{I}' = Y^{-1} \mathscr{I}$, where $\mathscr{I}$ is the $\sigma$-algebra of $\theta$-invariant sets in $\Omega^+$. We could also call a set $B' \in \mathscr{A}'$ invariant mod $P$ if there exists an $A \in \mathscr{A}$ for which $P(B' \Delta \{(Y_n, Y_{n+1}, \ldots) \in A\}) = 0$ is true for all $n$; but then, again, we could use the same arguments to show that these are just sets $B'$ which differ from an invariant $B$ on a set of $P$-measure 0.

Once we have a concept of an invariant set for a stationary process, we also have a concept of ergodicity. $Y$ is called *ergodic* if any invariant set $B \in \mathscr{A}'$ satisfies $P(B) = 0$ or $P(B^c) = 0$.

**Proposition 4.3.** *If $Y = (Y_0, Y_1, \ldots)$ is stationary and ergodic and $f: \Omega^+ \to \tilde{E}$ is measurable, then the process $\tilde{Y} = (\tilde{Y}_0, \tilde{Y}_1, \ldots)$ defined by $\tilde{Y}_i = f(Y_i, Y_{i+1}, \ldots)$ is ergodic.*

*Proof.* We have $\tilde{Y} = Z \circ Y$. If $\tilde{Y}$ is not ergodic there exists an invariant $B$ for the $\tilde{Y}$-process such that $0 < P(B) < 1$. For $B$ there exists a shift-invariant set $\tilde{A}$ in $\tilde{\Omega}^+$ with $B = \tilde{Y}^{-1} \tilde{A}$. Put $A = Z^{-1} \tilde{A}$. It is straightforward to check that the invariance of $\tilde{A}$ under the shift in $\tilde{\Omega}^+$ implies the $\theta$-invariance of $A$. Therefore $B = Y^{-1} A$ is invariant for the $Y$-process, a contradiction to the ergodicity of $Y$. □

Birkhoff's ergodic theorem, spelled out for stationary processes instead of endomorphisms $\tau$, now has the following form:

**Theorem 4.4.** *If $Y_0, Y_1, \ldots$ is a stationary real valued process and $Y_0$ is integrable, then* $\lim_{n \to \infty} n^{-1} \sum_{k=0}^{n-1} Y_k = E(Y_0 | \mathscr{I}')$.

This follows in the same way from Birkhoff's theorem as Kolmogorov's strong law.

We have discussed stationarity for a process $Y = (Y_j, j \in \mathbb{Z}^+)$ defined on an arbitrary probability space. Now note that $X = (X_j, j \in \mathbb{Z}^+)$ is the identity on $\Omega^+$. Therefore the distribution $\mu = P \circ Y^{-1}$ agrees with the distribution of $X$. Hence, all probability statements on $Y$ can be expressed in terms of $X$. For this reason $X$ is called the *canonical representation* of processes with distribution $\mu$.

**2. Remote $\sigma$-algebras.** The $\sigma$-algebra $\mathscr{A}_\infty = \bigcap_{n=0}^{\infty} \mathscr{A}(n, \infty)$ in $\Omega^+$ and in $\Omega = E^{\mathbb{Z}}$ is called the (right) *remote $\sigma$-algebra*. (Also the term *tail $\sigma$-algebra* is in use). In $\Omega$ we can also define the left remote $\sigma$-algebra $\mathscr{A}_{-\infty} = \bigcap_{n=0}^{\infty} \mathscr{A}(-\infty, -n)$. $\mathscr{A}'(m, n) = Y^{-1}\mathscr{A}(m, n)$ is the smallest $\sigma$-algebra in $\Omega'$ in which the random variables $Y_j$ $(m \leqq j \leqq n, j \neq \pm\infty)$ are measurable.

$$\mathscr{A}'_\infty = \bigcap_{n=0}^{\infty} \mathscr{A}'(n, \infty) \, (= Y^{-1}\mathscr{A}_\infty) \quad \text{and} \quad \mathscr{A}'_{-\infty} = \bigcap_{n=0}^{\infty} \mathscr{A}'(-\infty, -n)$$

are the right and left remote $\sigma$-algebras of $Y$. Events like $\{\omega' \in \Omega': \lim_{n\to\infty} n^{-1} \sum_{k=0}^{n-1} Y_k(\omega') \geqq \alpha\}$ or $\{\omega' \in \Omega': Y_k(\omega') \in B_k$ for infinitely many $k\}$ belong to $\mathscr{A}'_\infty$.

The $\sigma$-algebra $\mathscr{I}$ in $\Omega^+$ is contained in $\mathscr{A}_\infty$ because of $A \in \mathscr{I} \Rightarrow A \in \mathscr{A} = \mathscr{A}(0, \infty) \Rightarrow A = \theta^{-n}A \in \mathscr{A}(n, \infty)$, $(n \geqq 0)$. Similarly $\mathscr{I}' \subset \mathscr{A}'_\infty$ for processes with index set $\mathbb{Z}^+$.

We call a $\sigma$-algebra *trivial* (with respect to a given measure) if it contains only nullsets and their complements. Ergodicity means that the $\sigma$-algebra of invariant sets is trivial. Therefore a stationary process $Y = (Y_j, j \in \mathbb{Z}^+)$ for which $\mathscr{A}'_\infty$ is trivial must be ergodic. If $\nu$ is a measure on a $\sigma$-algebra $\mathscr{B}$ and $\mathscr{B}_0$, $\mathscr{B}_1$ are sub-$\sigma$-algebras, $\mathscr{B}_0 \subset \mathscr{B}_1 \pmod \nu$ shall mean that for all $B_0 \in \mathscr{B}_0$ there exists a $B_1 \in \mathscr{B}_1$ with $\nu(B_0 \Delta B_1) = 0$.

**Proposition 4.5.** *If $(Y_j, j \in \mathbb{Z})$ is a bilateral stationary process, we have $\mathscr{I}' \subset \mathscr{A}'_\infty$* (mod $P$) *and $\mathscr{I} \subset \mathscr{A}_\infty$* (mod $\mu$).

*Proof.* Because of $\mathscr{I} \subset \mathscr{A} = \mathscr{A}(-\infty, +\infty)$ there exists for any $A \in \mathscr{I}$ and for any $\varepsilon > 0$ an $n_\varepsilon \in \mathbb{N}$ and an $A_\varepsilon \in \mathscr{A}(-n_\varepsilon, +\infty)$ with $\mu(A_\varepsilon \Delta A) < \varepsilon$. For all $n$, $A = \theta^{-n}A$ and $\mu = \mu \circ \theta^{-n}$ imply $\mu(A \Delta \theta^{-n} A_\varepsilon) < \varepsilon$. But $\theta^{-n}A_\varepsilon \in \mathscr{A}(-n_\varepsilon + n, \infty)$, so that $A$ can be approximated arbitrarily well by sets in $\mathscr{A}(m, \infty)$ where $m$ is arbitrarily large. This implies $A \in \mathscr{A}_\infty \bmod \mu$. Applying $Y^{-1}$ also $\mathscr{I}' \subset \mathscr{A}'_\infty$ mod $P$ follows. □

These observations yield an alternative proof of the ergodicity of Bernoulli shifts.

One simply has to apply *Kolmogorov's zero-one-law* which asserts that $\mathscr{A}'_\infty$ is trivial for any process $(Y_j, j \in \mathbb{Z}$ or $\mathbb{Z}^+)$ consisting of independent random variables $Y_j$. The following theorem of Blackwell and Freedman [1964] shows that the triviality of the remote $\sigma$-algebra is, in fact, equivalent to an asymptotic independence condition:

**Theorem 4.6.** *If $(Y_j, j \in \mathbb{Z}^+)$ is any sequence of random variables defined on $(\Omega', \mathscr{A}', P)$, the remote $\sigma$-algebra $\mathscr{A}'_\infty$ is trivial if and only if each $B \in \mathscr{A}'$ satisfies*

$$(4.5) \qquad \lim_{n\to\infty} \sup_{A \in \mathscr{A}'(n,\infty)} |P(A \cap B) - P(A)\,P(B)| = 0\,.$$

*Proof.* For any $A \in \mathscr{A}'_\infty$ and any $\varepsilon > 0$ there exists an $n_\varepsilon$ and an $A_\varepsilon \in \mathscr{A}(n_\varepsilon, \infty)$ with $P(A \Delta A_\varepsilon) < \varepsilon$. As $A$ belongs to each $\mathscr{A}'(n, \infty)$ we can apply (4.5) to $B = A_\varepsilon$ to get $P(A \cap A_\varepsilon) = P(A)\,P(A_\varepsilon)$. Passing with $\varepsilon$ to zero we arrive at $P(A \cap A) = P(A)^2$, so that $P(A) \in \{0, 1\}$. Conversely, suppose $\mathscr{A}'_\infty$ is trivial and fix $B \in \mathscr{A}'$. By the backward martingale convergence theorem (see Doob [1953: thm. 4.2, 382]) the sequence $E_P(1_B|\mathscr{A}'(n,\infty))$ of conditional expectations converges a.e. and in $L_1$-norm to $E_P(1_B|\mathscr{A}'_\infty)$, which equals $P(B)$ because of the triviality of $\mathscr{A}'_\infty$. For any $A \in \mathscr{A}'(n, \infty)$,

$$\begin{aligned} |P(A \cap B) - P(A)\,P(B)| &= |\int_A (1_B - P(B))\,dP| \\ &= |\int_A (E_P(1_B|\mathscr{A}'(n,\infty)) - P(B))\,dP| \\ &\leqq \|E_P(1_B|\mathscr{A}'(n,\infty)) - P(B)\|_1. \end{aligned}$$

Hence (4.5) follows. □

**3. Recurrence times.** Let $Y = (Y_j, j \in \mathbb{Z}^+)$ be a stationary process and let $A'$ be of the form $A' = \{Y \in A\} = Y^{-1}A$. For example, $A' = \{Y_0 \in A_0\}$ or $A' = \{Y_0 > Y_1 + Y_2\}$. The set $A$ may depend on infinitely many coordinates. We assume $P(A') > 0$.

Let $P_A(B') = P(A')^{-1}\,P(A' \cap B')$ denote the conditional probability of $B'$ given $\{Y \in A\}$. $P_A \circ Y^{-1}$ is the normalized restriction of $\mu = P \circ Y^{-1}$ to $A$.

By the recurrence theorem $\mu$-a.e. point of $A$ belongs to the set $A_{inf}$ of points $\omega \in A$ returning to $A$ infinitely often under $\theta$. We delete the $P$-nullset $Y^{-1}(A \setminus A_{inf})$ from $\Omega'$. This does not affect the distribution of $Y$. Then, for all $\omega' \in A'$ there are infinitely many $n \in \mathbb{N}$ with $(Y_n(\omega'), Y_{n+1}(\omega'), \ldots) \in A$, so that the random variables $R_0 = 0$,

$$R_{i+1}(\omega') = \inf\{n > R_i(\omega') \colon (Y_n(\omega'), Y_{n+1}(\omega'), \ldots) \in A\}, \qquad (i \geqq 0)$$

are finite in $A'$. The *recurrence times* are given by $T_i = R_i - R_{i-1}$, $(i \geqq 1)$. If $r_A(\omega) = \inf\{n \geqq 1 \colon \theta^n \omega \in A\}$ is the return time studied in section 1.3 and $\omega = Y(\omega')$, then $R_i(\omega')$ is the time of the $i$-th return of $\omega$ to $A$. Formally, we have $T_i(\omega')$

$= r_A(\theta^{R_{i-1}(\omega')}\omega)$. Therefore $(Y_{R_i}, Y_{R_i+1}, \ldots)(\omega')$ is the point $\theta_A^i(\omega)$, where $\theta_A$ is the transformation induced by the shift on $A$. Combining Theorem 3.6, proposition 3.8, corollary 4.2 and proposition 4.3 we have proved:

**Theorem 4.7.** *Let $Y = (Y_i, i \geqq 0)$ be an E-valued stationary process, and let f be a measurable map from $\Omega^+$ into $\tilde{E}$. Assume $A \in \mathscr{A}$ and $\mu(A) > 0$. Then the process*

$$U_i = f(Y_{R_i}, Y_{R_i+1}, \ldots) \quad (i \geqq 0)$$

*on $(A', A' \cap \mathscr{A}', P_A)$ is stationary. The process $(U_i, i \geqq 0)$ is ergodic if Y is ergodic. In particular the processes $T = (T_i, i \geqq 1)$ and $V = (V_i, i \geqq 0)$ with $V_i = Y_{R_i}$ are stationary, and they are ergodic if Y is ergodic.*

It is clear that an analogous theorem holds for bilateral processes.

**4. Bilateral extensions of unilateral processes.** In a way, bilateral stationary processes are simpler than unilateral processes because the bilateral shift is invertible. It is therefore of interest that in all probabilistic questions on stationary processes we may assume that the process is bilateral if $E$ satisfies very mild regularity assumptions.

Call $(E, \mathscr{F})$ a *Borel space* if there exists a bijective map $\psi$ of $E$ onto a Borel subset $E' \subset \mathbb{R}^1$ which is measurable in both directions. E.g., *Polish spaces* (i.e., complete separable metric spaces) and Borel subsets of such spaces with their Borel $\sigma$-algebra are Borel spaces.

**Theorem 4.8.** *If $Y = (Y_j, j \in \mathbb{Z}^+)$ is a unilateral stationary process defined on a probability space $(\Omega', \mathscr{A}', P)$ and taking values in a Borel space $(E, \mathscr{F})$, then there exists a $\theta$-invariant probability measure $\mu$ on $\Omega = E^{\mathbb{Z}}$ such that the distribution of $(X_j, j \geqq 0)$ in $(\Omega, \mathscr{A}, \mu)$ agrees with that of Y.*

*Proof.* For any set $A \in \mathscr{A}(-n, +n)$ there exists a measurable subset $A^{(n)}$ of $E^{2n+1}$ such that $A = \{\omega \in \Omega: (X_{-n}(\omega), \ldots, X_n(\omega)) \in A^{(n)}\}$. If we put $\mu(A) = P(\{\omega' \in \Omega': (Y_{k-n}(\omega'), \ldots, Y_{k+n}(\omega')) \in A^{(n)}\})$, then the stationarity of $Y$ implies that this definition is independent of $k$ as long as $k \geqq n$, and also that the same number is assigned to $\mu(A)$ if $A$ is considered as an element of the $\sigma$-algebra $\mathscr{A}(-(n+1), n+1)$. Thus these definitions of the marginal distributions of $(X_{-n}, \ldots, X_n)$ are consistent. An application of the Daniell-Kolmogorov extension theorem (see e.g. Jacobs [1978]), completes the proof. □

It is a simple consequence of proposition 4.5 that the ergodicity of a unilateral stationary process is equivalent to the ergodicity of its bilateral extension.

**5. Further examples of stationary processes.** A class of examples of great interest in probability theory and in prediction theory is provided by the stationary

*Gaussian processes.* A real valued process $(Y_j, j \in \mathbb{Z})$ is called Gaussian if the distribution of $(Y_{j_1}, \ldots, Y_{j_n})$ is a multivariate normal distribution for arbitrary $j_1, \ldots, j_n \in \mathbb{Z}$. Such distributions are completely determined by the expectations $E(Y_{j_k})$ and the covariances $\operatorname{Cov}(Y_{j_l}, Y_{j_k})$. In the stationary case the expectations are identical and the covariances depend only on the differences $j_l - j_k$, so that the distribution of the entire process is determined by the sequence $r(n) = \operatorname{Cov}(Y_n, Y_0)$ together with $E(Y_0)$. $r(n)$ can be an arbitrary non negative definite sequence. Small $r(n)$ corresponds to approximate independence of $Y_k$ and $Y_{k+n}$. Therefore Gaussian processes are suited for studies of the effect of various speeds of asymptotic independence. Another advantage is that their finite dimensional distributions have a density which allows explicit computations involving only few parameters.

Another important class of stationary processes arises in the theory of *Markov chains*. Let $E \neq \emptyset$ be a finite or countably infinite set, and let $Y_0, Y_1, \ldots$ be a sequence of $E$-valued random variables. We can think of $E$ as the set of possible states in which a system can be, and $Y_n$ can be considered as the state of the system at time $n$. If $P(A|B) = P(A \cap B)/P(B)$ denotes the *conditional probability* of $A$ given $B$, the conditional probabilities $P(Y_{n+1} = i_{n+1} | Y_0 = i_0, \ldots, Y_n = i_n)$ will in general depend on the whole sequence $i_0, i_1, \ldots, i_{n+1}$. We say that $Y_0, Y_1, \ldots$ has the *Markov property* and call $Y_0, Y_1, \ldots$ a *Markov chain*, if, for all $n \geqq 1$ and all $i_0, i_1, \ldots, i_{n+1} \in E$, this conditional probability is independent of $i_0, \ldots, i_{n-1}$ and just depends on $i_n$ and $i_{n+1}$; this is formally expressed by

$$P(Y_{n+1} = i_{n+1} | Y_0 = i_0, \ldots, Y_n = i_n) = P(Y_{n+1} = i_{n+1} | Y_n = i_n).$$

Intuitively, this means that the probability of a transition from state $i_n$ at time $n$ to state $i_{n+1}$ at time $n+1$ is not influenced by the states in which the system was before time $n$. (Strictly speaking $P(A|B)$ is defined for events $B$ of *positive* probability, so that $P(Y_{n+1} = i_{n+1} | Y_0 = i_0, \ldots, Y_n = i_n)$ may be undefined for some $i_0, \ldots, i_n$. But this is no reason to be alarmed because we can then define the conditional probabilities in accordance with the Markov property).

We say that a Markov chain has *stationary transition probabilities* if $P(Y_{n+1} = j | Y_n = i)$ is independent of $n$. We then write $p_{ij}$ for this probability of passing from $i$ to $j$. Clearly, a Markov chain which is a stationary process must have stationary transition probabilities.

The theory of stationary Markov chains has many applications and is highly developed. Many of the results extend to the case of more general state spaces which need not be countable. The ergodic theoretic aspects are treated in Chapter 3 and in the supplement. Here we just show how Markov chains lead in a natural way to stationary processes, and to shift transformations with a $\sigma$-finite *infinite* invariant measure.

A family $(p_{ij}: i, j \in E)$ is called a *stochastic matrix* if the $p_{ij}$ are non negative and $\sum_{j \in E} p_{ij} = 1$ for all $i$. Given any finite or $\sigma$-finite measure $\varrho = (\varrho_i: i \in E)$ on $E$ and any stochastic matrix we can define a measure $\mu_\varrho$ on $\Omega^+$ by

(4.6) $$\mu_\varrho(X_0 = i_0, X_1 = i_1, \ldots, X_n = i_n) = \varrho_{i_0} \prod_{t=0}^{n-1} p_{i_t, i_{t+1}}.$$

(As these marginal distributions are consistent, $\mu_\varrho$ can be extended to $\mathscr{A}$).

If $\pi = (\pi_i)$ with $\pi_i = P(Y_0 = i)$ is the *initial distribution* of a Markov chain with stationary transition probabilities $p_{ij}$, one easily checks that $P \circ Y^{-1}$ is the measure $\mu_\pi$ on $\Omega^+$. Conversely, if $\varrho$ is a probability measure on $E$, the process $X_0, X_1, \ldots$ on $(\Omega^+, \mathscr{A}, \mu_\varrho)$ is a Markov chain with initial distribution $\varrho$ and stationary transition probabilities $p_{ij}$. $\sigma$-finite measures $\varrho$ have no interpretation as probabilities, but come up naturally in the theory of Markov chains.

If $\mu_\varrho$ is invariant under the shift $\theta$, we have $\varrho_j = \mu_\varrho(X_0 = j) = \mu_\varrho(\theta^{-1}\{X_0 = j\}) = \mu_\varrho(X_1 = j)$ for all $j$. But $\mu_\varrho(X_1 = j) = \sum_{i \in E} \mu_\varrho(X_0 = i, X_1 = j) = \sum_{i \in E} \varrho_i p_{ij}$. Call $\varrho$ *invariant* under $(p_{ij})$ when

(4.7) $$\sum_{i \in E} \varrho_i p_{ij} = \varrho_j \quad (j \in E)$$

holds. For such a $\varrho$, we have for all $n$ and $i_0, \ldots, i_n \in E$

$$\begin{aligned}
&\mu_\varrho(X_1 = i_0, X_2 = i_1, \ldots, X_{n+1} = i_n) \\
&= \sum_i \mu_\varrho(X_0 = i, X_1 = i_0, \ldots, X_{n+1} = i_n) \\
&= \sum_i \varrho_i p_{i i_0} \prod_{t=0}^{n-1} p_{i_t i_{t+1}} = \mu_\varrho(X_0 = i_0, \ldots, X_n = i_n),
\end{aligned}$$

and this suffices to establish the $\theta$-invariance of $\mu_\varrho$.

We have proved:

**Proposition 4.9.** *The shift in $(\Omega^+, \mathscr{A}, \mu_\varrho)$ defined with the help of a measure $\varrho$ on $E$ and a stochastic matrix $(p_{ij})$ is measure preserving if and only if $\varrho$ is invariant under $(p_{ij})$.*

A measure preserving shift in $(\Omega^+, \mathscr{A}, \mu_\varrho)$ is called a (unilateral) *Markov shift*. Observe that the procedure described in the proof of theorem 4.8 also permits to extend $\sigma$-finite shift invariant measures from $\Omega^+$ to $E^{\mathbb{Z}} = \Omega$. With this extension of $\mu_\varrho$ the shift in $\Omega$ is called a bilateral Markov shift.

If $\pi$ is *any* initial distribution and $\varrho$ a $(p_{ij})$ – invariant measure on $E$ with $\pi \ll \varrho$, then $\mu_\pi \ll \mu_\varrho$. Therefore any assertion proved for $\mu_\varrho$-a.e. $\omega \in \Omega^+$ with the help of the $\theta$-invariance of $\mu_\varrho$ will be true for $\mu_\pi$ -a.e. $\omega \in \Omega^+$ even when $\pi$ is not invariant under $(p_{ij})$. In this way ergodic theoretic results on endomorphisms in $\sigma$-finite measure spaces can lead to results on Markov chains $(Y_0, Y_1, \ldots)$ defined on a probability space.

A state $i$ is called *recurrent*, if the probability of a return to $i$ given a start in $i$ is 1, and *transient* otherwise. If all states are recurrent, there exists a $\sigma$-finite strictly positive invariant $\varrho$ and the shift is conservative. If all states are transient and such a $\varrho$ exists, $\theta$ is dissipative; see the Notes of § 3.1.

**6. Stationarity in the wide sense.** If $Y = (Y_i, i \in \mathbb{Z}^+)$ is a real or complex valued stochastic process on $(\Omega', \mathscr{A}', P)$ there is a different notion of stationarity which can be used to apply the ergodic theorem of von Neumann. $Y$ is called *stationary in the wide sense* if all $Y_i$ belong to $L_2(\Omega', \mathscr{A}', P)$ and $E(Y_i \bar{Y}_j) = E(Y_0 \bar{Y}_{j-i})$ is true for all $i, j \in \mathbb{Z}^+$ with $i \leqq j$. Here $\bar{a}$ is the complex conjugate of a. Recall that $\langle f, g \rangle = E(f\bar{g}) = \int f\bar{g}dP$ is the scalar product in $L_2$. If $Y$ is stationary in the wide sense, and $V = \sum_{i=0}^{n} \alpha_i Y_i$ put $SV = \sum_{i=0}^{n} \alpha_i Y_{i+1}$. Then

$$\begin{aligned} \|SV\|_2 &= \langle \sum_{i=0}^{n} \alpha_i Y_{i+1}, \sum_{j=0}^{n} \alpha_j Y_{j+1} \rangle \\ &= \sum_{i=0}^{n} \sum_{j=0}^{n} \alpha_i \bar{\alpha}_j E(Y_{i+1} \bar{Y}_{j+1}) \\ &= \sum_{i=0}^{n} \sum_{j=0}^{n} \alpha_i \bar{\alpha}_j E(Y_i \bar{Y}_j) = \|V\|_2 . \end{aligned}$$

If $V = \sum_{k=0}^{m} \beta_k Y_k$ is a different representation of $V$, $\| \sum_{i=0}^{n} \alpha_i Y_i - \sum_{k=0}^{m} \beta_k Y_k \|_2 = 0$ implies $\| \sum_{i=0}^{n} \alpha_i Y_{i+1} - \sum_{k=0}^{m} \beta_k Y_{k+1} \|_2 = 0$. Hence $S$ is defined unambiguously. $S$ acts isometrically on the linear span $\mathfrak{H}_0$ of $\{Y_i, i \geqq 0\}$, and extends to an isometry $S$ of the Hilbert space $\mathfrak{H} = cl\,\mathfrak{H}_0$. When we apply von Neumann's theorem to $S$ and $Y_0$, we see that $n^{-1} \sum_{i=0}^{n-1} Y_i$ converges in $L_2$-norm.

A stationary process $Y = (Y_i, i \geqq 0)$ is stationary in the wide sense if $Y_0$ belongs to $L_2$. Apart from this boundedness condition stationarity in the wide sense is a much weaker property than stationarity. However, if $Y$ is a real valued Gaussian process which is stationary in the wide sense and satisfies $E(Y_i) = 0$ for all $i$, then $Y$ is stationary, because the $n$-dimensional normal distribution with given expectation is determined by the covariances.

For bilateral processes stationarity in the wide sense is defined in the same way, and $S$ is unitary.

**7. Asymptotic mean stationarity.** Consider a measurable transformation $\tau: \Omega \to \Omega$ of a probability space $(\Omega, \mathscr{A}, \mu)$. $\mu$ is called *asymptotically mean stationary* (with respect to $\tau$) if

$$(4.8) \qquad \bar{\mu}(B) := \lim n^{-1} \sum_{i=0}^{n-1} \mu(\tau^{-i} B)$$

exists for all $B \in \mathscr{A}$. Clearly $\bar{\mu}$ is $\tau$-invariant. $\bar{\mu}$ is called the *stationary mean* of $\mu$.

The Vitali-Hahn-Saks theorem implies that $\bar{\mu}$ is a probability measure. The following theorem has been proved by Mrs. Dowker [1951] for invertible null preserving $\tau$, and by Gray and Kieffer [1980] in the general form:

**Theorem 4.10.** *$\mu$ is asymptotically mean stationary iff $A_n(\tau)f$ converges $\mu$-a.e. for all bounded measurable $f$.*

*Proof.* If $A_n(\tau)1_B$ converges a.e. for $B \in \mathscr{A}$, then (4.8) is obtained by integrating. For the converse, consider the set $B_f$ of points $\omega$ for which $A_n(\tau)f(\omega)$ converges. The $\tau$-invariance of $B_f$ implies $\mu(B_f) = \bar{\mu}(B_f)$. Birkhoff's theorem yields $\bar{\mu}(B_f) = 1$. □

Let us say that $\mu'$ *asymptotically dominates* $\mu$ (with respect to $\tau$) if $\mu'(B) = 0$ implies $\lim \mu(\tau^{-n}B) = 0$.

**Proposition 4.11.** *If a $\tau$-invariant probability measure $\mu'$ asymptotically dominates $\mu$, then $\mu$ is asymptotically mean stationary.*

*Proof.* Birkhoff's theorem implies $\mu'(B_f) = 1$. Hence $\lim \mu(\tau^{-n}B_f) = 1$. But the sets $\tau^{-n}B_f$ are identical. Hence $\mu(B_f) = 1$. □

An $E$-valued process $Y = (Y_0, Y_1, \ldots)$ on a probability space $(\Omega', \mathscr{A}', P)$ is called asymptotically mean stationary if $P \circ Y^{-1}$ is asymptotically mean stationary under the shift in $\Omega^+$. This means that $\lim n^{-1} \sum_{i=0}^{n-1} P((Y_i, Y_{i+1}, Y_{i+2}, \ldots) \in B)$ exists for measurable $B \subset \Omega^+$. Theorem 4.10 asserts that this is equivalent to the convergence $P$-a.e. of $n^{-1} \sum_{i=0}^{n-1} f(Y_i, Y_{i+1}, Y_{i+2}, \ldots)$ for all bounded measurable $f$ on $\Omega^+$.

The point of these considerations is that asymptotically mean stationary processes come up in a natural way in several examples. E.g., asymptotic mean stationarity is stable under conditioning, while stationarity is not.

## Notes

Borel [1909] proved the strong law of large numbers for independent $Y_1, Y_2, \ldots$ with $p = P(Y_i = 1) = 1 - P(Y_i = 0)$. Kolmogorov [1930] proved the almost sure convergence to 0 of $a_n^{-1} \sum_{i=1}^{n} (Y_i - EY_i)$ when $Y_1, Y_2, \ldots$ are independent random variables with

$$\sum_{i=1}^{\infty} a_i^{-2} \operatorname{Var}(Y_i) < \infty$$

and $0 < a_n \uparrow \infty$. This is used in the proof of his strong law for independent identically distributed random variables with finite expectations, of which the first statement seems to be that in the book of Kolmogorov [1933].

Doob [1953] has given an account of the relations between measure preserving transformations and stationary processes.

Theorem 4.7 is a generalization of results of Mrs. Moy [1959] and Ryll-Nardzewski

[1961], who proved that the process $(T_i)$ of successive time differences between visits to $A$ is stationary and ergodic under $P_A$, when $A$ is of the form $\{Y_0 \in B\}$.

Geman and Horowitz [1976] constructed ergodic stationary proceses $(X_n)$ for which the return time $N(\omega) = \inf\{n \geqq 1: X_n(\omega) = X_0(\omega)\}$ is finite a.e. although $X_0$ has a continuous distribution.

The recurrence of $S_n = X_1 + X_2 + \ldots + X_n$ received a lot of attention recently. Call $(S_n)$ recurrent if $P(\exists n > 1: S_n \in U) = 1$ for each neighborhood $U$ of 0. Atkinson [1976] showed for ergodic $(X_n)$ with integrable $X_1$ that $(S_n)$ is recurrent iff $EX_1 = 0$. Berbee [1981] showed for $\mathbb{R}^d$-valued stationary processes that $\Omega$ is the union of the sets $\{\omega: \|S_n(\omega)\| \to \infty\}$ and $\{\omega$: each $S_m(\omega)$ is a limit point of $(S_n(\omega))\}$. Related references are Dekking [1982], Westman [1980]. Aaronson-Keane [1982] analysed the asymptotic growth for $N \to \infty$ of the number of visits to 0 of $(S_n)$ up to time $N$, when $(X_i)$ stems from certain rotations of $[0, 1[$.

There exists an extensive literature on Gaussian processes; see Neveu [1968], Ibragimov-Rozanov [1978]. Ergodic and mixing properties of Gaussian processes have been explored by K. Ito [1944], [1944a], [1952], Maruyama [1949], Grenander [1952], Leonov [1960], Nawrotzki [1969], Totoki [1964], [1970], Newton [1966], [1968], and Versik [1962]. For a Gaussian automorphism $\theta$ ergodicity, weak mixing and $r$-fold mixing are equivalent to $r(n) \to 0$. $\theta$ is isomorphic to a Bernoulli shift iff the spectral measure $m$ (with Fourier transform $(r(n))$ is absolutely continuous with respect to Lebesgue measure. The books of Chung [1967], Freedman[1971], and Kemeny-Snell-Knapp [1966] are well known references on Markov chains.

Wide sense stationary processes are of great interest in prediction theory; see Urbanik [1967] and Rozanov [1967].

Under various mixing conditions one can derive central limit theorems (and stronger results) for stationary processes; see Ibragimov-Linnik [1971] and Philipp-Stout [1975].

Stationarity plays an important role in studies of point processes; see the books of Kerstan-Matthes-Mecke [1974], and Franken-König-Arndt-Schmidt [1981], and the lecture notes of Neveu [1976], and see § 6.2. Rolski [1981] discusses stationary random processes associated with problems in queuing theory. (He also considers asymptotically mean stationary processes, calling them *stable*.)

Ambrose [1941] and Ambrose-Kakutani [1942] have established a basic link between discrete and continuous time by their representation of flows by flows under a function. If $\{\tau_t: t \in \mathbb{R}\}$ is a (proper) measure preserving flow in a probability space $(\Omega, \mathscr{A}, \mu)$, there exists a cross-section $\Omega_0 \subset \Omega$ such that each orbit meets $\Omega_0$ during a discrete sequence of times. Papangelou [1974] has connected this to the theory of stationary point processes. De Sam Lazaro and Meyer [1975] have worked out a whole theory of flows, stationary processes and point processes. Related work in a probabilistic setting appears also in papers of Prizva [1972] and Nawrotzki [1982].

Krengel [1969] and Kubo [1969] extended the representation to nonsingular flows, and Krengel [1969], [1971a] to semiflows and "filtered" flows, where the representation has to respect additional structure. (These papers were overlooked by de Sam Lazaro and Meyer, who derived a representation of filtered flows in their setting).

Rudolph [1976] showed that, for ergodic measure preserving flows, $\Omega_0$ can be found such that the times between two visits to $\Omega_0$ assume only two values. This was slightly improved and extended to nonsingular flows by Krengel [1976a]. Arques and Gabriel [1977] obtained this for filtered flows.

Kieffer and Rahe [1981] have applied asymptotic mean stationary processes in information theory. They also proved the convergence a.e. of $n^{-1} \sum_{i=0}^{n-1} X_1 X_2 \ldots X_n$, when $(X_i)$ is an asymptotically mean stationary process with values in the set of $b \times b$ stochastic matrices.

In a sense *all* ergodic automorphisms in probability spaces satisfying mild regularity conditions can be represented as shifts with finite or countable state space $E$. This follows from the theory of *generators* of measure preserving transformations. We refer to the surveys of Krieger [1975] and Sujan [1983] and to the book of Denker-Grillenberger-Sigmund [1976]. Grillenberger and Krengel [1976a] showed that one can even prescribe, for fixed $n$, the distribution of $(X_0, \ldots, X_n)$, if the automorphism does not have too large entropy.

## § 1.5 Kingman's subadditive ergodic theorem and the multiplicative ergodic theorem of Oseledeč

In 1968 an important new impetus has been given to the study of ergodic theorems for measure preserving transformations by Kingman's proof of this subadditive ergodic theorem, which opened up an impressive number of new applications.

We give a proof of Kingman's pointwise convergence theorem via a maximal inequality which extends Wieners maximal inequality to the superadditive case. We then sketch some applications. Following Raghunathan [1979] and Ruelle [1979] we show that Kingman's theorem can be used to derive the multiplicative ergodic theorem of Oseledeč [1968], which is of considerable interest in the study of differentiable dynamical systems.

**1. Discrete parameter subadditive processes.** Let $\tau$ be an endomorphism of $(\Omega, \mathscr{A}, \mu)$. Put $Q = \{(i,k) \in \mathbb{Z}^2 : 0 \leqq i < k\}$. A family $F = \{F_{i,k}: (i,k) \in Q\}$ of integrable functions is called a *subadditive process* if

(i) $F_{i,k} \circ \tau = F_{i+1,k+1} \quad ((i,k) \in Q)$,

(ii) $F_{i,l} \leqq F_{i,k} + F_{k,l} \quad ((i,k),(k,l) \in Q)$,

and

(iii) $\gamma(F) = \inf\{n^{-1} \int F_{0,n} d\mu : n \in \mathbb{N}\} > -\infty$.

$F$ is a *superadditive process* if $\{-F_{i,k}: (i,k) \in Q\}$ is subadditive and an *additive* process if it is both subadditive and superadditive. An additive process simply has the form $F_{i,k} = \sum_{j=i}^{k-1} F_{0,1} \circ \tau^j$. Birkhoff's theorem asserts the a.e.-convergence of $n^{-1} F_{0,n}$ for additive processes and Kingman's theorem extends this to the subadditive case. $\gamma(F)$ is called the *time constant* of the process $F$.

Put $g_n = \int F_{0,n} d\mu$. As $\tau$ is measure preserving, (i) implies $\int F_{i,k} d\mu = \int F_{i+1,k+1} d\mu$. Using (ii) we therefore see that the sequence $g_1, g_2, \ldots$ obtained from a subadditive $F$ is a *subadditive sequence* of real numbers, i.e., it satisfies

(5.1) $\quad g_{n+m} \leqq g_n + g_m, \quad (n, m \in \mathbb{N})$.

The following elementary lemma is classical:

**Lemma 5.1.** *If* $(g_n)_{n \geqq 1}$ *is a subadditive sequence of real numbers then* $n^{-1} g_n$ *converges to* $\inf \{n^{-1} g_n : n \in \mathbb{N}\} =: \gamma$.

*Proof.* Given $N$ we may write $n = k_n N + r_n$ with $1 \leqq r_n \leqq N$ and $n^{-1} k_n \to N^{-1} (n \to \infty)$. We find that $\gamma \leqq n^{-1} g_n \leqq n^{-1} (k_n N N^{-1} g_N + g_{r_n}) \leqq N^{-1} g_N + \varepsilon$ for large $n$ because $n^{-1} g_{r_n}$ tends to 0. As $N$ and $\varepsilon > 0$ were arbitrary the lemma is proved. □

To obtain a maximal inequality it turns out that it is better to change the sign and to consider superadditive processes. For them the time constant is given by $\gamma(F) = \sup \{n^{-1} g_n : n \in \mathbb{N}\} = \lim_{n \to \infty} n^{-1} g_n$.

We denote the number of elements of a set $A$ by card $(A)$. The following inequality is due to Akcoglu-Krengel [1981]:

**Theorem 5.2.** *If* $F = \{F_{i,k} : (i, k) \in Q\}$ *is a non negative superadditive process,* $\alpha > 0$, *and* $E = \{\omega : \sup \{n^{-1} F_{0,n}(\omega) : n \in \mathbb{N}\} > \alpha\}$ *then* $\mu(E) \leqq \alpha^{-1} \gamma(F)$.

*Proof.* Let $N$ be a fixed integer, and let

$$E_N = \{\omega : \sup \{n^{-1} F_{0,n}(\omega) : 1 \leqq n \leqq N\} > \alpha\}.$$

Let $K > N$ be another integer, and for each $\omega \in \Omega$ define

$$A(\omega) = \{k : 0 \leqq k < K - N, \tau^k \omega \in E_N\}.$$

Let $k_1 = k_1(\omega)$ be the smallest element of $A(\omega)$. There is some $n_1 = n_1(\omega)$ with $1 \leqq n_1 \leqq N$ and $F_{0,n_1}(\tau^{k_1} \omega) > \alpha n_1$.

Let $k_2 = k_2(\omega)$ be the smallest element of $A(\omega)$ with $k_2 \geqq k_1 + n_1$. Find $1 \leqq n_2 = n_2(\omega) \leqq N$ with $F_{0,n_2}(\tau^{k_2} \omega) > \alpha n_2$. Continuing in this way one finds finitely many $k_1, \ldots, k_r, n_1, \ldots, n_r$ such that $A(\omega)$ is contained in the union of the disjoint intervals $[k_i, k_i + n_i[$ and $F_{0,n_i}(\tau^{k_i} \omega) > \alpha n_i$, $(i = 1, \ldots, r = r(\omega))$. The superadditivity and nonnegativity of $F$ and the fact that all these intervals are contained in $[0, K[$, yield

$$\begin{aligned} F_{0,K}(\omega) &\geqq \sum_{i=1}^{r} F_{k_i, k_i + n_i}(\omega) \\ &= \sum_{i=1}^{r} F_{0,n_i}(\tau^{k_i} \omega) \geqq \alpha \sum_{i=1}^{r} n_i \geqq \alpha \operatorname{card}(A(\omega)). \end{aligned}$$

Integrating over $\Omega$ and noticing that

$$\int \operatorname{card}(A(\omega))\, d\mu = \int \sum_{k=0}^{K-N-1} 1_{E_N}(\tau^k \omega)\, d\mu = (K - N)\mu(E_N)$$

we obtain $\alpha\mu(E_N)(K-N) \leqq g_K$. Dividing by $K$ and letting $K \to \infty$, $\mu(E_N) \leqq \alpha^{-1}\gamma(F)$ results. Then let $N$ tend to $\infty$. □

**Theorem 5.3** (Kingman). *Let $\tau$ be an endomorphism of a measure space $(\Omega, \mathscr{A}, \mu)$, and $F = \{F_{i,k}: (i,k) \in Q\}$ a subadditive process for $\tau$. Then $n^{-1}F_{0,n}$ converges a.e. to a $\tau$-invariant integrable limit $\bar{f}$. If $\mu$ is finite, $n^{-1}F_{0,n}$ converges also in $L_1$-norm to $\bar{f}$, and $\gamma(F) = \int \bar{f}\,d\mu$ holds.*

*Proof.* Passing to $\{-F_{i,k}\}$ we may assume that $F$ is superadditive. Subtracting the additive process $F'_{i,k} = \sum_{j=i}^{k-1} F_{0,1}\circ\tau^j$, for which the assertion follows from Birkhoff's theorem, we may further assume $F_{ik} \geqq 0$.

Let $\bar{f}$ and $f$ be, respectively, the pointwise lim sup and lim inf of $n^{-1}F_{0,n}$ as $n \to \infty$. We know that $f$ is integrable because of Lemma 5.1 and Fatou's lemma. If $m \geqq 1$ is a fixed integer then $\bar{f}$ and $f$ are also the lim sup and lim inf of $(km)^{-1}F_{0,km}$ as $k \to \infty$, because $F_{0,n}$ is monotonely increasing in $n$.

Let $\alpha > 0$ be given and let $E = \{\omega: \bar{f}(\omega) - f(\omega) > \alpha\}$. We now show $\mu(E) = 0$. Let $\varepsilon > 0$ be given. By lemma 5.1 there exists an integer $m$ with $m^{-1}g_m > \gamma(F) - \varepsilon$. The process $H^m = \{H^m_{k,l}\}$ with $H^m_{k,l} = \sum_{j=k}^{l-1} F_{0,m}\circ\tau^{mj}$ is additive, and the process $F^m = \{F^m_{k,l}\}$ with $F^m_{k,l} = F_{km,lm} - H^m_{k,l}$ is superadditive and non negative. We have $\gamma(F^m) = m\gamma(F) - \gamma(H^m) = m\gamma(F) - g_m \leqq \varepsilon m$. By Birkhoff's theorem $k^{-1}H^m_{0,k}$ converges a.e.. Therefore

$$\bar{f} - f \leqq \sup_{k\geqq 1} \frac{1}{km} F^m_{0,k}.$$

We now apply theorem 5.2 and obtain

$$\mu(E) \leqq \mu(\sup\{k^{-1}F^m_{0,k}: k \geqq 1\} > \alpha m) \leqq (am)^{-1}\gamma(F^m) \leqq \alpha^{-1}\varepsilon.$$

We have proved the a.e.-convergence of $n^{-1}F_{0,n}$ to an integrable $\bar{f} = f$.

If $h^m$ is the limit of $(nm)^{-1}H^m_{0,n}$ we can use the $\tau^m$-invariance of $h^m$ and the estimate $0 \leqq \bar{f} - h^m \leqq \sup\{(km)^{-1}F^m_{0,k}: k \geqq 1\}$ to conclude that

$$\mu(|\bar{f} - \bar{f}\circ\tau^m| > 2\alpha) \leqq \mu(\bar{f} - h^m > \alpha) + \mu(\bar{f}\circ\tau^m - h^m > \alpha) < \frac{2\varepsilon}{\alpha}.$$

If $m$ was large enough $(m+1)^{-1}g_{m+1} > \gamma(F) - \varepsilon$ holds as well, and the same argument gives us $\mu(|\bar{f} - \bar{f}\circ\tau^{m+1}| > 2\alpha) \leqq 2\varepsilon/\alpha$. Combining these estimates we see that $\mu(|\bar{f} - \bar{f}\circ\tau| > 4\alpha) = \mu(|\bar{f}\circ\tau^m - \bar{f}\circ\tau^{m+1}| > 4\alpha) < 4\varepsilon/\alpha$. As $\varepsilon > 0$ and $\alpha > 0$ were arbitrary the $\tau$-invariance of $\bar{f}$ follows.

If $\mu$ is finite the limits $0 \leqq h^m$ exist also in the $L_1$-sense, and the integral of $h^m$ equals $m^{-1}g_m \leqq \gamma(F)$. The superadditivity of $F$ implies $h^{2m} \geqq h^m$. Therefore the sequence $h^{2^i}$ is increasing and tends in $L_1$ to a limit $h^\infty$ with $\int h^\infty d\mu = \gamma(F)$. Given

$\eta > 0$ we may fix an $m$ of the form $2^i$ such that $\|h^\infty - h^m\|_1 < \eta$ and $|m^{-1} g_m - \gamma(F)| < \eta$.

Writing $n \geqq m$ in the form $n = k_n m + r_n$ with $0 \leqq r_n < m$ we observe that $0 \leqq F_{0,n} - H^m_{0,k_n} \leqq F_{0,(k_n+1)m} - H^m_{0,k_n}$. This implies

$$\int (F_{0,n} - H^m_{0,k_n})\, d\mu \leqq g_{(k_n+1)m} - k_n g_m \leqq (k_n+1) m\gamma(F) - k_n g_m \leqq k_n m\eta + m\gamma(F).$$

Now $n^{-1}(k_n m) \to 1$and

$$\|(k_n m)^{-1} F_{0,n} - h^\infty\|_1 \leqq (k_n m)^{-1} \|F_{0,n} - H^m_{0,k_n}\|_1 + \|(k_n m)^{-1} H^m_{0,k_n} - h^m\|_1 + \|h^m - h^\infty\|_1$$

yield the $L_1$-convergence of $n^{-1} F_{0,n}$ to $h^\infty = \tilde{f}$. □

The assertion of a.e.-convergence in theorem 5.3 can be proved under slightly weaker assumptions when $\mu$ is finite:

**Theorem 5.4.** *Let $\tau$ be an endomorphism of a finite measure space $(\Omega, \mathscr{A}, \mu)$ and $F = \{F_{i,k}: (i,k) \in Q\}$ a family of real valued measurable functions satisfying* (i), (ii) *and $\|F_{0,1}{}^+\|_1 < \infty$. Then $n^{-1} F_{0,n}$ convergences* a.e. *to a $\tau$-invariant limit $f^\infty$ with $\int f^\infty d\mu = \inf\{\int n^{-1} F_{0,n} d\mu: n \geqq 1\}$.*

($\|F_{0,1}{}^+\|_1 < \infty$ implies $\|F_{i,k}{}^+\|_1 < \infty$ for all $(i,k) \in Q$; therefore the integrals $\int F_{i,k} d\mu$ are well defined, although we may have $\|F_{i,k}{}^-\|_1 = \infty$.)

*Proof.* It suffices to consider the case $\inf\{\int n^{-1} F_{0,n} d\mu\} = -\infty$. The processes $F^N$ defined by $F^N_{i,k} = \text{Max}\,(F_{i,k}, -N(k-i))$ satisfy all three properties (i)–(iii) for all $N \geqq 1$. Therefore the limits $f^N = \lim_{n\to\infty} n^{-1} F^N_{0,n}$ exist a.e. and in $L_1$. Let $f^\infty$ be the limit of the decreasing sequence $f^1, f^2, \ldots$ It is easy to check $n^{-1} F_{0,n} \to f^\infty$. For all $N$, $\int f^\infty d\mu \leqq \int f^N d\mu = \gamma(F^N) = \inf\{\int n^{-1} F^N_{0,n} d\mu\} \leqq \int n^{-1} F^N_{0,n} d\mu$. Letting $N \to \infty$ it follows that $\int f^\infty d\mu \leqq \int n^{-1} F_{0,n} d\mu$ for all $n$. □

We remark that it is possible to construct a process $F$ for an endomorphism of a $\sigma$-finite measure space such that (i), (ii), and $\|F_{0,1}{}^+\|_1 < \infty$ are satisfied, but $n^{-1} F_{0,n}$ does not converge a.e..

The case $\mu(\Omega) = 1$ of the subadditive ergodic theorem admits also a probabilistic formulation: Notice that by (i) and the $\tau$-invariance of $\mu$ the *joint* distribution of the random variables $\{F_{i,k}: (i,k) \in Q\}$ is the same as that of the random variables $\{F_{i+1,k+1}: (i,k) \in Q\}$. In other words: The family $\{F_{i,k}\}$ of random variables satisfies

(i*) $$\mu(F_{i_1,k_1} \in B_1, \ldots, F_{i_s,k_s} \in B_s) = \mu(F_{i_1+1,k_1+1} \in B_1, \ldots, F_{i_s+1,k_s+1} \in B_s)$$

for all $s \in \mathbb{N}$, all $(i_1, k_1), \ldots, (i_s, k_s) \in Q$ and all Borel sets $B_1, \ldots, B_s$.

We say that a family $F = \{F_{i,k}: (i, k) \in Q\}$ of integrable real valued random variables defined on a probability space $(\Omega, \mathscr{A}, \mu)$ is a *probabilistic subadditive process* if (i*), (ii), (iii) hold. Given such a process we can construct a shift on a slightly more complicated space than in section 1.4 as follows: Let $\Omega' = \mathbb{R}^Q$ and let $X_{(i,k)}$ for $(i, k) \in Q$ be the coordinate variables. Let $\tilde{F}_{i,k}$ be the restriction of the coordinate variable $X_{(i,k)}$ to the set

$$\tilde{\Omega} = \{\omega \in \Omega': X_{(i,l)}(\omega) \leqq X_{(i,k)}(\omega) + X_{(k,l)}(\omega) \quad ((i, k), (k, l) \in Q)\}.$$

Let $\tilde{\mathscr{A}}$ be the restriction of the product-$\sigma$-algebra in $\mathbb{R}^Q$ to $\tilde{\Omega}$, and let $\tilde{\mu}$ on $\tilde{\mathscr{A}}$ be the measure with

$$\tilde{\mu}(F_{i_1,k_1} \in B_1, \ldots, F_{i_s,k_s} \in B_s) = \mu(F_{i_1,k_1} \in B_1, \ldots, F_{i_s,k_s} \in B_s).$$

Let $\theta: \tilde{\Omega} \to \tilde{\Omega}$ be the shift defined by $\tilde{F}_{i,k}(\theta\omega) = \tilde{F}_{i+1,k+1}(\omega)$. Then $\theta$ is an endomorphism of $(\tilde{\Omega}, \tilde{\mathscr{A}}, \tilde{\mu})$, $\tilde{F} = \{\tilde{F}_{i,k}\}$ is a subadditive process for $\theta$, and the joint distribution of $\{\tilde{F}_{i,k}\}$ under $\tilde{\mu}$ is the same as that of $\{F_{i,k}\}$ under $\mu$. Thus, any result for subadditive processes immediately implies the same result for probabilistic subadditive processes. We can now drop the term "probabilistic" again and need not distinguish these notions.

We now discuss some examples of applications:

(a) *Percolation*. The notion of a subadditive process as studied here is due to Kingman. Hammersley and Welsh [1965] had introduced a similar notion in their study of percolation processes, requesting (i*) only for $s = 1$. There is now a vast literature on percolation; see Kesten [1982]. We just give a typical example.

Let $S = \mathbb{Z}^2$ and let $\mathscr{N}$ be the family of pairs $(u, v) \in S^2$ which are neighbors, i.e., which have euclidean distance 1. Assume that for any $(u, v) \in \mathscr{N}$ there is a positive random variable $T_{u,v}$ representing the time needed to travel from $u$ to $v$. A sequence $u_0, u_1, \ldots, u_r \in S$ is called a path from $u$ to $w$ if $u_0 = u$, $u_r = w$ and $(u_{i-1}, u_i) \in \mathscr{N}$ for $i = 1, \ldots, r$. For $u, w \in S$, the travel time $U(u, w)$ along the fastest route from $u$ to $w$ is the infimum over all pathes from $u$ to $w$ of the sums $\sum_{i=1}^{r} T_{u_{i-1},u_i}$. It is easy to see that the process $F_{i,k} = U((i,0), (k, 0))$ always satisfies (ii). If the random variables $T_{u,v}$ are integrable, independent and identically distributed it also satisfies (i*) and (iii).

(b) *The range of a random walk*. Let $X_1, X_2, \ldots$ be independent identically distributed random variables taking values in a topological group $G$, and let $S_n = \prod_{i=1}^{n} X_i$ be the corresponding random walk. $R_{i,k}(\omega) = \text{card}(\{S_j(\omega): i < j \leqq k\})$ is the number of distinct points of $G$ visited in the time interval $[i+1, k]$. The process $\{R_{i,k}: (i, k) \in Q\}$ is subadditive. The subadditive ergodic theorem in this case generalizes a result of Kesten, Spitzer and Whitman treated in the book of Spitzer [1964].

(c) *Random products in Banach algebras.* If we convert the multiplicative inequality $\|A_1 A_2\| \leqq \|A_1\| \, \|A_2\|$ valid in a Banach algebra into an additive one by taking logarithms we obtain

**Theorem 5.5.** *Let $\tau$ be an endomorphism of a finite measure space $(\Omega, \mathscr{A}, \mu)$, and let $T\colon \Omega \to \mathfrak{B}$ be a measurable map of $\Omega$ into a Banach algebra $\mathfrak{B}$ for which $\log^+ \|T(\cdot)\|$ is integrable. Put $T_{i,k}(\omega) = \prod_{\nu=i}^{k-1} T(\tau^\nu \omega)$. Then there exists a $\tau$-invariant measurable function $\chi\colon \Omega \to \mathbb{R} \cup \{-\infty\}$ with $\chi^+ \in L_1$ such that*

$$\lim_{n\to\infty} n^{-1} \log \|T_{0,n}(\omega)\| = \chi(\omega) \quad \text{a.e.},$$

*and*

$$\lim_{n\to\infty} n^{-1} \int \log \|T_{0,n}(\omega)\| \, d\mu = \inf_n \{n^{-1} \int \log \|T_{0,n}(\omega)\| \, d\mu\}$$
$$= \int \chi(\omega) \, d\mu .$$

The special case where $\mathfrak{B}$ is the algebra of $(m \times m)$ matrices endowed with any matrix norm is a result of Furstenberg and Kesten [1960], which required deep arguments as long as Kingman's theorem was not available. It will serve as the basis for the proof of the multiplicative ergodic theorem.

**2. Continuous parameter subadditive processes.** While the derivation of the continuous parameter version of Birkhoff's theorem from its discrete counterpart is very easy, one must be more cautious in the case of additive or subadditive processes. There are two difficulties which do not appear in the study of $\int_0^t f(\tau_s \omega) \, ds$.

The first one is a technical problem well known in probability theory. It is due to the fact that $t$ ranges through an uncountable set and two processes which agree for each fixed $t$ almost everywhere may then have a different limit behaviour.

Let $\{\tau_s, s \geqq 0\}$ be a measurable measure preserving semiflow in a $\sigma$-finite measure space $(\Omega, \mathscr{A}, \mu)$. $Q_c = \{(s,t) \in \mathbb{R}^2\colon 0 \leqq s < t\}$ will be the new parameter set. A family $F = \{F_{s,t}\colon (s,t) \in Q_c\}$ of integrable real valued functions $F_{s,t}$ is called a (continuous parameter) *subadditive process* if the conditions

(i)$_c$ $\quad F_{s,t} \circ \tau_u = F_{s+u,t+u} \quad ((s,t) \in Q_c,\ u \geqq 0)$,

(ii)$_c$ $\quad F_{s,u} \leqq F_{s,t} + F_{t,u} \quad ((s,t), (t,u) \in Q_c)$,

(iii)$_c$ $\quad \inf \{t^{-1} \int F_{0,t} \, d\mu\colon t > 0\} > -\infty$

are satisfied, and $F_{(s,t)}(\omega)$ is a measurable map of $Q_c \times \Omega \to \mathbb{R}$ with respect to the product-$\sigma$-algebra in $Q_c \times \Omega$.

The following example shows that there exist two subadditive processes $F, F'$ in this sense for which $F_{s,t} = F'_{s,t}$ a.e. is true for every fixed $(s,t) \in Q_c$, yet $\lim_{t\to\infty} t^{-1} F_{0,t} = 0$ holds everywhere and $t^{-1} F'_{0,t}$ diverges everywhere as $t \to \infty$: Simply take $\Omega = [0, 1[$ with Lebesgue measure and with addition mod 1 and $\tau_t \omega = \omega + t$. If $f\colon \Omega \to \mathbb{R}$ is a function which equals 0 on the irrationals and is unbounded on the rationals we may use $F_{s,t}(\omega) \equiv 0$ and $F'_{s,t}(\omega) = f(\omega + t) - f(\omega + s)$.

Basically, the difficulty is that measure theory permits taking limits, suprema etc. only along countable index sets. Usually continuity in $(s, t)$ is too strong a requirement. A process $F$ is called *separable* if there exists a countable subset $S \subset Q_c$ and a set $N \in \mathscr{A}$ with $\mu(N) = 0$ such that for all $\omega \notin N$ and all $(s, t) \in Q_c$ $F_{s,t}(\omega)$ belongs to the closure of each of the sets $\{F_{u,v}(\omega)\colon (u, v) \in S \cap U\}$ with $(s, t) \in U$ and $U$ relatively open in $Q_c$.

Separability is a very weak requirement. Moreover, if $F'$ is any given process there exists a separable process $F$ equivalent to $F'$ in the sense that $\mu(F_{s,t} \neq F'_{s,t}) = 0$ for all $(s, t)$. This is a variant of a general theorem on processes with a continuous parameter due to Doob. (See e.g. Borges [1966]). It does not disturb if $F$ satisfies $(\mathrm{i})_c$ and $(\mathrm{ii})_c$ only in an a.e.-sense because limits and suprema for $F$ may be taken along the countable set $S$: Note that $S$ may be replaced by any larger countable set, so that we can assume $(0, t) \in S$ for $(s, t) \in S$, etc. It is easy to see that, for $\omega \notin N$, $\sup \{F_{s,t}(\omega)\colon (s, t) \in U\} = \sup \{F_{s,t}(\omega)\colon (s, t) \in U \cap S\}$ holds for all relatively open sets $U$, and the same is true for inf. Consequently, the existence of the limit of $t^{-1} F_{0,t}(\omega)$ as $t \to \infty$ follows from the existence of the limit as $t \to \infty$ with $(0, t) \in S$.

The second difficulty requires a more restrictive condition. Kingman [1973] has shown that for any positive increasing function $\Gamma(t)$ there exists a separable additive process $F$ on a probability space such that, for all $\omega \in \Omega$, $F_{0,t}(\omega)$ as a function of $t$ has derivatives of all orders and yet $\lim\sup_{t\to\infty} \Gamma(t)^{-1} F_{0,t}(\omega) = \infty$. Intuitively, what happens is that there are some rare sudden high increases in the process which are almost instantly followed by decreases of the same size. Thus, as these excursions are very short, they practically have no influence on the integrals.

To state a condition which avoids such pathologies one can define the maximal oscillation on an interval $I$ by

$$\Omega_I = \sup\{|F_{s,t}|\colon s < t;\ s, t \in I\}.$$

**Theorem 5.6.** *If $F$ is a subadditive separable process for which $\Omega_{[0,1]}$ is integrable, $t^{-1} F_{0,t}$ converges a.e. as $t \to \infty$, and the limit is $\tau_s$-invariant for all $s$.*

*Proof.* We may assume $s = 1$. If $n$ is the integer part of $t$ we have

$$F_{0,n+1} - F_{t,n+1} \leqq F_{0,t} \leqq F_{0,n} + F_{n,t},$$

which implies

$$F_{0,n+1} - \Omega_{[n,n+1]} \leqq F_{0,t} \leqq F_{0,n} + \Omega_{[n,n+1]}.$$

By Birkhoff's theorem $n^{-1}\Omega_{[n,n+1]} = n^{-1}\Omega_{[0,1]} \circ \tau_n$ tends to 0, and the result follows from theorem 5.3. □

If the condition (ii)$_c$ is satisfied everywhere and not only a.e. with exceptional sets depending on $u$, $s$, $t$, the only way the separability enters now is that it guarantees the measurability of $\Omega_{[0,1]}$. It can then be replaced by the assumption that $\Omega_I$ is bounded by an integrable function. The separability has a slightly more important function when one gives the probabilistic definition of a subadditive process in which no semiflow is specified. The condition (i)$_c$ is then replaced by
(i*)$_c$ The joint distribution of $\{F_{s,t}: (s,t) \in Q_c\}$ is the same as that of $\{F_{s+r,t+r}: (s,t) \in Q_c\}$ for any $r$.

Now separability ensures that all $\Omega_{[n,n+1]}$ are well defined and have the same distribution. $n^{-1}\Omega_{[n,n+1]} \to 0$ a.e. follows because

$$\sum_{n=1}^{\infty} \mu(\Omega_{[n,n+1]} > \varepsilon n) = \sum_{n=1}^{\infty} \mu(\Omega_{[0,1]} > \varepsilon n) \leqq \varepsilon^{-1} \int \Omega_{[0,1]} d\mu < \infty$$

converges for all $\varepsilon > 0$.

Of course no assumption like separability or integrability of $\Omega_{[0,1]}$ is needed when one extends the $L_1$-norm convergence assertion in theorem 5.3 to continuous parameters.

**3. The multiplicative ergodic theorem of Oseledeč.** If $\omega$ moves along an orbit $\omega, \tau\omega, \tau^2\omega, \ldots$ and a visit in $\tau^k\omega$ is associated with a linear transformation described by a matrix $A(\tau^k\omega)$, then the composition of the transformations performed until time $n-1$ is described by $P_n(A,\omega) = A(\tau^{n-1}\omega) \cdot A(\tau^{n-2}\omega) \cdots A(\omega)$. The theorem of Furstenberg and Kesten, obtained above as a consequence of Kingman's theorem, tells us that $n^{-1} \log \|P_n(A,\omega)\|$ converges a.e. under the integrability condition $\log^+ \|A(\cdot)\| \in L_1$. The multiplicative ergodic theorem provides more precise information on the limit behaviour of $P_n(A,\omega)$.

For some fixed natural number $r$, $\mathfrak{E}$ denotes the space $\mathbb{R}^r$ endowed with the euclidean norm, and $\mathbf{M} = \mathbf{M}(r)$ denotes the set of $r \times r$ matrices with elements from $\mathbb{R}$. For $A \in \mathbf{M}$ we use the norm $\|A\| = \sup\{\|Au\|: u \in \mathfrak{E}, \|u\| \leqq 1\}$. $A^*$ is the transposed matrix of $A$.

**Theorem 5.7** (Oseledeč). *Let $\tau$ be an endomorphism of a probability space $(\Omega, \mathscr{A}, \mu)$ and $A(\cdot)$ a measurable map $\Omega \to \mathbf{M}$, $(\omega \to A(\omega))$, with $\log^+ \|A(\cdot)\| \in L_1(\mu)$. There exists a $\tau$-invariant subset $\Omega'$ of $\Omega$ with $\mu(\Omega') = 1$ such that for $\omega \in \Omega'$ the following is true:*

*(i)* $\lim_{n\to\infty} (P_n^*(A,\omega) P_n(A,\omega))^{1/2n} =: \Lambda(\omega) \in \mathbf{M}$ *exists.*

*(ii) Let* $\exp \lambda_1(\omega) < \exp \lambda_2(\omega) < \ldots < \exp \lambda_s(\omega)$ *be the distinct eigenvalues of* $\Lambda(\omega)$ *in increasing order. (Their number $s = s(\omega)$ may depend on $\omega$, $\lambda_1(\omega)$ may be $-\infty$). Let $\mathfrak{F}_\nu(\omega)$ be the eigenspace corresponding to* $\exp \lambda_\nu(\omega)$, $m_\nu(\omega)$

$= \dim(\mathfrak{F}_\nu(\omega))$ *the multiplicity of* $\exp \lambda_\nu(\omega)$, *and* $\mathfrak{G}_\nu(\omega) = \mathfrak{F}_1(\omega) + \ldots + \mathfrak{F}_\nu(\omega)$, $\mathfrak{G}_0 = \{0\}$. *Then we have*

$$\lim_{n\to\infty} \frac{1}{n} \log \| P_n(A, \omega) u \| = \lambda_\nu(\omega)$$

*for* $u \in \mathfrak{G}_\nu(\omega) \setminus \mathfrak{G}_{\nu-1}(\omega)$, $(\nu = 1, \ldots, s)$.

*(iii) The functions* $\omega \to m_\nu(\omega)$ *and* $\omega \to \lambda_\nu(\omega)$ *are* $\tau$*-invariant.*

*(iv) If* $\tau$ *is ergodic,* $\det(A(\omega)) \equiv 1$, *and* $\limsup \int \frac{1}{n} \log \| P_n(A, \omega) \| \, d\mu > 0$, *then* $\lambda_1$ *is negative and* $\lambda_s$ *is positive.*

Before we start with the proof we recall some auxiliary results from linear and multilinear algebra, and we discuss the meaning of the assertions.

$B \in \mathbf{M}$ is called *positive semidefinite* if scalar product $\langle Bu, u \rangle$ is non negative for all $u \in \mathfrak{E}$. For any $A \in \mathbf{M}$ the matrices $AA^*$ and $A^*A$ are symmetric and positive semidefinite and have roots of all orders which again are symmetric and positive semidefinite. The polar decomposition (see Gantmacher [1958: § 14, 263] says that $A$ may be written in the form $A = (AA^*)^{1/2} C' = C'' (A^*A)^{1/2}$ with orthogonal matrices $C', C'' \in \mathbf{M}$. As a symmetric, positive semidefinite $B$ may be written in the form $C^{-1} DC$ with $C$ orthogonal and $D$ a diagonal matrix with non negative entries, we see that any $A \in \mathbf{M}$ can be written in the form $A = \tilde{C} DC$ with orthogonal matrices $C, \tilde{C}$. Moreover, we may assume that the diagonal elements of $D = (d_{ij})$ appear in increasing order, i.e., we have $0 \leqq d_{11} \leqq d_{22} \leqq \ldots, \ldots d_{rr}$. They are the eigenvalues of $(A^*A)^{1/2}$. $d_{rr}$ is the norm $\|A\|$.

Applying the polar decomposition to $P_n = P_n(A, \omega)$ we see that $P_n$ is of the form $C_n'' (P_n^* P_n)^{1/2}$. Thus (i) means that, for fixed $\omega$, $P_n$ is essentially a power $\Lambda^n$ of a positive semidefinite matrix $\Lambda$, followed by an orthogonal transformation. As the latter is an isometry, the length of the vectors $P_n u$ is of the order $\|\Lambda^n u\|$. If $u$ is an eigenvector of $\Lambda$ for the eigenvalue $\exp \lambda_\nu$, we have $\|\Lambda^n u\| = \|u\| \cdot \exp(n\lambda_\nu)$. If $u$ is a linear combination of eigenvectors, the component belonging to the largest eigenvalue is ultimately dominant. (iv) treats the case where each $A(\omega)$ preserves the volume in $\mathfrak{E}$. The positivity of $\limsup \int n^{-1} \log \|P_n\| \, d\mu$ guarantees that there are vectors $u$ for which $\|P_n u\|$ grows exponentially fast. There must then also be vectors $u'$ for which $\|P_n u'\|$ decreases exponentially fast. This splitting of the space into expansive and contractive components is of great importance in the theory of differentiable dynamical system. The numbers $\lambda_\nu(\omega)$ are called *characteristic exponents*; with the *multiplicities* $m_\nu(\omega)$ they constitute the *spectrum* of $(\tau, A(\cdot))$ at $\omega$. $\mathfrak{G}_1(\omega) \subset \mathfrak{G}_2(\omega) \subset \ldots$ is called the *associated filtration* of $\mathbb{R}^r$.

We need a bit of multilinear algebra: Recall that, for $1 \leqq k \leqq r$, the $k$-fold *exterior power* $\bigwedge_k \mathfrak{E}$ of $\mathfrak{E}$ is the set of all formal expressions $\sum_{i=1}^{m} c_i u_{i1} \wedge u_{i2} \ldots \wedge u_{ik}$ with $c_i \in \mathbb{R}$, $m \in \mathbb{N}$, and $u_{ij} \in \mathfrak{E}$, when we compute with the following conventions:

(i) $u_1 \wedge \ldots \wedge (u_j + u_j') \wedge \ldots \wedge u_k = u_1 \wedge \ldots \wedge u_j \wedge \ldots \wedge u_k + u_1 \wedge \ldots \wedge u_j' \wedge \ldots \wedge u_k$;

(ii) $c(u_1 \wedge \ldots \wedge u_j \wedge \ldots \wedge u_k) = u_1 \wedge \ldots \wedge cu_j \wedge \ldots \wedge u_k$;

(iii) for any permutation $(\pi 1, \ldots, \pi k)$ of $(1, \ldots, k)$: $u_{\pi 1} \wedge \ldots \wedge u_{\pi k} = \operatorname{sign}(\pi)\, u_1 \wedge \ldots \wedge u_k$.

For more detail, see e.g. Kowalsky [1967].

A scalar product is given in $\bigwedge_k \mathfrak{E}$ by

$$\langle u_1 \wedge \ldots \wedge u_k, v_1 \wedge \ldots \wedge v_k \rangle = \det\left((\langle u_i, v_j \rangle)_{i,j=1,\ldots,k}\right).$$

In particular, the norm $\|u_1 \wedge \ldots \wedge u_k\|$ is the square root of the Gram-determinant $\det\left((\langle u_i, u_j \rangle)_{i,j=1,\ldots,k}\right)$. Thus $\|u_1 \wedge \ldots \wedge u_k\|$ is the volume of the $k$-dimensional parallelepiped spanned by the vectors $u_1, \ldots, u_k$; see Greub [1967: Ch. VII, § 3].

The $k$-fold exterior power $A^{\wedge k}$ of a matrix $A \in \mathbf{M}$ is the map $\bigwedge_k \mathfrak{E} \to \bigwedge_k \mathfrak{E}$ given by $u_1 \wedge \ldots \wedge u_k \to Au_1 \wedge \ldots \wedge Au_k$. As above, write $A = \tilde{C}DC$ with orthogonal $C, \tilde{C}$, and $0 \leqq d_{11} \leqq \ldots \leqq d_{rr}$. Because of $\langle \tilde{C}Du_i, \tilde{C}Du_j \rangle = \langle Du_i, Du_j \rangle$, and the fact that the set of $k$-vectors $Cu_1 \wedge Cu_2 \ldots \wedge Cu_k$ with $\|u_1 \wedge \ldots \wedge u_k\| \leqq 1$ agrees with the set of $k$-vectors $u_1 \wedge \ldots \wedge u_k$ with $\|u_1 \wedge \ldots \wedge u_k\| \leqq 1$, we have

$$\|A^{\wedge k}\| = \sup\left\{\det\left((\langle Du_i, Du_j \rangle)_{i,j=1,\ldots k}\right) : \|u_1 \wedge \ldots \wedge u_k\| \leqq 1\right\}.$$

Putting $u_1 = (0, \ldots, 0, 1)^*$, $u_2(0, \ldots, 0, 1, 0)^*, \ldots$, we see that

$$\|A^{\wedge k}\| \geqq \prod_{r-k<i\leqq r} d_{ii}.$$

Using a stepwise reduction of the dimension of the parallelepipeds (formula 7.33 in Greub [1967]), it can be checked that $\leqq$ holds as well.

*Proof of theorem 5.7.* $\|A^{\wedge k}\| \leqq \|A\|^k$ implies that $\log^+ \|A^{\wedge k}(\cdot)\|$ is integrable. Applying the Birkhoff theorem to the function $\log^+ \|A(\cdot)\|$ and the Furstenberg-Kesten theorem to the maps $\omega \to A^{\wedge k}(\omega)$, $(1 \leqq k \leqq r)$, we see that there is a $\tau$-invariant set $\Omega'$ of full measure such that for all $\omega \in \Omega'$, the averages $n^{-1} \sum_{i=0}^{n-1} \log^+ \|A(\tau^i \omega)\|$ tend to a finite limit, and $n^{-1} \log \|P_n(A^{\wedge k}, \omega)\|$ tends to a limit $\chi(k, \omega) \in \mathbb{R} \cup \{-\infty\}$ for $1 \leqq k \leqq r$. The limits are $\tau$-invariant. We fix $\omega \in \Omega'$.

There exist orthogonal matrices $\tilde{C}_n(\omega)$, $C_n(\omega)$ and diagonal matrices $D_n(\omega) = (d_n(i, j, \omega))_{ij}$ with $P_n(A, \omega) = \tilde{C}_n(\omega) D_n(\omega) C_n(\omega)$, and $0 \leqq d_n(1, 1, \omega) \leqq d_n(2, 2, \omega) \leqq \ldots \leqq d_n(r, r, \omega)$. Mostly we shall suppress $\omega$ in the notation and write $\tilde{C}_n$, $d_n(i, j)$, etc..

Because of $\|P_n(A^{\wedge k})\| = \prod_{r-k<i\leqq r} d_n(i, i)$ and the monotonicity of the $d_n(i, i)$ in $i$, there exist $-\infty \leqq \varrho_1 \leqq \varrho_2 \leqq \ldots \leqq \varrho_r < \infty$ with $n^{-1} \log d_n(i, i) \to \varrho_i$. We may split the "interval" $\{1, 2, \ldots, r\} = [1, r+1[ \cap \mathbb{N}$ into finitely many intervals $[i_\nu, i_{\nu+1}[ \cap \mathbb{N}$, $(\nu = 1, \ldots, s;\ i_1 = 1, i_{s+1} = r+1)$, such that $\varrho_i = \varrho_j$ iff $i$ and $j$ belong to the same interval.

Let $\mathfrak{E}_\nu$ be the subspace of $\mathfrak{E}$ spanned by the $i_{\nu+1}-1$ unit vectors $e_i = (\delta_{1i}, \delta_{2i}, \ldots, \delta_{ri})^*$, $(1 \leqq i < i_{\nu+1})$. In particular $\mathfrak{E}_0 = \{0\}$, $\mathfrak{E}_s = \mathfrak{E}$. We want to show that the subspaces $\mathfrak{E}_\nu^n = C_n^{-1}\mathfrak{E}_\nu$ converge to a limit $\mathfrak{G}_\nu$.

To do this, first note that the convergence of $n^{-1}\sum_{i=0}^{n-1}\log^+\|A(\tau^i\omega)\|$ to a finite limit readily implies that, for any $\varepsilon > 0$, $\|A(\tau^n\omega)\| < \exp(n\varepsilon)$ holds for all sufficiently large $n$.

**Lemma 5.8.** *Assume* $\varrho_{i_\nu} > -\infty$. *Given* $\varepsilon > 0$ *there exists* $N_\varepsilon$ *with the following property: If* $v \in \mathfrak{E}_\nu^n$ *has length* $\|v\| = 1$, *and there are numbers* $b_i \in \mathbb{R}$ *and vectors* $v' \in \mathfrak{E}_\nu^{n+1}$ *with*

(5.2) $$v = v' + \sum_{i \geqq i_{\nu+1}} b_i C_{n+1}^{-1} e_i,$$

*then* $|b_i| < \exp\{-n(\varrho_i - \varrho_{i_\nu} - \varepsilon)\}$ *for* $n \geqq N_\varepsilon$.

*Proof.* For large $n$, $\log d_n(j,j) < (\varrho_{i_\nu} + \varepsilon)n$ holds for all $j \leqq i_{\nu+1} - 1$. Using $P_{n+1}(A,\omega) = A(\tau^n\omega)\cdot P_n(A,\omega)$ and $v \in \mathfrak{E}_\nu^n$ this implies

$$\|P_{n+1}(A,\omega)v\| < \exp(n\varepsilon)\exp(n(\varrho_{i_\nu} + \varepsilon)).$$

Similarly, for large $n$

$$\begin{aligned}\|P_{n+1}(A,\omega)v\| &\geqq |b_i|\,\|P_{n+1}(A,\omega)\,C_{n+1}^{-1}e_i\| = |b_i|\,d_{n+1}(i,i)\\ &\geqq |b_i|\exp((n+1)(\varrho_i - \varepsilon)).\end{aligned}$$

As the $\varrho_i$ $(i \geqq i_{\nu+1})$ lie in a bounded interval and $\varepsilon$ may be assumed $\leqq 1$ we may write $\exp((n+1)(\varrho_i - \varepsilon)) \geqq \exp(n(\varrho_i - 2\varepsilon))$ for large $n$. Putting everything together we obtain $|b_i| \leqq \exp(n(\varrho_{i_\nu} - \varrho_i + 4\varepsilon))$. □

In the case $\nu = 1$, $\varrho_1 = -\infty$, the conclusion in this lemma is that, for any $\varrho > -\infty$, $|b_i| \leqq \exp(n\varrho)$ holds for sufficiently large $n$. We shall assume $\varrho_1 > -\infty$ in the sequel, because the proof in the case $\varrho_1 = -\infty$ is quite similar.

Let $\pi_\nu^m$ be the orthogonal projection of $\mathfrak{E}$ into $\mathfrak{E}_\nu^m$ and $\bar\pi_\nu^m$ the orthogonal projection of $\mathfrak{E}$ into the orthogonal complement $\bar{\mathfrak{E}}_\nu^m$ of $\mathfrak{E}_\nu^m$. The lemma implies

$$\|\bar\pi_\sigma^{n+1}u\| \leqq r\|u\|\exp\{-n(\varrho_{i_{\sigma+1}} - \varrho_{i_\nu} - \varepsilon)\}$$

for large $n$ and $u \in \mathfrak{E}_\nu^n$, $\sigma \geqq \nu$. For $k > 1$, we can use the decomposition

$$\begin{aligned}\bar\pi_\nu^{n+k} &= \bar\pi_\nu^{n+k}\circ\bar\pi_\nu^{n+1} + \bar\pi_\nu^{n+k}\circ\pi_\nu^{n+1}\\ &= \bar\pi_\nu^{n+k}\circ\bar\pi_\nu^{n+1} + \bar\pi_\nu^{n+k}\circ\bar\pi_\nu^{n+2}\circ\pi_\nu^{n+1} + \bar\pi_\nu^{n+k}\circ\pi_\nu^{n+2}\circ\pi_\nu^{n+1} = \ldots\end{aligned}$$

to derive the estimate

(5.3) $$\|\bar{\pi}_\nu^{n+k} u\| \leqq r \sum_{j=0}^{k-1} \|u\| \exp\{-(n+j)(\varrho_{i_{\nu+1}} - \varrho_{i_\nu} - \varepsilon)\}$$
$$\leqq K_1 \|u\| \exp\{-n(\varrho_{i_{\nu+1}} - \varrho_{i_\nu} - \varepsilon)\},$$

with a constant $K_1$ independent of $u \in \mathfrak{E}_\nu^n$, $n$ and $k$.

Using similar decompositions into projections on the spaces $\mathfrak{E}_\nu^m$, $\bar{\mathfrak{E}}_{\nu+1}^m$ and the spaces spanned by $C_m^{-1} e_i$, $(i \in [i_{\nu+1}, i_{\nu+2}[)$, we can show that

$$\|\bar{\pi}_{\nu+1}^{n+k} u\| \leqq \sum_{j=0}^{k-1} r \cdot \|u\| \exp(-(n+j)(\varrho_{i_{\nu+2}} - \varrho_{i_\nu} - \varepsilon))$$
$$+ \sum_{j=0}^{k-1} K_1 \|u\| \exp(-n(\varrho_{i_{\nu+1}} - \varrho_{i_\nu} - \varepsilon)) \cdot r \cdot \exp(-(n+j)(\varrho_{i_{\nu+2}} - \varrho_{i_{\nu+1}} - \varepsilon))$$
$$\leqq K_2 \|u\| \exp\{-n(\varrho_{i_{\nu+2}} - \varrho_{i_\nu} - 2\varepsilon)\}.$$

Continuing in this way we see that there are constants $K_l$ such that, for all large $n$ (say $n \geqq N'_\varepsilon$), $u \in \mathfrak{E}_\nu^n$, $k \geqq 1$, $l = 0, \ldots, s - \nu - 1$,

(5.4) $$\|\bar{\pi}_{\nu+l}^{n+k} u\| \leqq K_{l+1} \|u\| \exp\{-n(\varrho_{i_{\nu+l+1}} - \varrho_{i_\nu} - (l+1)\varepsilon)\}.$$

Redefining $N'_\varepsilon$ we may assume $K_l = 1$.

By (5.3) each sequence $\pi_\nu^{n+k} C_n^{-1} e_i$, $(k = 1, 2, \ldots)$, with $i < i_{\nu+1}$, is a Cauchy sequence. If $n$ is fixed but large, it stays close to $C_n^{-1} e_i$, and we may assume that the vectors $v^i = \lim_{k\to\infty} \pi_\nu^{n+k} C_n^{-1} e_i$ are linearly independent. The space $\mathfrak{G}_\nu$ spanned by $v^i$ $(i < i_{\nu+1})$ may be considered as the limit of the spaces $\mathfrak{E}_\nu^k$. It is clearly independent of the fixed $n$.

Our next aim is to study the limit behavior of $n^{-1} \log \|P_n(A, \omega) v\|$ for $v \in \mathfrak{G}_\nu$.

Put $f(i) = l$ for $i \in [i_l, i_{l+1}[$, $(l = 1, \ldots, s)$.

**Lemma 5.9.** *Let $B = (b_{ij})$ be an $r \times r$-orthogonal matrix. Assume that there are numbers $0 \leqq \alpha_1 < \alpha_2 < \ldots < \alpha_s$ with $|b_{ij}| \leqq \alpha_{f(i)}/\alpha_{f(j)}$. Then $B^{-1} = (b_{ij}^*)$ satisfies*

$$|b_{ij}^*| \leqq (r-1)!\, \alpha_{f(i)}/\alpha_{f(j)}.$$

*Proof.* If $\psi$ is a permutation of $\{1, 2, \ldots, r\}$ with $\psi(j) = i$, and we write $\psi$ as a product of cycles, one of the cycles has the form $(i, \psi(i), \ldots, \psi^q(i) = j)$. Using $|b_{ij}| \leqq 1$ we observe

$$|\prod_{k \neq j} b_{k,\psi(k)}| \leqq \prod_{h=0}^{q-1} |b_{\psi^h(i), \psi^{h+1}(i)}| \leqq \alpha_{f(i)}/\alpha_{f(j)}.$$

The claim now follows from $|\det B| = 1$ and

$$b_{ij}^* = (\det B)^{-1} \sum_\psi \pm \prod_{k \neq j} b_{k,\psi(k)}. \quad \square$$

If $B$ is the matrix $C_n C_{n+k}^{-1}$, we have $C_n^{-1} e_i = \sum_{j=1}^r b_{ij} C_{n+k}^{-1} e_j$, and (5.4) with constants $K_l = 1$ implies that $B$ satisfies the assumptions of the lemma with $\alpha_h = \exp(n(\varrho_{i_h} - h\varepsilon))$. Because of $C_{n+k}^{-1} e_i = \sum_{j=1}^{r} b_{ij}^* C_n^{-1} e_j$, the lemma now implies that any $v \in \mathfrak{E}_\nu^{n+k}$ with $\|v\| \leqq 1$ is of the form $v = v_n' + v_n''$ with $v_n' \in \mathfrak{E}_\nu^n$, $\|v_n'\| \leqq 1$, $v_n'' = \sum_{j=i_{\nu+1}}^{r} \gamma_{j,n} C_n^{-1} e_j$, and

$$|\gamma_{j,n}| \leqq \text{const} \exp(-n(\varrho_j - \varrho_{i_\nu} - r\varepsilon)).$$

As this is true uniformly in $k$ it remains true for $v \in \mathfrak{G}_\nu$.

Now for $i < i_{\nu+1}$

$$\limsup n^{-1} \log \|P_n(A, \omega)\, C_n^{-1} e_i\| = \limsup n^{-1} \log d_n(i, i) = \varrho_i \leqq \varrho_{i_\nu},$$

and hence $\limsup n^{-1} \log \|P_n(A, \omega) v_n'\| \leqq \varrho_{i_\nu}$. For $j \geqq i_{\nu+1}$,

$$\limsup n^{-1} \log \|P_n(A, \omega) \gamma_{j,n} C_n^{-1} e_j\| \leqq$$

$$\limsup n^{-1} \log \{\text{const} \exp(-n(\varrho_j - \varrho_{i_\nu} - r\varepsilon)) \cdot d_n(j, j)\} = -\varrho_j + \varrho_{i_\nu} + r\varepsilon + \varrho_j.$$

As $\varepsilon > 0$ was arbitrarily small

$$\limsup n^{-1} \log \|P_n(A, \omega)\, v\| \leqq \varrho_{i_\nu}$$

holds for all $v \in \mathfrak{G}_\nu$.

On the other hand, if $v \in \mathfrak{E}$ does not belong to $\mathfrak{G}_\nu$, there is a $c > 0$ with $\|\bar{\pi}_\nu^n v\| > c$ for large $n$. It follows then that

$$\liminf n^{-1} \log \|P_n(A, \omega)\, v\| \geqq \varrho_{i_{\nu+1}}.$$

Hence $\lim n^{-1} \log \|P_n(A, \omega)\, v\| = \varrho_{i_\nu}$ holds for $v \in \mathfrak{G}_\nu \setminus \mathfrak{G}_{\nu-1}$, $(\nu \geqq 1)$.

By the construction of $D_n$ and $C_n$, $C_n^{-1} e_i$ is an eigenvector for $(P_n(A)^* P_n(A))^{1/2}$ with eigenvalue $d_n(i, i)$ and an eigenvector for $(P_n(A)^* P_n(A))^{1/2n}$ with eigenvalue $d_n(i, i)^{1/n}$.

As the spaces spanned by the vectors $C_n^{-1} e_i$, $(i \in [i_\nu, i_{\nu+1}[)$, are the orthogonal complements of $\mathfrak{E}_{\nu-1}^n$ in $\mathfrak{E}_\nu^n$ and converge to the orthogonal complements $\mathfrak{F}_\nu$ of $\mathfrak{G}_{\nu-1}$ in $\mathfrak{G}_\nu$, and as $d_n(i, i)^{1/n}$ converges to $\exp \varrho_i$, the matrices $(P_n(A)^* P_n(A))^{1/2n}$ must converge to the matrix $\Lambda$ which has $\mathfrak{F}_\nu$ as eigenspace of $\exp \lambda_\nu = \exp \varrho_{i_\nu}$. This proves (i) and (ii). The assertion (iii) follows from the $\tau$-invariance of $\lim_{n \to \infty} n^{-1} \log \|P_n(A^{\wedge k}, \omega)\|$. It remains to prove (iv). The assumption $\det(A(\omega)) \equiv 1$ implies $\det P_n(A, \omega) \equiv 1$ and, hence, $\|P_n(A, \omega)\| \geqq 1$. In particular, $\log^+ \|P_n(A, \omega)\|$ agrees with $\log \|P_n(A, \omega)\|$, and $\log \|P_1(A, \omega)\|$ is integrable. It follows that $F_{i,k}^P(\omega) := \log \|P_{k-i}(A, \tau^i \omega)\|$ is a subadditive process, and the assumption $\limsup \int n^{-1} \log \|P_n(A, \omega)\| \, d\mu > 0$ means that its time constant $\gamma(F^P)$ is strictly positive. As $\tau$ is ergodic, $\lim n^{-1} \log \|P_n(A, \omega)\| = \gamma(F^P) > 0$ a.e..

$\|P_n(A, \omega)\| = d_n(r, r, \omega)$ and $\varrho_i = \lim n^{-1} \log d_n(i, i)$ now imply $\varrho_r = \gamma(F^P) > 0$. Next observe that

$$d_n(1, 1, \omega) \cdot d_n(2, 2, \omega) \ldots d_n(r, r, \omega) = \det(P_n(A, \omega)) \equiv 1$$

and $d_n(1,1) \leqq d_n(2,2) \leqq \ldots \leqq d_n(r,r)$ yield $d_n(r,r)^{-r} \leqq d_n(1,1) \leqq d_n(r,r)^{-1/(r-1)}$. Hence $-r\varrho_r \leqq \varrho_1 \leqq -(1/(r-1))\varrho_r$, and $\varrho_1$ is a finite negative number. □

## Notes

**Subadditive processes.** Kingman's proof of the subadditive ergodic theorem was based on a decomposition theorem: A subadditive $F = \{F_{ik}\}$ always can be written as a sum of an additive process and a non negative subadditive process with time constant 0, see § 3.5. The present proof of pointwise convergence, which follows Akcoglu-Krengel [1981], has the advantage that the same maximal inequality can be used to derive also Birkhoff's theorem (as in theorem 7.3). The argument used to show the invariance of the limit and $L_1$-convergence is taken from Derriennic-Krengel [1981].

There are several other proofs of pointwise convergence for $\mu(\Omega) = 1$: Derriennic [1975] used the following maximal lemma: If $E = \{\liminf n^{-1} F_{0,n} < 0\}$, then $\lim n^{-1} \int_E F_{0,n}\, d\mu \leqq 0$. Steele [1984] has a proof without explicit use of maximal inequalities. Smeltzer [1977] used upcrossing inequalities generalizing those of Bishop: For $a < b$ let $W_N(\omega)$ be the maximum of all $n$, for which there exist $0 \leqq u_1 < v_1 < \ldots < u_n < v_n \leqq N$ with $v_i b - u_i a \leqq F_{u_i, v_i}(\omega)$, $(1 \leqq i \leqq n)$, and $F_{v_i, u_{i+1}}(\omega) \leqq u_{i+1} a - v_i b$, $(1 \leqq i \leqq n-1)$. Then

$$\int W_N\, d\mu \leqq \frac{1}{b-a} \int (F_{0,1} - a)^+\, d\mu.$$

This quantitative estimate has independent interest. (The proof appears in the unpublished thesis written at Yale). Neveu [1983] has the following generalization of the superadditive maximal inequality: If $F$ is non negative superadditive, $g \in L_1^+$, and $E = \{\sup (F_{0,n} - S_n g) > 0\}$, then $\int_E g\, d\mu \leqq \gamma(F)$. However, his proof of the subadditive ergodic theorem (for ergodic $\tau$) uses only the special case $g \equiv \alpha$, the inequality in theorem 5.2.

Kingman gave a subadditive version of the maximal ergodic theorem:

$$\text{If } E_N = \{\sup_{n \leqq N} F_{0,n} > 0\}, \text{ then } \int_{E_N} F_{0,1}\, d\mu \geqq 0.$$

He remarked that this does not suffice to prove pointwise convergence.

The examples of applications given here are a sample from many in the survey of Kingman [1973]. Other applications can be found in the articles of Hammersley [1974], Kingman [1976], and Derriennic [1980]; see also Steele [1978], and Smythe-Wierman [1978].

A subadditive process $F$ on a probability space is called *independent*, if the random variables $F_{m_1,n_1}, F_{m_2,n_2}, \ldots$ are independent whenever the intervals $[m_1, n_1[, [m_2, n_2[, \ldots$ are disjoint. Let $\mathscr{F}(m, \infty)$ be the smallest $\sigma$-algebra in which all $F_{ik}$ with $m \leqq i < k < \infty$ are measurable. The intersection of the $\mathscr{F}(m, \infty)$ is the remote $\sigma$-algebra of $F$. If $F$ is independent, it is trivial. In particular, independent subadditive processes are ergodic. The following result of Kesten (contained in Kingman [1976], and in more general form in Hammersley [1974]) has been useful:

**Theorem 5.10.** *If $F$ is a non negative independent subadditive process with $\|F_{0,1}\|_2 < \infty$, then $\lim \|n^{-1} F_{0,n} - \gamma(F)\|_2 = 0$. The sequence $\|F_{0,n}\|_2$ is subadditive. If $V_n$ is the variance of $F_{0,n}$, we have*

$$\sum_{k=1}^{\infty} 2^{-2k} V_{2^k} < \infty.$$

Derriennic and Krengel [1981] proved that $n^{-1}F_{0,n}$ converges in $L_p$ $(1<p<\infty)$ for *any* non negative subadditive process with $\|F_{0,1}\|_p<\infty$. On the other hand, they showed that there exists a subadditive process with $\sup\|n^{-1}F_{0,n}\|_2<\infty$, for which $n^{-1}F_{0,n}$ fails to converge in $L_2$. Thus, the condition $F_{0,n}\geqq 0$ is important. In the same paper it is shown that the subadditive form of Weyl's theorem fails even for non negative continuous $F_{ik}$ on $[0,1[$.

Derriennic [1983a] considered processes for which the condition of subadditivity is weakened: He showed that $n^{-1}F_{0,n}$ converges in $L_1$ for any process $F=\{F_{i,k}\}$ on a probability space, satisfying $F_{i,k}\circ\tau=F_{i+1,k+1}$, $\inf\int n^{-1}F_{0,n}>-\infty$, and

$$\int(F_{0,n+k}-F_{0,n}-F_{n,n+k})^+\,d\mu\leqq c_k \text{ with } c_k/k\to 0.$$

Almost sure convergence holds if, in addition, $F_{0,n+k}-F_{0,n}-F_{n,n+k}\leqq h_k\circ\tau^n$ $(n,k\geqq 0)$, with $\{h_k\}$ an $L_1$-bounded sequence of functions. This can be used to prove the pointwise ergodic theorem for information. Moulin Olagnier [1983] proposed a different weakening of subadditivity, see § 9.2.

Wacker [1983], [1984] characterized the processes which can be decomposed into a difference of two subadditive processes. He also proved central limit theorems, invariance principles and other probabilistic results for subadditive processes satisfying suitable mixing conditions, and for mixing processes which can be approximated by additive processes. His results apply, e.g., to the range of random walks, and to products of random matrices, processes previously studied by Jain, Orey, Pruitt, Furstenberg, Kesten, Lange, and others. First steps towards an abstract treatment were taken by Ishitani [1977].

In the probabilistic formulation of Birkhoff's and Kingman's theorem the condition of stationarity can be weakened, replacing it by *superstationarity*: If $P'$, $P''$ are probability measures on a Polish (i.e., complete, separable, metric) space $\mathfrak{E}$ endowed with a partial order relation $\leqq$ which is closed (i.e. $\{(a,b)\colon a\leqq b\}$ is closed in $\mathfrak{E}\times\mathfrak{E}$), then $P'$ is called *stochastically larger* than $P''$ if $\int\varphi\,dP'\geqq\int\varphi\,dP''$ holds for all bounded, increasing, measurable $\varphi\colon\mathfrak{E}\to\mathbb{R}$. If $(X_1,X_2,\dots)$ is a real valued stochastic process on $(\Omega,\mathscr{A},P)$ the distribution of $(X_n,X_{n+1},\dots)$ is the measure $P_n$ on $\mathfrak{E}=\mathbb{R}^{\mathbb{N}}$ with $P_n(B)=P(\{\omega\colon(X_n(\omega),X_{n+1}(\omega),\dots)\in B\})$. The process is called superstationary if $P_1$ is stochastically larger than $P_2$. Intuitively, a process is superstationary if – compared to stationarity – there is a tendency towards smaller values. It has been shown by Krengel [1976] that $n^{-1}\sum_{i=1}^{n}X_i$ converges a.e. if $(X_1,X_2,\dots)$ is superstationary and $X_1^+$ integrable. Abid [1978] has a common generalization of this result and Kingman's theorem.

Schürger [1983] has results of this type for random convex sets; see § 9.4. Hachem [1981] has maximal inequalities and dominated estimates for superstationary processes and a continuous parameter form of Abid's result.

Krawczak [1985] has proved upcrossing inequalities for superstationary subadditive processes. In the continuous parameter case the conditions which suffice for convergence do not suffice for the integrability of the number of upcrossings.

Liggett [1985] has weakened the assumptions in the subadditive ergodic theorem in yet another direction.

**The multiplicative ergodic theorem.** Oseledeč [1948] proved the multiplicative ergodic theorem for invertible transformations and matrices, using a reduction to triangular matrices. The argument of Raghunathan [1979] works in the noninvertible case, and, in fact, for matrices with elements from a local field. Some steps omitted in his paper were sketched in the article of Ruelle [1979], who used these results in the proof of a stable manifold theorem for diffeomorphisms of a compact manifold. Pesin [1977] derived a stable manifold theorem a.e. with respect to a smooth invariant measure. Ledrappier

[1981] gave related applications. Also Zakharevich [1978] has a proof of the multiplicative ergodic theorem.

As pointed out by Ruelle, it is useful to isolate the main part of the proof of theorem 5.7: If $A_1, A_2, \ldots$ is a sequence of $r \times r$ matrices with $\limsup n^{-1} \log \|A_n\| \leqq 0$, and $P_n = A_n A_{n-1} \ldots A_1$, then the existence of $\lim n^{-1} \log \|P_n^{\wedge k}\|$ for $k = 1, \ldots, r$ implies the existence of $\lim (P_n^* P_n)^{1/2n} =: \Lambda$, and the assertion corresponding to theorem 5.7. (ii).

He also proved a perturbation theorem in the case $\det(\Lambda) \neq 0$: If $A'_1, \ldots, A'_n, \ldots$ is another sequence, $P'_n = A'_n, \ldots, A'_1$, $\eta > 0$, and $\sup(\|P_n - P'_n\| e^{\eta n})$ is small enough, then $\Lambda' = \lim (P'^*_n P'_n)^{1/2n}$ exists and has the same eigenvalues.

*The invertible case.* Now let $\tau$ be an automorphism, and let the matrices $A(\omega)$ be invertible. Assume that $\log^+ \|A(\cdot)\|$ and $\log^+ \|A^{-1}(\cdot)\|$ are integrable. Put $P_{-n}(\omega) = A^{-1}(\tau^{-n}\omega) \ldots A^{-1}(\tau^{-1}\omega)$. Then there is a $\tau$-invariant $\Omega'' \subset \Omega$ with $\mu(\Omega'') = 1$, and a measurable splitting $\mathbb{R}^r = W^{(1)}(\omega) \oplus \ldots \oplus W^{(s)}(\omega)$ with

$$\lim_{n \to \pm\infty} n^{-1} \log \|P_n(\omega) u\| = \lambda_\nu(\omega) \quad \text{for } 0 \neq u \in W^{(\nu)}(\omega).$$

The spaces $W^{(\nu)}$ are called *Oseledeč spaces*; they need not be orthogonal. We sketch the construction following Ruelle: Put $\hat{P}_n = A^*(\tau^{-n+1}\omega) \ldots A^*(\tau^{-1}\omega) A^*(\omega)$, $\check{P}_n = A^*(\omega) \ldots A^*(\tau^{n-1}\omega)$, $\tilde{P}_n = A^{*-1}(\tau^{n-1}\omega) \ldots A^{*-1}(\omega)$. Since the spectrum of $\hat{\Lambda}(\omega) = \lim (\hat{P}_n^* \hat{P}_n)^{1/2n}$ is $\tau$-invariant, it is also the limit of the spectra of $(\check{P}_n^* \check{P}_n)^{1/2n}$. But the spectrum of $\check{P}_n^* \check{P}_n$ is the same as that of $\check{P}_n \check{P}_n^* = P_n^* P_n$. Hence, the spectrum of $\hat{\Lambda}(\omega)$ is the same as that of $\Lambda(\omega)$. In other words, the spectrum of $(\tau^{-1}, A^*(\cdot))$ is that of $(\tau, A(\cdot))$.

Now $\tilde{P}_n^* \tilde{P}_n = (P_n^* P_n)^{-1}$. Hence, the spectrum of $(\tau, A^{*-1}(\cdot))$ is obtained by changing the sign of the spectrum of $(\tau, A(\cdot))$. The filtration associated with $(\tau, A^{*-1}(\cdot))$ is the one orthogonal to that associated with $(\tau, A)$: $\tilde{\mathfrak{G}}_\nu(\omega) = \mathfrak{G}^\perp_{s-\nu}(\omega)$. Combining these observations we see that the spectrum of $(\tau^{-1}, A^{-1} \circ \tau^{-1})$ at $\omega$ consists of the numbers $-\lambda_s(\omega) \ldots < -\lambda_1(\omega)$ with multiplicities $m_s(\omega), \ldots, m_1(\omega)$.
Let $\mathfrak{G}_{-s}(\omega) \subset \mathfrak{G}_{-(s-1)}(\omega) \subset \ldots \subset \mathfrak{G}_{-1}(\omega)$ be the associated filtration. Put $W_\nu(\omega) = \mathfrak{G}_r(\omega) \cap \mathfrak{G}_{-r}(\omega)$. It is possible to check $\mathfrak{G}_{\nu-1}(\omega) \cap \mathfrak{G}_{-\nu}(\omega) = \{0\}$ and $\mathfrak{G}_{\nu-1}(\omega) + \mathfrak{G}_{-\nu}(\omega) = \mathbb{R}^r$. Then the desired assertions follow. One has $A(\omega) W_\nu(\omega) = W_\nu(\tau\omega)$.

Ruelle [1982] has extended the multiplicative ergodic theorem and the other related results to the case of operators in Hilbert space under some compactness condition.

*Continuous parameters.* Let $\{\tau_s, s \geqq 0\}$ be a measurable measure preserving semiflow in a probability space $(\Omega, \mathscr{A}, \mu)$. A map $(\omega, t) \to A(\omega, t)$ of $\Omega \times \mathbb{R}^+$ into $\mathbf{M} = \mathbf{M}(r)$ is called *cocycle* if $A(\omega, t+s) = A(\tau_s \omega, t) A(\omega, s)$ holds for $s, t \geqq 0$ and $\omega \in \Omega$. Assume that this map is measurable and that $\sup\{\log^+ \|A(\omega, u)\|: 0 \leqq u \leqq 1\}$ and $\sup\{\log^+ \|A(\tau_u \omega, 1 - u)\|: 0 \leqq u \leqq 1\}$ are integrable. The continuous parameter multiplicative ergodic theorem asserts the existence of $\lim_{t \to \infty} (A(\omega, t)^* A(\omega, t))^{1/2t} = \Lambda(\omega)$, and the assertions analogous to (ii), (iii), (iv). Oseledeč deduced such a theorem from the discrete theorem. One can also adapt the proof of the discrete case. This has been done by Crauel [1981], who gives applications to linear stochastic differential equations.

## § 1.6 Relatives of the maximal ergodic theorem

In a famous paper, Hardy and Littlewood [1930] have introduced the *maximal function* $M_\infty f = f^* = \sup A_n(\tau) f$ for the case where $\tau$ is a translation in $\mathbb{R}$. They obtained estimates $\|f^*\|_p \leqq \dfrac{p}{p-1}\|f\|_p$ for $1 < p < \infty$, and a related estimate for $\|f^*\|_1$ when $f$ belongs to *Zygmund's class* $L \log L$ consisting of all measurable $f$, for which $|f|(\log|f|)^+$ is integrable.

We shall now apply the maximal ergodic theorem to derive Wiener's [1939] dominated ergodic theorem which extends this to the case where $\tau$ is a general measure preserving transformation. We shall also discuss Ornstein's converse of the $L \log L$ – result for ergodic $\tau$, some identities involving sets like $A_f = \{\omega: S_n f(\omega) > 0$ for all $n \geqq 1\}$, and continuous parameter semigroups.

Some of these results will be given for a class of positive contractions in $L_1$. Those interested only in measure preserving transformations may disregard our comments on operators and read $T^k f$ as $f \circ \tau^k$.

**1. Dominated estimates.** A linear operator $T$ in $L_1$ is called an $L_1 - L_\infty$-*contraction* if it is a contraction in $L_1$ and satisfies

$$\|T\|_{1,\infty} := \sup\{\|Tf\|_\infty : f \in L_1 \cap L_\infty, \|f\|_\infty \leqq 1\} \leqq 1 .$$

If $T$ is a positive contraction in $L_1$, we may define $Tf$ for measurable $f$ with $f^- \in L_1$ by $Tf = \lim Th_n$, where $h_n$ is an increasing sequence in $L_1$ tending to $f$ a.e.. It is an exercise to show that this definition is unique. In the same way we may extend $T$ to the class of measurable functions $f$ with $f^+ \in L_1$ using a decreasing sequence $h_m$ tending to $f$.

Hopf's maximal ergodic theorem then implies

$$\int_{\{M_n h_{m+k} \geqq 0\}} h_m \, d\mu \geqq \int_{\{M_n h_{m+k} \geqq 0\}} h_{m+k} \, d\mu \geqq 0 .$$

If we first let $k$ tend to $\infty$ and then let $m$ tend to $\infty$ we find

$$\int_{\{M_n f \geqq 0\}} f \, d\mu \geqq 0 ,$$

and thus see that Hopf's maximal ergodic theorem holds even for all $f$ with $f^+ \in L_1$. (Also Garsia's proof goes through).

**Lemma 6.1.** *If $T$ is a positive $L_1 - L_\infty$-contraction the following inequalities hold for $f$ with $f^+ \in L_1$ and $\alpha > 0$:*

(6.1) $$\alpha\mu(M_n f \geqq \alpha) \leqq \int_{\{M_n f \geqq \alpha\}} f \, d\mu ,$$

*and*

(6.2) $$\alpha\mu(M_n f \geqq 2\alpha) \leqq \int_{\{f \geqq \alpha\}} f \, d\mu .$$

*Proof.* We apply Hopf's theorem to $g = f - \alpha$ and obtain

$$\int_{\{M_n g \geqq 0\}} (f - \alpha) \, d\mu \geqq 0 .$$

As $T$ contracts the $L_\infty$-norm, and as $T$ is positive, $T\mathbb{1} \leqq \mathbb{1}$, and hence $\{f \geqq \alpha\} \subset \{M_n f \geqq \alpha\} \subset \{M_n g \geqq 0\}$. It follows that $\int_{\{M_n f \geqq \alpha\}} (f-\alpha)\,d\mu \geqq 0$, and this implies (6.1). $\{M_n f \geqq 2\alpha\}$ is contained in $\{M_n(f-\alpha) \geqq \alpha\}$. Hence $\mu(M_n f \geqq 2\alpha)$ is bounded by

$$\alpha^{-1} \int_{\{M_n(f-\alpha) \geqq \alpha\}} (f-\alpha)\,d\mu \leqq \alpha^{-1} \|(f-\alpha)^+\|_1 . \quad \square$$

We say that two measurable functions $X, Y$: $\Omega \to \mathbb{R}^+$ are in a *maximal type relation* if

$$(6.3) \qquad \alpha\mu(Y \geqq \alpha) \leqq \int_{\{Y \geqq \alpha\}} X\,d\mu < \infty$$

holds for all $\alpha > 0$.

**Lemma 6.2.** *If $X, Y$ are in a maximal type relation and $\psi$: $\mathbb{R}^+ \to \mathbb{R}^+$ is non decreasing, right continuous with $\psi(0) = 0$, then*

$$(6.4) \qquad \int \psi(Y)\,d\mu \leqq \int (X(\omega) \int_{]0, Y(\omega)]} t^{-1}\psi(dt))\,\mu(d\omega).$$

*Proof.* The function $\psi$ defines a measure on $]0,\infty[$, also denoted by $\psi$, with $\psi(]0,t]) = \psi(t)$. If $v$ is the distribution of $Y$ on $]0,\infty]$, i.e., $v(]a,b]) = \mu(Y \in ]a,b])$, the left hand side of (6.4) equals

$$\begin{aligned} \iint 1_{\{(y,\alpha):\, y \geqq \alpha\}}\,\psi(d\alpha)\,v(dy) &= \int_{]0,\infty[} \mu(Y \geqq \alpha)\,\psi(d\alpha) \\ &\leqq \int_{]0,\infty[} (\alpha^{-1} \int_{\{Y \geqq \alpha\}} X\,d\mu)\,\psi(d\alpha) \\ &= \iint_{\{(\omega,\alpha):\, Y(\omega) \geqq \alpha > 0\}} \alpha^{-1} X(\omega)\,\mu(d\omega)\,\psi(d\alpha), \end{aligned}$$

and this is equal to the right hand side of (6.4). $\square$

We can now obtain the desired estimates for the *maximal operator*

$$M_\infty : f \to f^* = M_\infty f = \sup\{A_n f\colon n \geqq 1\}.$$

As usual, we shall use $x \log^+ x$ as a short notation for $x(\log x)^+$ for $x > 0$, and put $0 \log^+ 0 = 0$. log ist the natural logarithm.

**Theorem 6.3** (Dominated ergodic theorem). *If $T$ is a positive $L_1 - L_\infty$-contraction, the following inequalities hold for measurable $f \geqq 0$:*

$$(6.5) \qquad \|M_\infty f\|_p \leqq \frac{p}{p-1} \|f\|_p \quad (1 < p < \infty)$$

*and*

$$(6.6) \qquad \|M_\infty f\|_1 \leqq \frac{e}{e-1} (\mu(\Omega) + \int f \log^+ f\,d\mu).$$

*Proof.* It suffices to prove the inequalities for $M_n$ instead of $M_\infty$, and for bounded functions $f$.

To prove (6.5) we apply (6.4) to $\psi(t) = t^p$ $(t > 0)$ and to $Y = M_n f$, $X = f$. Using $t^{-1}\psi(dt) = t^{-1}dt^p = pt^{p-2}dt$ we obtain

$$\| Y \|_p^p = \int \psi(Y)\, d\mu \leqq \int f\Big[\int_0^Y p\, t^{p-2}\, dt\Big]\, d\mu$$
$$= \frac{p}{p-1} \int f \cdot Y^{p-1}\, d\mu.$$

By the Hölder inequality this is $\leqq \dfrac{p}{p-1} \|f\|_p \| Y \|_p^{p-1}$. The bound for $f$ is a bound for $Y$. Hence $\| Y \|_p < \infty$. We can simplify and obtain $\| Y \|_p \leqq \dfrac{p}{p-1} \|f\|_p$.

To prove (6.6) we shall need the inequality

(6.7) $\quad a \log b \leqq a \log^+ a - b/e, \quad (0 \leqq a,\ 0 < b).$

($\log x \leqq x - 1$ implies $\log x \cdot e \leqq x$, and hence $\log b \leqq b/e$; this implies (6.7) for $a \leqq 1$. For $a > 1$, $a \log b - a \log a = a \log(b/a) \leqq a \cdot b/a \cdot e = b/e$).

We may assume $\mu(\Omega) < \infty$. We apply (6.4) with the function $\psi(t)$ which equals $t$ for $t \geqq 1$ and is 0 for $t < 1$. With $Y = M_n f$ and $X = f$ we obtain

$$\int (Y - \mathbb{1})\, d\mu \leqq \int \psi(Y)\, d\mu \leqq \int f \cdot \Big(\int_0^Y t^{-1} \psi(dt)\Big)\, d\mu$$
$$= \int_{\{Y \geqq 1\}} f \cdot \Big(\int_1^Y t^{-1}\, dt\Big)\, d\mu = \int_{\{Y \geqq 1\}} f \log Y\, d\mu$$
$$\leqq \int f \log^+ f\, d\mu + e^{-1} \int Y\, d\mu.$$

Hence $(1 - 1/e)\, \| Y \|_1 \leqq \mu(\Omega) + \int f \log^+ f\, d\mu$ and (6.6) follows. □

Theorem 6.3 is due to Hardy and Littlewood for translations, to Wiener for measure preserving transformations, and to Dunford-Schwartz [1956] for operators. Dunford-Schwartz have also shown that $T$ need not be positive. This generalization follows via theorem 4.1.1.

If $L'$, $L''$ are subspaces of the space of equivalence classes of measurable functions, an operator $M$: $L' \ni f \to Mf \in L''$ mapping $L'$ into $L''$ is called *sublinear* if $|M(f_1 + f_2)| \leqq |Mf_1| + |Mf_2|$ is true for $f_1, f_2 \in L'$. $M$ is said to be of *weak type* $(p_1, p_2)$ if $L' = L_{p_1}$, $L'' = L_{p_2}$ and

$$\mu(|Mf| > \alpha) \leqq (c \cdot \alpha^{-1} \|f\|_{p_1})^{p_2}$$

holds for all $f \in L_{p_1}$ with a constant independent of $f$ and $\alpha > 0$. $M$ is said to be of *strong type* $(p_1, p_2)$ if

$$\|Mf\|_{p_2} \leqq c \|f\|_{p_1}$$

holds for all $f \in L_{p_1}$ with a constant $c$ independent of $f$. The maximal operator $M_\infty$

clearly is sublinear. As $T$ is an $L_\infty$-contraction, $M_\infty$ is of strong type $(\infty, \infty)$ with constant 1. Lemma 6.1 asserts that $M_\infty$ is of weak type $(1,1)$. By a special case of an *interpolation theorem* of Marcinkiewicz (see e.g. Zygmund [1935, vol. II]) these two facts suffice to prove that $M_\infty$ is of strong type $(p, p)$ for $1 < p < \infty$. This is the main content of (6.5).

Sometimes a variant of the arguments above is of use. Say that $X$, $Y$: $\Omega \to \mathbb{R}^+$ are in a *weak maximal type relation* if

$$(6.8)\qquad \alpha\mu(Y \geqq \alpha) \leqq 2 \int_{\{2X \geqq \alpha\}} X\,d\mu \quad (\alpha > 0).$$

Then the same argument as in the proof of (6.4) shows, that for non decreasing, right continuous $\psi$: $\mathbb{R}^+ \to \mathbb{R}^+$ with $\psi(0) = 0$

$$(6.9)\qquad \int \psi(Y)\,d\mu \leqq 2\int \Big(X(\omega) \int_0^{2X(\omega)} t^{-1}\psi(dt)\Big)\,\mu(d\omega).$$

Let us apply this with $\psi(t) = t\,(\log^+ t)^{m-1}$ and $m \geqq 2$. We have $d\psi/dt = (m-1)(\log^+ t)^{m-2} + (\log^+ t)^{m-1}$, and hence

$$\int_0^{2X(\omega)} t^{-1}\psi(dt) \leqq m(\log^+ 2X(\omega))^{m-1} \int_0^{2X(\omega)} t^{-1}\,dt = m(\log^+ 2X(\omega))^m.$$

Let $L\log^m L$ denote the class of all measurable functions $f$: $\Omega \to \mathbb{R}$ for which $|f|(\log^+|f|)^m$ is integrable. We obtain that, if $\mu$ is a finite measure, then $X \in L\log^m L$ implies $Y \in L\log^{m-1} L$. Now note that (6.2) means that $X = f$ and $Y = M_\infty f$ are in a weak maximal type relation. Together with (6.6) we have proved:

**Theorem 6.4.** *If $T$ is a positive $L_1 - L_\infty$-contraction in $L_1$ of a finite measure space, then $f \in L\log^m L$ implies $M_\infty|f| \in L\log^{m-1} L$, $(m \geqq 1)$.*

**2. A converse.** If $\tau$ is ergodic and $\mu(\Omega) = \infty$, $M_\infty f$ cannot be integrable for $0 \leqq f \not\equiv 0$, because this would imply that $A_n f$ converges in $L_1$-norm to the a.e.-limit 0. For ergodic $\tau$ and $\mu(\Omega) < \infty$ the following theorem of Ornstein [1971] shows that the sufficient condition $f \in L\log L$ is also necessary for the integrability of $f^*$:

**Theorem 6.5.** *If $\tau$ is an ergodic automorphism of a finite measure space $(\Omega, \mathcal{A}, \mu)$ and $f \geqq 0$, then $f^* \in L_1$ implies $f \in L\log L$.*

We adapt an argument of Derriennic [1973]:

**Lemma 6.6** (Moy [1960]). *If $\tau$ is an automorphism of a $\sigma$-finite measure space $(\Omega, \mathcal{A}, \mu)$, $A \in \mathcal{A}$, $r_A(\omega) = \inf\{n \geqq 1: \tau^n(\omega) \in A\}$ with $\inf\emptyset = \infty$, and $A^* = \bigcup_{i=0}^{\infty} \tau^{+i}A$, then $g \geqq 0$ implies*

(6.10) $$\int_{A^*} g\,d\mu = \int_A \sum_{k=0}^{r_A-1} g\circ\tau^k\,d\mu.$$

*Proof.* The sets $A_k = \tau^k\{\omega\in A: r_A(\omega)\geqq k+1\}$, $(1\leqq k<\infty)$, form a disjoint partition of $A^*\setminus A$. The identity therefore follows from

$$\int_A \sum_{k=0}^{r_A-1} g\circ\tau^k\,d\mu = \int_A g\,d\mu + \sum_{k=1}^{\infty}\int_{A\cap\{r_A>k\}} g\circ\tau^k\,d\mu = \int_A g\,d\mu + \sum_{k=1}^{\infty}\int_{A_k} g\,d\mu. \quad \square$$

**Lemma 6.7.** *If the automorphism $\tau$ is conservative and ergodic, $f\geqq 0$, and $\alpha>0$ is so large that $A=\{f^*<\alpha\}$ has positive measure, then*

(6.11) $$\int_{\{f^*\geqq\alpha\}} f\,d\mu \leqq 2\alpha\,\mu(f^*\geqq\alpha).$$

*Proof.* The assumptions imply $A^*=\Omega$. For $\omega\in A\subset\{f<\alpha\}$ one has

$$\sum_{k=1}^{r_A(\omega)-1} f(\tau^k\omega) \leqq 2\alpha(r_A(\omega)-1) \leqq 2\alpha \sum_{k=1}^{r_A(\omega)-1} 1,$$

because otherwise $f^*(\omega)\geqq r_A(\omega)^{-1}\sum_{k=0}^{r_A(\omega)-1} f(\tau^k\omega)\geqq\alpha$. Applying lemma 6.6 to $g=f$ and to $g\equiv 1$ we find

$$\int_{A^c} f\,d\mu = \int_A \sum_{k=1}^{r_A-1} f\circ\tau^k\,d\mu \leqq 2\alpha\int_A \sum_{k=1}^{r_A-1} 1\,d\mu = 2\alpha\,\mu(A^c). \quad \square$$

*Proof of theorem 6.5.* We may assume $f\geqq 1$ by passing to $f+1$. Let $\alpha_0 = \inf\{\alpha: \mu(f^*<\alpha)>0\}$. Now

$$\begin{aligned}
\int f\log f\,d\mu &= \int f(\omega)\int_1^{f(\omega)} \alpha^{-1}\,d\alpha\,\mu(d\omega)\\
&= \int_1^{\infty}\alpha^{-1}\int_{\{f\geqq\alpha\}} f(\omega)\,\mu(d\omega)\,d\alpha\\
&\leqq \int_1^{\infty}\alpha^{-1}\Big(\int_{\{f^*\geqq\alpha\}} f(\omega)\,\mu(d\omega)\Big)d\alpha \leqq 2\int_{\alpha_0}^{\infty}\mu(f^*\geqq\alpha)\,d\alpha\\
&\quad + \|f\|_1\int_1^{\alpha_0}\alpha^{-1}\,d\alpha \leqq 2\|f^*\|_1 + \|f\|_1\int_1^{\alpha_0}\alpha^{-1}\,d\alpha < \infty. \quad \square
\end{aligned}$$

The only instance where we have used the invertibility of $\tau$ has been the proof of Lemma 6.6. Therefore the following generalization of the Kac recurrence theorem ($g\equiv 1$) shows that Theorem 6.5 and lemma 6.7 hold also for endomorphisms:

**Proposition 6.8.** *If $\tau$ is a conservative endomorphism of a $\sigma$-finite measure space $(\Omega,\mathscr{A},\mu)$, then (6.10) holds for $A^*=\bigcup_{i=0}^{\infty}\tau^{-i}A$ and $g\geqq 0$.*

The case where $\tau$ is invertible is a special case of Lemma 6.6 because for conservative $\tau$ the definitions of $A^*$ coincide. We postpone the proof of the general case to section 3.3. (Theorem 6.5 for endomorphisms follows also from the invertible case by a passage to the bilateral extension).

If $\tau$ is a dissipative automorphism and $f \in L_1^+$, the inequality (6.11) holds for all $\alpha > 0$: Simply observe that $\sum_{i=-\infty}^{+\infty} f(\tau^i \omega) < \infty$ a.e.. Therefore, for almost all $\omega \in \{f^* \geqq \alpha\}$, there exists some $k \geqq 1$ with $\tau^{-k}\omega \in \{f^* < \alpha\}$. This is all that is needed to apply the argument used in the proof of lemma 6.7.

**3. The sign of partial sums.** For $f \in L_1$ we consider the sets

$$A_f = \{\omega\colon S_n f(\omega) > 0 \quad \text{for all } n \geqq 1\}$$

and

$$\underline{A}_f = \{\omega\colon S_n f(\omega) \geqq 0 \text{ for all } n \geqq 1\}.$$

$A_f^*$, $\underline{A}_f^*$, etc. are defined as $A^*$ in proposition 6.8. Put

$$S_* f = \inf_{n \geqq 1} S_n f.$$

The following identities have been observed by Marcus and Petersen [1979] for invertible ergodic $\tau$. Part of their result follows also from a result of Halász [1976], who proved that, for ergodic $\tau$ in a finite measure space, $\int f d\mu = 0$ implies $\mu(A_f) = 0$.

**Theorem 6.9.** *If $\tau$ is a conservative endomorphism of a $\sigma$-finite measure space $(\Omega, \mathscr{A}, \mu)$ and $f \in L_1$ then*

$$(6.12) \quad \int_{A_f^*} f d\mu = \int_{A_f} S_* f d\mu = \int_{\underline{A}_f} S_* f d\mu = \int_{\underline{A}_f^*} f d\mu.$$

*If $\int_I f d\mu$ is non negative for all $\tau$-invariant sets $I \in \mathscr{A}$, the four integrals agree also with $\int f d\mu$.*

*Proof.* The second identity in (6.12) is obvious since $S_* f(\omega) = 0$ holds on $\underline{A}_f \setminus A_f$. Let $r_A = r_A(\omega)$ be the first return time of $\omega$ to $A = \underline{A}_f$. We claim that

$$(6.13) \quad S_* f(\omega) = \sum_{i=0}^{r_A - 1} f(\tau^i \omega), \quad (\omega \in A).$$

As $\tau^{r_A}\omega$ belongs to $A$, all sums $\sum_{i=r_A}^{m} f(\tau^i \omega)$, $(m \geqq r_A)$, are non negative. Therefore, there must be some $k$ with $1 \leqq k \leqq r_A(\omega)$ and $S_* f(\omega) = S_k f(\omega)$. For all $m \geqq 1$, $S_m f(\tau^k \omega) = S_{m+k} f(\omega) - S_k f(\omega) = S_{m+k} f(\omega) - S_* f(\omega) \geqq 0$. This shows that $\tau^k \omega$ belongs to $A$. As $r_A$ was the time of the first return, $k = r_A$, and (6.13) is proved. The third identity in (6.12) now is a consequence of proposition 6.8. To prove the first identity we may assume $\Omega = A_f^*$, since $A_f^*$ is $\tau$-invariant. But then $A_f^*$ agrees with $\underline{A}_f^*$.

To prove the last assertion we may assume that $\underline{A}_f^*$ is empty. For $g = -f$ the set $\{M_\infty g \geqq 0\}$ then agrees with $\Omega$ and the maximal ergodic theorem implies $\int(-f)\,d\mu \geqq 0$. Together with the assumption the proof is complete. □

It is easy to show by example that (6.12) may fail for dissipative transformations.

Marcus and Petersen derive the following corollary using the ergodic decomposition of $\Omega$, but is is not necessary to use this deep tool.

**Corollary 6.10.** *If $\tau$ is conservative, $S^*f = \sup\{S_nf: n \geqq 1\}$, and $B_f = \{\omega: S_nf(\omega) < 0$ for all $n \geqq 1\}$, then*

$$\int f\,d\mu = \int_{A_f} S_*f\,d\mu + \int_{B_f} S^*f\,d\mu.$$

*Proof.* By an exhaustion argument we may split $\Omega$ into three invariant sets $I_+$, $I_-$ and $I_0$ such that, for any invariant $I$ of positive measure, $\int_I f\,d\mu$ is positive when $I \subset I_+$, negative when $I \subset I_-$, and zero when $I \subset I_0$. (6.13) implies the strict positivity of $S_*f$ on $A_f$. If $I$ is any invariant set with $\mu(A_f \cap I) > 0$, we obtain

$$0 < \int_{A_f \cap I} S_*f\,d\mu = \int_{(A_f \cap I)^*} f\,d\mu.$$

Hence $A_f$ is contained mod $\mu$ in $I_+$.

Applying theorem 6.9 to $I_+$ we find $\int_{A_f} S_*f\,d\mu = \int_{I_+} f\,d\mu$. By symmetry $\int_{B_f} S^*f\,d\mu = \int_{I_-} f\,d\mu$, and the part $I_0$ does not disturb since $\int_{I_0} f\,d\mu = 0$. □

Theorem 6.9 admits some interesting probabilistic conclusions: Consider a stationary process $X_0, X_1, \ldots$ taking values in $\{-1, 0, +1\}$. In the product space representation we may write $X_i = f \circ \tau^i$, where $\tau$ is the shift. $\mu$ is the probability measure $P$. $S_n = X_0 + \ldots + X_{n-1}$ is a "walk" on the integers. On $A_f = \{S_n \geqq 1$ for all $n\}$, $S_*f$ is $\equiv 1$. Therefore, if the process $(X_i)$ is ergodic and $EX_0$ non negative, the probability that the walk always stays above 0 is $EX_0$.

Even without the condition $EX_0 \geqq 0$ and ergodicity the corollary shows that the probability that the walk first returns to 0 from above is the same as the probability that it first returns to 0 from below. [We have $S_*f = 1$ on $A_f$, and $S^*f = -1$ on $B_f$. Hence $P(S_n \geqq 1$ for all $n) - P(S_n \leqq -1$ for all $n) = \mathrm{P}(\mathrm{X}_0 = 1) - \mathrm{P}(\mathrm{X}_0 = -1)$. But the two probabilities in question are $P(X_0 = 1) - P(S_n \geqq 1$ for all $n)$ and $P(X_0 = -1) - P(S_n \leqq -1$ for all $n)$.]

**4. Continuous parameters.** The continuous parameter maximal and dominated ergodic theorems can be deduced from their discrete counterparts by approximation. Again we treat the operator case right away.

A family $\mathscr{S} = \{T_t, t > 0\}$ of bounded linear operators in a Banach space $\mathfrak{X}$ is called a semigroup if $T_{t+s} = T_t T_s$ holds for all $t$, $s$. $\mathscr{S}$ is called strongly continuous at $t$ if $\| T_s f - T_t f \|$ tends to 0 as $s \to t$ for all $f \in \mathfrak{X}$, and strongly continuous if it is strongly continuous at all $t$. We also consider semigroups $\mathscr{S} = \{T_t, t \geqq 0\}$ with parameter set $[0, \infty[$. We do not request that $T_0$ is the identity, though this is the main case. $\mathscr{S}$ is called *locally bounded* if $\sup \{\| T_t \|: 0 < t \leqq a\}$ is finite for one and hence for all $a > 0$.

If $\mathscr{S} = \{T_t, t > 0\}$ is a locally bounded strongly continuous semigroup of linear operators in $L_1(\mu)$ we may define the integrals

$$(6.14) \qquad S_{0,a} f = \int_0^a T_t f \, dt, \quad (f \in L_1),$$

in several essentially equivalent ways: Put $f_n(t, \omega) = T^i_{1/n} f$ for $i/n \leqq t < (i+1)/n$. For given $\varepsilon > 0$ and $t_\varepsilon > 0$ we may find $n_\varepsilon$ such that $n \geqq n_\varepsilon$, $t > t_\varepsilon$ implies $\| f_n(t, \cdot) - T_t f \|_1 < \varepsilon$. Therefore there exists a subsequence $n_\nu$ and a measurable $f_\infty(t, \omega)$ on $]0, \infty[ \times \Omega$ such that $f_{n_\nu}(t, \cdot)$ converges in $L_1$ and a.e. to $f_\infty(t, \cdot)$. The $L_1$-convergence is uniform in $t \geqq t_\varepsilon$. $f_\infty(t, \cdot)$ is a representative of the equivalence class $T_t f$. When $\lambda$ denotes the Lebesgue measure, $f_\infty(t, \omega)$ is $(\lambda \times \mu)$integrable on $]0, a] \times \Omega$ for each $a < \infty$, and we may put $S_{0,a} f(\omega) = \int_0^a f_\infty(t, \omega) \, dt$.

If we choose another subsequence or other points of subdivision than $i/n$ we get a function $f'_\infty(t, \omega)$ agreeing with $f_\infty(\omega, t)$ $(\lambda \times \mu)$-a.e. and the corresponding integrals $S'_{0,a} f$ will agree with $S_{0,a} f$ except on a nullset which is independent of $a$. For fixed $a$, $S_{0,a} f$ is also the $L_1$-limit of the Riemann sums

$$R_a(n) f = \frac{a}{n} \sum_{i=0}^{n-1} T^i{}_{a/n} f.$$

We can now deduce a continuous time version of Hopf's inequality:

**Theorem 6.11.** *Let $(\Omega, \mathscr{A}, \mu)$ be a $\sigma$-finite measure space and $\mathscr{S} = \{T_t, t \geqq 0\}$ a strongly continuous semigroup of positive contractions in $L_1$. For $\alpha > 0$ and $f \in L_1$ put*

$$E(\alpha) = \{\omega \in \Omega: \sup \{S_{0,a} f(\omega): 0 < a < \alpha\} > 0\}.$$

*Then*

$$(6.15) \qquad \int_{E(\alpha)} T_0 f \, d\mu \geqq 0.$$

*Proof.* Let $D_n = \{k/2^n: k = 1, 2, \ldots\}$ be the set of positive dyadic rationals of order $n$ and $D = \bigcup_{n=1}^{\infty} D_n$. Applying Hopf's inequality to the function $T_0 f$ and the operator $T_{2^{-n}}$ we obtain, for $\alpha \in D_n$ and $m \geqq n$, the inequality

(6.16) $$\int_{E^{(m)}(\alpha)} T_0 f d\mu \geqq 0,$$

where

$$E^{(m)}(\alpha) = \{\sup \{\sum_{i=0}^{k-1} T^i_{2^{-m}} T_0 f\colon 0 \leqq k \leqq \alpha 2^m\} > 0\}$$
$$= \{\sup \{R_\beta(2^m) T_0 f\colon \quad 0 \leqq \beta \leqq \alpha,\ \beta \in D_m\} > 0\}.$$

$R_\beta(2^m) T_0 f$ tends to $S_{0,\beta} T_0 f = S_{0,\beta} f$ in $L_1$-norm. Passing to subsequences and finding a common subsequence by the diagonal procedure the convergence along the subsequence is also pointwise except on a $\mu$-nullset $N_0$ independent of $\alpha \in D$ and $\beta \in D$. Because of

$$\sup_{0<\beta<\alpha} S_{0,\beta} f = \sup_{\substack{0<\beta<\alpha \\ \beta \in D}} S_{0,\beta} f,$$

it follows that $\mu(E(\alpha) \setminus E^{(m_i)}(\alpha)) \to 0$ as $m_i \to \infty$ along the subsequence. Again by the strong continuity of the semigroup, the set $\{T_0 f > 0\}$ is contained in $E(\alpha)$ up to a $\mu$-nullset. (6.15) now follows from (6.16). □

It is now not difficult to proceed as in the beginning of this section to prove a continuous parameter version of Lemma 6.1. for strongly continuous semigroups $\{T_t, t \geqq 0\}$ of positive $L_1 - L_\infty$-contractions. A generalization of such a lemma giving a simultaneous estimate for several functions $f_u$ is proved in Vol. I of Dunford-Schwartz [1958; 690].

For $f \in L_p$, $(1 < p < \infty)$, and strongly continuous semigroups $\{T_t, t > 0\}$ of contractions in $L_p$, it is possible to define $S_{0,a} f$ essentially as above. The $p$-th power of $f_\infty(t, \omega)$ is $(\lambda \times \mu)$-integrable on $[0, a] \times \Omega$. Therefore, $f_\infty$ is integrable on each $[0, a] \times H$ with $\mu(H) < \infty$, so that the integrals $\int_0^a f(t, \omega) dt$ are well defined and finite for a.e. $\omega$.

It follows from the Riesz convexity theorem (cf. Dunford-Schwartz [1958]) that $L_1 - L_\infty$-contractions are also contractions in $L_p$, $(1 < p < \infty)$. (For positive operators a short proof is given in section 1.7). Using this fact, it is not hard to see that a strongly continuous semigroup $\{T_t, t > 0\}$ of $L_1 - L_\infty$-contractions is also strongly continuous in $L_p$.

The following theorem is the continuous parameter version of the dominated ergodic theorem 6.3. We leave the proof (by discrete approximation) to the reader:

**Theorem 6.12.** *If $\mathcal{T} = \{T_t, t > 0\}$ is a strongly continuous semigroup of positive $L_1 - L_\infty$-contractions in $L_1$ $(\Omega, \mathscr{A}, \mu)$ and $M_{0,\infty} f = \sup \{a^{-1} S_{0,a} f\colon 0 < a < \infty\}$, then the following inequalities hold for measurable $f \geqq 0$:*

(6.17) $$\|M_{0,\infty} f\|_p \leqq \frac{p}{p-1} \|f\|_p, \quad (1 < p < \infty),$$

*and*

(6.18) $$\|M_{0,\infty} f\|_1 \leqq \frac{e}{e-1}\left(\mu(\Omega) + \int f \log^+ f d\mu\right).$$

A similar theorem appears in Dunford-Schwartz [1958: thm. VIII 7.7]. (The condition of positivity of $T_t$ can be eliminated by theorem 4.1.1).

We now look at the special case of semigroups $\{T_t\}$ induced by measure preserving transformations: Let $\{\tau_t: t \geqq 0\}$ be a measurable measure preserving semiflow in $(\Omega, \mathscr{A}, \mu)$, and $T_t$ the composition with $\tau_t$. In order to apply one of the theorems stated above it remains to show that the measurability implies the strong continuity of the semigroup.

A mapping $t \to F(t)$ of a measure space $(C, \mathscr{C}, \nu)$ into a Banach space $\mathfrak{X}$ is called *weakly $\mathscr{C}$-measurable* if, for any $h$ from the dual space $\mathfrak{X}^*$, the map $t \to \langle F(t), h\rangle$ is $\mathscr{C}$-measurable. $F$ is said to be *finitely valued* if it assumes only finitely many values $x_j \in \mathfrak{X}$, and, for $x_j \neq 0$, the sets $\{t \in C: F(t) = x_j\}$ have finite $\nu$-measure. $F$ is called *strongly $\mathscr{C}$-measurable*, if there exists a sequence of finitely valued functions $F_n(t)$ strongly convergent to $F(t)$ for $\nu$-a.e. $t \in C$. $F$ is called *$\nu$-almost separably valued* if there exists a $\nu$-nullset $C_0$ such that $\{F(t): t \in C \setminus C_0\}$ is a separable subset of $\mathfrak{X}$. A theorem of Pettis asserts that $F$ is strongly $\mathscr{C}$-measurable iff it is weakly $\mathscr{C}$-measurable and $\nu$-almost separably valued; see Yosida [1974: 131].

We apply it with $C = [0, \infty[$, $\mathscr{C} = \mathscr{B}$, and $\nu = \lambda$, where $\lambda$ is the Lebesgue measure, and $\mathscr{B}$ the family of Lebesgue measurable sets. Let $E \in \mathscr{A}$ be a set with finite $\mu$-measure, and $F(t) = 1_E \circ \tau_t$. The measurability of the semiflow shows that, for all $h = 1_A$, $(A \in \mathscr{A})$, $\langle F(t), h\rangle = \mu(A \cap \tau_t^{-1} E)$ depends measurably on $t$. This is enough to imply the weak $\mathscr{C}$-measurability of $F$. Let $\tilde{\mathscr{A}}$ be the completion of $\mathscr{B} \otimes \mathscr{A}$ with respect to $\lambda \times \mu$. For any set $G \in \tilde{\mathscr{A}}$ there exists a countably generated sub-$\sigma$-algebra $\mathscr{A}_G$ of $\mathscr{A}$ such that $G$ belongs even to the completion of $\mathscr{B} \otimes \mathscr{A}_G$; (the family of sets $G$ with this property is a $\sigma$-algebra). Now take $G = \{(t, \omega): \tau_t \omega \in E\}$.

$F(t)$ belongs for almost all $t$ to the separable space $L_1(\Omega, \mathscr{A}_G, \mu)$. By the Pettis theorem $F$ is strongly $\mathscr{C}$-measurable.

By a theorem of N. Dunford (Dunford-Schwartz [1958], 8.1.3) a strongly $\mathscr{C}$-measurable family $F(t) = T_t f$ for a semigroup of bounded linear operators $T_t$ is strongly continuous at $t > 0$, i.e., we have $\|T_s f - T_t f\| \to 0$ as $s \to t > 0$. Approximating a general $f \in L_1$ with simple functions we conclude that our semigroup given by $T_t f = f \circ \tau_t$ is strongly continuous at $t > 0$. As the $T_t$ are isometries, we find, for $0 < r \to 0$, that $\|T_r f - f\|_1 = \|T_{r+t} f - T_t f\|_1 \to 0$, and the semigroup is also continuous at $t = 0$. Modulo the results of Pettis and Dunford we have proved

**Theorem 6.13.** *If $\{\tau_t: t \geqq 0\}$ is a measurable measure preserving semiflow in a $\sigma$-*

*finite measure space* $(\Omega, \mathscr{A}, \mu)$, *the semigroup* $\{T_t, t \geqq 0\}$ *defined by* $T_t f = f \circ \tau_t$ *is strongly continuous in each* $L_p$, $(1 \leqq p < \infty)$.

## Notes

A dominated estimate similar to (6.6) was given by Marcinkiewicz and Zygmund [1937] in the case of independent identically distributed random variables, called the *i.i.d. case* in the sequel. Wiener's larger constant in the estimate (6.5) was improved by Fukamiya [1940]. Notes concerning theorem 6.4 are given in section 6.1.

Ornstein's converse of the $L \log L$-result was already obtained by Stein [1969] for translations in $\mathbb{R}^d$. It was also motivated by a theorem of Burkholder [1962], saying that, in the i.i.d. case $X_i = f \circ \tau^i$, the integrability of $f^{**} := \sup_n |A_n f|$ implies $f \in L \log L$. R.L. Jones [1976] has shown that this stronger converse requires the i.i.d. assumption. B. Davis [1982] has characterized the distributions of $f$ for which $f^{**}$ can be integrable for an ergodic stationary process:

**Theorem 6.14.** *Let* $g$ *be non decreasing in* $[0,1]$ *and* $M(x) = \int_0^x g(t)\,dt + \int_{1-x}^1 g(t)\,dt$. *There exists a stationary ergodic sequence* $f \circ \tau^i$ *such that the distribution of* $f$ *agrees with that of* $g$ *(under Lebesgue measure) and* $f^{**}$ *is integrable, iff* $\int_0^{1/2} |M(x)/x|\,dx < \infty$.

For $g \geqq 0$ this reduces to the $L \log L$-criterion.

Derriennic [1973] proved a dominated estimate and a converse for ergodic ratios: Let $\tau$ be an ergodic conservative endomorphism of $(\Omega, \mathscr{A}, \mu)$, $f, g \in L_1^+$ and $g > 0$ a.e.. Put $s = \sup_n (S_n f / S_n g)$. If $\int f \log^+ (f/g)\,d\mu < \infty$, then $\int g s\,d\mu < \infty$. Conversely, if $\int g[s + s \circ \tau]\,d\mu < \infty$ then $\int f \log^+ (f/g)\,d\mu < \infty$.

Other related work was done by Gundy [1963] and R.L. Jones [1977]. B. Davis [1971] proved another strong converse in the i.i.d. case: Put $\tilde{f} = \sup \{n^{-1} f \circ \tau^n\}$. As $\tilde{f}$ is dominated by $M_\infty |f|$, it is clear that $\tilde{f}$ is integrable for $f \in L \log L$. Davis showed that for $f \geqq 0$ the integrability of $\tilde{f}$ implies $f \in L \log L$. Krengel and Sucheston [1978] gave an example showing that this does not extend to the general ergodic case.

Petersen [1979] proved a continuous parameter form of Ornstein's converse. In the continuous parameter setting the maximal inequality frequently holds even with equality:

**Theorem 6.15** (Marcus and Petersen [1979]). *If* $\{\tau_t, t \in \mathbb{R}\}$ *is a measurable ergodic measure preserving flow in a probability space,* $f \in L_1$, *and* $\alpha \geqq \int f\,d\mu$, *then*

$$(6.19) \qquad \int_{\{M_{0,\infty} f > \alpha\}} f\,d\mu = \alpha \mu(M_{0,\infty} f > \alpha).$$

They obtain this from continuous parameter results related to theorem 6.9; see also Petersen [1983]. (The fact that there is equality in (6.19) was known for translation in $\mathbb{R}$; see Hewitt-Stromberg [1969: 423]. Sato [1982] has proved maximal equalities for semiflows. Recently, Sato [1984a] obtained a generalization of theorem 6.14, which concerns the ratio maximal function for ergodic flows. Fefferman and Stein [1971] have maximal estimates for sequences of functions which unify the Hardy-Littlewood estimates with results of Marcinkiewicz and Carleson. Bru and Heinich [1981] studied dominated estimates in Orlicz spaces.

*Other dominated estimates.* Krengel, Röttger and Wacker [1983] proved: If $Y_0, Y_1, \ldots$ is a stationary sequence of positive valued random variables and $Z = \inf_n (Y_0 + Y_1 + \ldots Y_{n-1})/n$, then $E(1/Z) \leqq 4E(1/Y_0)$. If $N_t = \sup\{n \geqq 1: Y_0 + \ldots + Y_{n-1} \leqq t\}$, then $E(\sup N_t/t) \leqq E(1/Z)$. This allows to prove the $L_1$-norm convergence of $N_t/t$ in the case $E(1/Y_0) < \infty$. The integrability condition $E(1/Y_0) < \infty$ is in a sense optimal.

Motivated by analogies in martingale theory, several authors considered the *ergodic square function*

$$SQ(T)f = \left(\sum_1^\infty |A_n f - A_{n-1} f|^2\right)^{1/2}$$

(with $A_0 f = 0$). Jones [1977] showed for ergodic automorphisms $\tau$ that $SQ(T)$ is a bounded operator in $L_p$ for $1 < p < \infty$, and of weak type (1,1). De la Torre [1979] proved that the boundedness in $L_p$ remains true for positive contractions $T$ in $L_p$. Further generalizations were give by Yoshimoto [1980].

Sometimes one is interested in a $\tau$ preserving a measure $\mu$, and in dominated estimates in $L_p(\nu)$ where $\nu$ has a density $w = d\nu/d\mu$. For this we refer to Muckenhoupt [1972], Coifman and Fefferman [1974], Atencia and de la Torre [1982], de la Torre [1982], and the references given there.

This is closely related to the existence of the *ergodic Hilbert transform.* Cotlar [1955] obtained an ergodic generalization of the classical Hilbert transform (for translations) as follows:

**Theorem 6.16.** *Let $\tau$ be an automorphism. Then $f^{HT} := \lim_{n\to\infty} \sum_{k=1}^n (f \circ \tau^k/k - f \circ \tau^{-k}/k)$ exists a.e. for $f \in L_1$. Similarly, if $\{\tau_t: -\infty < t < \infty\}$ is a measurable measure preserving flow, then $\lim_{\varepsilon\to 0} \int_{I_\varepsilon} f \circ \tau_t/t\, dt$ exists a.e., where $I_\varepsilon = \{t: \varepsilon \leqq |t| \leqq 1/\varepsilon\}$.*

Petersen [1983] gives a thorough discussion and references.

Teicher [1967] has used Wieners dominated ergodic theorem to deduce in the i.i.d. case: if $\int f d\mu = 0$ and $\|f\|_p < \infty$, $(p > 2)$, or $\||f|^p \log^+|f|\|_1 < \infty$, $(p = 2)$, and if $c_n$ is a positive decreasing sequence with $c_n = O(n^{-r/2})$ and $\sum_{n=1}^\infty n^{[r]-1} c_n{}^{2[r]/r} < \infty$, then $\sup\{c_n |S_n f|^r\}$ is integrable. Related results for $U$-statistics appear in Ahmad [1981].

Teicher [1971] proved: If $X_0, X_1, \ldots$ are i.i.d. with $EX_i = 0$, then $\sup_{n \geqq e^e} |\sum_1^n X_i|^r / n \log\log n)^{r/2}$ is integrable iff $E|X_1|^r < \infty$, (for $r > 2$), resp. iff $X_1^2 \log|X_1| 1_{\{|X_1| > e^e\}} / \log\log|X_1|$ is integrable, (for $r = 2$). The case where $r$ is an integer is due to Siegmund [1969].

G. Baxter [1965] has given a subtle refinement of Hopf's maximal ergodic theorem. To state it we abreviate "for infinitely many values of $k$" by i.m.k, and "for finitely many values of $k$" by f.m.k:

**Theorem 6.17.** *If $T$ is a positive contraction in $L_1$, $f \in L_1$, and $1 \leqq m \leqq n$, then*

$$\int_{\{M_n^S f > M_m^S f\}} S_m f\, d\mu \geqq - \int_{\{M_n^S f = M_m^S f\}} M_m f\, d\mu$$

*and*

$$\int_{\{S_k f > M_m^S f \text{ i.m.k}\}} S_m f\, d\mu \geqq - \int_{\{S_k f > M_m^S f \text{ f.m.k}\}} M_m f\, d\mu.$$

*In particular*

$$\int_{\{S_k f > 0 \ i.m.k\}} f d\mu \geqq 0 .$$

Continuous parameter analogues of these inequalities have been established by Berk [1968].

## § 1.7 Some general tools and principles

In this section we discuss the Banach principle, the ergodic theorem for sequences of functions, and, in the Notes, the necessity of maximal inequalities.

$\Lambda$ is a *directed* set if there is a partial order $\leqq$ (satisfying $\lambda \leqq \lambda$ and transitivity, but not necessarily $\lambda_1 \leqq \lambda_2 \leqq \lambda_1 \Rightarrow \lambda_1 = \lambda_2$), and if for $\lambda, \lambda' \in \Lambda$ there exists $\lambda'' \geqq \lambda, \lambda'$. E.g., $\Lambda = \mathbb{Z}^+$ or $\Lambda = (\mathbb{Z}^+)^d$ with partial order $(u_i) \leqq (v_i)$ iff $u_i \leqq v_i$ for $i = 1, \ldots, d$. In the proof of von Neumann's theorem we have implicitly used:

**Proposition 7.1.** *Let $\Lambda$ be a directed set, and let $T_\lambda$, $(\lambda \in \Lambda)$, be a family of linear operators mapping a Banach space $\mathfrak{X}$ into a Banach space $\mathfrak{Y}$. If the $T_\lambda$ are uniformly bounded, (i.e., if $\sup \| T_\lambda \| < \infty$), and if $T_\lambda x$ converges in norm for all $x$ in a dense subset of $\mathfrak{X}$, then $T_\lambda x$ converges in norm for all $x \in \mathfrak{X}$.*

The proof is obvious. There also is a converse: If the $T_\lambda$ are bounded and $T_\lambda x$ converges for all $x$, then the norms $\| T_\lambda x \|$ are bounded for each fixed $x$, and, by the uniform boundedness principle, the $T_\lambda$ are uniformly bounded. $T_\infty x := \lim_\lambda T_\lambda x$ then defines a linear operator with $\| T_\infty \| \leqq \sup \| T_\lambda \|$, (theorem of Banach Steinhaus).

One of the most useful tools in the study of pointwise ergodic theorems is *Banach's principle*, which, similarly, allows to deduce almost sure convergence for all $x \in \mathfrak{X}$ from almost sure convergence for all $x$ in a dense subset. In this principle, $\mathfrak{Y}$ is a space of equivalence classes of functions and the uniform boundedness of the $T_\lambda$ is replaced by a pointwise condition stated below.

Let $(\Omega, \mathscr{A}, \mu)$ be a finite measure space and $\mathfrak{Y}$ a linear space of equivalence classes of real or complex valued measurable functions on $\Omega$. A map $T: \mathfrak{X} \to \mathfrak{Y}$ is called *continuous in measure* if, for any sequence $x_n \in \mathfrak{X}$ converging in norm to some $x$, and for any $\varepsilon > 0$, $\mu(|Tx_n - Tx| > \varepsilon) \to 0$. For a family $\mathscr{T} = \{T_\lambda, \lambda \in \Lambda\}$ we employ the notation

$$M^{\mathscr{T}} x = \sup_{\lambda \in \Lambda} \{|T_\lambda x|\} \quad M_\lambda^{\mathscr{T}} x = \sup \{|T_\varrho x| : \varrho \leqq \lambda\} .$$

$\Lambda$ is called *locally finite* if $\{\varrho \in \Lambda : \varrho \leqq \lambda\}$ is finite for all $\lambda \in \Lambda$. Clearly, if $\Lambda$ is countable and locally finite, the a.e.-convergence of $T_\lambda x$ implies the finiteness a.e. of $M^{\mathscr{T}} x$.

We shall only need the easy half (a) of the following theorem in the sequel:

**Theorem 7.2** (Banach principle). *(a) Let $\Lambda$ be a countable directed set and $\mathcal{T} = \{T_\lambda, \lambda \in \Lambda\}$ a family of linear operators from $\mathfrak{X}$ into $\mathfrak{Y}$. If there exists a positive decreasing function $C(\alpha)$ with $\lim_{\alpha \to \infty} C(\alpha) = 0$ and*

$$(7.1) \qquad \mu(M^{\mathcal{T}} x > \alpha \|x\|) \leqq C(\alpha) \quad \forall \alpha > 0, x \in \mathfrak{X},$$

*then the $T_\lambda$ are continuous in measure, and the set of elements $x \in \mathfrak{X}$ for which $T_\lambda x$ converges a.e. is closed.*

*(b) Assume $\Lambda$ countable and locally finite. If the $T_\lambda$ are continuous in measure and $M^{\mathcal{T}} x$ is finite a.e. for all $x \in \mathfrak{X}$, then there is a positive decreasing function $C(\alpha)$, $(\alpha > 0)$, tending to 0 as $\alpha \to \infty$, and satisfying* (7.1).

*Proof.* (a) The continuity in measure of the $T_\lambda$ follows from $\mu(|T_\lambda x_n - T_\lambda x| > \varepsilon) \leqq \mu(M^{\mathcal{T}}(x_n - x) > \varepsilon) \leqq C(\varepsilon \|x_n - x\|^{-1}) \to 0$, as $\|x_n - x\| \to 0$. Now let $z$ be a limit of elements $z_n$ of $\mathfrak{X}$ for which $\lim_\lambda T_\lambda z_n$ exists a.e.. Put $\Delta(\omega, x) := \limsup_{\lambda, \varrho \in \Lambda} |T_\lambda x(\omega) - T_\varrho x(\omega)|$ We have $|\Delta(\cdot, z) - \Delta(\cdot, z_n)| \leqq \Delta(\cdot, z - z_n) \leqq M^{\mathcal{T}}(z - z_n)$. Hence, for any $\varepsilon > 0$, $\mu\{\omega: |\Delta(\omega, z)| > \varepsilon\} \leqq \mu(M^{\mathcal{T}}(z - z_n) > \varepsilon) \leqq C(\varepsilon \|z - z_n\|^{-1}) \to 0$. Hence, $\Delta(\cdot, z) = 0$, and $\lim_\lambda T_\lambda z$ exists a.e..

(b) For any $x \in \mathfrak{X}$ and $\varepsilon > 0$ the finiteness of $M^{\mathcal{T}} x$ and of $\mu$ implies the existence of an $n$ with $\mu(M^{\mathcal{T}} x > n) \leqq \varepsilon$. Thus $\mathfrak{X}$ is the union of the sets $B_n = \{x: \mu(M^{\mathcal{T}} x > n) \leqq \varepsilon\}$. For each $n$, $B_n$ is the intersection of the sets $B_{n,\lambda} = \{x: \mu(M_\lambda^{\mathcal{T}} x > n) \leqq \varepsilon\}$. $M_\lambda^{\mathcal{T}}$ is continuous in measure because $\Lambda$ is locally finite. Hence, each of the sets $B_{n,\lambda}$, and therefore also each $B_n$ is closed. By the Baire category thorem at least one of the sets $B_n$ must have a non empty interior, and thus must contain a closed sphere. If $x_0$ is the center, and $r > 0$ the radius of this sphere, we have

$$\mu(M^{\mathcal{T}} x > n) \leqq \varepsilon \quad \text{for all } x \text{ with } \|x - x_0\| \leqq r.$$

For this we may write

$$\mu\{\omega: M^{\mathcal{T}}(x_0 + rz)(\omega) > n\} \leqq \varepsilon \quad \text{for all } z \text{ with } \|z\| \leqq 1.$$

Using

$$M^{\mathcal{T}} z \leqq M^{\mathcal{T}}(x_0 + rz)/r + M^{\mathcal{T}} x_0/r$$

we obtain, for $\|z\| \leqq 1$, the estimate

$$\mu\left(M^{\mathcal{T}} z \geqq \frac{2n}{r}\right) \leqq \mu(M^{\mathcal{T}}(x_0 + rz) > n) + \mu(M^{\mathcal{T}} x_0 > n) \leqq 2\varepsilon.$$

Thus, for any $\alpha \geqq 2n/r$, we have $\mu(M^{\mathcal{T}} z > \alpha) \leqq 2\varepsilon$. Hence, $C(\alpha) := \sup\{\mu(M^{\mathcal{T}} z > \alpha): \|z\| \leqq 1\}$ tends to 0 as $\alpha \to \infty$. To prove (7.1), we may assume $x \neq 0$ and pass to $z = x/\|x\|$. □

In applications the operators $T_\lambda$ usually are bounded linear operators in a space $L_p$, $(1 \leqq p < \infty)$. Such operators clearly are continuous in measure. The measure $\mu$ may then also be $\sigma$-finite, because the theorem can then be applied with $\mathfrak{X} = L_p(\Omega)$ and $\mathfrak{Y} = L_p(A)$, where $A$ is a subset of $\Omega$ having finite measure.

As an example of an application we sketch the proof of an extension of Birkhoff's theorem proved by Hopf [1954] (for $T\mathbb{1} = \mathbb{1}$) and Dunford-Schwartz [1956]. When $T$ is the composition with an endomorphism we obtain an alternative simple proof of Birkhoff's theorem:

**Theorem 7.3.** *If $T$ is a positive $L_1 - L_\infty$-contraction in $L_1$ of a finite measure space, the averages $A_n f$ converge a.e. for $f \in L_1$.*

*Proof.* $T$ is also a contraction in $L_2$. This is clear when $T$ is the composition with $\tau$, and it follows from the Riesz convexity theorem or from the subsequent lemma in the general case. By von Neumann's theorem the set of functions $f = g + (Th - h)$ with $Tg = g \in L_2$ and $h \in L_2$ is dense in $L_2$. We may assume $h \in L_\infty$ replacing $h$ by $(-K) \vee h \wedge K$ for large enough $K$. It follows that the set of such $f$ with bounded $h$ is also dense in $L_1$. For these $f$ we clearly have $\lim A_n f = g$ a.e.. As Lemma 6.1 implies (7.1) with $C(\alpha) = 1/\alpha$, the proof is complete. □

Later, we shall encounter various generalizations of this theorem.

The following lemma was shown to me by M. Lin. It provides a simple proof of the fact that positive $L_1 - L_\infty$ contractions are also contractions in $L_p$ for $1 < p < \infty$:

**Lemma 7.4.** *If $T$ is a positive linear operator in a space of equivalence classes of measurable functions, $1 < p < \infty$, and $p^{-1} + q^{-1} = 1$, then*

(7.2) $$|T(fg)| \leqq [T(|f|^p)]^{1/p} [T(|g|^q)]^{1/q} \quad \text{a.e..}$$

*In particular, if $T$ contracts the $L_\infty$-norm,* $|Tf|^p \leqq T(|f|^p)$ a.e..

*Proof.* It is known from the proof of Hölder's inequality (see Hewitt-Stromberg [1969]) that

$$ab \leqq p^{-1} a^p + q^{-1} b^q, \quad (a, b \geqq 0).$$

We may assume $f, g \geqq 0$. For any $c, d > 0$ we obtain

$$\frac{fg}{cd} \leqq \frac{f^p}{c^p p} + \frac{g^q}{d^q q},$$

and hence

(7.3) $$T(fg)/cd \leqq T(f^p)/c^p p + T(g^q)/d^q q \quad \text{a.e..}$$

Fix representatives in the equivalence classes $T(fg)$, $T(f^p)$, $T(g^q)$. There is a fixed nullset $N$ such that (7.3) holds for all rational $c, d > 0$ in $N^c$, and – by continuity – for all $c, d > 0$ in $N^c$. Now set $c = c(\omega) = (T(f^p))(\omega)^{1/p}$ and $d = d(\omega) = (T(g^q)(\omega)^{1/q}$, and use $1/p - 1/q = 1$, to obtain (7.2) on $\{\omega \in N^c: c(\omega) > 0$ and $d(\omega) > 0\}$. (7.2) follows on $\{\omega: c(\omega) = 0\}$ from $fg \leqq \lim_n n f^p$, and on $\{\omega: d(\omega) = 0\}$ from $fg \leqq \lim_m m g^q$. □

We now discuss the ergodic theorem for sequences of functions. Maker [1940] proved a variant of

**Theorem 7.5.** *Let $\tau$ be an endomorphism of $(\Omega, \mathscr{A}, \mu)$ and $(f_{ni})$ a family of measurable functions such that $\sup_{i,n} |f_{n,i}|$ is integrable and $f_{ni} \to f$ a.e., $(n, i \to \infty)$. Then the limit of the sequence $h_n = n^{-1} \sum_{i=0}^{n-1} f_{ni} \circ \tau^i$ exists a.e., and is equal to the limit of $A_n(\tau) f$.*

*Proof.* We may assume $f = 0$. Put $g_k = \sup \{|f_{ni}|: n, i \geqq k\}$.

As the $f_{mi}$ with $i \leqq k$ and $m \leqq k$ do not matter for the limit behaviour of $h_n$, we have $\limsup_n |h_n| \leqq \lim_n A_n(\tau) g_k$ for each $k$. The conclusion now follows from $\int \lim A_n(\tau) g_k \leqq \int g_k \to 0$. □

Sucheston [1983] has given a similar result for operators in certain Orlicz spaces. To avoid various definitions we assume that the spaces $L, L'$ below are spaces $L_p$, $(1 \leqq p < \infty)$, or $L \log^m L$ of a finite measure space.

**Theorem 7.6.** *Let $\Lambda$ be a countable directed set, and let $L \subset L'$ be two spaces as above. Let $\{T_\lambda, \lambda \in \Lambda\}$ be a family of positive (and hence bounded) linear operators $T_\lambda: L \to L'$. Assume that, for each $f \in L$, $T_\infty f := \lim_\lambda T_\lambda f$ exists a.e. and is in $L'$. If $f_n$ is a sequence in $L$ with $\sup |f_n| \in L$ and $f_n \to f$ a.e., then*

$$\lim_{\substack{\lambda \in \Lambda \\ n \in \mathbb{N}}} T_\lambda f_n = T_\infty f \quad \text{a.e.}.$$

*Proof.* The sequence $g_n := \sup_{k \geqq n} |f_k - f|$ is decreasing with limit 0. For each $n$ we have

$$\begin{aligned} \limsup_{\lambda, k} |T_\lambda f_k - T_\infty f| &\leqq \limsup_{\lambda, k} |T_\lambda (f_k - f)| + \limsup_{\lambda, k} |T_\lambda f - T_\infty f| \\ &\leqq \limsup_{\lambda} T_\lambda g_n = T_\infty g_n. \end{aligned}$$

Now $g_n \downarrow 0$ implies $T_\infty g_n \downarrow 0$. □

## Notes

The idea of the Banach principle was conceived by Banach [1926]; see also Saks [1927], and Mazur-Orlicz [1933]. Its importance for ergodic theory was realized by Yosida

[1940]. We have partly followed Garsia [1970], whose book gives a splendid introduction to the Banach principle and to several related results.

Yosida [1940a], [1974] has a lattice theoretic version of the first part of theorem 7.2, further generalized by von Weizsäcker [1974]. A special case of his results is: Assume $\{T_\lambda, \lambda \in \Lambda\}$ is a set of continuous maps of a Banach space $\mathfrak{X}$ into a Banach lattice $\mathfrak{Y}$ in which order convergence implies norm convergence. (E.g. $\mathfrak{Y} = L_p$, $(1 \leqq p < \infty)$). Assume $T_\lambda(\alpha x) = \alpha T_\lambda x$, $(\alpha \geqq 0)$, and $T_\lambda(x + x') \leqq T_\lambda x + T_\lambda x'$, $(x, x' \in \mathfrak{X})$. If the maps $x \to \limsup T_\lambda x$ are well defined on $\mathfrak{X}$, they are continuous.

The papers of Klimko and Sucheston [1969], Klimko [1969], Mesiar [1984], and Burke [1965] are references related to the ergodic theorem for sequences of functions.

**The necessity of maximal inequalities.** The maximal inequality (6.1) says that (7.1) holds even with $C(\alpha) = \alpha^{-1}$ for the operators $T_n = A_n(T)$, when $T$ is a positive $L_1 - L_\infty$-contraction. This is much more than the statement $C(\alpha) \to 0$, $(\alpha \to \infty)$. Similarly, (6.5) implies that the same family $\mathscr{T} = \{T_n\}$ satisfies an estimate

$$\mu(M^{\mathscr{T}} f \geqq \alpha \|f\|_p) \leqq \left(\frac{p}{p-1}\right)^p \alpha^{-p}.$$

Hence, the family $\mathscr{T}$ of operators in $L_p$, $(1 < p < \infty)$, fulfills (7.1) with $C(\alpha) = (p/(p-1))^p \alpha^{-p}$, and again we have a quantitative result.

We now report briefly on some results about the necessity of such stronger estimates for rather general sequences of operators $T_n$. Apart from their intrinsic interest such results can be useful tools for the construction of counterexamples. For specific sequences $T_n$ of linear operators in $L_p$, $(1 \leqq p < \infty)$, it is frequently much simpler to show that there is no constant $K > 0$ with

(7.4) $\quad \mu(M_\infty^{\mathscr{T}} f > \lambda \|f\|_p) \leqq K/\lambda^p, \quad \forall f \in L_p$

than to exhibit an $f \in L_p$ for which the sequence $T_n f$ diverges on a set of positive measure.

The first results on inequalities of the form (7.4) as necessary conditions for a.e.-convergence arose in Fourier analysis. Kolmogorov [1925] studied the operators

$$T_n f(\omega) = \int_{|t| > 1/n} f(\omega + t)/t \, dt$$

for integrable functions vanishing outside a finite interval and deduced an inequality (7.4) with $p = 1$. Calderon proved that the a.e.-convergence of $T_n f$ for all square integrable $f$ on the unit circle is equivalent to (7.4) with $p = 2$, when $T_n f$ is the $n$-th partial sum of the Fourier series for $f$. E.M. Stein [1961] generalized this considerably. He considered a compact group $G$ and a homogeneous space $\Omega$ of $G$, i.e., there are maps $\tau_g$: $\Omega \to \Omega$, $(g \in G)$, with $\tau_{gh} = \tau_g \tau_h$ and each orbit $\{\tau_g \omega : g \in G\}$ is all of $\Omega$. There is a measure $\mu$ in $\Omega$ invariant under all "translations" $\tau_g$. (E.g. $\Omega = G$, $\mu =$ Haar measure). In this setting he proved that for any $p$ with $1 \leqq p \leqq 2$ and any sequence of bounded linear operators $T_n$ in $L_p(\mu)$ which commute with all translations $((T_n f) \circ \tau_g = T_n(f \circ \tau_g))$ the a.e.-finiteness of all $M^{\mathscr{T}} f$ implies the existence of a $K < \infty$ for which (7.4) holds.

As many problems in ergodic theory and in probability theory do not fit into this setting, Burkholder [1964], Sawyer [1966], and Garsia [1970] introduced several other conditions allowing similar conclusions. The basic idea is as follows: If there is no constant $K < \infty$ with (7.4), then there are functions $f^n$ for which $M^{\mathscr{T}} f^n$ is very large on small sets. If we take many independent copies of each $f^n$ or "sufficiently independent" translates they will have this bad behaviour on other small sets and the union of these sets will have a fairly large measure. But if $f$ is a convex combination of the $f^n$ then $M^{\mathscr{T}} f$ will be large on this union.

Let us make this precise in the situation discussed by Burkholder: Let **D** be the set of sequences $X=(X_1, X_2, \ldots)$ of non negative random variables on a probability space $(\Omega, \mathscr{A}, \mu)$. $X \sim Y$ means that $X$ and $Y=(Y_1, Y_2, \ldots)$ have the same joint distribution. A family $\mathbf{C} \subset \mathbf{D}$ is called *stochastically convex* if for each sequence $X^n=(X_1^n, X_2^n, \ldots)$, $(n \geqq 1)$, in **D** there exists a sequence $Y^n$ in **D** with $Y^n \sim X^n$ such that $Y^1, Y^2, \ldots$ are stochastically independent and, for each $a_1, a_2, \ldots \geqq 0$ with $\sum_{k=1}^{\infty} a_n = 1$, there is a $Z=(Z_1, Z_2, \ldots)$ in **C** with $Z \sim (\sum_{n=1}^{\infty} a_n Y_1^n, \sum_{n=1}^{\infty} a_n Y_2^n, \ldots)$. For $X \in \mathbf{C}$ put $X^* = \sup\{X_i\colon 1 \leqq i < \infty\}$. For $p = 1$ Burkholder proved:

**Theorem 7.7.** *Suppose that* **C** *is stochastically convex and that* $\mu\{X^* < \infty\} > 0$ *holds for all* $X \in \mathbf{C}$. *Then there exists a* $K < \infty$ *such that* $\mu\{X^* > \alpha\} \leqq K\alpha^{-1}$ *holds for all* $\alpha > 0$ *and all* $X \in \mathbf{C}$.

*Proof.* If there is no such $K$ and we put

$$M(\alpha) = \sup\{\alpha\mu\{X^* > \alpha\}\colon X \in \mathbf{C}\},$$

then $\limsup_{\alpha \to \infty} M(\alpha) = \infty$. It is not hard to see that this implies the existence of sequences $(a_n)$ and $(\alpha_n)$ satisfying $a_n > 0$, $\sum_{n=1}^{\infty} a_n = 1$, $\alpha_n \to \infty$, and $\sum_{n=1}^{\infty} (a_n/\alpha_n) M(\alpha_n/a_n) = \infty$. Let $X^n \in \mathbf{C}$ satisfy

$$\mu((X^n)^* > \alpha_n/a_n) > (a_n/\alpha_n) M(\alpha_n/a_n) - 2^{-n},$$

$(n=1,2,\ldots)$. Take $Y^n$ as in the definition of stochastic convexity. Then $\mu((Y^n)^* > \alpha_n/a_n) = \mu((X^n)^* > \alpha_n/a_n)$ yields $\sum_{n=1}^{\infty} \mu((Y^n)^* > \alpha_n/a_n) = \infty$. As the processes $Y^n$ are independent, the Borel-Cantelli lemma shows that $(Y^n)^*(\omega) > \alpha_n/a_n$ almost surely holds for infinitely many $n$. Hence $\limsup_{n\to\infty} a_n (Y^n)^*(\omega) = \infty$ a.e.. Again by stochastic convexity there exists $Z \in \mathbf{C}$ with $Z \sim (\sum_{n=1}^{\infty} a_n Y_1^n, \sum_{n=1}^{\infty} a_n Y_2^n, \ldots)$. As all terms are non negative

$$\sup_i \sum_{n=1}^{\infty} a_n Y_i^n \geqq \sup_i \sup_n a_n Y_i^n = \sup_n a_n (Y^n)^* = \infty \quad \text{a.e..}$$

We obtain $Z^* = \infty$ a.e., a contradiction to the assumption that $\mu(X^* < \infty) > 0$ holds for all $X \in \mathbf{C}$. □

There is a similar theorem for $0 < p < \infty$. Burkholder gave applications to ergodic theory, probability theory, and orthogonal series. E.g. the family of non negative stationary processes on the unit interval $\Omega$ and the family of non negative submartingales with bounded integrals on the same $\Omega$ are stochastically convex. If **C** is stochastically convex and $\bar{\mathbf{C}}$ is the family of processes $\bar{X} = (\bar{X}_1, \bar{X}_2, \ldots)$ for which there exists an $X \in \mathbf{C}$ with $\bar{X}_n = n^{-1} \sum_{i=1}^{n} X_i$, then it is easy to see that $\bar{\mathbf{C}}$ is stochastically convex.

Garsia [1970] worked in an ergodic theoretic setting with rather weak assumptions. We refer the reader to his book for the following *continuity principle*:

**Theorem 7.8.** *Let* $\mathscr{T} = \{T_n\}$ *be a sequence of linear operators in* $L_p$, $(1 \leqq p \leqq 2)$, *of a finite measure space* $(\Omega, \mathscr{A}, \mu)$, *which are continuous in measure. Let* $\mathscr{E}$ *be a family of endomor-*

*phisms* $\tau$ *with*

(7.5) $\quad (M^{\mathcal{T}} f) \circ \tau \leqq M^{\mathcal{T}} (f \circ \tau), \quad \forall f \in L_p.$

*Suppose that for each* $A, B \in \mathcal{A}$ *and each* $\alpha > 1$ *there exists a* $\tau \in \mathcal{E}$ *with* $\mu(A \cap \tau^{-1} B)\, \mu(\Omega) \leqq \alpha \mu(A)\, \mu(B)$. *Then the condition*

$$M^{\mathcal{T}} f < \infty \text{ a.e.} \quad \forall f \in L_p$$

*is equivalent to the existence of a* $K < \infty$ *with* (7.4)

The condition (7.5) is satisfied if $(T_n f) \circ \tau = T_n(f \circ \tau)$ holds for all $f \in L_p$, $\tau \in \mathcal{E}$ and $T_n$. Sawyer [1966] has proved the continuity principle for $1 \leqq p < \infty$, but then the operators $T_n$ must also be positive.

As we mentioned before, the Banach principle, the continuity principle, and similar results can be powerful aids for the construction of counterexamples. In this way, del Junco and Rosenblatt [1979] have obtained a unified approach to many negative results. Their paper contains also some interesting variations of the main theme. E.g. in some instances where theorem 7.8 allows to infer that there is an $f \in L_1$ with $\limsup T_n f = \infty$ they can obtain an indicator function $f = 1_A$ with $\limsup T_n f = 1$ and $\liminf T_n f = 0$.

# Chapter 2: Mean ergodic theory

The mean ergodic theorem deals with averages taken along the orbits of elements of a topological vector space under semigroups of operators. In the Banach space case we also deal with uniform convergence of the averages of operators. It can be obtained for quasi-compact operators. Then we discuss weak mixing and multiple recurrence. The theorem of Jacobs-Deleeuw-Glicksberg extends the spectral mixing theorem to semigroups in Banach spaces.

## § 2.1 The mean ergodic theorem

We are now concerned with questions of weak and strong convergence of averages $n^{-1} \sum_{i=0}^{n-1} T^i x$, where $T$ is a continuous linear operator in a Banach space $\mathfrak{X}$ with norm $\|\cdot\|$. We then proceed to the study of more general semigroups of operators in more general spaces.

**1. Operators in a Banach space.** Let us fix some notation: If $A$ is a subset of a linear space, $\operatorname{co} A$ or $\operatorname{co}(A)$ is the *convex hull* $\{\sum_{i=1}^{m} \alpha_i x_i \colon x_i \in A, 0 \leqq \alpha_i, \sum_{i=1}^{m} \alpha_i = 1, m \in \mathbb{N}\}$ of $A$. $\operatorname{lin} A$ denotes the *linear span* of $A$, the smallest linear space containing $A$. We write $A + B$ for $\{x + y \colon x \in A, y \in B\}$; if $A$ and $B$ are linear spaces and $A \cap B = \{0\}$, we say that $A + B$ is a *direct sum* and write $A \oplus B$. This means that the representation $x + y$ is unique.

Recall that, for a space $\mathfrak{Y}_2$ of linear functionals on a space $\mathfrak{Y}_1$, the $\sigma(\mathfrak{Y}_1, \mathfrak{Y}_2)$-topology is the coarsest topology on $\mathfrak{Y}_1$ with respect to which all functionals in $\mathfrak{Y}_2$ are continuous. The *weak topology* is the $\sigma(\mathfrak{X}, \mathfrak{X}^*)$-topology, where $\mathfrak{X}^*$ is the space of continuous functionals on $\mathfrak{X}$. The *$w^*$-topology* is the $\sigma(\mathfrak{X}^*, \mathfrak{X})$-topology on $\mathfrak{X}^*$. lim denotes the limit in the strong(= norm) topology, w-lim in the weak topology and w*-lim in the w*-topology. $\operatorname{cl} A$ is the closure of $A \subset \mathfrak{X}$, and w-cl $A$ the closure in the weak topology. As usual, $\overline{\operatorname{co}}\, A$ denotes $\operatorname{cl} \operatorname{co} A$. Note that this agrees with w-$\overline{\operatorname{co}}\, A$ = w-cl co $A$, (Dunford-Schwartz [1958: 422]).

$\langle x, h \rangle$ is the value of the functional $h \in \mathfrak{X}^*$ in $x \in \mathfrak{X}$. Recall that $\mathscr{L}(\mathfrak{X})$ is the space of bounded linear operators $T$ in $\mathfrak{X}$, and that $T^0 = I$ is the identity operator. The *adjoint* or *dual* operator of $T$ is the operator $T^* \colon \mathfrak{X}^* \to \mathfrak{X}^*$ with $\langle Tx, h \rangle = \langle x, T^* h \rangle$ for $x \in \mathfrak{X}$, $h \in \mathfrak{X}^*$.

$T$ is called *power bounded* if the norms of the powers $T^k$, $(k \geqq 0)$, are uniformly bounded ($\sup_k \|T^k\| < \infty$), and *Cesàro bounded* if the norms of the averages $A_n = A_n(T) = n^{-1} \sum_{i=0}^{n-1} T^i$ are uniformly bounded, a clearly weaker condition. Because of the identity

$$(1.1) \qquad n^{-1} T^{n-1} = A_n - \frac{n-1}{n} A_{n-1}$$

the condition

$$(1.2) \qquad \lim n^{-1} T^{n-1} x = 0$$

is necessary for the convergence of $A_n x$. It is satisfied for all $x$ when $T$ is power bounded.

**Theorem 1.1** (Mean ergodic theorem). *Let $T$ be a Cesàro bounded linear operator in a Banach space $\mathfrak{X}$. For any $x \in \mathfrak{X}$ satisfying* (1.2), *and any $y \in \mathfrak{X}$ the following assertions are equivalent:*

*(i) $Ty = y$ and $y \in \overline{\text{co}}\{x, Tx, T^2x, \dots\}$,*

*(ii) $y = \lim_n A_n x$,*

*(iii) $y = \text{w-lim}\, A_n x$,*

*(iv) $y$ is a weak cluster point of the sequence $(A_n x)$.*

*Proof.* The implications (ii) $\Rightarrow$ (iii) $\Rightarrow$ (iv) are trivial. (i) $\Rightarrow$ (ii): Set $M := \sup_n \|A_n\|$. For $\varepsilon > 0$, (i) implies the existence of an operator $S \in \text{co}\{T^0, T^1, T^2, \dots\}$ with $\|y - Sx\| < \varepsilon$. For any $k$ and any $n$ we have

$$A_n T^k x - A_n x = n^{-1} \sum_{i=0}^{k-1} T^{n+i} x - n^{-1} \sum_{i=0}^{k-1} T^i x.$$

Using (1.2) and $(i+n)/n \to 1$ we see that $\|A_n T^k x - A_n x\| < \varepsilon$ holds for large enough $n$. As $S$ is a convex combination of finitely many $T^k$ and $\{z \in \mathfrak{X}: \|z\| < \varepsilon\}$ is convex, we may find an $N$ such that $\|A_n Sx - A_n x\| < \varepsilon$ holds for $n \geqq N$. As $y$ is a fixed point of $T$ we have $A_n y = y$ for all $n$. For $n \geqq N$ we therefore obtain

$$\|y - A_n x\| \leqq \|A_n(y - Sx)\| + \|A_n Sx - A_n x\| \leqq M\varepsilon + \varepsilon.$$

(iv) $\Rightarrow$ (i): We now need Mazur's theorem which says that any closed convex subset of $\mathfrak{X}$ is also weakly closed. The weak cluster point $y$ of the sequence $A_n x$ of convex combinations of $x, Tx, \dots$ therefore belongs to $\overline{\text{co}}\{x, Tx, \dots\}$.

To prove $y = Ty$ take any $h \in \mathfrak{X}^*$ and $\varepsilon > 0$. By the argument used above we know $TA_n x - A_n x \to 0$. For large enough $n$

$$|\langle TA_n x - A_n x, h\rangle| < \varepsilon.$$

As $y$ is a weak cluster point of $(A_n x)$ there exist arbitrarily large values of $n$ with

$$|\langle y, h\rangle - \langle A_n x, h\rangle| < \varepsilon \quad \text{and} \quad |\langle y, T^* h\rangle - \langle A_n x, T^* h\rangle| < \varepsilon.$$

Combining these estimates we arrive at $|\langle y, h\rangle - \langle Ty, h\rangle| < 3\varepsilon$. But $\varepsilon$ and $h$ had been arbitrary. □

The present formulation of the mean ergodic theorem is a special case of results of Eberlein [1949], but the main assertions emerged already with the work of F. Riesz [1938] for $\mathfrak{X} = L_p$, and independently with the work of K. Yosida [1938] and S. Kakutani [1938] for general Banach spaces; see also Yosida-Kakutani [1941]. The following important special case was proved by E. Lorch [1939] independently:

**Theorem 1.2.** *If $T$ is a power bounded linear operator in a reflexive Banach space $\mathfrak{X}$, the averages $A_n x$ converge in norm to a $T$-invariant limit for all $x \in \mathfrak{X}$.*

*Proof.* The power-boundedness implies (1.2), and each norm bounded set in a reflexive Banach space has a weakly compact closure. By the Eberlein-Šmulian theorem, it is then weakly sequentially compact. □

Note that the spaces $L_p$, $(1 < p < \infty)$, are reflexive. Theorem 1.1 also contains a mean ergodic theorem in $L_1$: If $T$ is a contraction in $L_1$ and there exists a strictly positive integrable function $p$ such that $|Tf| \leqq p$ holds for all $f$ with $|f| \leqq p$, then $A_n g$ converges in $L_1$-norm for all $g \in L_1$. (*Proof.* For any $\varepsilon > 0$ there is a $c$ and a splitting $g = g_c + g_c'$ with $|g_c| \leqq cp$ and $\int |g_c'| \, d\mu < \varepsilon$. This implies $\int (|A_n g| - cp)^+ \, d\mu < \varepsilon$ for all $n$. Thus, the sequence $(A_n g)$ is uniformly integrable (Def. 3.3.13), and therefore weakly sequentially compact; see Dunford-Schwartz [1958].) The application of theorem 1.1 to $C(K)$ is treated in § 5.1.

Our next aim is a splitting theorem for

$$\mathfrak{X}_{me} = \mathfrak{X}_{me}(T) := \{x \in \mathfrak{X} \colon \lim A_n x \text{ exists}\}.$$

Clearly, if $T$ is Cesàro bounded, $\mathfrak{X}_{me}$ is a closed linear subspace of $\mathfrak{X}$. $T$ is called *mean ergodic* if $\mathfrak{X} = \mathfrak{X}_{me}$. We shall use the notation

$$F = F(T) := \{x \in \mathfrak{X} \colon Tx = x\}, \quad N := \{x - Tx \colon x \in \mathfrak{X}\} = (I - T)\mathfrak{X}$$
$$F_* = F_*(T) := \{h \in \mathfrak{X}^* \colon T^* h = h\}, \quad N_* := (I - T^*)\mathfrak{X}^*.$$

Most of the next theorem is due to Yosida [1938]:

**Theorem 1.3.** *Let $T$ be Cesàro bounded, and assume that* (1.2) *holds for all $x \in \mathfrak{X}$. Then $\mathfrak{X}_{me} = F \oplus \operatorname{cl} N$. The operator $P$ assigning to $x \in \mathfrak{X}_{me}$ the limit $Px := \lim A_n x$ is the projection of $\mathfrak{X}_{me}$ onto $F$. We have $P = P^2 = TP = PT$. For any $z \in \mathfrak{X}$ the assertions*

*(i)* $\lim A_n z = 0$,

*(ii)* $\langle z, h\rangle = 0$ *for all* $h \in F_*$,

*(iii)* $z \in \operatorname{cl} N$

*are equivalent.*

*Proof.* Clearly $F$ and $\operatorname{cl} N$ are linear subspaces. Let us verify $F \cap \operatorname{cl} N = \{0\}$: For $\varepsilon > 0$ and $z \in F \cap \operatorname{cl} N$ there exists $u$ with $\|z - (u - Tu)\| < \varepsilon$. Hence $\|A_n(z - (u - Tu))\| < M\varepsilon$. Using $A_n z = z$ and $A_n(u - Tu) \to 0$, we find $\|z\| < M\varepsilon + \varepsilon$. A similar argument shows $F_* \cap \operatorname{cl} N_* = \{0\}$.

For $x \in \mathfrak{X}_{me}$, theorem 1.1 implies $Px \in F$. Thus $z = x - Px$ satisfies (i). (i) $\Rightarrow$ (ii): $h \in F_*$ implies $h = A_n^* h$ for all $n$ and hence $\langle z, h\rangle = \langle z, A_n^* h\rangle = \langle A_n x, h\rangle \to 0$.

(ii) $\Rightarrow$ (iii): If $z$ does not belong to $\operatorname{cl} N$ there exists, by Hahn-Banach, an $h \in \mathfrak{X}^*$ with $\langle z, h\rangle \neq 0$ and with $\langle y, h\rangle = 0$ for all $y \in \operatorname{cl} N$. In particular, $\langle u - Tu, h\rangle = 0$ for $u \in \mathfrak{X}$. Hence $\langle u, h - T^* h\rangle = 0$ for all $u$. This implies $h \in F_*$, a contradiction to $\langle z, h\rangle \neq 0$.

We have proved $\mathfrak{X}_{me} \subset F \oplus \operatorname{cl} N$. As $F \subset \mathfrak{X}_{me}$ is trivial, the opposite inclusion will follow from (iii) $\Rightarrow$ (i): For any $u \in \mathfrak{X}$ we have $A_n(u - Tu) = n^{-1}(u - T^n u)$ and this tends to 0. Thus, all $z \in N$ satisfy $A_n z \to 0$. But the set of $z$ with this property is closed because $T$ is Cesàro bounded. As $P$ is the projection of $F \oplus \operatorname{cl} N$ on $F$ and the elements of $F$ are fixed under $T$ the identities $P = P^2 = TP$ are clear. $P = PT$ follows from $A_n(x - Tx) \to 0$. □

Recall that a linear operator $Q$ is called *projection* if $Q = Q^2$. A projection with $Q = QT = TQ$ will be called *T-absorbing*. Thus, $P$ is a $T$-absorbing projection.

Occasionally, the following criterion of Sine [1970] for mean ergodicity is useful:

**Theorem 1.4.** *Let $T$ be Cesàro bounded and assume* (1.2) *for all $x$. $T$ is mean ergodic iff $F$ separates $F_*$.*

*Proof.* First assume convergence. If $h_1, h_2 \in F_*$ are distinct there exists an $x \in \mathfrak{X}$ with $\langle x, h_1\rangle \neq \langle x, h_2\rangle$. Now $Px$ belongs to $F$ and separates the $h_i$ because of $h_i = P^* h_i$ and $\langle Px, h_i\rangle = \langle x, P^* h_i\rangle$. Now assume that $F$ separates $F_*$. If $F \oplus \operatorname{cl} N$ is not all of $\mathfrak{X}$ there exists a non zero $h \in \mathfrak{X}^*$ with $\langle y, h\rangle = 0$ for all $y \in F \oplus \operatorname{cl} N$. In particular, $\langle x - Tx, h\rangle = 0$ for all $x$ shows $h \in F_*$. As $F$ separates $h$ and 0, there exists $y \in F$ with $\langle y, h\rangle \neq 0$, a contradiction. □

As the case of reflexive $\mathfrak{X}$ suffices for many applications we end this subsection with a sketch of Lorch's elegant argument. Assume $T$ power bounded in a reflexive space $\mathfrak{X}$:

For $E \subset \mathfrak{X}$ set $E^\perp = \{h \in \mathfrak{X}^*: \langle x, h\rangle = 0\ \forall x \in E\}$. Using Hahn-Banach and $\mathfrak{X} = \mathfrak{X}^{**}$ one easily checks $(N^\perp)^\perp = \operatorname{cl} N$. From $h \in N^\perp \Leftrightarrow \langle x - Tx, h\rangle = 0\ \forall x \Leftrightarrow \langle x, h\rangle = \langle Tx, h\rangle = \langle x, T^* h\rangle\ \forall x \Leftrightarrow T^* h = h$ we see $N^\perp = F_*$, and hence $F_*^\perp = \operatorname{cl} N$. By duality $F^\perp = \operatorname{cl} N_*$.

To establish $\mathfrak{X} = F \oplus \operatorname{cl} N$ it remains to make sure that any $h \in (F \oplus \operatorname{cl} N)^\perp$ vanishes. But this now follows from $h \in F^\perp = \operatorname{cl} N_*$ and $h \in N^\perp = F_*$.

**2. $\mathscr{S}$-ergodic nets in topological vector spaces.** We shall now derive similar results as above for more general semigroups, and we allow that they act in a locally convex topological vector space.

We recall just a few concepts: A real valued function $p$ on a real or complex vector space $\mathfrak{X}$ is called *seminorm* if it is *subadditive* and *positively homogeneous*, i.e., it satisfies $p(x+y) \leqq p(x) + p(y)$ for all $x, y \in \mathfrak{X}$, and $p(\alpha x) = |\alpha| p(x)$ for all $x \in \mathfrak{X}$ and all scalars $\alpha$. A family $\{p_\gamma: \gamma \in \Gamma\}$ of seminorms satisfies the *axiom of separation if, for any* $x \neq 0$, there exists some $\gamma$ with $p_\gamma(x) \neq 0$. $V$ is a neighbourhood of $x$ if there are finitely many $\gamma_i$ and $\varepsilon_i$ such that $V$ contains all $y$ with $p_{\gamma_i}(y-x) < \varepsilon_i$. With this topology, $\mathfrak{X}$ is a *topological vector space*, i.e., the mappings $(x, y) \to x + y$, $(\alpha, x) \to \alpha x$ are continuous. $\mathfrak{X}$ is called a *locally convex* topological vector space (l.c.t.v.s) if there exists a family $\{p_\gamma: \gamma \in \Gamma\}$ of seminorms which satisfies the axiom of separation and generates the topology in $\mathfrak{X}$. For more detail and for the functional analytic tools needed in the sequel we refer to Yosida [1974] and Schaefer [1966].

Let $\mathscr{N}(y)$ be the family of neighbourhoods of $y$. A family $\{T_k\}$ of linear operators in $\mathfrak{X}$ is called *equi-continuous* if, for any $V \in \mathscr{N}(0)$, there exists some $W \in \mathscr{N}(0)$ such that $T_k W \subset V$ holds for all $k$. E.g., a set $\mathscr{S}$ of linear operators $T$ in a Banach space is equi-continuous iff it is *bounded*, i.e., there exists an $M < \infty$ with $\|T\| \leqq M$ for all $T \in \mathscr{S}$.

We say that a semigroup $\mathscr{S}$ of continuous linear operators $T$ in $\mathfrak{X}$ admits a *right $\mathscr{S}$-ergodic net* $\{A_\lambda: \lambda \in \Lambda\}$ if $\Lambda$ is a directed set and

(E1) each $A_\lambda$ is a linear operator in $\mathfrak{X}$,
(E2) for each $x \in \mathfrak{X}$ and all $\lambda$, $A_\lambda x \in \overline{\text{co}}\, \mathscr{S} x$,
(E3) the $A_\lambda' s$ are equi-continuous, and
(E4r) for every $x \in \mathfrak{X}$ and $T \in \mathscr{S}$, $\lim_\lambda (A_\lambda T x - A_\lambda x) = 0$.

We say that the net is *left $\mathscr{S}$-ergodic* if (E4r) is replaced by
(E4l) for every $x \in \mathfrak{X}$ and $T \in \mathscr{S}$, $\lim_\lambda (T A_\lambda x - A_\lambda x) = 0$.

It is called *$\mathscr{S}$-ergodic* if it is right and left $\mathscr{S}$-ergodic. If lim in condition (E4r) is replaced by w-lim the net is called *weakly right $\mathscr{S}$-ergodic*, etc.

Perhaps the most important example is a multiparameter semigroup. Let $T_1, \ldots, T_d$ be commuting power bounded operators in a Banach space. Put $\Lambda = \mathbb{N}^d$. $\mathscr{S}$ is the semigroup generated by $T_1, T_2, \ldots, T_d$. For $\lambda = (\lambda_1, \ldots, \lambda_d)$ we may take

$$A_\lambda = \Big(\prod_{\nu=1}^{d} \lambda_\nu\Big)^{-1} \sum_{i_1=0}^{\lambda_1-1} \cdots \sum_{i_d=0}^{\lambda_d-1} T_1^{i_1} \ldots T_d^{i_d}.$$

We may also take averages over convex sets like spheres instead of rectangles.

It is less obvious that $\mathscr{S}$-ergodic nets exist for arbitrary Abelian equicontinuous semigroups $\mathscr{S}$: We may then take $\Lambda = \text{co}\, \mathscr{S}$ and $A_\lambda = \lambda$. A partial order in $\Lambda$ is defined by $T \geqq S$ iff there exists an $R \in \Lambda$ with $T = RS$. It follows from $RS = SR$ that $\Lambda$ is a directed set. The net clearly has the properties (E1)–(E3). (E4):

For $A_\lambda \geqq A_n(T)$, i.e., for all large enough $\lambda$, there exists an $R \in \operatorname{co} \mathscr{S}$ with $A_\lambda = RA_n(T)$, and then $A_\lambda - A_\lambda T = R(n^{-1}(T^0 - T^n))$. This clearly yields (E4r), and by commutativity (E4l).

In most applications the question of existence of a net does not arise because the interest lies in convergence for a particular given net.

**Theorem 1.5** (Eberlein [1949]). *If $\mathscr{S}$ is a semigroup of continuous linear operators in a* l.c.t.v.s. $\mathfrak{X}$, *and admits an $\mathscr{S}$-ergodic net $\{A_\lambda: \lambda \in \Lambda\}$, then, for any $x, y \in \mathfrak{X}$, the following conditions are equivalent:*

*(i) $Ty = y$ for all $T \in \mathscr{S}$ and $y \in \overline{\operatorname{co}}\, \mathscr{S} x$,*

*(ii) $y = \lim_\lambda A_\lambda x\ (=: Px)$,*

*(iii) $y = w\text{-}\lim_\lambda A_\lambda x$,*

*(iv) $y$ is a weak cluster point of $\{A_\lambda x: \lambda \in \Lambda\}$.*

The proof of theorem 1.1 has been written in such a way that it carries over to the present more general situation with minor modifications (e.g. $\varepsilon$-neighbourhoods are replaced by topological neighbourhoods).

It seems to be unknown if the equivalence of (i)–(iv) persists if one of the conditions (E4r), (E4l) is dropped. (E4r) was needed for (i) $\Rightarrow$ (ii), and (E4l) for (iv) $\Rightarrow$ (i). If we assume only that the net in theorem 1.6 is *weakly* $\mathscr{S}$-ergodic, the conditions (i), (iii), (iv) remain equivalent.

We mention two applications of Eberlein's theorem:

(1) *Existence of the mean for continuous almost periodic functions.* Let $\mathfrak{X}$ be the Banach space of bounded continuous functions $x: \mathbb{R} \to \mathbb{R}$ with the usual norm $\|x\| = \sup |x(t)|$. We consider the group $\mathscr{S} = \{T_t, t \in \mathbb{R}\}$ of translations $(T_t x)(\cdot) = x(\cdot + t)$, and the averages

$$A_\lambda x(\cdot) = \lambda^{-1} \int_0^\lambda T_t x(\cdot)\, dt.$$

Taking von Neumann's definition of almost periodicity, $x$ is *almost periodic* if $\mathscr{S} x$ is conditionally compact. Then $\overline{\operatorname{co}}\, \mathscr{S} x$ is compact. It is rather straightforward to check that $\{A_\lambda: \lambda > 0\}$ is an $\mathscr{S}$-ergodic net. As $A_\lambda x$ belongs to $\overline{\operatorname{co}}\, \mathscr{S} x$ for all $\lambda > 0$, the equivalence of (iv) and (ii) in Eberlein's theorem yields $\|A_\lambda x - Px\| \to 0$, $(\lambda \to \infty)$. Because of $T_t P = P$, $(t \in \mathbb{R})$, the function $Px$ must be constant.

(2) *Fejér's theorem.* As the equicontinuity has only been required for the operators $A_\lambda$, and not for $\mathscr{S}$, we may view Fejér's theorem as an ergodic theorem:

This time $\mathfrak{X}$ shall be the space of continuous real valued functions on $\mathbb{R}$ of period $2\pi$. We again use $\|x\| = \sup |x(t)|$. Let $S_n x$ be the $n$-th partial sum of the Fourier expansion of $x$. Set $D_n(t) = (1/2) + \sum_{\nu=1}^{n} \cos \nu t$. Then

$$S_n x(\cdot) = 1/\pi \int_{-\pi}^{\pi} x(\cdot + t)\, D_n(t)\, dt.$$

The semigroup $\mathscr{S}$ shall consist of the identity $I$ and the transformations $U_n = I - S_n$. The semigroup property follows from $U_n U_m = U_k$ with $k = \max(m, n)$. As the norm of $S_n$ is the Lebesgue constant $L_n = \pi^{-1} \int_{-\pi}^{\pi} |D_n(u)|\,du$ and $L_n = (4/\pi^2) \log n$, (see Zygmund [1935: 67]), the semigroup $\mathscr{S}$ is not equicontinuous.

Now put $A_\lambda = (n+1)^{-1} \sum_{\nu=0}^{n} U_\nu$, $(\lambda \in \Lambda = \mathbb{N})$. It is not hard to check (E1), (E2), and (E4). Let $K_n$ be the Fejér kernel:

$$K_n(t) = (n+1)^{-1} \sum_{\nu=0}^{n} D_\nu(t).$$

A simple computation shows

$$A_\lambda x(\cdot) = x(\cdot) - \pi^{-1} \int_{-\pi}^{\pi} x(\cdot + t) K_\lambda(t)\,dt.$$

As $K_n$ has the properties $K_n \geqq 0$, and $\pi^{-1} \int_{-\pi}^{\pi} K_n(t)\,dt = 1$, (see e.g. Zygmund [1935: 88]) we clearly have $\|A_\lambda\| \leqq 2$ for all $\lambda$. Thus, (E3) holds, too. The uniform continuity of $x$ and the properties of $K_\lambda$ imply the equicontinuity of the functions $A_\lambda x$, $(\lambda \in \Lambda)$. By Arzelà-Ascoli $\{A_\lambda x \colon \lambda \in \Lambda\}$ is conditionally compact. Eberlein's theorem therefore yields the convergence of $A_\lambda x$ to an element $y \in \mathfrak{X}$ with $U_k y = y$ for all $k$. As $y = 0$ is the only element of $\mathfrak{X}$ fixed under $\mathscr{S}$ we obtain $\|A_\lambda x\| \to 0$, for all $x \in \mathfrak{X}$. This is equivalent to $\|n^{-1} \sum_{\nu=0}^{n-1} S_\nu x - x\| \to 0$, the assertion of Fejér's theorem.

**3. The splitting theorem for semigroups.** Next we shall derive a semigroup generalization of theorems 1.3 and 1.4. We assume that $\mathfrak{X}$ is a *complete* l.c.t.v.s. [Recall that a sequence $(x_\varrho)$ indexed by the elements of a directed set is called a *Cauchy net* if, for any $V \in \mathscr{N}(0)$, $x_\varrho - x_\sigma$ belongs to $V$ for all large enough $\varrho$, $\sigma$. $\mathfrak{X}$ is complete if each Cauchy net converges to an element of $\mathfrak{X}$]. Set

$$F = \{x \in \mathfrak{X} \colon Tx = x \quad \text{for all } T \in \mathscr{S}\},$$
$$N = \operatorname{lin} \{x - Tx \colon x \in \mathfrak{X}, T \in \mathscr{S}\},$$
$$F_* = \{h \in \mathfrak{X}^* \colon T^* h = h \quad \text{for all } T \in \mathscr{S}\}.$$

The elements of $F$ are the *fixed points* under $\mathscr{S}$, and the elements of $F_*$ are the fixed points under $\mathscr{S}^* = \{T^* \colon T \in \mathscr{S}\}$.

We assume only the existence of a weakly right $\mathscr{S}$-ergodic net $(A_\lambda \colon \lambda \in \Lambda)$. Put

$$D = \{x \in \mathfrak{X} \colon w\text{-}\lim_\lambda A_\lambda x \text{ exists}\},$$
$$Px = w\text{-}\lim_\lambda A_\lambda x, \ (x \in D),$$
$$D(0) = \{x \in D \colon Px = 0\},$$
$$D(F) = \{x \in D \colon Px \in F\}.$$

**Proposition 1.6.** *If $\mathfrak{X}$ is a complete* l.c.t.v.s. *and $(A_\lambda\colon \lambda \in \Lambda)$ a weakly right $\mathcal{S}$-ergodic net for the semigroup $\mathcal{S}$, then*

*(a) $D$ is a closed linear subspace of $\mathfrak{X}$, and $TD \subset D$ holds for all $T \in \mathcal{S}$,*

*(b) $PD \subset D$, and $P$ is linear and continuous on $D$,*

*(c) $PT = P$ on $D$ for all $T \in \mathcal{S}$.*

*Proof.* It is clear that $D$ is a linear subspace. We show that it is complete and hence closed. Let $(x_\varrho)$ be a Cauchy net in $D$, and $x$ its limit. The equicontinuity of the $A_\lambda$'s implies the continuity of $P$. Therefore $(Px_\varrho)$ is a Cauchy net. Let $y$ be its limit.

If $U$ is an arbitrary weak convex neighbourhood of 0, there exists $\varrho_0$ such that

$$Px_{\varrho 0} - y \in (1/3)\,U \quad \text{and} \quad A_\lambda(x - x_{\varrho 0}) \in (1/3)\,U$$

holds for all $\lambda$. Because of $Px_{\varrho 0} = w\text{-lim}\, A_\lambda x_{\varrho 0}$, there exists a $\lambda_0$ with

$$A_\lambda x_{\varrho 0} - Px_{\varrho 0} \in (1/3)\,U \quad \text{for all } \lambda \geqq \lambda_0 .$$

But then

$$A_\lambda x - y = [A_\lambda(x - x_{\varrho 0})] + [A_\lambda x_{\varrho 0} - \mathrm{P} x_{\varrho 0}] + [Px_{\varrho 0} - y]$$

belongs to $U$ for $\lambda \geqq \lambda_0$. As $U$ was arbitrarily small we see that $y = w\text{-lim}\, A_\lambda x$ exists and (a) is proved.

By (a) the set $\overline{\mathrm{co}}\, \mathcal{S} x$ is contained in $D$ for $x \in D$. (E2) therefore yields $A_\lambda D \subset D$ for all $\lambda$, and (b) follows because $D$ is weakly closed. (c) is a consequence of the weak right $\mathcal{S}$-ergodicity of the net. □

**Corollary 1.7.** (a) $F$, $D(F)$, and $D(0)$ are closed subspaces of $D$ with $F = PD(F) \subset D(F) \subset D$;

(b) $TD(F) \subset D(F)$ *for all* $T \in \mathcal{S}$, *and* $A_\lambda D(F) \subset D(F)$ *for all* $\lambda$;

(c) on $D(F)$ *we have* $TP = P = PT = P^2$ *for all* $T \in \mathcal{S}$.

*Proof.* Obvious. (It seems to be unknown whether $D = D(F)$ holds.)

**Lemma 1.8.** $D(0) = \mathrm{cl}\, N$.

*Proof.* The equicontinuity of the $A_\lambda$'s shows that $D(0)$ is closed. It contains the elements $x - Tx$ because of $w\text{-lim}\,(A_\lambda Tx - A_\lambda x) = 0$. Therefore $\mathrm{cl}\, N$ is contained in $D(0)$. If some $x_0 \in D(0)$ does not belong to $\mathrm{cl}\, N$ there exists an $h \in \mathfrak{X}^*$ with $\langle x_0, h\rangle \neq 0$ and $\langle z, h\rangle = 0$ for all $z \in \mathrm{cl}\, N$. As $\langle x - Tx, h\rangle = 0$ is valid for all $x$ and $T$, $h$ must belong to $F_*$. Hence $\langle x_0, h\rangle = \langle x, h\rangle$ holds for all $x \in \mathrm{co}\, \mathcal{S} x_0$, and by (E2) for all $x = A_\lambda x_0$. But then $\langle A_\lambda x_0, h\rangle$ does not tend to 0, contradicting $x_0 \in D(0)$. □

An operator $Q$ is called *$\mathcal{S}$-absorbing projection* if $Q = Q^2 = TQ = QT$ holds for all $T \in \mathcal{S}$.

**Theorem 1.9.** (Koliha-Nagel-Sato). *Let $\mathfrak{X}$ be a* l.c.t.v.s. *and $(A_\lambda, \lambda \in \Lambda)$ a weakly right $\mathcal{S}$-ergodic net.*

*(a) $D(F)$ is a closed linear subspace with $\mathcal{S} D(F) \subset D(F)$, $D(F) = F \oplus \mathrm{cl}\, N$, and $P$: $D(F) \to F$ is an $\mathcal{S}$-absorbing projection.*

*(b) $D(F) = \mathfrak{X}$ iff $F$ separates $F_*$.*

*(c) If $\mathfrak{Y}$ is a closed linear subspace of $\mathfrak{X}$ with $\mathcal{S}\mathfrak{Y} \subset \mathfrak{Y}$, and $Q$ an $\mathcal{S}$-absorbing projection on $\mathfrak{Y}$ with $Qx \in \overline{\mathrm{co}}\, \mathcal{S}x$ for all $x$, then $\mathfrak{Y} \subset D(F)$, and $Q$ agrees with $P$ on $\mathfrak{Y}$.*

*(d) If the net is even weakly $\mathcal{S}$-ergodic, $D = D(F)$.*

*(e) If the net is right $\mathcal{S}$-ergodic,* $\lim A_\lambda x = Px$ *for $x \in D(F)$.*

*Proof.* (a) follows from proposition 1.6, corollary 1.7 and lemma 1.8.

(b) Assume that $F$ separates $F_*$. If $D(F) = \mathfrak{X}$ fails, there exists $0 \neq h \in \mathfrak{X}^*$ with $\langle x, h \rangle = 0$ for all $x \in D(F)$. By lemma 1.8 and $D(0) \subset D(F)$ we obtain $\langle y - Ty, h \rangle = 0$ for all $y \in \mathfrak{X}$ and $T \in \mathcal{S}$, and hence $h \in F_*$. As $F$ separates $F_*$, there exists some $x \in F$ with $\langle x, h \rangle \neq 0$. This contradicts $F \subset D(F)$.

Next assume $D(F) = \mathfrak{X}$. We have to show that, for any $0 \neq h \in F_*$, there exists some $z \in F$ with $\langle z, h \rangle \neq 0$. Because of $h \neq 0$ there exists an $x \in F$ with $0 \neq \langle x, h \rangle$. $h \in F_*$ implies $\langle x, h \rangle = \langle x, S^* h \rangle$ for all $S \in \mathrm{co}\, \mathcal{S}$. In particular, $\langle x, h \rangle = \langle x, A_\lambda^* h \rangle = \langle A_\lambda x, h \rangle$, and hence $\langle x, h \rangle = \langle Px, h \rangle$. By $P = TP$ the element $Px$ belongs to $F$, and we can take $z = Px$.

(c) Restricting everything to $\mathfrak{Y}$ we may assume $\mathfrak{X} = \mathfrak{Y}$ and must show $\mathfrak{X} = D(F)$ and $P = Q$. Take $h, x$ as in the previous paragraph. Then $\langle x, h \rangle = \langle Sx, h \rangle$ for all $S \in \mathrm{co}\, \mathcal{S}$, and the assumption $Qx \in \overline{\mathrm{co}}\, \mathcal{S}x$ implies $\langle x, h \rangle = \langle Qx, h \rangle$. By $Q = TQ$ the element $Qx$ belongs to $F$. As $0 \neq h \in F_*$ was arbitrary, $F$ separates $F_*$, and $\mathfrak{X} = D(F)$. It follows from $QT = TQ = Q$ that $A_\lambda Q = QA_\lambda = Q$ for all $\lambda$ and hence $PQ = QP = Q$. Similarly $P\mathcal{S} = P$ and $Qx \in \overline{\mathrm{co}}\, \mathcal{S}x$ yield $PQ = P$. Hence $P = Q$.

(d) If $(A_\lambda)$ is $\mathcal{S}$-ergodic $TPx = Px$ holds on $D$.

(e) Passing to $x - Px$ we may assume $x \in D(0) = \mathrm{cl}\, N$. Clearly, the set $\{x: \lim A_\lambda x = 0\}$ is closed by the equicontinuity of the $A_\lambda$'s. It contains $N$ because $\lim (A_\lambda y - A_\lambda Ty) = 0$ holds for all $y$. □

Note that the description of $D(F)$ as a direct sum $F \oplus \mathrm{cl}\, N$ is independent of $(A_\lambda)$. In particular, if there are two nets $(A_\lambda)$, $(A'_\varrho)$, convergence to fixed points holds for the same set of elements, and the limits are the same.

**4. General semigroups.** The results above depend on the existence of weakly right $\mathcal{S}$-ergodic nets. The question of existence of such nets leads to the study of right amenable semigroups; see § 6.4. However, a different argument yields fixed points for general semigroups:

**Theorem 1.10.** (Alaoglu-Birkhoff). *If $\mathcal{S}$ is a semigroup consisting of contractions in a Banach space $\mathfrak{X}$ with uniformly convex norm, and if the norm of $\mathfrak{X}^*$ is strictly convex, then, for each $x$, $\overline{\text{co}}\,\mathcal{S}x$ contains a unique fixed point $Px$.*

*Proof.* Let $(x_n)$ be a sequence in $C := \overline{\text{co}}\,\mathcal{S}x$ with $\|x_n\| \to \alpha := \inf\{\|x\|: x \in C\}$. The uniform convexity of the norm implies that $(x_n)$ is a Cauchy sequence. Thus, $C$ contains a point $x'$ of minimal norm, and $x'$ is $\mathcal{S}$-fixed.

Assume $x''$ is another fixed point in $C$. Find $h \in \mathfrak{X}^*$ with $\langle x' - x'', h\rangle = \|x' - x''\|$. Using the $w^*$-compactness of $\overline{\text{co}}\,\mathcal{S}^*h$ we find a point $h'$ of minimal norm in $\overline{\text{co}}\,\mathcal{S}^*h$. From $x', x'' \in C$ and the fact that $h'$ is fixed under co $\mathcal{S}^*$ we may infer $\langle x, h'\rangle = \langle x', h'\rangle = \langle x'', h'\rangle$. Hence $\langle x' - x'', h'\rangle = 0$. As $x' - x''$ is fixed for co $\mathcal{S}$, and $h' \in \overline{\text{co}}\,\mathcal{S}^*h$, we have $\langle x' - x'', h'\rangle = \langle x' - x'', h\rangle$, and therefore $x' = x''$. □

**Remarks.** Similarly, one checks that $P^*h = h'$ in $\overline{\text{co}}\,\mathcal{S}^*h$ is unique, that $\mathfrak{X} = P\mathfrak{X} \oplus \ker(P)$, where $\ker(P)$ is the *kernel* $\{x: Px = 0\}$, and $\mathfrak{X}^* = P^*\mathfrak{X}^* \oplus \ker(P^*)$. $P\mathfrak{X}$ and $\ker(P^*)$, and also $\ker(P)$ and $P^*\mathfrak{X}^*$ are orthogonal to each other; see Jacobs [1960: theorem 1.2.3.]. Alaoglu-Birkhoff have shown by example that $x' \neq x''$ may happen above if the condition that $\mathfrak{X}^*$ has a strictly convex norm is deleted.

For the interpretation of their theorem Alaoglu and Birkhoff say that a sequence $x_\lambda$ indexed by the elements $\lambda$ of a set $\Lambda$ which is only partially ordered (and not necessarily directed) converges to $y$, if for each neighborhood $U$ of $y$ and each $\lambda$ there exists $\zeta \geqq \lambda$ such that $x_\eta$ belongs to $U$ for all $\eta \geqq \zeta$. They consider the partial order in $\Lambda = \text{co}\,\mathcal{S}$ defined by $\lambda \leqq \zeta$ iff there exists $\xi \in \Lambda$ with $\zeta = \xi \circ \lambda$. ($\Lambda$ was directed in the Abelian case.) It is an exercise to show that the convergence of $x_\lambda = \lambda x$ *for all* $x$ is equivalent to the existence of a unique fixed point in $\overline{\text{co}}\,\mathcal{S}x$ *for all* $x$.

Theorem 1.9(c) and theorem 1.10 motivate the following result which will be used in section 9.1. Recall that the *weak and strong operator topologies* are defined in $\mathcal{L}(\mathfrak{X})$ by wo-lim $T_\lambda = T$, (so-lim $T_\lambda = T$), iff w-lim $T_\lambda x = Tx$, (lim $T_\lambda x = Tx$), for all $x$; with $\lambda \in \Lambda$, a directed set. The *$w^*$-operator topology* is defined in $\mathcal{L}(\mathfrak{X}^*)$ by $w^*$o-lim $T'_\lambda = T'$ iff $w^*$-lim $T'_\lambda h = T'h$ for all $h \in \mathfrak{X}^*$. $\overline{\text{co}}\,\mathcal{S}$ denotes the closure of co $\mathcal{S}$ in the strong operator topology. It agrees with the closure in the weak operator topology (Dunford-Schwartz [1958: 477]).

**Theorem 1.11** (Nagel). *Let $\mathcal{S}$ be a bounded semigroup of linear operators in a Banach space $\mathfrak{X}$. Then the following properties are equivalent:*

*(i)* $\overline{\text{co}}\,\mathcal{S}$ *contains an $\mathcal{S}$-absorbing projection $P$;*

*(ii)* $w^*$o-$\overline{\text{co}}\,\mathcal{S}^*$ *contains an $\mathcal{S}^*$-absorbing projection $P'$ which is $w^*$-continuous;*

*(iii)* $\overline{\text{co}}\,\mathcal{S}x \cap F \neq \emptyset$ *for all* $x \in \mathfrak{X}$, *and* $\overline{\text{co}}\,\mathcal{S}^*h \cap F_* \neq \emptyset$ *for all* $h \in \mathfrak{X}^*$;

*(iv)* $F$ *separates* $F_*$, *and* $\overline{\text{co}}\,\mathcal{S}^*h \cap F_* \neq \emptyset$ *for all* $h \in \mathfrak{X}^*$.

*Proof.* The equivalence of (i) and (ii) follows because the $w^*$-continuous linear operators in $\mathfrak{X}^*$ are the adjoints of the elements of $\mathcal{L}(\mathfrak{X})$. We may take $P' = P^*$. $P \in \overline{\text{co}}\,\mathcal{S}$ is equivalent to $P^* \in w^*\text{o-}\overline{\text{co}}\,\mathcal{S}^*$.

(i) $\Rightarrow$ (iii): $Px$ belongs to $F \cap \overline{\text{co}}\,\mathcal{S}x$, and $P^*h$ to $F_* \cap \overline{\text{co}}\,\mathcal{S}^*h$.

(iii) $\Rightarrow$ (iv): We argue as before. For $0 \neq h \in F_*$ there exists $x$ with $\langle x, h\rangle \neq 0$. Take $x' \in F \cap \overline{\text{co}}\,\mathcal{S}x$. For any $T \in \text{co}\,\mathcal{S}$ we have $\langle Tx, h\rangle = \langle x, T^*h\rangle = \langle x, h\rangle$. Hence $\langle x', h\rangle = \langle x, h\rangle \neq 0$.

(iv) $\Rightarrow$ (ii): If $h', h''$ are two elements in $F_* \cap \overline{\text{co}}\,\mathcal{S}^*h$, there exists $x \in F$ with $\langle x, h' - h''\rangle \neq 0$. $x \in F$ and $h', h'' \in \overline{\text{co}}\,\mathcal{S}^*h$ again implies $\langle x, h'\rangle = \langle x, h\rangle = \langle x, h''\rangle$, a contradiction. Thus $F_* \cap \overline{\text{co}}\,\mathcal{S}^*h$ contains exactly one element, which we denote by $P'h$.

Put $M = \sup\{\|T\|: T \in \mathcal{S}\}$. To show $P'(h_1 + h_2) = P'h_1 + P'h_2$, observe that for any $x_1, \dots, x_n, \varepsilon$ there exists $R' \in \text{co}\,\mathcal{S}^*$ with $|\langle x_i, R'h_1 - P'h_1\rangle| < \varepsilon$. There exists $h' \in \overline{\text{co}}\,\mathcal{S}^*R'h_2 \cap F_*$. By the uniqueness of $P'h_2$ we have $h' = P'h_2$. Hence, there exists $S' \in \text{co}\,\mathcal{S}^*$ with $|\langle x_i, P'h_2 - S'R'h_2\rangle| < \varepsilon$. Now $S'P'h_1 = P'h_1$ yields

$$|\langle x_i, S'R'(h_1 + h_2) - P'h_1 - P'h_2\rangle| \leqq |\langle x_i, S'R'h_1 - S'P'h_1\rangle| + |\langle x_i, S'R'h_2 - P'h_2\rangle| \leqq M\varepsilon + \varepsilon.$$

Thus, each $w^*$-neighborhood of $P'h_1 + P'h_2$ contains an element of $\text{co}\,\mathcal{S}^*(h_1 + h_2)$. Hence $P'h_1 + P'h_2 = P'(h_1 + h_2)$. It follows that $P'$ is linear. It is simple to check that $P'$ is an $\mathcal{S}^*$-absorbing projection. Now consider a net $(h_\lambda)$ with w*-lim $h_\lambda = 0$. As the norm of any element of $\text{co}\,\mathcal{S}^*$ is $\leqq M$, we have $|\langle x, P'h_\lambda\rangle| \leqq M|\langle x, h_\lambda\rangle| \to 0$. Hence, $P'$ is $w^*$-continuous, and (ii) is proved. □

**5. More on fixed points.** An operator $T$ in a vector space $\mathfrak{X}$ is called *affine* if $T(\alpha x + (1 - \alpha)y) = \alpha Tx + (1 - \alpha)Ty$ holds for all $x, y \in \mathfrak{X}$ and $0 \leqq \alpha \leqq 1$.

**Theorem 1.12** (Fixed point theorem of Markov-Kakutani). *Let $K$ be a convex weakly compact subset of a topological vector space $\mathfrak{X}$. Let $\mathcal{S}$ be a commuting family of continuous affine maps $T: \mathfrak{X} \to \mathfrak{X}$ with $TK \subset K$. Then $K$ contains a point $p$ fixed under all $T \in \mathcal{S}$.*

*Proof.* The operators $A_n(T)$ map $K$ into $K$ and commute. Therefore, the sets $A_{n_1}(T_1)A_{n_2}(T_2)\dots A_{n_k}(T_k)K$ with $n_i \in \mathbb{N}$, $T_i \in \mathcal{S}$, form a decreasingly filtered family of non empty weakly compact sets, and there exists a $p$ in the intersection of these sets. For any $T \in \mathcal{S}$ and $n \in \mathbb{N}$ there exists $q_n \in K$ with $Tp - p = TA_n(T)q_n - A_n(T)q_n = n^{-1}(T^n - I)q_n \in n^{-1}(K - K)$. As $\{\langle u, h\rangle: u \in K\}$ is bounded for each $h \in \mathfrak{X}^*$, we must have $\langle Tp - p, h\rangle = 0$ for all $h$, and hence $Tp = p$. □

Next, let us look at the fixed points of convex combinations of commuting operators. We need the following lemma of Kakutani:

**Lemma 1.13.** *The identity $I$ is an extreme point in the convex set consisting of all contractions in a Banach space $\mathfrak{X}$.*

*Proof.* If $I$ is not an extreme point there exist contractions $T_1 \neq T_2$ with $(T_1 + T_2)/2 = I$. For $T = T_1 - I$ we then have $T_1 = I + T$, $T_2 = I - T$ and $T \neq 0$. For any $h \in \mathfrak{X}^*$ put $h_i = T_i^* h$. As $T_i^*$ is a contraction, $\|h_i\| \leqq \|h\|$. If $h$ is an extreme point of $B = \{h \in \mathfrak{X}^* : \|h\| \leqq 1\}$, the identity $h = (h_1 + h_2)/2$ shows $h = h_1 = h_2$, i.e., $T^* h = 0$. By Krein-Milman $B$ is the closed convex hull of the set of extreme points. It follows that $T^* h = 0$ holds for all $h \in B$. But $T^* = 0$ contradicts $T \neq 0$. □

Recall the notation $F(T) = \{x \in \mathfrak{X} : Tx = x\}$.

**Lemma 1.14** (Brunel-Falkowitz). *Let $\{\alpha_j : j \in J\}$ be strictly positive numbers with $\sum \alpha_j = 1$ and let $T_j$ be commuting contractions in a Banach space $\mathfrak{X}$. If $T = \sum \alpha_j T_j$ then*

$$F(T) = \bigcap_j F(T_j).$$

*Proof.* The inclusion $\supset$ is obvious. To prove $F(T) \subset F(T_j)$ put $S = (1 - \alpha_j)^{-1} \sum_{k \neq j} \alpha_k T_k$. For $x \in F(T)$ we have $T_j x = T_j T x = T T_j x$, and hence $T_j x \in F(T)$. Similarly, the contraction $S$ maps $F(T)$ into itself. But on $F(T)$, $T$ is the identity. Because of $T = \alpha_j T_j + (1 - \alpha_j) S$ the previous lemma shows that $T_j$ is the identity on $F(T)$. □

The next result is essentially due to Sine [1975]:

**Theorem 1.15.** *Convex combinations of finitely many commuting power bounded mean ergodic operators $T_1, \ldots, T_n$ in a Banach space $\mathfrak{X}$ are mean ergodic.*

*Proof.* Passing to the equivalent norm

$$|||x||| = \sup \{ \| T_1^{k_1} \ldots T_n^{k_n} x \| : k_i \geqq 0 \},$$

we may assume that the $T_i$ are contractions. We may also assume $n = 2$. By the previous lemma and by theorem 1.4 it is enough to prove that $F(T_1) \cap F(T_2)$ separates $F(T_1^*) \cap F(T_2^*)$. Let $h_1, h_2$ be two distinct elements of $F(T_1^*) \cap F(T_2^*)$. As $T_1$ is mean ergodic there exists an $x \in F(T_1)$ with $\langle x, h_1 \rangle \neq \langle x, h_2 \rangle$. Put $x_\infty = \lim A_n(T_2) x$. As $T_2 x = T_2 T_1 x = T_1 T_2 x$ shows $T_2 F(T_1) \subset F(T_1)$, we have $x_\infty \in F(T_1) \cap F(T_2)$. $x_\infty$ separates $h_1$ and $h_2$ because of $\langle x_\infty, h_i \rangle = \lim \langle A_n(T_2) x, h_i \rangle = \lim \langle x, A_n(T_2^*) h_i \rangle = \langle x, h_i \rangle$. □

## Notes

**Continuous parameter semigroups and resolvents.** If $\mathscr{S} = \{T_t, t > 0\}$ is a semigroup with a continuous parameter, ergodic theorems are frequently formulated for averages

$$A_{0,t}x = t^{-1}\int_0^t T_s x\,ds \quad \text{or} \quad \lambda R_\lambda x = \lambda \int_0^\infty e^{-\lambda t} T_t x\,dt$$

where $t$ tends to $\infty$ and $\lambda$ to $0+0$. For the sake of simplicity let us assume that the $T_t$ are contractions in a Banach space, and that $\mathscr{S}$ is continuous in the strong operator topology for $t > 0$. Ergodic theorems follow from theorem 1.5, see Eberlein [1976]. They can also easily be deduced by applying theorem 1.1 to $T_1$ and $A_{0,1}x$. The convergence of $\lambda R_\lambda x$ may be obtained by a reduction to the case of averages $A_{0,t}x$; see § 8.2.

It is more interesting that different conditions for convergence may be described in terms of the generator $A$ of a semigroup. Consider a semigroup $\mathscr{S} = \{T_t: t \geqq 0\}$ of contractions in a Banach space $\mathfrak{X}$, which is continuous in the strong operator topology. Assume $T_0 = I$. The set dom$(A)$ of points $x$ for which $Ax := \lim_{t \to 0+0} t^{-1}(T_t x - x)$ exists, is dense in $\mathfrak{X}$. It is clear that $F = \{x \in \mathfrak{X}: T_t x = x \text{ for all } t\}$ is contained in $\ker(A) = \{x \in \operatorname{dom}(A): Ax = 0\}$. It follows from semigroup theory that $F = \ker(A)$; see e.g. Davies [1980: 3]. If both $\mathscr{S}$ and the adjoint semigroup $\mathscr{S}^*$ are continuous in the strong operator topology, the condition that $F$ separates $F_*$ may therefore be expressed as:

(c′): $\ker(A)$ separates $\ker(A^*)$, where $A^*$ is the generator of $\mathscr{S}^*$.

Hille and Phillips [1957] have proved that also each of the following two conditions is equivalent to the norm convergence, for all $x$, of $A_{0,t}x$:

(HP1) For all $x$ in a weakly dense subset of $\mathfrak{X}$ the set $\{\lambda(\lambda I - A)^{-1}x: 0 < \lambda \leqq 1\}$ is conditionally weakly compact;

(HP2) $\operatorname{cl}(\ker(A) + A\operatorname{dom}(A)) = \mathfrak{X}$.

(See also Davies [1980]).

It may be of interest that the convergence of $A_{0,t}x$ for all $x$ does not imply the convergence of the discrete averages $A_n(T_1)x$ for all $x$. As an example let $\mathfrak{X}$ be the space of all Lebesgue integrable real valued functions on $\mathbb{R}$ with integral 0, and $(T_t x)(\cdot) = x(\cdot + t)$. Shaw [1980], Sato [1981], and Kataoka [1981] considered a sufficient condition for the mutual implication of discrete and continuous parameter convergence.

Some authors (e.g. Cohen [1940], Hille [1945]) have dealt with theorems asserting that convergence holds for some averaging method (e.g. $A_{0,t}x$) iff it holds for some other method (e.g. $\lambda R_\lambda x$). Such results usually follow from Eberlein's theorem because the first condition there does not involve the $A_\lambda$'s. However, it may happen that the boundedness conditions are distinct. Call $T$ *Abel bounded* if $\sup_{\lambda > 0} \|\lambda W_\lambda\|$ is finite, where

$$W_\lambda x = \sum_{n=0}^\infty (\lambda + 1)^{-(n+1)} T^n x.$$

Cesàro bounded operators are Abel bounded, and the converse holds for positive $T$ in a Banach lattice; see Emilion [1984].

A family $(V_\lambda, \lambda > 0)$ of elements of $\mathscr{L}(\mathfrak{X})$ is said to satisfy the *resolvent equation* if $V_\lambda - V_\mu = (\mu - \lambda) V_\lambda V_\mu, (\lambda, \mu > 0)$. Then $(\lambda V_\lambda, \lambda > 0)$, or sometimes $(V_\lambda)$, is called *pseudo resolvent*. The most important examples are the *resolvents* $(\lambda R_\lambda)$ of semigroups. Yosida [1974] has mean ergodic theorems for pseudo resolvents. There is also a local version:

**Proposition 1.16.** *If $(\lambda V_\lambda)$ is a pseudo resolvent in a Banach space $\mathfrak{X}$, $\sup\|\lambda V_\lambda\| < \infty$,* *and* $(\lambda_n)$ *a sequence with* $\lambda_n \to \infty$, *then the existence of* w-lim $\lambda_n V_{\lambda_n} x = y$ *implies* $x \in \mathfrak{X}_0 := \{z: \lim_{\lambda\to\infty} \lambda V_\lambda z$ *exists*$\}$.

*Proof.* $\lambda V_\lambda \lambda_n V_{\lambda_n} x = \lambda\lambda_n(\lambda-\lambda_n)^{-1}(V_{\lambda_n} - V_\lambda)x$ and the boundedness condition shows that this converges to $\lambda_n V_{\lambda_n} x$ for $\lambda \to \infty$. Hence $\lambda_n V_{\lambda_n} x \in \mathfrak{X}_0$. As $\mathfrak{X}_0$ is weakly closed, $y \in \mathfrak{X}_0$. Now

$$\begin{aligned}\lambda V_\lambda(I - \lambda_n V_{\lambda_n})x &= \lambda V_\lambda x - \lambda\lambda_n(\lambda-\lambda_n)^{-1}(V_{\lambda_n} - V_\lambda)x \\ &= (1 + \lambda_n(\lambda-\lambda_n)^{-1})\lambda V_\lambda x - \lambda\lambda_n(\lambda-\lambda_n)^{-1} V_{\lambda_n} x\end{aligned}$$

tends to 0 in norm for $n \to \infty$. Hence $\lambda V_\lambda(x-y) = 0$ and $\lambda V_\lambda x = \lambda V_\lambda y$ for all $\lambda$. It follows that $x \in \mathfrak{X}_0$. □

This also proves that $T_0 x := \lim_{\lambda\to\infty} \lambda V_\lambda x$ satisfies $V_\lambda T_0 x = V_\lambda x$, which yields $T_0^2 = T_0$. Similarly, $T_0 V_\lambda x = V_\lambda x$. Frequently $\mathfrak{X} = \mathfrak{X}_0$ (e.g., for reflexive $\mathfrak{X}$). Then the Hille-Yosida theorem implies that $(\lambda V_\lambda)$ on $T_0\mathfrak{X}$ is the resolvent of a semigroup. Pseudo resolvents are then compositions of a projection on a subspace and a resolvent acting on the subspace.

Proposition 1.16 is taken from Emilion [1984]. Masani [1976] has such a result for resolvents.

Necessary and sufficient conditions for the convergence of $A_{0,t}x$ for all $x$ have been given for *general* strongly continuous semigroups by Lin, Montgomery and Sine [1977].

Yu [1972] has a mean ergodic theorem for "harmonizable" processes.

**Cyclic semigroups** $\{\mathbf{T}^n\}$. If $T$ is a contraction in a Banach space, the sequence $s_n = \|S_n(T)x\|$ is subadditive. Hence $\lim s_n/n$ exists. Derriennic [1976] has shown that the limit is $\sup\{|\langle x, h\rangle|: h = T^*h, \|h\| \leqq 1\}$. He also proved $\lim \|T^n x\| = \sup\{|\langle x, h\rangle: h \in \bigcap_{n=0}^{\infty} T^{*n}S\}$ with $S = \{h \in \mathfrak{X}^*: \|h\| \leqq 1\}$.

A sequence $F_n$ in a Banach lattice is called subadditive for a positive contraction $T$ if $F_{n+k} \leqq F_n + T^n F_k$ holds for all $n, k \geqq 1$. The convergence of $n^{-1}F_n$ has been studied in the paper of Derriennic and Krengel [1981]. E.g. it is shown that the existence of a weak cluster point $f = Tf$ of $(n^{-1}F_n)$ implies $\|(n^{-1}F_n - f)^+\| \to 0$. In $L_p$, $(1 \leqq p < \infty)$, the existence of a weak cluster point of $(n^{-1}F_n)$ for non negative $(F_n)$ is equivalent with norm convergence. If $T$ is an isometry in $L_p$, $(1 < p < \infty)$, and $\sup\{n^{-1}\|F_n\|_p\} < \infty$, then $n^{-1}F_n$ converges weakly, but it need not converge strongly.

If $T$ is a contraction and $A_n(T^2)x$ converges for all $x$, then (almost obviously) $A_n(T)x$ converges for all $x$, but the converse is not true. For an example let $\mathfrak{X}$ be the Banach space of sequences $x = (x_i: i \in \mathbb{Z})$ of real numbers with $\|x\| = \sum|x_i| < \infty$ and $\sum x_i = 0$, and take $(Tx)_i = x_{i+1}$. However, if $T$ is positive and $\mathfrak{X}$ a Banach lattice with order continous norm, then the converse above does hold; see Derriennic-Krengel [1981]. Sine [1976] has constructed a uniquely ergodic transformation $\tau$ in a compact metric space $K$ (for which $T: x \to x \circ \tau$ is mean ergodic in $\mathfrak{X} = C(K)$), such that $T^2$ is not mean ergodic.

Butzer and Westphal [1971] have proved the following (essentially negative) result on the speed of convergence in the mean ergodic theorem for power bounded linear operators $T$ in a reflexive Banach space $\mathfrak{X}$:
$\|A_n x - Px\| = o(n^{-1})$ implies $x \in F$, and $\|A_n x - Px\| = O(n^{-1})$ implies $x \in F \oplus N$. (In both cases the converse is obvious). For a simplified argument, see Lin-Sine [1983]. For a strongly continuous bounded semigroup $\{T_t, t \geqq 0\}$ with generator $A$, Butzer and Westphal show that $\|\lambda R_\lambda x - Px\| = o(\lambda)$, $(\lambda \to 0+0)$, implies $x \in \ker(A) = F$, and $\|\lambda R_\lambda x - Px\| = O(\lambda)$ implies $x \in F \oplus A$ dom$(A)$. For related results, see Butzer and Westphal [1972], Leviatan and Westphal [1973], Leviatan [1974], Goldstein, Radin and Showalter [1978], and Butzer [1980].

The papers of Sato [1979], [1979a], Shreider [1967], Rubinov [1977], Lloyd [1976], and Anzai [1977] are related to Sine's criterion for mean ergodicity of $T$.

Sato [1975] noticed that Cesàro boundedness can be replaced by Cesàro boundedness along the subsequence $n_k$ with w-lim $A_{n_k}x = y$ in theorem 1.1. Emilion [1984] considered weakened boundedness conditions. E.g., he showed: A positive Abel bounded $T$ in a reflexive Banach lattice is mean ergodic. By Heinich [1983] the reflexivity is necessary here. Emilion's paper also contains an example of Assani of a non positive Cesàro bounded $T$ which does not satisfy $T^n/n \to 0$. For further related work, see Assani [1984a].

Haïnis [1977] remarked that in a Banach algebra $n^{-1}\sum_{1}^{n} y_k \to y$ implies $(e + y_1/n)\,(e + y_2/n)\ldots(e + y_n/n) \to \exp(y)$. This yields "multiplicative" ergodic theorems. Sarymsakov [1964], [1966] has mean ergodic theorems for vector spaces over a semifield. Atalla [1976] and Shaw [1983] derived a criterion for mean ergodicity in a Grothendieck space.

**General semigroups.** Dunford [1939a], and Day [1942] proved mean ergodic theorems for $d$-parameter and Abelian semigroups. Koliha [1973], and Nagel [1973] have a splitting theorem in Banach spaces. We largely followed Sato [1978] in theorem 1.9. C. Ionescu Tulcea [1980] has a generalization of Eberlein's theorem in uniform spaces.

The uniform convexity argument originated with Wiener [1939] and was used by Garrett Birkhoff [1939] for a single $T$ before L. Alaoglu and G. Birkhoff [1940] obtained their theorem for general semigroups. Jacobs [1954] proved the existence of a unique fixed point in $\overline{\text{co}}\,\mathscr{S}x$ for bounded groups $\mathscr{S}$ in Hilbert space and in $L_p$, $(1 < p < \infty)$.

Now consider the space $L_1$ of a *finite* measure space $(\Omega, \mathscr{A}, \mu)$. If $\mathscr{T}$ is a semigroup of measure preserving transformations in $\Omega$ and $\mathscr{S}$ the semigroup of operators $Tf = f \circ \tau$, $(\tau \in \mathscr{T})$, then the operators in co $\mathscr{S}$ are also contractions in $L_2 \subset L_1$, and $L_2$ is dense in $L_1$. Although theorem 1.10 is not directly applicable to $L_1$, it is not hard to see that $P$ may be extended from $L_2$ to a projection in $L_1$, also denoted by $P$, such that $Pf$ belongs, for all $f \in L_1$, to $\overline{\text{co}}\,\mathscr{S}f$. (Use the positivity of $T$ and the property $T\mathbb{1} \leqq \mathbb{1}$). $P$ again satisfies $TP = PT = P = P^2$ for all $T$.

This was proved (for groups) by Aribaud [1970], and extended to more general operators in Banach lattices by Nagel [1973]. Theorem 1.11 of Nagel [1973] includes complements due to Kümmerer-Nagel [1979].

**Affine operators and the equation $(I - T)x = y$.** Iterative methods for the approximate solution of this functional equation, investigated by Browder, Petryshyn, Koliha, and others, have motivated the following generalization of Eberlein's theorem by Dotson [1971]; (having a similar proof):

**Theorem 1.17.** *Let $\mathscr{S} = \{T_n\}$ be any collection of affine operators in a* l.c.t.v.s. $\mathfrak{X}$, *and $\{U_\lambda, \lambda \in \Lambda\}$ a net of affine operators in $\mathfrak{X}$ which are equicontinuous. Assume $U_\lambda z \in \overline{\text{co}}\,\mathscr{S}z$ for all $\lambda$, $z$. For fixed $x$ assume*

$$\lim_{\lambda} (U_\lambda T_u x - U_\lambda x) = \lim_{\lambda} (T_u U_\lambda x - U_\lambda x) = 0$$

*for all $T_u \in \mathscr{S}$. If $y$ is a weak cluster point of $\{U_\lambda x\}$ then $y = T_u y$ for all $T_u$, $y \in \overline{\text{co}}\,\mathscr{S}x$ and* $\lim U_\lambda x = y$.

Dotson, Groetsch [1975/76], Sato [1979b], and others have given applications.

Lin and Sine [1983] showed for Cesàro bounded mean ergodic $T$ with so-lim $T^n/n = 0$ that (i) $y \in (I - T)\mathfrak{X}$, (ii) $x_n = n^{-1}\sum_{k=1}^{n}\sum_{j=0}^{k-1} T^j y$ has a weakly convergent subsequence, and

(iii) $\lim x_n = x$ with $y = (I - T)x$, are equivalent. The paper settles the problem of existence of $x$ also for dual operators and for many Markov operators.

Krengel and Lin [1984] study the analogous problems for semigroups $\{T_t, t \geqq 0\}$, i.e., the problem when $y$ belongs to the range of the generator $A$. The necessary condition $\sup \|\int_0^t T_s y\, ds\| < \infty$ is also sufficient for contractions in $L_1$ and for dual semigroups. For mean ergodic semigroups the existence of $\lim \alpha^{-1} \int_0^\alpha [\int_0^\beta T_s y\, ds]\, d\beta$ is necessary and sufficient. Davies [1982] studies conditions for the range of $A$ to be dense.

**Fixed points.** There is now an extensive literature on fixed points, see e.g. Istratescu [1981], Takahashi [1980]. We mention the following fixed point theorem of Ryll-Nardzweski [1967]:

**Theorem 1.18.** *Let $\mathscr{S}$ be a semigroup of affine continuous transformations in a* l.c.t.v.s. $\mathfrak{X}$. *Assume that $\mathscr{S}$ is distal, i.e., for any $x \neq y$, 0 does not belong to* cl $\{Sx - Sy : S \in \mathscr{S}\}$. *If $K$ is a weakly compact convex subset of $\mathfrak{X}$ with $\mathscr{S}K \subset K$, then $K$ contains a fixed point for $\mathscr{S}$.*

We refer to Glasner [1976], and Asplund-Namioka [1967] for proofs; see also Day [1973], Petersen [1983].

Brunel [1973] and Falkowitz [1973] arrived independently at lemma 1.14. We followed Nagel-Palm-Derndinger [1984]. The lemma also follows easily from the following lemma of Foguel and Weiss [1973]:

**Lemma 1.19.** *If $P_1$, $P_2$ are commuting elements of a Banach algebra with $\|P_1\| = \|P_2\| = 1$ and $Q = \alpha P_1 + (1 - \alpha) P_2$, $(0 < \alpha < 1)$, then $\|Q^n (P_1 - P_2)\| \leqq K\alpha(1 - \alpha) n^{-1/2}$, where $K$ is a constant.*

## § 2.2 Uniform convergence

**1. The uniform ergodic theorem.** Let $S_n$, $S$, $T$, ... be continuous linear operators in a Banach space $\mathfrak{X}$. The sequence $(S_n)$ is said to *converge uniformly* (or in the *uniform operator topology*) to $S$, if $\|S_n - S\|$ tends to 0. $T$ is called *uniformly ergodic* if $A_n = A_n(T)$ converges uniformly.

This is much more than mean ergodicity. E.g. if $\mathfrak{X} = L_p$, and $T$ is the composition with a measure preserving transformation, then $T$ is uniformly ergodic iff $T$ is periodic with bounded period. Yet, some interesting uniformly ergodic operators do arise in probability theory and functional analysis. We begin with those results which do not require spectral theory.

By (1.1) any uniformly ergodic $T$ must satisfy

(2.1) $$\|n^{-1} T^n\| \to 0.$$

As any uniformly ergodic $T$ is mean ergodic, theorem 1.3 implies that for uniformly ergodic $T$ the space $\mathfrak{X}$ is a direct sum $F \oplus \mathrm{cl}\, N$ with $F = F(T)$

$= \{x \in \mathfrak{X}: Tx = x\}$ and $N = (I - T)\mathfrak{X}$. The following theorem, due to Lin [1974], shows even $\mathfrak{X} = F \oplus N$:

**Theorem 2.1.** *If $T$ satisfies* (2.1), *then $T$ is uniformly ergodic iff $N$ is closed.*

*Proof.* Let $T$ be uniformly ergodic. $Y = \mathrm{cl}\, N$ is invariant under $T$, and the restriction $S$ of $T$ to $Y$ satisfies $\|A_n(S)\| \to 0$. For any $n$ with $\|A_n(S)\| < 1$, $I - A_n(S)$ is invertible. The invertibility of $I - S$ therefore follows from the identity

$$(I-S)\left(\frac{n-1}{n} I + \frac{n-2}{n} S + \ldots + \frac{1}{n} S^{n-2}\right) = I - A_n(S).$$

Hence $Y = (I-S)Y = (I-T)Y \subset (I-T)\mathfrak{X} = N$, and $Y = N$.

Conversely, if $N$ equals $Y$, the open mapping theorem asserts that $(I-T)U$ is open in $Y$ for any open $U \subset \mathfrak{X}$. Hence there exists a $K > 0$ such that, for any $y \in Y$, there is a $z \in \mathfrak{X}$ with $(I-T)z = y$ and $\|z\| \leqq K\|y\|$. (Otherwise there is a sequence $y_n$ converging to 0 and disjoint to $(I-T)\{z: \|z\| < 1\}$.) From

$$\|A_n(T)y\| = \|A_n(T)(I-T)z\| \leqq n^{-1}\|I - T^n\|\,\|z\| \leqq Kn^{-1}\|I - T^n\|\,\|y\|$$

we see that the restriction $S$ of $T$ to $Y$ is uniformly ergodic. It follows as above that $I - S$ is invertible in $Y$, and $(I-T)\mathfrak{X} = Y = (I-S)Y = (I-T)Y$. Therefore there exists, for any $x \in \mathfrak{X}$, an element $y \in Y$ with $(I-T)x = (I-T)y$, and by the invertibility of $(I-S)$ we may assume $\|y\| \leqq K'\|(I-T)x\|$ with some $K' < \infty$ independent of $x$. We now write $x = (x-y) + y$. As $(x-y)$ is fixed under $T$ we find

$$\|A_n(T)x - (x-y)\| = \|A_n(T)y\| \leqq n^{-1}KK'\|I - T^n\|\,\|I - T\|\,\|x\|,$$

i.e., the convergence of $A_n(T)$ to the projection $P$ with $Px = x - y$ is uniform. □

It is an exercise to deduce a multiparameter extension: If $T_1, T_2, \ldots, T_d$ are commuting bounded linear operators in $\mathfrak{X}$ with $\|n^{-1} T_i^n\| \to 0$ and we define $A_\lambda$ for $\lambda = (\lambda_1, \ldots, \lambda_d) \in \Lambda = \mathbb{N}^d$ by $A_\lambda = A_{\lambda_1}(T_1) \ldots A_{\lambda_d}(T_d)$, then $A_{\lambda(k)}$ converges uniformly for all increasing sequences $\lambda(k) \in \Lambda$ iff $(I - T_i)\mathfrak{X}$ is closed for $i = 1, \ldots, d$. (For "only if" take $\lambda_i(k) = k$ and $\lambda_j(k) = 1$, $(j \neq i)$).

We treat the next criterion right away for a general bounded semigroup $\mathscr{S}$ of linear operators in $\mathfrak{X}$. A family $\{A_\lambda: \lambda \in \Lambda\}$ is called a uniformly $\mathscr{S}$-ergodic net, if $\Lambda$ is a directed set, and the $A_\lambda$'s are linear operators belonging to the norm closure $\|\cdot\|$-$\overline{\mathrm{co}}\,\mathscr{S}$ of $\mathrm{co}\,\mathscr{S}$, such that

(2.2) $\quad \lim_\lambda \|A_\lambda T - A_\lambda\| = 0 \quad$ and $\quad \lim_\lambda \|TA_\lambda - A_\lambda\| = 0$

holds for all $T \in \mathscr{S}$.

**Theorem 2.2.** *Let $\{A_\lambda: \lambda \in \Lambda\}$ be a uniformly $\mathscr{S}$-ergodic net for a bounded semigroup $\mathscr{S}$. Then $\lim_\lambda A_\lambda$ exists in the uniform operator topology iff there exists a bounded linear operator $P$ in $\|\cdot\|$-$\overline{\text{co}}\,\mathscr{S}$ with $TP = PT = P$ for all $T \in \mathscr{S}$.*

*Proof.* If $(A_\lambda)$ converges uniformly to some $P$, then obviously $P$ has the desired properties. Conversely, assume that $P$ exists.

Let $K$ be a bound for the norms of the operators in $\mathscr{S}$ and hence for those in $\|\cdot\|$-$\overline{\text{co}}\,\mathscr{S}$. For any $\varepsilon > 0$ we may find a convex combination $T_\alpha$ of finitely many operators $T_i \in \mathscr{S}$ with $\|T_\alpha - P\| < \varepsilon$. Clearly $TP = P$ holds even for all $T$ in $\|\cdot\|$-$\overline{\text{co}}\,\mathscr{S}$, and in particular for $T = A_\lambda$. For large enough $\lambda$ we have $\|A_\lambda T_i - A_\lambda\| < \varepsilon$ and therefore $\|A_\lambda T_\alpha - A_\lambda\| < \varepsilon$. As $\varepsilon > 0$ was arbitrary the estimate $\|A_\lambda - P\| = \|A_\lambda - A_\lambda P\| \leqq \|A_\lambda - A_\lambda T_\alpha\| + \|A_\lambda T_\alpha - A_\lambda P\| \leqq \varepsilon + K\|T_\alpha - P\| \leqq (K+1)\varepsilon$ completes the proof. □

**Remarks.** 1) The theorem also follows from theorem 1.9 because $\mathscr{S}$ acts on the Banach space $\tilde{\mathfrak{X}} = \mathscr{L}(\mathfrak{X})$ by left multiplication. 2) The theorem implies that the uniform convergence of $(A_\lambda)$ for some net implies the uniform convergence for any other net. E.g. if $T$ is power bounded, the uniform convergence of $A_n(T)$ for $n \to \infty$ is equivalent to the uniform convergence of $(\lambda - 1) \sum_{m=0}^{\infty} \lambda^{-(m+1)} T^m$ for $\lambda \downarrow 1$; see also Lin [1974].

**2. Quasi-compact operators.** Kryloff and Bogolioùboff [1937], [1937a] introduced an important class of operators, for which uniform convergence can be obtained:

**Definition 2.3.** $T$ is called *(weakly) compact* if the image under $T$ of the unit sphere of $\mathfrak{X}$ is conditionally (weakly) compact. $T$ is called *(weakly) quasi-compact* if there exists an integer $m$ and a (weakly) compact operator $Q$ with $\|T^m - Q\| < 1$.

We shall use the fact that linear combinations and uniform limits of (weakly) compact operators are (weakly) compact, and that the product of a (weakly) compact operator and a bounded linear operator is (weakly) compact. (See e.g. Dunford-Schwartz [1958]).

**Lemma 2.4.** *$T$ is (weakly) quasi-compact iff there exists a sequence $Q_n$ of (weakly) compact operators with $\|T^n - Q_n\| \to 0$.*

*Proof.* For (weakly) quasi-compact $T$ put $\Delta = T^m - Q$. For any integer $n \geqq 0$ find integers $r, s$ with $n = ms + r$ and $0 \leqq r < m$, and put $Q_n = T^n - T^r \Delta^s$. Then we have $Q_n = T^r[T^{ms} - \Delta^s] = T^r[T^{ms} - (T^m - Q)^s]$. The only term in the expansion of $(T^m - Q)^s$ which does not contain $Q$ as a factor is $T^{ms}$. Therefore, $Q_n$ is (weakly) compact. $\|\Delta\| < 1$ implies $\|T^n - Q_n\| = \|T^r \Delta^s\| \leqq \|\Delta\|^s \cdot \sup\{\|T^r\|: 0 \leqq r < m\} \to 0$. □

The lemma implies that powers of (weakly) quasi-compact operators are (weakly) quasi-compact. Weak quasi-compactness can be used to prove a mean ergodic theorem:

**Theorem 2.5.** *Let $\{A_\lambda: \lambda \in \Lambda\}$ be an $\mathscr{S}$-ergodic net for a semigroup $\mathscr{S} \subset \mathscr{L}(\mathfrak{X})$. If $\|\cdot\|$-$\overline{\text{co}}\,\mathscr{S}$ contains a weakly quasi-compact operator $T$, then $Px := \lim_\lambda A_\lambda x$ exists for all $x \in \mathfrak{X}$, and $P$ is weakly compact. If $T$ is even quasi-compact, then $P$ is compact.*

*Proof.* Again put $\Delta = T^m - Q$. It is clear from $\|\Delta\| < 1$ that $(I - \Delta)^{-1} = \sum_{k=0}^{\infty} \Delta^k$ exists, and it is simple to verify the identity

(2.3) $$A_\lambda = (I-\Delta)^{-1} Q A_\lambda + (I-\Delta)^{-1}(A_\lambda - T^m A_\lambda).$$

Fix any $x$. By the definition of an $\mathscr{S}$-ergodic net $\lim_\lambda (A_\lambda - SA_\lambda)x = 0$ holds for all $S \in \mathscr{S}$, and hence for all $S \in \|\cdot\|$-$\overline{\text{co}}\,\mathscr{S}$. Powers of elements of $\text{co}\,\mathscr{S}$ belong to $\text{co}\,\mathscr{S}$. Therefore $T \in \|\cdot\|$-$\overline{\text{co}}\,\mathscr{S}$ implies $T^m \in \|\cdot\|$-$\overline{\text{co}}\,\mathscr{S}$, and we obtain $\lim_\lambda (A_\lambda - T^m A_\lambda)x = 0$. By the weak compactness of $(I-\Delta)^{-1} Q$ and the boundedness of $\{A_\lambda x: \lambda \in \Lambda\}$ there exists a weak cluster point of $(I-\Delta)^{-1} Q A_\lambda x$. It then follows from (2.3) that $A_\lambda x$ has a weak cluster point. By theorem 1.5 this is enough to prove convergence.

As the net is $\mathscr{S}$-ergodic $SP = P$ holds for all $S \in \mathscr{S}$. If $T_\alpha z$ is a convex combination of elements $S_i z$ with $S_i \in \mathscr{S}$, we therefore have $T_\alpha P z = Pz$. As $A_\lambda z$ belongs to $\overline{\text{co}}\,\mathscr{S} z$ for all $z$, we obtain $A_\lambda P = P$, and a similar argument shows $T^m A_\lambda P = P$. Now use $P = A_\lambda P$ and (2.3) to prove $P = (I-\Delta)^{-1} Q A_\lambda P$. The (weak) compactness of $Q$ yields the (weak) compactness of $P$. □

If $\mathscr{S}$ is an arbitrary bounded semigroup such that $\|\cdot\|$-$\overline{\text{co}}\,\mathscr{S}$ contains a quasi-compact $T$, an application of theorem 2.5 to $A_n(T)$ shows that $F(T)$ and hence also the fixed space of $\mathscr{S}$ is finite dimensional: We have $F(T) = \{x: Px = x\} = P\mathfrak{X}$ and $\{x \in F(T): \|x\| \leqq 1\} = P\{x: \|x\| \leqq 1\}$ is compact.

**Theorem 2.6.** *Let $\{A_\lambda: \lambda \in \Lambda\}$ be a uniformly $\mathscr{S}$-ergodic net for a semigroup $\mathscr{S}$. If $\|\cdot\|$-$\overline{\text{co}}\,\mathscr{S}$ contains a quasi-compact $T$, then $\|A_\lambda - P\|$ tends to 0.*

*Proof.* With the notation used above $Q$ is now compact. Using $T^m = \Delta + Q$ and $(I-\Delta)^{-1} = \sum_0^{\infty} \Delta^k$ it is simple to verify

(2.4) $$A_\lambda = A_\lambda Q (I-\Delta)^{-1} + (A_\lambda - A_\lambda T^m)(I-\Delta)^{-1}.$$

As the net is uniformly $\mathscr{S}$-ergodic $\|A_\lambda - A_\lambda S\| \to 0$ holds for all $S \in \mathscr{S}$ and hence also for $S = T^m \in \|\cdot\|\text{-}\overline{\text{co}}\,\mathscr{S}$. It is therefore sufficient to show that $A_\lambda Q(I-\Delta)^{-1}$ converges uniformly, or that $A_\lambda$ converges uniformly on the compact set $C := \text{cl}(Q(I-\Delta)^{-1}\{x: \|x\| < 1\})$. Let $K$ be a bound for all $\|A_\lambda\|$ and for $\|P\|$. For any $\varepsilon > 0$ there exist finitely many $c_1, \ldots, c_k \in C$ such that any $z \in C$ has distance $\leqq \varepsilon/K$ from some $c(z) \in \{c_1, \ldots, c_k\}$.

As $\|A_\lambda x - Px\| \to 0$ holds for each $x$ by theorem 2.5, we have $\|A_\lambda c_i - Pc_i\| < \varepsilon$ for $i = 1, \ldots, k$ for large enough $\lambda$. Then $\|A_\lambda z - Pz\| \leqq \|A_\lambda c(z)\| + \|A_\lambda c(z) - Pc(z)\| + \|Pc(z) - Pz\| < 3\varepsilon$ holds for all $z \in C$. □

Theorem 2.5 and 2.6 are Eberlein's [1949] generalizations of results of Yosida-Kakutani [1941], who treated the case $A_\lambda = A_n(T)$. Yosida and Kakutani proved even a structure theorem for power bounded quasi-compact $T$ (theorem 2.8 below). Its proof depends on some spectral theory.

We refer to Dunford-Schwartz [1958] for an introduction to spectral theory, but we summarize the main definitions here for the convenience of the reader:

Let $T$ be a bounded linear operator in a complex Banach space $\mathfrak{X}$. The *resolvent set* $\varrho(T)$ is the set of complex numbers $\lambda$ for which the *resolvent* $R(\lambda, T) = (\lambda I - T)^{-1}$ exists as a bounded linear operator with domain $\mathfrak{X}$. The function $\lambda \to R(\lambda, T)$ defined on $\varrho(T)$ is holomorphic. (The theory of vector valued holomorphic functions is analogous to that of complex valued holomorphic functions). The *spectrum* $\sigma(T) := \mathbb{C} \setminus \varrho(T)$ is a non empty compact subset of $\mathbb{C}$. $r(T) := \sup\{|\lambda|: \lambda \in \sigma(T)\}$ is called the *spectral radius* of $T$. It may be computed from $r(T) = \lim \|T^n\|^{1/n}$. For $|\lambda| > r(T)$ we have $R(\lambda, T) = \sum_{n=0}^{\infty} T^n/\lambda^{n+1}$.

$\lambda_0$ is called a *pole of order n*, $(n \geqq 1)$, if $R(\lambda, T)$ has a Laurent expansion

$$R(\lambda, T) = \sum_{k=-n}^{\infty} B_k(\lambda - \lambda_0)^k, \quad (B_{-n} \neq 0)$$

near $\lambda_0$. $B_{-1}$ is the *residue* of $R(\lambda, T)$ at $\lambda_0$.

We can now continue with our study of uniform ergodicity:

**Theorem 2.7.** *A power bounded linear operator $T$ in a Banach space $\mathfrak{X}$ is uniformly ergodic iff either* 1 *belongs to the resolvent set $\varrho(T)$ or* 1 *is a pole of first order of the resolvent $R(\lambda, T)$.*

*Proof.* If $T$ is uniformly ergodic $N = (I - T)\mathfrak{X}$ is closed and $\mathfrak{X} = F \oplus N$ by theorem 2.1. If $S$ is the restriction of $T$ to $N$, and $P$ the projection of $\mathfrak{X}$ onto $F$, we can verify

$$(2.4) \qquad R(\lambda, T) = [(\lambda - 1)^{-1} P + R(\lambda, S)(I - P)] \quad \text{for } 1 \neq \lambda \in \varrho(S)$$

by showing that $(\lambda I - T)[\ldots]x = x$ holds both for $x \in F$ and for $x \in N$. We know from the proof of theorem 2.1 that $I - S$ is invertible. Thus 1 belongs to

$\varrho(S)$, and $R(\lambda, S)$ is holomorphic in a neighborhood of 1. In the case $F \neq \{0\}$, the projection $P$ is non degenerate and 1 is a pole of first order of $R(\lambda, T)$. In the case $F = \{0\}$ we have $\mathfrak{X} = N$ and $T = S$ and 1 belongs to $\varrho(S) = \varrho(T)$.

Conversely, the uniform ergodicity of $T$ follows in the case $1 \in \varrho(T)$ because then the invertibility of $I - T$ shows that $(I - T)\mathfrak{X}$ is $\mathfrak{X}$, and hence closed. Finally, if 1 is a pole of first order of $R(\lambda, T)$, we have an expansion $R(\lambda, T) = \sum_{n=-1}^{\infty} B_n(\lambda - 1)^n$ for $1 \neq \lambda$ in some neighborhood of 1. Spectral theory asserts that $B_{-1}$ is a projection onto $F$. Let $S$ be the restriction of $T$ to $Y = \mathrm{cl}\, N$. For $\lambda$ with $|\lambda| > 1 \geqq r(T) \geqq r(S)$, $R(\lambda, S)$ is just the restriction of $R(\lambda, T) = \sum_{k=0}^{\infty} T^k/\lambda^{k+1}$ to $Y$. As $B_{-1}$ and $T$ commute, we have $B_{-1}Tx = TB_{-1}x = B_{-1}x$ for all $x$. Therefore the restriction of $B_{-1}$ to $Y$ vanishes and $R(\lambda, S)$ has the form $\sum_{n=0}^{\infty} B_n(\lambda - 1)^n$ for $\lambda$ with $|\lambda| > 1$ in a neighborhood of 1. This implies that 1 belongs to the resolvent set of $S$, and $Y = (I - S)Y$. Thus $Y$ is contained in $(I - T)\mathfrak{X} = N$, and $N = Y$ is closed. By theorem 2.1 $T$ is uniformly ergodic. □

If $T$ is power bounded and quasi-compact, then $\xi T$ is quasi-compact, and hence uniformly ergodic for each $\xi$ on the unit circle. It therefore follows from $R(\lambda, T) = (\lambda I - T)^{-1} = \xi(\xi\lambda I - \xi T)^{-1} = \xi R(\xi\lambda, \xi T)$ that $R(., T)$ can have only poles of first order on the unit circle, and no other singularities. Consequently, there can be only finitely many poles $\lambda_1, \dots, \lambda_k$ with $|\lambda_i| = 1$. (There may be none). $\lambda_i$ is a pole of $R(., T)$ iff 1 is a pole of $R(., \lambda_i^{-1}T)$. Set $T_i = \lambda_i^{-1}T$. The proof of theorem 2.7 shows that the residue $B_{-1}(T_i)$ in the expansion of $R(., T_i)$ in a neighborhood of 1 is the projection of $\mathfrak{X}$ onto the fixed space $F(T_i)$, and that $\lambda_i$ is a pole iff $F(T_i) = \{x\colon Tx = \lambda_i x\}$ differs from $\{0\}$, i.e., iff $\lambda_i$ is an eigenvalue of $T$.

As $T_i$ is quasi-compact, $A_n(T_i)$ converges strongly to a projection $P_i$ with fixed space $F(T_i)$ and with $T_iP_i = P_iT_i = P_i = P_i^2$. $F(T_i)$ is finite dimensional by the remark after Theorem 2.5. As $A_m(T_i)$ and $A_n(T_j)$ commute, the projections $P_i$ commute, too. From $P_iP_jx = P_jP_ix \in F(T_i) \cap F(T_j)$ we infer that $P_iP_j = 0$ holds for $i \neq j$. Set

$$S := T - \sum_{i=1}^{k} \lambda_i P_i.$$

Using the relations above we find $P_iS = SP_i = 0$ for $i = 1, \dots, k$. We can now state the uniform ergodic theorem of Yosida-Kakutani:

**Theorem 2.8.** *Let $T$ be a power bounded, quasi-compact linear operator in $\mathfrak{X}$. Then each power $T^n$ has a representation*

$$(2.5) \qquad T^n = \sum_{i=1}^{k} \lambda_i^n P_i + S^n$$

*with $P_i$, $S$ as above. There exist constants* $0 < \varrho < 1$, $M > 0$ *with* $\|S^n\| \leqq M\varrho^n$, $(n = 1, 2, \ldots)$.

*Proof.* (2.5) follows by induction from the definition of $S$. As the terms $\lambda_i^n P_i$ are norm bounded, $S$ is power bounded. Since the range $F(T_i)$ of $P_i$ is finite dimensional, $P_i$ is compact. Hence $S$ is quasi-compact. If $x$ is an eigenvector of $S$ with eigenvalue $\lambda$ having moduls 1, then $\lambda^2 x = S^2 x = TSx = T\lambda x$ follows from $P_i S = 0$. $\lambda$ must then be some $\lambda_l$ and $x \in F(T_l)$ but then $\lambda_l x = Sx = Tx - \lambda_l P_l x = \lambda_l x - \lambda_l x = 0$ yields a contradiction. Thus $S$ has no eigenvalues of modulus 1 and the entire unit circle belongs to the resolvent set of $S$. As $S$ is power bounded we have $r(S) \leqq 1$ and therefore $r(S) < 1$, which is equivalent to the existence of $M$ and $\varrho$ in the statement of the theorem. □

**Remarks.** As $N^{-1} \sum_{n=0}^{N-1} \lambda_i^n$ tends to 0 for $\lambda_i \neq 1$, the uniform convergence of $A_N(T)$ follows. (We did not use theorem 2.6). The definition of $P_i$ and $S$ above could be replaced by an application of the spectral projections corresponding to the spectral sets $\{\lambda_1\}, \ldots, \{\lambda_k\}, \sigma(T) \setminus \{\lambda_1, \ldots, \lambda_k\}$.

If $T$ is a linear operator having a representation as in theorem 2.8, $T$ clearly is quasi-compact. Looking at the proof we therefore see that a power bounded $T$ is quasi-compact iff $\{z \in \sigma(T)\colon |z| = 1\}$ containes only poles of $R(\lambda, T)$ and the corresponding eigenspaces are finite dimensional.

The study of quasi-compact operators was largely motivated by applications to Markov operators, which will be treated in the supplement on Harris processes.

## Notes

Lin [1974a], [1977] has also characterized uniform ergodicity for strongly continuous semigroups $\{T_t, t \geqq 0\}$ of bounded linear operators in $\mathfrak{X}$ with $T_0 = I$ and $\|T_t/t\| \to 0$, $(t \to \infty)$. With the notation of the previous section the following conditions are shown to be equivalent:

(i) There exists a bounded operator $P$ with $\|A_{0,t} - P\| \to 0$, $(t \to \infty)$;
(i) the infinitesimal generator $A$ has closed range;
(iii) $S_{0,1} = \int_0^1 T_t dt$ is uniformly ergodic;
(iv) there exists a projection $P$ on $\{x\colon T_t x = x \forall t > 0\}$ with $\|\lambda R_\lambda - P\| \to 0$ for $\lambda \to 0$;
(v) $(\lambda R_\lambda)^n$ converges uniformly for some (every) $\lambda > 0$.

This generalizes results of Hille and Phillips [1957].

Dunford-Schwartz [1958] proved a uniform ergodic theorem for quasi-compact $T$ with wo-lim $T^n/n = 0$.

Brunel and Revuz [1974a] proved that any bounded linear operator is quasi-compact iff it is the sum of an operator $S$ with $r(S) < 1$ and an operator $V$ with finite dimensional range.

An interesting class of quasi-compact operators in spaces of Lipschitz functions has been introduced by C.T. Ionescu Tulcea and G. Marinescu [1950]. They consider a Banach space $\mathfrak{X}$ which is a subset of a Banach space $\mathfrak{Y}$. The norm $|||\cdot|||$ in $\mathfrak{X}$ may differ from the norm $\|\cdot\|$ in $\mathfrak{Y}$, but it is assumed that $x_n \in \mathfrak{X}$, $|||x_n||| \leqq K$, $\|x_n - x\| \to 0$ implies $x \in \mathfrak{X}$ and $|||x||| \leqq K$. It is proved that any bounded linear operator $T$ *in* $\mathfrak{X}$ with the following three properties is power bounded and quasi-compact: (i) $\sup\{\|T^n x\|: n \in \mathbb{N}, x \in \mathfrak{X}, \|x\| \leqq 1\} < \infty$; (ii) there exists $R > 0$ and $0 < r < 1$ such that $|||Tx||| \leqq r|||x||| + R\|x\|$ holds for all $x \in \mathfrak{X}$, (iii) the image under $T$ of any $|||\cdot|||$-bounded subset of $\mathfrak{X}$ is conditionally compact in $\mathfrak{Y}$.

This axiomatic formulation is fitted for the following situation: $\mathfrak{Y}$ is the space of continuous functions on a compact metric space $(C, \varrho)$ and $\|y\| = \sup\{|y(c)|: c \in C\}$. $\mathfrak{X}$ is the subspace of functions $x$, satisfying, for some $0 < d \leqq 1$, a Lipschitz-condition $L(x) = \sup\{|x(t) - x(s)|\,\varrho(t,s)^{-d}: t \neq s\} < \infty$, and $|||x|||$ is given by $\|x\| + L(x)$. We assume that $(W, \mathscr{W})$ is a measurable space, and $P$: $(c, A) \to P(c, A)$ is a given map $C \times \mathscr{W} \to \mathbb{R}$, which is Borel measurable for any fixed $A$, and a signed measure of total variation $\leqq 1$ for any fixed $c$. If $\tau$: $C \times W \to C$ is a map with $\varrho(\tau(c_1, w), \tau(c_2, w)) \leqq r\varrho(c_1, c_2)$, which is Borel measurable for any fixed $c$, we can define $T$ by

$$(Tx)(c) = \int_W x(\tau(c, w))\, P(c, dw).$$

Such operators have originally been introduced for the study of chains with complete connections, but there are now also other applications; see e.g. Hofbauer-Keller [1982].

**Positive operators.** We now consider positive (necessarily bounded) linear operators in the complexification $\mathfrak{X}$ of a Banach lattice. For example $\mathfrak{X}$ may be a complex $L_p$-space. Karlin [1959] proved that $T$ is uniformly ergodic iff $r(T) \leqq 1$ holds and 1 is a pole of $R(\lambda, T)$ of order $\leqq 1$, (i.e., $1 \in \varrho(T)$ or 1 is a pole of order 1). Another equivalent condition is: $r(T) \leqq 1$ holds and $(\lambda - 1) R(\lambda, T)$ converges uniformly for $\lambda \downarrow 1$; (cf. Lotz [1981]).

The spectral theory of positive operators has profound implications for the relation of uniform ergodicity and quasi-compactness. The key result is a theorem of Niiro-Sawashima [1966] in the form of Lotz and Schaefer [1968]: If $r(T) = 1$ is a pole of $R(\lambda, T)$ and the range of the residue $P$ of $R(\lambda, T)$ at $\lambda = 1$ is of finite dimension $k$, then $\{\lambda \in \sigma(T): |\lambda| = 1\}$ consists entirely of poles of $R(\lambda, T)$. We refer to Schaefer [1974: 331] for the proof. Lin [1978] showed for the same class of operators that the range of the residue in each pole on the unit circle is of dimension $\leqq k$. It then follows that for positive $T$ the following conditions are equivalent:

(a) $T$ is uniformly ergodic and $F(T)$ is finite dimensional;

(b) $T$ is quasi-compact and $\{(\lambda - 1) R(\lambda, T): \lambda > 1\}$ is uniformly bounded;

(c) $T$ is quasi-compact and $n^{-1} T^n$ converges to 0 in the weak operator topology.

[The positivity of $T$ is crucial: There exists a uniformly ergodic non positive contraction $T$ in $\ell_2$ with $\dim(F(T)) < \infty$, which is not quasi-compact: If $e_1, e_2, \ldots$ is an ON-basis and $\lambda_1, \lambda_2, \ldots$ a sequence on the unit circle converging to $\lambda \neq 1$, then $T$ can be given by $Te_k = \lambda_k e_k$.]

Lotz [1981] has explored conditions for uniform ergodicity of positive linear operators $T$ in $C(X)$ with $T\mathbb{1} = \mathbb{1}$ (Markov operators), where $X$ is a compact Hausdorff space. In the following theorem he gives sufficient conditions for quasi-compactness which are also necessary when $T$ is irreducible.

**Theorem 2.9.** *A Markov operator $T$ on $C(X)$ is quasi-compact if the following conditions* (i)–(iii) *are satisfied:*

*(i) $T^*$ is mean ergodic;*

*(ii) $F(T^*)$ has a weak order unit, i.e., there exists a $\mu \in F(T^*)$ with $\nu \ll \mu$ for all $\nu \in F(T^*)$;*

*(iii) every probability $\mu \in F(T^*)$ has non meager support.*

This has several interesting consequences. E.g. using the representation of $L_\infty$ as a space $C(X)$ one can deduce that every positive irreducible contraction $T$ in $L_\infty$ is uniformly ergodic. ($T$ is called irreducible if there exists no closed $T$-invariant ideal. An ideal is a subspace $I$ with $f \in I$, $|g| \leqq |f| \Rightarrow g \in I$. Ando [1968] had proved that such $T$ are mean ergodic).

Apparently the *peripheral spectrum* $\sigma(T) \cap \{\lambda \in \mathbb{C}: |\lambda| = r(T)\}$ is particularly important for questions of uniform ergodicity. For this we refer to Schaefer [1974]. Lotz [1968] proved that positive operators $T$ in a Banach lattice have cyclic peripheral spectrum if they satisfy a growth condition, satisfied, e.g., when $(\lambda - r(T))\, R(\lambda, T)$ is uniformly bounded for $\lambda > r(T)$.

Uniform ergodic theory for positive operators in Banach lattices is also studied in the thesis of Axmann [1980] and in the book of Nagel-Palm-Derndinger [1984].

Deshpande and Padhye [1979] showed: If $\{T_t, t \geqq 0\}$ is a semigroup of self-adjoint operators such that $T_1$ is compact and $1 \in \sigma(T_1)$, then $T_t$ converges in norm for $t \to \infty$.

We also mention that Istratescu [1974], [1977] has considered generalizations of the class of quasi-compact operators using measures of non-compactness. R. O'Brien [1978] has an application of quasi-compact operators to control theory.

Deshpande and Padhye [1977] have proved ergodic theorems for quasi-compact operators in barrelled spaces.

## § 2.3 Weak mixing, continuous spectrum and multiple recurrence

We first prove the spectral mixing theorem for contractions $T$ in a complex Hilbert space $\mathfrak{X}$: The vectors $x$ in the orthogonal complement $\mathfrak{X}_{fl}$ of the closed space $\mathfrak{X}_{uds}$ spanned by the eigenvectors belonging to eigenvalues $\lambda$ with modulus $|\lambda| = 1$ are exactly those which satisfy $n^{-1} \sum_{i=0}^{n-1} |\langle T^i x, y\rangle| = 0$ for all $y \in \mathfrak{X}$. The results for general $T$ will be applied to the isometries arising from endomorphisms in a probability space. We obtain the principal results about weak mixing. Finally, we prove Furstenberg's ergodic theorem for generic measures and give a sample of an application.

The splitting of $\mathfrak{X}$ into a direct sum of a space $\mathfrak{X}_{uds}$ on which $T$ has "unimodular discrete spectrum" and a space $\mathfrak{X}_{fl}$ of "flight vectors" will be extended to more general Banach spaces and to more general semigroups in the next section. But here the tools are simpler. Actually, further obvious simplification is possible for the case of isometries, which is sufficient for the study of endomorphisms.

A sequence $(a_n)_{n \in \mathbb{Z}}$ of complex numbers is called *non negative definite* if $\sum_l \sum_m z_l \bar{z}_m a_{l-m} \geqq 0$ holds for every finite sequence $(z_l: |l| \leqq n)$ of complex numbers. Here $\bar{z}$ is the complex conjugate of $z$. Put

$$T_n = T^n, \ (n \geqq 0), \quad \text{and} \quad T_n = T^{*|n|}, \ (n < 0).$$

**Proposition 3.1.** *For any contraction $T$ in a Hilbert space $\mathfrak{X}$, and for any $x \in \mathfrak{X}$, the* *sequence* $\langle T_n x, x\rangle$ *is non negative definite. There exists a unique finite measure* $v_x$ *on* $\mathbb{R}/\mathbb{Z} = [0,1[$, *the spectral measure of* $x$, *with*

$$\langle T_n x, x\rangle = \int \exp(2\pi \operatorname{int}) v_x(dt), \quad (n \in \mathbb{Z}).$$

*Proof.* The second assertion follows from the first and a well known theorem of Herglotz [1911], sometimes called Bochner's theorem. The first assertion is simple when $T$ is an isometry:

$$\begin{aligned}\sum_{l,m=-n}^{n} z_l \bar{z}_m \langle T_{l-m} x, x\rangle &= \sum_l \sum_m z_l \bar{z}_m \langle T_{l+n} x, T_{m+n} x\rangle \\ &= \|\sum_l z_l T_{l+n} x\|^2 \geqq 0.\end{aligned}$$

As the first identity in this calculation may fail for contractions, the proof in the general case involves more effort. For $0 < r < 1$ and $t \in [0,1[$ put

$$U(r, t) = \sum_{k \geqq 0} r^k \exp(2\pi ikt) T^k$$

and

$$V(r, t) = \sum_{k \in \mathbb{Z}} r^{|k|} \exp(2\pi ikt) T_k = -I + U(r, t) + U(r, t)^*.$$

With $y = U(r, t)x$, $\|y - x\| \leqq \|y\|$ follows from $y - x = r\exp(2\pi it) Ty$, and this yields

$$\begin{aligned}\langle V(r, t)x, x\rangle &= -\langle x, x\rangle + \langle y, x\rangle + \langle x, y\rangle \\ &= \langle y, y\rangle - \langle y - x, y - x\rangle \geqq 0.\end{aligned}$$

For complex $(z_l\colon |l| \leqq n)$ and $0 < r < 1$ we then obtain

$$\begin{aligned}&\sum_{l,m} z_l \bar{z}_m r^{|l-m|} \langle T_{l-m} x, x\rangle \\ &\quad = \sum_{l,m} \sum_k z_l \bar{z}_m r^{|k|} \langle T_k x, x\rangle \int_0^1 \exp(2\pi i(l - m - k)t)\, dt \\ &\quad = \int \sum_{l,m} z_l \bar{z}_m \exp(2\pi i(l - m)t) \langle V(r, t)x, x\rangle\, dt \\ &\quad = \int |\sum_l z_l \exp(2\pi ilt)|^2 \langle V(r, t)x, x\rangle\, dt \geqq 0.\end{aligned}$$

When $r$ tends to 1 the desired inequality follows. □

The *upper density* $D^*(M)$ of a subset $M \subset \mathbb{N} \cup \{0\}$ is defined by

$$D^*(M) = \limsup_{n \to \infty} n^{-1} \operatorname{card}(M \cap [1, n])$$

and the *lower density* $D_*(M)$ is the corresponding lim inf. We say that $M$ has *density* $D(M)$ if $D^*(M) = D_*(M) = D(M)$. A sequence $(a_n)_{n \geqq 0}$ is said to *converge in density* to a, if there exists a subset $M \subset \mathbb{N}$ with $D(M) = 0$ and

$\lim_{n\to\infty, n\notin M} a_n = a$. Equivalently, for any $\varepsilon > 0$ the set of integers $n$ with $|a_n - a| > \varepsilon$ has density 0, We then write $D$-lim $a_n = a$. It is an exercise to prove that for *bounded* sequences $(a_n)$ one has

$$\text{(3.1)} \qquad D\text{-lim}\, a_n = a \Leftrightarrow \lim n^{-1} \sum_{k=0}^{n-1} |a_n - a| = 0\,.$$

**Lemma 3.2** (Wiener). *Let* $\nu$ *be a finite measure on* $[0,1[$ *and* $\hat{\nu}(n) = \int \exp(2\pi \text{int})\,\nu(dt)$ *its Fourier transform. Then* $\hat{\nu}(n)$ *converges in density to* 0 *iff* $\nu$ *has no atoms,* i.e., $\nu(\{t\}) = 0$ *for all* $t$.

*Proof.* $\hat{\nu}(n)$ converges in density to 0 iff $\hat{\nu}(n)^2$ converges in density to 0. We have

$$\begin{aligned} n^{-1} \sum_{k=0}^{n-1} |\hat{\nu}(k)|^2 &= n^{-1} \sum_{k=0}^{n-1} \int \exp(2\pi ikt)\,\nu(dt) \int \exp(-2\pi iks)\,\nu(ds) \\ &= \int [n^{-1} \sum_{k=0}^{n-1} \exp(2\pi ik(t-s))]\,\nu(dt)\,\nu(ds)\,. \end{aligned}$$

The last term tends to

$$\int_{\{t=s\}} \nu(dt)\,\nu(ds) = \sum_{0 \leqq t < 1} \nu(\{t\})^2$$

as $n \to \infty$ because the integrands $[\ldots]$ are everywhere bounded by 1 and converge to $1_{\{t=s\}}$. □

**Lemma 3.3** (Akcoglu-Sucheston). *If* $S$ *is a contraction in a Hilbert space* $\mathfrak{X}$, *then* $\|x\|^2 - \|Sx\|^2 < \varepsilon^2$ *implies that for each* $y \in \mathfrak{X}$

$$\text{(3.2)} \qquad |\langle x, y\rangle - \langle Sx, Sy\rangle| \leqq \varepsilon \|y\|\,.$$

*Proof.* We have

$$|\langle x, y\rangle - \langle Sx, Sy\rangle| = |\langle x, y\rangle - \langle S^* Sx, y\rangle| \leqq \|x - S^* Sx\|\,\|y\|.$$

Therefore (3.2) follows from $\|x - S^* Sx\|^2 = \|x\|^2 - 2\|Sx\|^2 + \|S^* Sx\|^2 \leqq \|x\|^2 - \|Sx\|^2 \leqq \varepsilon^2$. □

**Theorem 3.4** (Spectral mixing theorem of Koopman-von Neumann [1932]). *Let* $T$ *be a contraction in a Hilbert space* $\mathfrak{X}$. *Then the following statements about an element* $x \in \mathfrak{X}$ *are equivalent:*

*(a)* $x \in \mathfrak{X}_{fl}$;
*(b)* $\nu_x$ *has no atoms;*
*(c)* D-lim $\langle T^n x, x\rangle = 0$;
*(d)* D-lim $\langle T^n x, y\rangle = 0$ for all $y \in \mathfrak{X}$.

*Proof.* (a) ⇒ (b): Let $x \in \mathfrak{X}_{fl}$ and assume $\nu_x(\{t\}) > 0$ for some $t$.

Applying von Neumann's mean ergodic theorem to the contraction $\exp(-2\pi it)T$ we find that the averages $A_n(\exp(-2\pi it)T)x$ converge to an element $y \in \mathfrak{X}$ with $\exp(-2\pi it)Ty = y$. We have

$$\begin{aligned}\langle y, x\rangle &= \lim_{n\to\infty} n^{-1}\sum_{k=0}^{n-1} \exp(-2\pi itk)\langle T^k x, x\rangle \\ &= \lim_{n\to\infty} \int n^{-1}\sum_{k=0}^{n-1} \exp(2\pi i(s-t)k)\nu_x(ds) = \nu_x(\{t\}) > 0.\end{aligned}$$

Hence $y \neq 0$, and $y$ is an eigenvector with eigenvalue $\exp(2\pi it)$ which is not orthogonal to $x$. This contradicts our assumption.

(b) ⇒ (c) follows from Lemma 3.2.

(c) ⇒ (d) By (3.1) the set of elements $y \in \mathfrak{X}$ with D-lim $\langle T^n x, y\rangle = 0$ is a closed linear subspace of $\mathfrak{X}$. For every $y$ which is orthogonal to all $T^k x$, $(k \geqq 0)$, we even have $\langle T^n x, y\rangle \equiv 0$. It therefore remains to show D-lim $\langle T^n x, T^k x\rangle = 0$. As $\|T^n x\|$ decreases we have $\|T^n x\| - \|T^{n+k}x\| < \varepsilon^2$ for all large $n$. Lemma 3.3 then implies $|\langle T^n x, x\rangle - \langle T^{n+k}x, T^k x\rangle| < \varepsilon$.

As $\varepsilon > 0$ was arbitrary, (c) yields D-$\lim_n \langle T^{n+k}x, T^k x\rangle = 0$, which is equivalent to D-$\lim_n \langle T^n x, T^k x\rangle = 0$.

(d) ⇒ (a) If $w$ is an eigenvector of $T$ having an eigenvalue $\lambda$ with $|\lambda| = 1$, then $S = \lambda^{-1}T$ is a contraction and $w$ is fixed under $S$. (d) says that $n^{-1}\sum_{i=0}^{n-1} |\langle T^i x, y\rangle|$ tends to 0 for all $y$. Because of $|\langle S^i x, y\rangle| = |\langle T^i x, y\rangle|$ this implies the weak and therefore also the strong convergence of $A_n(S)x$ to 0. By theorem 1.3 and lemma 1.1.2, $x$ is orthogonal to $w = Sw$. As $w$ was arbitrary we obtain $x \in \mathfrak{X}_{uds}^{\perp} = \mathfrak{X}_{fl}$. □

The terminology "spectral mixing theorem" is motivated by the application to endomorphisms $\tau$ of a probability space $(\Omega, \mathscr{A}, \mu)$. $\tau$ is said to be *weakly mixing* iff

$$\text{(3.3)} \qquad \text{D-}\lim_{n\to\infty} \mu(A \cap \tau^{-n}B) = \mu(A)\mu(B)$$

holds for all $A, B \in \mathscr{A}$.

Comparing (3.3) with proposition 1.1.8 we see that weak mixing implies ergodicity. The isometry $f \to Tf = f \circ \tau$ in $\mathfrak{X} = L_2$ always has the constant function $\mathbb{1}$ as an eigenvector with eigenvalue 1.

**Theorem 3.5.** *For an endomorphism $\tau$ of a probability space $(\Omega, \mathscr{A}, \mu)$ the following conditions are equivalent:*

*(a) $\tau$ is weakly mixing;*

*(b)* D-$\lim_n \langle T^n f, g\rangle = \langle f, \mathbb{1}\rangle \cdot \langle \mathbb{1}, g\rangle$ *for all* $f, g \in L_2$;

*(c)* D-$\lim_n \langle T^n h, h\rangle = 0$ *for all* $h \in \{\mathbb{1}\}^{\perp}$;

*(d) $T$ has continuous spectrum (i.e., no eigenvectors) in $\{\mathbb{1}\}^{\perp}$.*

*Proof.* (3.3) is equivalent to the validity of (b) for $f = 1_A$ and $g = 1_B$. Thus (b) implies (a). Approximating general $f, g \in L_2$ by linear combinations of indicator functions we also obtain (a) $\Rightarrow$ (b). (b) $\Rightarrow$ (c) is trivial. As no eigenvalues of modulus less than 1 exist, the spectral mixing theorem shows that (c) implies (d), and that (d) yields D-lim $\langle T^n f', g' \rangle = 0$ for all $f', g' \in \{\mathbb{1}\}^\perp$. To prove that this implies (b), put $f' = f - \langle f, \mathbb{1} \rangle$, $g' = g - \langle g, \mathbb{1} \rangle$ and check

$$\langle T^n f, g \rangle = \langle T^n f' + \langle f, \mathbb{1} \rangle \mathbb{1}, g' + \langle g, \mathbb{1} \rangle \mathbb{1} \rangle = \langle T^n f', g' \rangle + \langle f, \mathbb{1} \rangle \langle \mathbb{1}, g \rangle. \quad \square$$

Another useful characterization of weak mixing can be given via *direct products* of endomorphisms: If $\tau_i$, $(i = 1, 2)$ are endomorphisms of probability spaces $(\Omega_i, \mathscr{A}_i, \mu_i)$ we define the direct product $\tau_1 \times \tau_2$ in the product $(\Omega_1 \times \Omega_2, \mathscr{A}_1 \otimes \mathscr{A}_2, \mu_1 \times \mu_2)$ by

$$\tau_1 \times \tau_2(\omega_1, \omega_2) = (\tau_1 \omega_1, \tau_2 \omega_2).$$

It is easy to check that $\tau_1 \times \tau_2$ is an endomorphism. The definition of $\tau_1 \times \tau_2 \times \ldots \times \tau_n$ is analogous.

**Theorem 3.6.** *(a) If $\tau_1$ is weakly mixing and $\tau_2$ ergodic, then $\tau_1 \times \tau_2$ is ergodic;*
*(b) If $\tau_1$ and $\tau_2$ are weakly mixing, then $\tau_1 \times \tau_2$ is weakly mixing;*
*(c) If $\tau \times \tau$ is ergodic, then $\tau$ is weakly mixing.*

*Proof.* (a) Put $\tilde{\tau} = \tau_1 \times \tau_2$, $\tilde{\mu} = \mu_1 \times \mu_2$. For any $A_1, B_1 \in \mathscr{A}_1$ and $A_2, B_2 \in \mathscr{A}_2$ we have

$$\tilde{\mu}((A_1 \times A_2) \cap \tilde{\tau}^{-i}(B_1 \times B_2)) = \mu_1(A_1 \cap \tau_1^{-i} B_1)\, \mu_2(A_2 \cap \tau_2^{-i} B_2).$$

The first factor on the righthand side tends to $\mu_1(A_1)\mu_1(B_1)$ except along a sequence of density 0. Taking Cesàro averages, and using

$$\mu_1(A_1)\mu_1(B_1)\,\mu_2(A_2)\,\mu_2(B_2) = \tilde{\mu}(A_1 \times A_2)\,\tilde{\mu}(B_1 \times B_2)$$

and

$$n^{-1} \sum_{i=0}^{n-1} \mu_2(A_2 \cap \tau_2^{-i} B_2) \to \mu_2(A_2)\mu_2(B_2)$$

we find that

$$n^{-1} \sum_{i=0}^{n-1} \tilde{\mu}(\tilde{A} \cap \tilde{\tau}^{-i} \tilde{B}) \to \mu(\tilde{A})\,\mu(\tilde{B}) \tag{3.4}$$

holds for sets of the type $\tilde{A} = A_1 \times A_2$, $\tilde{B} = B_1 \times B_2$. But then (3.4) remains true for finite disjoint unions $\tilde{A}, \tilde{B}$ of such "rectangles" and, by approximation, for all $\tilde{A}, \tilde{B} \in \mathscr{A}_1 \otimes \mathscr{A}_2$. Thus, $\tilde{\tau}$ is ergodic.

(b) Apply the same argument to the convergence in density of $\tilde{\mu}(\tilde{A} \cap \tilde{\tau}^{-i} \tilde{B})$ and observe that finite unions of sequences of density 0 have density 0.

(c) $\tau$ must be ergodic. Assume $f \neq 0$ is an eigenvector of $T$, then the eigenvalue has modulus 1 because $T$ is an isometry, and thus $|f|$ is $\tau$-invariant and hence constant. Now $g(\omega_1, \omega_2) = f(\omega_1)/f(\omega_2)$ satisfies $g(\tau\omega_1, \tau\omega_2) = \lambda f(\omega_1)/\lambda f(\omega_2) = g(\omega_1, \omega_2)$. As $\tau \times \tau$ is ergodic $g$ is constant on $\Omega \times \Omega$ and hence $f$ is constant on $\Omega$. As the only eigenvectors of $T$ are constants $\tau$ is weakly mixing by theorem 3.5. □

**A glimpse into Furstenberg's multiple recurrence theory.** The defining properties of mixing and weak mixing involve only *pairs* of sets. Similarly, ergodicity is characterized by the convergence of the Cesàro averages of $\mu(A \cap \tau^{-i} B)$ to $\mu(A)\,\mu(B)$ for all pairs $A$, $B$. It is not clear how this could be used for a proof of limiting results for expressions involving three or more sets. While the question whether mixing implies $\mu(A \cap \tau^{-i} B \cap \tau^{-2i} C) \to \mu(A)\,\mu(B)\,\mu(C)$ remains unsolved, the multiple recurrence theory yields

$$\text{(3.5)} \qquad \operatorname*{D-lim}_n \mu(A_0 \cap \tau_1^{-n} A_1 \ldots \cap \tau_k^{-n} A_k) = \mu(A_0) \ldots \mu(A_k)$$

for commuting weakly mixing $\tau_1, \ldots, \tau_k$. The Furstenberg-Katznelson theorem asserts

$$\text{(3.6)} \qquad \liminf_n n^{-1} \sum_{i=0}^{n-1} \mu(A_0 \cap \tau_1^{-i} A_1 \cap \ldots \cap \tau_k^{-i} A_k) > 0$$

for arbitrary commuting automorphisms $\tau_1, \ldots, \tau_k$ of a probability space and arbitrary $A_0, \ldots, A_k$ of positive measure. Furstenberg has shown that results like (3.6) are equivalent to deep theorems in combinatorial number theory. E.g. the case $\tau_i = \tau^i$ yields a theorem of Szemerédi, saying that any subset $M \subset \mathbb{N}$ with positive upper density contains arbitrarily long arithmetic progressions.

As the recent monograph of Furstenberg [1981] gives a beautiful systematic account, a brief introduction shall suffice here. We begin with Furstenberg's ergodic theorem for generic measures.

**Definition 3.7.** Let $\tau$ be an endomorphism of a probability space $(\Omega, \mathscr{A}, \mu)$. If $\mathscr{L}$ is an algebra of bounded, complex valued, $\mathscr{A}$-measurable functions which is closed with respect to complex conjugation and satisfies $g \circ \tau \in \mathscr{L}$ for all $g \in \mathscr{L}$, we call a probability measure $\nu$ on $(\Omega, \mathscr{A})$ *generic* with respect to $\mathscr{L}$ in $(\Omega, \mathscr{A}, \mu, \tau)$ when

$$\text{(3.7)} \qquad \int A_n(\tau) f\,d\nu \to \int f\,d\mu$$

holds for all $f \in \mathscr{L}$.

The measure $\nu$ need not be invariant! The terminology is motivated by an example in topological dynamics: If $\tau$ is a continuous transformation of a compact metric space $\Omega$, a point $\omega$ is called *generic* for $\mu$ if $(A_n f)(\omega)$ tends to $\int f\,d\mu$ for all $f$ in the algebra $\mathscr{L} = C(\Omega)$. Thus $\omega$ is generic iff the point mass in $\omega$ is generic.

For $f \in \mathscr{L}$ also $|f|^2$ belongs to $\mathscr{L}$ because $\mathscr{L}$ is a conjugation closed algebra. If $f$ is $\tau$-invariant, (3.7) shows

$$\int |f|^2 d\mu = \lim n^{-1} \sum_{i=0}^{n-1} \int |f|^2 \circ \tau^i d\nu = \int |f|^2 d\nu.$$

The $L_2(\mu)$-norm $\|f\|_{2,\mu}$ then agrees with $\|f\|_{2,\nu}$. If $\mathscr{L}$ contains a set of $\tau$-invariant elements of $L_2(\mu)$ which is dense in the space $L_2(\mu, \mathscr{I}_\tau)$ of $\tau$-invariant elements of $L_2(\mu)$, then this subset and therefore all of $L_2(\mu, \mathscr{I}_\tau)$ is embedded isometrically in $L_2(\nu)$. Let $P_\tau$ be the projection in $L_2(\mu)$ onto $L_2(\mu, \mathscr{I}_\tau)$.

**Theorem 3.8.** *Let $\nu$ be generic with respect to $\mathscr{L}$ in $(\Omega, \mathscr{A}, \mu, \tau)$. Assume that $\mathscr{L}$ contains a dense subset of $L_2(\mu, \mathscr{I}_\tau)$. Then $A_n f$ tends to $P_\tau f$ in the norm $\|\cdot\|_{2,\nu}$ of $L_2(\nu)$ for all $f \in \mathscr{L}$.*

*Proof.* Let $\varepsilon > 0$ be given. Let $g$ be an element of $\mathscr{L} \cap L_2(\mu, \mathscr{I}_\tau)$ with $\|P_\tau f - g\|_{2,\mu} < \varepsilon$. As the inequality is strict, von Neumann's mean ergodic theorem and the $\tau$-invariance of $g$ imply the existence of a $K$ with $\int |A_K(f-g)|^2 d\mu < \varepsilon^2$. Put $h = f - g$. By the properties of $\mathscr{L}$ the function $|A_K h|^2$ again belongs to $\mathscr{L}$. For all large $n$, (3.7) therefore yields

$$\int A_n |A_K h|^2 d\nu < \varepsilon^2.$$

Applying the Cauchy-Schwarz inequality $(n^{-1} \sum_{i=0}^{n-1} a_i)^2 \leqq n^{-1} \sum_{i=0}^{n-1} a_i^2$ to $a_i = |A_K h|(\tau^i \omega)$ we find $|A_n A_K h|^2 \leqq A_n |A_K h|^2$, and hence $\int |A_n A_K h|^2 d\nu < \varepsilon^2$. For large $n$ the function $A_n A_K h$ is close to $A_n h$: Indeed, when $c > 0$ is a bound for $h$, we have

$$|n^{-1} K^{-1} \sum_{i=0}^{n-1} \sum_{j=0}^{K-1} h \circ \tau^{i+j} - n^{-1} \sum_{l=0}^{n-1} h \circ \tau^l| \leqq 2n^{-1} c.$$

Consequently, $\int |A_n h|^2 d\nu < \varepsilon^2$ holds for large enough $n$. Now $\|A_n f - P_\tau f\|_{2,\nu} < 2\varepsilon$ follows from $\|A_n h\|_{2,\nu} = \|A_n f - g\|_{2,\nu} < \varepsilon$ and $\|P_\tau f - g\|_{2,\nu} = \|P_\tau f - g\|_{2,\mu} < \varepsilon$. □

The example of chief interest is the diagonal measure in a product space. Let $\tau$ be an endomorphism in a probability space $(\Omega, \mathscr{A}, \mu)$. Let $\sigma$ be the product $\tau \times \tau^2$ in $(\Omega^2, \mathscr{A}^2, \mu^2) = (\Omega \times \Omega, \mathscr{A} \otimes \mathscr{A}, \mu \times \mu)$. The diagonal mesure $\mu_\Delta$ on $\mathscr{A}^2$ is the unique measure with

$$\int f(\omega_1, \omega_2) d\mu_\Delta(\omega_1, \omega_2) = \int f(\omega_1, \omega_1) d\mu(\omega_1)$$

for all bounded $\mathscr{A}^2$-measurable $f$: $\Omega^2 \to \mathbb{R}$. For $f, g \in L_\infty(\mu)$ we denote by $f \otimes g$ the function on $\Omega^2$ with $(f \otimes g)\ (\omega_1, \omega_2) = f(\omega_1) g(\omega_2)$. Let $\mathscr{L} = L_\infty(\mu) \otimes L_\infty(\mu)$ be the algebra of finite sums of such functions $f \otimes g$.

**Proposition 3.9.** *If $\tau$ is ergodic, then $\mu_\Delta$ is generic for $\mathscr{L}$ in $(\Omega^2, \mathscr{A}^2, \mu^2, \sigma)$.*

*Proof.* $\mathscr{L}$ has the properties required in definition 3.7, and

$$\int n^{-1}\sum_{i=0}^{n-1}(f\otimes g)\circ\sigma^i\,d\mu_\Delta=\int n^{-1}\sum_{i=0}^{n-1}f(\tau^i\omega_1)\,g(\tau^{2i}\omega_2)\,d\mu_\Delta(\omega_1,\omega_2)$$

$$=n^{-1}\sum_{i=0}^{n-1}\int f(\tau^i\omega)\,g(\tau^{2i}\omega)\,d\mu(\omega)$$

$$=n^{-1}\sum_{i=0}^{n-1}\int f(\omega)\,g(\tau^i\omega)\,d\mu(\omega)$$

$$\to\left(\int f\,d\mu\right)\left(\int g\,d\mu\right)=\int(f\otimes g)\,d\mu^2.\quad\square$$

A sample application shall illustrate the spirit of the method: If $\tau$ is weakly mixing, $\sigma$ is ergodic, and $L_2(\mu^2,\mathscr{I}_\sigma)$ consists only of the constants. In particular, $P_\sigma(f\otimes g)=(\int f\,d\mu)(\int g\,d\mu)\mathbb{1}$.

For $f,g\in L_\infty(\mu)$, Theorem 3.8 implies the strong convergence in $L_2(\mu_\Delta)$ of $A_n(\sigma)(f\otimes g)$ to $P_\sigma(f\otimes g)$. As strong convergence implies weak convergence, this yields

$$\langle h\otimes\mathbb{1},A_n(\sigma)(f\otimes g)\rangle\to\langle h\otimes\mathbb{1},P_\sigma(f\otimes g)\rangle$$

for $h\in L_\infty(\mu)$, or more explicitly

$$n^{-1}\int\sum_{i=0}^{n-1}h(\omega)f(\tau^i\omega)\,g(\tau^{2i}\omega)\,d\mu\to\left(\int h\,d\mu\right)\left(\int f\,d\mu\right)\left(\int g\,d\mu\right).\tag{3.8}$$

We apply this result to $\tau$ and $h=1_A, f=1_B, g=1_C$, and also to the weakly mixing transformation $\tau\times\tau$ with $h=1_A\otimes 1_A$, $f=1_B\otimes 1_B$, $g=1_C\otimes 1_C$, and obtain

$$n^{-1}\sum_{i=0}^{n-1}\mu(A\cap\tau^{-i}B\cap\tau^{-2i}C)\to\mu(A)\,\mu(B)\,\mu(C)$$

and

$$n^{-1}\sum_{i=0}^{n-1}\mu^2(A\times A\cap\tau^{-i}B\times\tau^{-i}B\cap\tau^{-2i}C\times\tau^{-2i}C)\to(\mu(A)\,\mu(B)\,\mu(C))^2.$$

If we put $\alpha_i=\mu(A\cap\tau^{-i}B\cap\tau^{-2i}C)$ and $\alpha=\mu(A)\,\mu(B)\,\mu(C)$, we therefore have $n^{-1}\sum_{i=0}^{n-1}\alpha_i^2\to\alpha^2$ and $n^{-1}\sum_{i=0}^{n-1}\alpha_i\to\alpha$. Then $n^{-1}\sum_{i=0}^{n-1}(\alpha_i-\alpha)^2=n^{-1}\sum_{i=0}^{n-1}\alpha_i^2-2\sum_{i=0}^{n-1}\alpha_i\alpha+\alpha^2$ tends to 0. By (3.1) this means D-lim $(\alpha_i-\alpha)^2=0$ and hence D-lim $(\alpha_i-\alpha)=0$.

We have proved

**Theorem 3.10.** *If $\tau$ is a weakly mixing endomorphism we have*

$$\lim n^{-1}\sum_{i=0}^{n-1}|\mu(A\cap\tau^{-i}B\cap\tau^{-2i}C)-\mu(A)\,\mu(B)\,\mu(C)|=0$$

*for all $A,B,C\in\mathscr{A}$.*

(As even (3.5) can be handled without the generic ergodic theorem, its main use is the deeper ergodic case).

## Notes

If $T$ is a contraction in a Hilbert space $\mathfrak{X}$, then $\mathfrak{X}$ also decomposes into the direct sum of $\mathfrak{X}_d$, the closure of the space spanned by *all* eigenvectors, and its orthogonal complement $\mathfrak{X}_c$. $T$ is said to have *discrete spectrum* if $\mathfrak{X} = \mathfrak{X}_d$ and *continuous spectrum* if $\mathfrak{X} = \mathfrak{X}_c$. For isometries we clearly have $\mathfrak{X}_d = \mathfrak{X}_{uds}$ and $\mathfrak{X}_c = \mathfrak{X}_{fl}$. These subspaces have also been characterized geometrically: Call $0 \neq f \in \mathfrak{X}$ a *weakly wandering vector*, if there exists a sequence $k_1 < k_2 < \dots$ of integers such that the vectors $T^{k_i} f$ are orthogonal to each other. It is simple to see that for any isometry $T$ the weakly wandering vectors are orthogonal to $\mathfrak{X}_d$. Krengel [1972] has proved that they are dense in $\mathfrak{X}_c$. $\mathfrak{X}_c$ therefore is the closed subspace spanned by the weakly wandering vectors. An alternative (unpublished) proof has been given by A. Bellow, and her method has been extended to the case of compactly generated, Abelian, locally compact groups of unitary operators by V. Graham [1974].

Several further characterizations are available for weak mixing: England and Martin [1968] proved that an automorphism $\tau$ of a probability space is weakly mixing iff, for any $A, B$ of positive measure, the set of integers $n \geqq 0$ with $\mu(A \cap \tau^{-n} B) > 0$ has density 1. Related results for Markov processes appear in Falkowitz [1973a]. Call a sequence $(B_k)$ *remotely trivial* if the intersection of the $\sigma$-algebras $\mathscr{B}_n$ generated by $\{B_k, k \geqq n\}$ contains only nullsets and their complements. Krengel [1972] proved that $\tau$ is weakly mixing iff every sequence $(\tau^{-k} A)$, $(A \in \mathscr{A})$, contains a remotely trivial subsequence, and that the operator $f \to f \circ \tau$ has discrete spectrum iff there exists no $A$ with $0 < \mu(A) < 1$ for which $(\tau^{-k} A)$ has a remotely trivial subsequence. (This is closely related to L. Sucheston's [1963] characterization of mixing: A sequence $(A_k, k \geqq 0)$ is mixing (i.e., it satisfies $\lim_n (\mu(A_n \cap A_k) - \mu(A_n)\mu(A_k)) = 0$ for all $k \geqq 0$) iff every subsequence contains a further subsequence which is remotely trivial; see also Rényi [1963] and Jones [1972a]). A step in Krengel's proof which has turned out to be useful is:

**Theorem 3.11.** *If $\tau$ is an endomorphism in a probability space, and $\mathscr{A}_{uds}$ the smallest $\sigma$-algebra in which the eigenvectors of the operator $Tf = f \circ \tau$ in $\mathfrak{X} = L_2$ are measurable, then $\mathfrak{X}_{uds}$ consists exactly of the $\mathscr{A}_{uds}$-measurable elements of $L_2$, and $\tau^{-1}\mathscr{A}_{uds} = \mathscr{A}_{uds}$.*

Call $A$ *weakly independent* if there exists a subsequence $k_1 < k_2 < \dots$ for which the sets $\tau^{-k_i} A$ are independent. It follows from the results of Krengel [1972] together with Furstenberg's theorem 3.10 above that $\tau$ is weakly mixing iff the family of weakly independent sets is dense in $\mathscr{A}$, (and also iff the family of weakly independent partitions is dense in the family of all finite partitions of $\Omega$). For a refinement of this, see Friedman [1979].

If the $\sigma$-algebra $\mathscr{A}$ is countably generated and $\tau$ weakly mixing there exists a fixed sequence $M \subset \mathbb{N}$ of density 1 such that $\lim_{n \in M} \mu(A \cap \tau^{-n} B) = \mu(A)\mu(B)$ holds for all $A, B \in \mathscr{A}$. This has been observed by Jones [1972].

(*Proof.* Let $A_1, A_2, \dots$ be dense in $\mathscr{A}$. Find $M_k$ of density 1 such that the limits exist along $M_k$ for $A, B \in \{A_1, A_2, \dots, A_k\}$. If $n_1 < n_2 < \dots$ increases fast enough the union $M$ of the sets $[n_{k-1}, n_k[ \cap (M_1 \cap \dots \cap M_k)$ has density 1 and $M_k$ contains $\{m \in M : m \geqq n_k\}$.).

The theory of weakly mixing flows largely parallels the theory for a single $T$, see Hopf [1937] or Cornfeld-Fomin-Sinai [1982].

Let us mention some relatives of the notion of weak mixing: Furstenberg and

Weiss [1977] call $\tau$ *mildly mixing* if every $f \in L_2$, for which there exists $n_1 < n_2 < \dots$ with $\|f \circ \tau^{n_k} - f\|_2 \to 0$, is constant. They call such $f$ *rigid*. As every eigenfunction is rigid, mild mixing implies weak mixing. $\tau$ is mildly mixing iff $\tau \times \tau'$ is ergodic for all ergodic automorphisms $\tau'$ of $\sigma$-finite measure spaces. The space spanned by the rigid functions $f$ was first considered by Foguel [1964]. A notion slightly stronger than mild mixing was defined by Friedman and Ornstein [1971]. They called $\tau$ *partially mixing* if there exists $\beta > 0$ such that $A, B \in \mathscr{A}$ implies $\mu(A \cap \tau^{-n} B) \geqq \beta \mu(A) \mu(B)$; see also Friedman [1984] and Aaronson-Lin-Weiss [1979].

Beck [1960] generalized the notion of eigenvalue: A unitary operator $V$ on a Hilbert space $\mathscr{H}$ is called *eigenoperator* for $\tau$ if there is a measurable $f: \Omega \to \mathscr{H}$ with $V(f(\omega)) = f(\tau\omega)$ $\mu$-a.e.. Flytzanis [1980], [1978] has studied rigidity and related concepts in this context.

Natarajan and Viswanath [1967] called $\tau$ *weakly stable* if D-lim $\mu(A \cap \tau^{-n} B)$ exists for all $A, B$. It is shown that this holds iff 1 is the only eigenvalue. In the ergodic case weak stability and weak mixing are equivalent. Kallenberg [1980] has investigated the analogous concept for flows.

A comprehensive treatment of the recurrence theory initiated by Furstenberg [1977] can be found in Furstenberg [1981], Furstenberg-Katznelson [1978], and Furstenberg-Weiss [1978]. An introduction has been given by Furstenberg [1981a]. Furstenberg-Katznelson-Ornstein [1982] have given an exposition of a streamlined proof of the "ergodic Szemerédi theorem" (statement (3.6) for $\tau_i = \tau^i$). Lesigne [1981] has proved convergence for averages of type (3.6) in some cases. He treats the convergence in the $L_1$-mean of $n^{-1} \sum_{k=0}^{n-1} g(\tau_1^k \omega) f(\tau_2^k \omega)$ for $f, g \in L_2$, where the $\tau_i$ are commuting homeomorphisms of a compact metric space. For totally ergodic $\tau$ he also shows $L_p$-convergence of $n^{-1} \sum_{k=0}^{n-1} (f \circ \tau^k)(g \circ \tau^{2k})(h \circ \tau^{3k})$ with $f, g, h \in L_\infty$.

## § 2.4 The splitting theorem of Jacobs-Deleeuw-Glicksberg

A semigroup $\mathscr{S}$ of continuous linear operators in a Banach space $\mathfrak{X}$ is called *weakly almost periodic* if for any $x \in \mathfrak{X}$ the orbit $\mathscr{S}x = \{Tx: T \in \mathscr{S}\}$ is conditionally weakly compact. In this section we show that for Abelian weakly almost periodic $\mathscr{S}$ the space $\mathfrak{X}$ splits into a direct sum of the closed space $\mathfrak{X}_{uds}$ spanned by the eigenvectors with unimodular eigenvalues and the space $\mathfrak{X}_{fl}$ of "flight vectors", having 0 in their weak orbit closure. We discuss the non Abelian case in the Notes.

We begin with some auxiliary results. If $S$ is an abstract semigroup endowed with some topology we say that multiplication in $S$ is *separately continuous* if for each $s \in S$ the maps $t \to st$ and $t \to ts$ are continuous. The multiplication is said to be *jointly continuous* if the map $(s, t) \to st$ of $S \times S$ into $S$ is continuous. We always assume that $S$ contains a *unit* $e$, i.e., an element that satisfies $et = te = t$ for all $t \in S$. A *semitopological (topological)* semigroup is a semigroup with unit which is a Hausdorff space in which multiplication is separately (jointly) continuous. A sub-semigroup $J \subset S$ is called a *left ideal* if $SJ = \{s \cdot t: s \in S, t \in J\}$ is contained in $J$, a *right ideal* if $JS \subset J$, and an *ideal* if it is a left and right ideal. A *minimal* left ideal is a left ideal containing no other left ideal of $S$.

**Theorem 4.1.** *Any compact Abelian semitopological semigroup $S$ contains a unique minimal ideal $K$, the kernel of $S$. $K$ is contained in any ideal, and*

$$(4.1) \qquad K = \bigcap_{t \in S} tS.$$

*$K$ is a group. If $q$ is the unit of this group, $K = qS$.*

*Proof.* If $J_1, \ldots, J_k$ are ideals, their product $J_1 \cdot J_2 \cdot \ldots J_k$ is non empty, and is equal to their intersection. Consequently, the family of closed ideals of $S$ has the finite intersection property. As $S$ is compact the intersection $K$ of all closed ideals is non empty.

If $J$ is any ideal, and $t \in J$, then $tS$ is a closed ideal contained in $J$. Hence, $K$ is the intersection of *all* ideals. As each $tS$ is an ideal and each ideal contains an ideal of the form $tS$, (4.1) follows. Since $K$ is the intersection of a family of ideals, $K$ is an ideal. Clearly K is minimal and unique.

For any $s \in K$, $sK$ is an ideal contained in $K$ and, therefore, equal to $K$ by the minimality of $K$. Therefore there exists $q \in K$ with $sq = s$. By $Ks = sK = K$ there exists, for any $t \in K$, an $r \in K$ with $rs = t$. Hence $tq = rsq = rs = t$ shows that $tq = t$ holds for all $t \in K$, and $q$ is a unit. Finally. $sK = K$ and $q \in K$ implies the existence of an inverse of $s$ in $K$. The last assertion follows from $qs \subset KS \subset K$ and (4.1). □

Let $\bar{\mathscr{S}} = \text{wo-cl}(\mathscr{S})$ denote the closure of $\mathscr{S} \subset \mathscr{L}(\mathfrak{X})$ in the weak operator topology.

If $\{T_\alpha\}$ is a generalized sequence and $T_\alpha$, $R$, $T$ belong to $\mathscr{L}(\mathfrak{X})$, then wo-lim $T_\alpha = T$ implies $\langle T_\alpha R x_i, h_j\rangle \to \langle TRx_i, h_j\rangle$ and $\langle RT_\alpha x_i, h_j\rangle = \langle T_\alpha x_i, R^* h_j\rangle \to \langle Tx_i, R^* h_j\rangle = \langle RTx_i, h_j\rangle$. Therefore, multiplication in $\mathscr{L}(\mathfrak{X})$ is separately continuous with respect to the weak operator topology. If $\mathscr{S}$ is a semigroup in $\mathscr{L}(\mathfrak{X})$ we have $\mathscr{S}\mathscr{S} \subset \mathscr{S}$ and thus $\mathscr{S}\bar{\mathscr{S}} \subset \bar{\mathscr{S}}$ by separate continuity. Applying separate continuity once more and using $\text{wo-cl}(\bar{\mathscr{S}}) = \bar{\mathscr{S}}$ we find $\bar{\mathscr{S}}\bar{\mathscr{S}} \subset \bar{\mathscr{S}}$, i.e., $\bar{\mathscr{S}}$ is a semigroup. It is equally easy to check that $\bar{\mathscr{S}}$ is Abelian, if $\mathscr{S}$ is Abelian.

If $M < \infty$ is a bound for the norms of the operators in $\mathscr{S}$ we have $|\langle Tx, h\rangle| \leq M\|x\|\,\|h\|$ for all $T \in \mathscr{S}$, all $x \in \mathfrak{X}$ and all $h \in \mathfrak{X}^*$. Therefore this inequality holds also for all $T \in \bar{\mathscr{S}}$ and $M$ must be a bound for the norms of the operators in $\bar{\mathscr{S}}$.

We remark that multiplication in $\mathscr{L}(\mathfrak{X})$ in general is not jointly continuous with respect to the weak operator topology. As an example let $\mathfrak{X}$ be a Hilbert space spanned by a family $\{e_i, i \in \mathbb{Z}\}$ of orthonormal vectors. If $T$ is the unitary operator determined by $Te_i = e_{i+1}$, $(i \in \mathbb{Z})$, and $T_0$ the zero operator $T_0 x \equiv 0$ we have wo-lim $T^n = T_0$ and wo-lim $T^{-n} = T_0$, but wo-lim $T^n \cdot (T^{-n}) = I \neq T_0 T_0$.

**Lemma 4.2.** *A subset $\mathscr{S} \subset \mathscr{L}(\mathfrak{X})$ is relatively wo-compact iff $\mathscr{S}x$ is relatively weakly compact in $\mathfrak{X}$ for each $x \in \mathfrak{X}$. In this case $\mathscr{S}$ is bounded and $\bar{\mathscr{S}}x = \text{w-cl}(\mathscr{S}x)$ holds for all $x$.*

*Proof.* If $\mathscr{S}$ is relatively wo-compact, $\bar{\mathscr{S}}$ is wo-compact, and $\bar{\mathscr{S}}x$ is weakly compact as the continuous image of a compact set. Conversely, assume that $\mathscr{S}x$ is relatively weakly compact for all $x$. Then $\mathscr{S}x$ is norm bounded for all $x$, and by the uniform boundedness principle $\mathscr{S}$ is bounded. For each $x$ let $E_x$ be the weak closure of $\mathscr{S}x$ with the weak topology and $E$ the Cartesian product of the spaces $E_x$ with the product topology. As each $E_x$ is compact, $E$ is compact by Tychonoff's theorem. Now let $\{T_\alpha\}$ be an arbitrary net in $\mathscr{S}$. Then $\{(T_\alpha x)_{x \in \mathfrak{X}}\}$ may be regarded as a net in $E$ and by the compactness of $E$ there exists a subnet $\{(T_\beta x)_{x \in \mathfrak{X}}\}$ converging in $E$. This means that w-lim $T_\beta x$ exists for all $x$. Define an operator $T$ by $Tx = \text{w-lim}\, T_\beta x$. Clearly, $T$ is linear. If $M$ is a bound for the norms of the operators in $\mathscr{S}$ we have $\|T\| \leqq M$ as above. Thus $T$ belongs to $\bar{\mathscr{S}}$, and $T = \text{wo-lim}\, T_\beta$. As $\{T_\alpha\}$ was an arbitrary net we have proved that $\mathscr{S}$ is relatively wo-compact.

In the last assertion the inclusion $\supset$ follows from the weak compactness of $\bar{\mathscr{S}}x$, and the inclusion $\subset$ from the definition of the weak operator topology. □

We remark that the lemma and its proof remain valid when the weak operator topology is replaced by the strong operator topology and the weak topology in $\mathfrak{X}$ by the norm topology.

**Definition 4.3.** Let $\mathscr{S} \subset \mathscr{L}(\mathfrak{X})$ be a semigroup. A vector $x$ is called *reversible* if for any $T \in \bar{\mathscr{S}}$ there exists an $R \in \bar{\mathscr{S}}$ with $RTx = x$. A vector $x$ is called a *flight vector* if there exists an $S \in \bar{\mathscr{S}}$ with $Sx = 0$.

Let $\mathfrak{X}_{rev}$ denote the set of reversible vectors and $\mathfrak{X}_{fl}$ the set of flight vectors. Clearly, $\mathfrak{X}_{rev} \cap \mathfrak{X}_{fl} = \{0\}$. We shall only be interested in the case when $\mathscr{S}$ is weakly almost periodic. By the previous lemma this means that $\mathscr{S}$ is relatively wo-compact, and the identity $\bar{\mathscr{S}}x = \text{w-cl}(\mathscr{S}x)$ then yields

$$\mathfrak{X}_{rev} = \{x \in \mathfrak{X} \colon y \in \text{w-cl}(\mathscr{S}x) \Rightarrow x \in \text{w-cl}(\mathscr{S}y)\},$$
$$\mathfrak{X}_{fl} = \{x \in \mathfrak{X} \colon 0 \in \text{w-cl}(\mathscr{S}x)\}.$$

If $\mathscr{S}$ is an Abelian weakly almost periodic semigroup in $\mathscr{L}(\mathfrak{X})$, $\bar{\mathscr{S}}$ is an Abelian compact semitopological semigroup with the weak operator topology and we can apply theorem 4.1.

**Theorem 4.4** (Jacobs-Deleeuw-Glicksberg). *Let $\mathscr{S}$ be an Abelian weakly almost periodic semigroup in $\mathscr{L}(\mathfrak{X})$ and $Q$ the unit in the kernel $K = K(\bar{\mathscr{S}})$ of $\bar{\mathscr{S}}$. Then $\mathfrak{X}_{rev} = Q\mathfrak{X}$ and $\mathfrak{X}_{fl} = Q^{-1}(0) = (I - Q)\mathfrak{X}$. In particular, $\mathfrak{X}$ is the direct sum of the closed invariant subspaces $\mathfrak{X}_{rev}$ and $\mathfrak{X}_{fl}$. The restriction of $\bar{\mathscr{S}}$ to $\mathfrak{X}_{rev}$ is a group.*

*Proof*. If $x$ is reversible there exists $R \in \bar{\mathscr{S}}$ with $RQx = x$. Hence $QRx = x$ and $x \in Q\mathfrak{X}$. For the converse observe that for any $T \in \bar{\mathscr{S}}$ there exists an $S \in K$ with $STQ = Q$, because $K$ is a group and $TQ \in K$. If $x$ is of the form $x = Qy$ for some $y$, we therefore obtain $STx = STQy = Qy = x$. This shows that all elements of $Q\mathfrak{X}$ are reversible.

Obviously all elements of $Q^{-1}(0) = \{x: Qx = 0\}$ are flight vectors. On the other hand, if $x$ is a flight vector, there exists an $R$ with $Rx = 0$. Hence $QRx = 0$. As $QR$ belongs to $K$ there exists $S$ with $SQR = Q$. Thus, $Qx = SQRx = 0$.

$Q^{-1}(0) = (I - Q)\mathfrak{X}$ follows from the projection property $Q^2 = Q$. As any $x$ can be written in the form $x = Qx + (I - Q)x$ we find $x = x_1 + x_2$ with $x_1 \in \mathfrak{X}_{rev}$, $x_2 \in \mathfrak{X}_{fl}$. Conversely, if $x$ is of this form we have $Qx = Qx_1 + 0$ and $x_1 = Qy_1$ for some $y_1$. Hence $Qx = Q^2 y_1 = Qy_1 = x_1$. The splitting is therefore unique. The group property of $\bar{\mathscr{S}}$ in $\mathfrak{X}_{rev}$ follows from the group property of $K$. □

A vector $x \neq 0$ is called an *eigenvector* for $\mathscr{S}$ if there is a map $\lambda: \mathscr{S} \to \mathbb{C}$ with $Tx = \lambda(T)x$, $(T \in \mathscr{S})$. If $\{T_\alpha\}$ is a net and $T = \text{wo-lim}\, T_\alpha$ the numbers $\lambda(T_\alpha)$ must converge to some $\lambda(T)$. Therefore eigenvectors for $\mathscr{S}$ are eigenvectors for $\bar{\mathscr{S}}$ and conversely. $x$ is called an *eigenvector with unimodular eigenvalues* if $|\lambda(T)| = 1$ holds for all $T$. Let $\mathfrak{X}_{uds}$ be the closure of the subspace of $\mathfrak{X}$ spanned by all eigenvectors with unimodular eigenvalues. We may say that $\mathscr{S}$ has unimodular discrete spectrum in $\mathfrak{X}_{uds}$.

**Theorem 4.5** (Jacobs-Deleeuw-Glicksberg). *If $\mathscr{S}$ is an Abelian weakly almost periodic semigroup in $\mathscr{L}(\mathfrak{X})$, then $\mathfrak{X}_{rev} = \mathfrak{X}_{uds}$.*

*Proof*. It is easy to show the inclusion $\supset$: For $x \in \mathfrak{X}_{uds}$ and $T \in \bar{\mathscr{S}}$ the identity $T^n x = \lambda(T)^n x$ shows the existence of a sequence $n_i$ with $T^{n_i} x \to x$. It follows that, for any $h_1, \ldots, h_m \in \mathfrak{X}^*$ and $\varepsilon > 0$, the set $\{S \in \bar{\mathscr{S}}: |\langle STx - x, h_i\rangle| \leqq \varepsilon, i = 1, \ldots, m\}$ is non empty, and any finite intersection of such sets is non empty. By the compactness of $\bar{\mathscr{S}}$ in the weak operator topology the intersection of all sets of this type is non empty. Therefore there exists an $S \in \bar{\mathscr{S}}$ such that $|\langle STx - x, h\rangle| \leqq \varepsilon$ holds for all $\varepsilon > 0$ and all $h \in \mathfrak{X}^*$. But then $STx = x$. As $T \in \bar{\mathscr{S}}$ was arbitrary $x$ is reversible.

The proof of the opposite inclusion depends on some abstract harmonic analysis and on the theorem of Ellis which asserts that any compact semitopological group is a topological group; see Ellis [1957].

By this theorem $K = K(\bar{\mathscr{S}})$ with the weak operator topology is a compact Abelian group. Let $\Gamma$ be its character group and let $\varrho$ denote the normalized Haar measure on $K$.

For $\gamma \in \Gamma$ let $\bar{\gamma}(T)$ be the complex conjugate of $\gamma(T)$. We now describe how "weak integration" can be used to define operators $T_\gamma$ on $\mathfrak{X}$. First an element $T_\gamma x$ of $\mathfrak{X}^{**}$ can be defined by

(4.2) $$\langle T_\gamma x, h\rangle = \int_K \langle \bar\gamma(T)\, Tx, h\rangle\, \varrho(dT).$$

If we show that $T_\gamma x$ belongs to the canonical embedding $\kappa\mathfrak{X}$ of $\mathfrak{X}$ in $\mathfrak{X}^{**}$, then $T_\gamma x$ can be identified with an element of $\mathfrak{X}$, and (4.2) will be the meaning of

(4.3) $$T_\gamma x = \int_K \bar\gamma(T)\, Tx\, \varrho(dT).$$

As the map $T \to \bar\gamma(T)\, Tx$ is weakly continuous on $K$, the set $\{\bar\gamma(T)\, Tx\colon T \in K\}$ is weakly compact. By the Krein-Shmulian theorem (see Dunford-Schwartz [1958: 434]) its closed convex hull $\overline{\mathrm{co}}\,\{\bar\gamma(T)\, Tx\colon T \in K\} =: A$ is weakly compact. Hence $\kappa A$ is compact in the w*-topology of $\mathfrak{X}^{**}$. Approximating the integrals in (4.2) by Riemann sums we notice that $T_\gamma x$ belongs to the w*-closure of $\kappa A$ and therefore to $\kappa A$. It is now clear that (4.3) defines a linear operator in $\mathfrak{X}$. Its norm is bounded by $\sup\{\|T\|\colon T \in K\} < \infty$.

For $R \in K$, $x \in \mathfrak{X}$, $h \in \mathfrak{X}^*$ we find

$$\begin{aligned}\langle RT_\gamma x, h\rangle &= \langle T_\gamma x, R^* h\rangle = \int \langle \bar\gamma(T)\, Tx, R^* h\rangle\, \varrho(dT)\\ &= \int \langle \bar\gamma(T)\, RTx, h\rangle\, \varrho(dT) = \gamma(R) \int \langle \bar\gamma(RT)\, RTx, h\rangle\, \varrho(dRT)\\ &= \gamma(R)\, \langle T_\gamma x, h\rangle.\end{aligned}$$

This, and a similar computation using $RT = TR$, shows

(4.4) $$RT_\gamma = \gamma(R)\, T_\gamma = T_\gamma R, \quad (R \in K).$$

As $Q$ is the unit in $K$, $\gamma(Q) = 1$ and $QT_\gamma = T_\gamma$. Applying (4.4) to $SQ \in K$ we obtain $ST_\gamma = SQT_\gamma = \gamma(SQ)\, T_\gamma = \gamma(S)\, T_\gamma$. This means that $T_\gamma \mathfrak{X}$ consists of eigenvectors with unimodular eigenvalues $\lambda(T) = \gamma(T)$.

To finish the proof it now remains to show that $\mathfrak{X}_{rev}$ is contained in the space $U$ spanned by the spaces $T_\gamma \mathfrak{X}$, $(\gamma \in \Gamma)$. If $h \in \mathfrak{X}^*$ vanishes on $U$, then

(4.5) $$\int_K \bar\gamma(T)\, \langle Tx, h\rangle\, \varrho(dT)$$

vanishes for all $\gamma \in \Gamma$ and all $x \in \mathfrak{X}$. Assume $\langle Qx, h\rangle \neq 0$ for some $x$. As $T \to \langle Tx, h\rangle$ is continuous on $K$, this implies $\int 1_F \langle Tx, h\rangle\, \varrho(dT) \neq 0$ for some neighbourhood $F$ of $Q$. As the characters form an orthonormal basis in $L_2(K, \varrho)$ (see e.g. Dunford-Schwartz [1963: 944]) there are linear combinations of characters arbitrarily close to $1_F$. But then the integrals (4.5) cannot vanish for all $\gamma$, a contradition. We have proved $\langle Qx, h\rangle = 0$ for all $x$ and for all $h$ vanishing on U. Thus $Q\mathfrak{X} = \mathfrak{X}_{rev}$ is contained in $U$. □

**Remarks.** If $x_\gamma \neq 0$ is an eigenvector with unimodular eigenvalues $\{\lambda(T)\colon T \in \mathcal{S}\}$, then $\lambda$ is continuous on $\mathcal{S}$, and $\lambda(TS)\, x_\lambda = TSx_\lambda = T\lambda(S)\, x_\lambda = \lambda(T)\, \lambda(S)\, x_\lambda$ implies that $\lambda$ is a character. Thus the unimodular eigenvalues are exactly the characters $\gamma$ with $T_\gamma \neq 0$. For any $x \in \mathfrak{X}$, $h \in \mathfrak{X}^*$ we have

$$\langle T_\gamma^2 x, h\rangle = \int \langle \bar{\gamma}(T)\, T T_\gamma x, h\rangle\, \varrho(dT)$$
$$= \int \langle \bar{\gamma}(T)\, \gamma(T)\, T_\gamma x, h\rangle\, \varrho(dT) = \langle T_\gamma x, h\rangle .$$

Therefore $T_\gamma$ is a projection onto the eigenspace corresponding to $\gamma$.

If $\lambda$ is a unimodular eigenvalue for $\mathscr{S}$, then $T_\lambda \neq 0$ and $T_\lambda^* h_0 \neq 0$ for some $h_0$. For all $y \in \mathfrak{X}$ one has $\langle y, T^* T_\lambda^* h_0\rangle = \langle T_\lambda T y, h_0\rangle = \langle \lambda(T)\, T_\lambda y, h_0\rangle = \langle y, \bar{\lambda}(T)\, T_\lambda^* h_0\rangle$. Hence $T_\lambda^* h_0$ is an eigenvector of $\mathscr{S}^* = \{T^* \colon T \in \mathscr{S}\}$ with eigenvalue $\lambda^*$ defined by $\lambda^*(T^*) = \bar{\lambda}(T)$. Conversely, if $h$ is an eigenvector of $\mathscr{S}^*$ with unimodular eigenvalue $\lambda^*$, then $\langle y, \lambda^*(T_1^* T_2^*) h\rangle = \langle y, T_1^* T_2^* h\rangle = \langle y, T_1^* \lambda^*(T_2^*) h\rangle = \langle y, \lambda^*(T_1^*)\, \lambda^*(T_2^*) h\rangle$ holds for all $y$. Thus $\lambda(T) := \bar{\lambda}^*(T^*)$ defines a character of $\mathscr{S}$. It follows from $\langle T_\lambda x, h\rangle = \int \bar{\lambda}(T) \langle Tx, h\rangle\, d\varrho = \int \bar{\lambda}(T) \langle x, T^* h\rangle\, d\varrho = \int \bar{\lambda}(T)\, \lambda(T) \langle x, h\rangle\, d\varrho = \langle x, h\rangle$ that $T_\lambda \neq 0$. Thus $\lambda$ is a unimodular eigenvalue of $\mathscr{S}$. The relation $\lambda^*(T^*) = \bar{\lambda}(T)$ defines a 1-1-correspondence of the unimodular eigenvalues of $\mathscr{S}$ and $\mathscr{S}^*$.

**Proposition 4.6.** *If $\mathscr{S}$ is an Abelian weakly almost periodic group in $\mathscr{L}(\mathfrak{X})$, then $x \in \mathfrak{X}$ is a flight vector iff $\langle x, h\rangle = 0$ holds for all eigenvectors $h$ of $\mathscr{S}^*$ having unimodular eigenvalues.*

*Proof.* If $h$ is an eigenvector with unimodular eigenvalue $\lambda^*$, then $|\langle Tx, h\rangle| = |\langle x, T^* h\rangle| = |\bar{\lambda}^*(T^*) \langle x, h\rangle| = |\langle x, h\rangle|$ is constant on $\mathscr{S}$. If $x$ is a flight vector 0 belongs to $\bar{\mathscr{S}} x$, and $\langle x, h\rangle = 0$.

For any $\lambda$ and any $f \in \mathfrak{X}^*$, $T_\lambda^* f$ is $= 0$ or an eigenvector of $\mathscr{S}^*$ with $\bar{\lambda}^*$. If $x$ is orthogonal to all eigenvectors of $\mathscr{S}^*$ with unimodular $\lambda$, then $0 = \langle x, T_\lambda^* f\rangle = \langle T_\lambda x, f\rangle$ holds for all $f$. Hence $T_\lambda x = 0$. As $\lambda$ was arbitrary $x$ has no component in $\mathfrak{X}_{uds}$ and is a flight vector. □

To complete the generalization of the spectral mixing theorem we still have to find an extension of the assertion D-lim $\langle T^n x, y\rangle = 0$ for flight vectors $x$. This will be done under a separability condition.
A linear functional $L$ on the space of bounded complex valued functions $g$ on $\mathscr{S}$ is called an *invariant mean* if $g \geqq 0$ implies $L(g) \geqq 0$, $L(\mathbb{1}) = 1$, and $U \in \mathscr{S}$ implies $L(g \circ U) = L(g)$ for all $g$. $g \circ U$ is the (left) translate $(g \circ U)(T) = g(UT)$; (we need not distinguish right and left for Abelian $\mathscr{S}$).

**Theorem 4.7** (Lin). *Let $\mathscr{S}$ be an Abelian weakly almost periodic semigroup in $\mathscr{L}(\mathfrak{X})$ and $x \in \mathfrak{X}$. Assume that the closed linear subspace $\mathfrak{X}(x)$ of $\mathfrak{X}$ spanned by $\mathscr{S} x$ is separable. Then the following are equivalent:*

*(a) $x \in \mathfrak{X}_{fl}$;*

*(b) there exists a sequence $T_i \in \mathscr{S}$ with $T_i x \to 0$ weakly;*

*(c) for every $h \in \mathfrak{X}^*$ and every invariant mean $L$ we have $L(g) = 0$, where $g(T) = |\langle Tx, h\rangle|$. (We may write $L(|\langle Tx, h\rangle|) = 0$ for short).*

*Proof.* (a) $\Rightarrow$ (b): As $\overline{\mathscr{S}}x$ is weakly compact, the Krein-Shmulian theorem shows that $\overline{\text{co}}(\mathscr{S}x)$ is weakly compact and hence weakly sequentially compact by the Eberlein-Shmulian theorem. By the Hahn-Banach theorem $\overline{\text{co}}(\mathscr{S}x)$ is also sequentially compact in the weak topology of $\mathfrak{X}(x)$ since the limits are in $\overline{\text{co}}(\mathscr{S}x) \subset \mathfrak{X}(x)$. But the weak topology of a weakly compact subset of a separable Banach space is metric; see theorem V.6.3 in Dunford-Schwartz [1958].

(b) $\Rightarrow$ (c): The closed unit sphere $B$ in $\mathfrak{X}^*$ is w*-compact. Define continuous functions $f_i$ on $B$ by $f_i(h) = |\langle T_i x, h\rangle|$. Then $f_i \to 0$ pointwise and the $f_i$ are uniformly bounded by $\|x\| M$ where $M = \sup\{\|T\|: T \in \mathscr{S}\} < \infty$. By Lebesgue's theorem $f_i \to 0$ weakly in $C(B)$. By Mazur's theorem (Yosida [1974: 120]) there exists, for any $\varepsilon > 0$, a convex combination $\sum_{i=1}^{n} \alpha_i f_i$ of norm $< \varepsilon$. This means that $\sum_{i=1}^{n} \alpha_i |\langle T_i x, h\rangle| < \varepsilon$ holds for $h \in B$. Hence

$$\sum_{i=1}^{n} \alpha_i |\langle T T_i x, h\rangle| < \varepsilon M$$

for all $T \in \mathscr{S}$, and using the invariance under translations by $T_i$:

$$\begin{aligned} L(|\langle Tx, h\rangle|) &= \sum \alpha_i L(|\langle Tx, h\rangle|) = \sum \alpha_i L(|\langle T T_i x, h\rangle|) \\ &= L(\sum \alpha_i |\langle T T_i x, h\rangle|) \leqq \varepsilon M. \end{aligned}$$

As $\varepsilon > 0$ was arbitrary (c) follows for $h \in B$ and hence for all $h$.

(c) $\Rightarrow$ (a): If $h$ is an eigenvector of $\mathscr{S}^*$ with unimodular $\lambda^*$ we have $|\langle Tx, h\rangle| = |\bar{\lambda}^*(T^*)\langle x, h\rangle| = |\langle x, h\rangle|$ for all $T$. Hence (c) implies that $x$ is orthogonal to $h$. (a) now follows from proposition 4.6. □

Now consider the special case of a semigroup $\mathscr{S} = \{T^0, T^1, \dots\}$ generated by a single power bounded operator $T$. If $L'$ is a Banach limit on $\ell_\infty(\{0, 1, \dots\})$ we can define an invariant mean on $\mathscr{S}$ by $L(g) = L'(g')$ with $g'(k) = g(T^k)$. By theorem 3.4.1 the condition (c) in the previous theorem implies

$$\lim_n \sup_j n^{-1} \sum_{i=j}^{j+n-1} |\langle T^i x, h\rangle| = 0, \quad (h \in \mathfrak{X}^*),$$

which is slightly stronger than D-lim $\langle T^i x, h\rangle = 0$. However, a simple argument will show that (b) implies a yet stronger condition: the convergence is even uniform on $B$.

**Proposition 4.8** (Jones-Lin) *Let $T$ be a power bounded linear operator in a Banach space $\mathfrak{X}$ and $x \in \mathfrak{X}$. If there exists a sequence $(n_i)$ with $T^{n_i} x \to 0$ weakly, then*

$$(4.6) \qquad \lim_{N \to \infty} \left(\sup\left\{N^{-1} \sum_{i=0}^{N-1} |\langle T^i x, h\rangle| : h \in B\right\}\right) = 0.$$

*Proof.* First assume $\|T\| \leqq 1$. $B$ is w*-compact. $Af(h) := f(T^* h)$ defines a contraction on $C(B)$. The function $f(h) = |\langle x, h\rangle|$ satisfies $A^{n_i} f(h) \to 0$, hence $A^{n_i} f \to 0$ weakly in $C(B)$ by the dominated convergence theorem. For every (signed) measure $\mu \in C(B)^*$ with $\mu = A^* \mu$ we have $\langle f, \mu\rangle = \langle A^{n_i} f, \mu\rangle \to 0$ and hence $\langle \mu, f\rangle = 0$. The Hahn-Banach theorem now yields $f \in \mathrm{cl}\,((I-A)\,C(B))$. But then

$$\sup\{N^{-1}\sum_{i=0}^{N-1} |\langle T^i x, h\rangle| \colon h \in B\} = \|N^{-1}\sum_{i=0}^{N-1} A^i f\| \to 0.$$

If $\sup\{\|T^n\| \colon n \geqq 0\} = M > 1$, we introduce a new norm by $|||x||| = \sup \|T^n x\|$. Then $\|x\| \leqq |||x||| \leqq M\|x\|$ and $T$ is a contraction for the new norm. For $h \in \mathfrak{X}^*$

$$|||h||| = \sup\{|\langle x, h\rangle| \colon |||x||| \leqq 1\} \leqq \|h\|.$$

Hence $B$ is contained in the unit ball of $\mathfrak{X}^*$ under the new norm and the assertion follows from the contraction case. □

**Proposition 4.9** (Jones-Lin). *Let $T$ be power bounded in a Banach space $\mathfrak{X}$ and $x \in \mathfrak{X}$, then* (4.6) *is equivalent to*

$$\text{(4.7)} \qquad \lim_{N\to\infty} N^{-1}\sum_{i=0}^{N-1} |\langle T^i x, h\rangle| = 0 \quad \forall h \in \mathfrak{X}^*.$$

*Proof.* Clearly (4.6) implies (4.7). The argument for the converse is nearly identical to that of the previous proof: Now $N^{-1}\sum_{i=0}^{N-1} A^i f$ converges to 0 pointwise and weakly and $\mu = A^*\mu$ implies $\langle f, \mu\rangle = \langle N^{-1}\sum_{i=0}^{N-1} A^i f, \mu\rangle \to 0$. □

## Notes

If $\mathscr{S}$ is allowed to be non Abelian, then $\mathfrak{X}_{rev}$ and $\mathfrak{X}_{fl}$ need no longer be linear subspaces of $\mathfrak{X}$. Jacobs has given the following simple example: Let $\mathscr{S}$ be the semigroup $\{I, T, S\}$ in $\mathbb{R}^2$ given by the matrices

$$T = \begin{pmatrix} 1 & 0 \\ 0 & 0 \end{pmatrix}, \quad S = \begin{pmatrix} 1 & 1 \\ 0 & 0 \end{pmatrix}.$$

Then $\mathfrak{X}_{rev} = \{(x_1, x_2) \colon x_2 = 0\}$, $\mathfrak{X}_{fl} = \{(x_1, x_2) \colon x_1 = 0\} \cup \{(x_1, x_2) \colon x_1 + x_2 = 0\}$. For the semigroup defined by the transposed matrices one obtains $\mathfrak{X}_{rev} = \{(x_1, x_2) \colon x_2 = 0\} \cup \{(x_1, x_2) \colon x_1 = x_2\}$, $\mathfrak{X}_{fl} = \{(x_1, x_2) \colon x_1 = 0\}$.

Yet, splitting theorems can be derived under suitable assumptions when the statement $\mathfrak{X}_{rev} = \mathfrak{X}_{uds}$ is weakened. A finite dimensional $\mathscr{S}$-invariant subspace $D$ of $\mathfrak{X}$ is called a *unitary subspace* of $\mathfrak{X}$ if the restriction of $\mathscr{S}$ to $D$ is contained in a bounded *group* of operators on $D$. ($D$ is a unitary subspace iff it is possible to choose an inner product on $D$ so that the restrictions of all operators of $\mathscr{S}$ to $D$ are unitary). Let $\mathfrak{X}_{ap}$ be the smallest closed subspace of $\mathfrak{X}$ containing all unitary subspaces. The vectors in $\mathfrak{X}_{ap}$ are called *almost periodic*. (This definition is equivalent to the more usual definition; see Maak [1950], Jacobs [1960]). Clearly $\mathfrak{X}_{uds} \subset \mathfrak{X}_{ap}$. In the Abelian case equality holds. We are interested in the splitting property

(Sp) $\mathfrak{X}_{fl}$ and $\mathfrak{X}_{rev}$ are closed $\mathscr{S}$-invariant linear subspaces of $\mathfrak{X}$, $\mathfrak{X}$ is their direct sum, and $\mathfrak{X}_{rev} = \mathfrak{X}_{ap}$.

To formulate the principal result of Deleeuw-Glicksberg [1961] we need some definitions: For general weakly almost periodic $\mathscr{S}$ the kernel $K(\bar{\mathscr{S}})$ is the intersection of all two sided ideals of $\bar{\mathscr{S}}$. It is the unique minimal ideal of $\bar{\mathscr{S}}$.

$C_b(\mathscr{S})$ is the space of bounded continuous functions on $\mathscr{S}$, and $C_{\mathfrak{X}}(\mathscr{S})$ is the smallest uniformly closed subalgebra closed under complex conjugation and containing the constant functions and all $f$ of the form $f(T) = \langle Tx, h\rangle$ with $x \in \mathfrak{X}$, $h \in \mathfrak{X}^*$.

An *invariant mean* on a closed subspace $D$ of $C_b(\mathscr{S})$ is an element $m$ of $D^*$ with $m(\mathbb{1}) = 1$, $f \geqq 0 \Rightarrow m(f) \geqq 0$, and invariant under left and right translations.

**Theorem 4.10** (Deleeuw-Glicksberg). *Let $\mathscr{S}$ be a weakly almost periodic semigroup of linear operators in a Banach space $\mathfrak{X}$. Then each of the following conditions is necessary and sufficient for the splitting property (Sp):*

*(a) $C_{\mathfrak{X}}(\mathscr{S})$ has an invariant mean;*

*(b) $K(\bar{\mathscr{S}})$ is a compact topological group;*

*(c) $K(\bar{\mathscr{S}})$ contains a unique projection.*

Apart from the Abelian case treated above and solved by Jacobs [1956] for reflexive $\mathfrak{X}$, the main case of interest seems to be the case of contractions in spaces having suitable convexity properties:

**Theorem 4.11** (Jacobs-Deleeuw-Glicksberg) *Assume that $\mathfrak{X}$ is a strictly convex Banach space for which $\mathfrak{X}^*$ is strictly convex. If $\mathscr{S}$ is a weakly almost periodic semigroup of contractions in $\mathfrak{X}$, (Sp) holds.*

(Jacobs [1957] gave the proof for $\mathfrak{X}$, $\mathfrak{X}^*$ reflexive and strictly convex, with one of them uniformly convex).

The splitting theorem for groups of unitary operators in Hilbert space has already been studied by Maak [1954]. (For a similar result see Godement [1948: 64]) The case of bounded groups in Hilbert space was settled by Jacobs [1955].

Our presentation of the Abelian case borrows from an unpublished manuscript of Lin [1974], which, in particular, contains proposition 4.6 and theorem 4.7. The book of Jacobs [1960] contains different complete proofs of theorems 4.4 and 4.5 and shows that the reflexivity assumption in the original paper of Jacobs [1956] was only needed to assure the weak almost periodicity of $\mathscr{S}$. Jacobs [1956] also gave a characterization of flight vectors resembling theorem 4.7.

Proposition 4.8 and 4.9 appear in Jones-Lin [1976]. Jones and Lin [1980] have shown that the sufficient condition w-lim $T^{n_i}x = 0$ for (4.6) and (4.7) (which clearly is also necessary when $\mathfrak{X}^*$ is separable, and also if $\{T^n x\}$ is weakly sequentially compact) is not necessary in general. If $\{T^n x\}$ is w* sequentially compact in $\mathfrak{X}^{**}$, then (4.6) is equivalent to each of the conditions

(4.8) *0 is a weak cluster point of $\{T^n x\}$;*

(4.9) *$\langle x, h\rangle = 0$ for all unimodular eigenvectors $h$ of $T^*$;*

(4.10) *for every $\lambda$ with $|\lambda| = 1$, $\|N^{-1} \sum_{0}^{N-1} \lambda^n T^n x\| \to 0$.*

For further related material, see section 8.1 and Nagel-Palm-Derndinger [1984]. Jamison [1964] gave a direct proof of the following consequence of the Jacobs-Deleeuw-Glicksberg theorems:

*If $\mathfrak{X}$ is a Banach space, $T \in \mathscr{L}(\mathfrak{X})$, and $\{Tx, T^2 x, \ldots\}$ conditionally compact in the norm topology for each x, then $\{T^n x\}$ converges in norm for all x iff there exists no eigenvector with eigenvalue $\lambda \neq 1$, $|\lambda| = 1$.*

Gundel [1979] studied mixing properties of semigroups of endomorphisms and the mutual implication of such properties in the Gaussian case. Zimmer [1976] has a general structure theory for semigroups $\{\tau_g, g \in G\}$, $G$ a locally compact group, including a generalization of theorem 3.11. The books of Berglund-Hofman [1967], Berglund-Junghenn-Milnes [1978], Burckel [1970] are references for the theory of semitopological semigroups.

The papers of Dye [1965], Kühne [1982] and Schmidt [1982] are further references concerning weak mixing for semigroups.

We remark that the results discussed here also yield mean ergodic theorems of the type discussed in § 2.1. If $\mathscr{S}$ is weakly almost periodic, also $\mathscr{T} = \mathrm{co}\,\mathscr{S}$ is weakly almost periodic. The space $\mathfrak{X}_{ap}$ for the semigroup $\mathscr{T}$ consists only of fixed points of $\mathscr{S}$, and the space $\mathfrak{X}_{fl}$ of $\mathscr{T}$ consists of vectors $x$ with $0 \in \overline{\mathrm{co}}\,\mathscr{S}x$. The results of this section, however, are intrinsically deeper.

# Chapter 3: Positive contractions in $L_1$

While mean ergodic theory has been developed for rather general spaces, pointwise ergodic theory for operators mainly deals with positive contractions in $L_1$ and closely related topics. This subject has flourished because it unifies ergodic theorems for measure preserving and nonsingular transformations with results from the theory of Markov processes.

In this chapter we discuss the basic theory, including the Hopf decomposition, the Chacon-Ornstein theorem, Brunel's lemma, and the existence of finite invariant measures. Generalizations and refinements are defered to subsequent chapters.

## § 3.1 The Hopf decomposition

Let us first describe some probabilistic notions motivating the study of contractions in $L_1$. Throughout, $(\Omega, \mathscr{A})$ will be a fixed measurable space.

A mapping $P: \Omega \times \mathscr{A} \to \mathbb{R}^+$, $(\omega, A) \to P(\omega, A)$, is called a *substochastic kernel* if

(k1) for each $A \in \mathscr{A}$, $P(\cdot, A): \Omega \to \mathbb{R}^+$ is measurable, and

(k2) for each $\omega \in \Omega$, $P(\omega, \cdot): \mathscr{A} \to \mathbb{R}^+$ is a measure with $P(\omega, \Omega) \leqq 1$.

$P(\cdot, \cdot)$ is called a *stochastic kernel*, if even $P(\omega, \Omega) = 1$ holds for all $\omega \in \Omega$. We may think of $\Omega$ as a set of "states" in which a system observed at time $0, 1, 2, \ldots$ can be, and of $P(\omega, A)$ as the probability of passing from state $\omega$ into the set $A$ of states in one time unit.

If $\tau: \Omega \to \Omega$ is a measurable transformation we may define a stochastic kernel $P_\tau(\omega, A) = 1_A(\tau\omega)$. This kernel describes the deterministic transition $\omega \to \tau\omega$.

Let $b\mathscr{A}$ denote the space of real valued bounded $\mathscr{A}$-measurable functions on $\Omega$, and let $\mathscr{M} = \mathscr{M}(\mathscr{A})$ be the real Banach space of signed measures $\nu$ with finite total variation $\|\nu\|$. A substochastic kernel induces important linear mappings in these spaces by

(1.1) $\quad T^* h(\omega) = \int h(\eta) P(\omega, d\eta), \quad (h \in b\mathscr{A})$,

and

(1.2) $\quad T\nu(A) = \int P(\omega, A)\, \nu(d\omega), \quad (\nu \in \mathscr{M})$.

If $h(\eta)$ is a reward one can receive in state $\eta$, then $T^* h(\omega)$ may be interpreted as the expected reward after one transition given a start in $\omega$. In particular $T^* 1_A(\omega) = P(\omega, A)$.

For an interpretation of $T$ one may think of a non negative $v$ as a distribution of matter. The matter in $\omega$ is scattered at random according to the transition probabilities. $Tv$ is the new distribution of the matter. If $v$ assumes also negative values, then $v = v^+ - v^-$. $v^-$ may be thought of as antimatter. Some of the matter and some of the antimatter may cancel each other after the transition. If $v(d\omega)$ is the probability that the system is in $d\omega$ at time $n$, then $Tv(A)$ is the probability that the system is in $A$ at time $n+1$.

$\langle v, h\rangle = \int h\, dv$ is a bilinear form on $\mathcal{M} \times b\mathcal{A}$. The Fubini theorem for kernels asserts that

$$\int [\int h(\eta) P(\omega, d\eta)]\, v(d\omega) = \int h(\eta) [\int P(\omega, d\eta)\, v(d\omega)].$$

It follows that

$$(1.3) \qquad \langle Tv, h\rangle = \langle v, T^* h\rangle, \quad (v \in \mathcal{M},\ h \in b\mathcal{A}).$$

For the study of ergodic theorems we need also a reference measure. Let $\mu$ be a $\sigma$-finite measure on $(\Omega, \mathcal{A})$, and

$$\tilde{L}_1(\mu) = \tilde{L}_1 = \{v \in \mathcal{M}\colon v \ll \mu\}.$$

We say that a kernel $P$ is *null preserving* if $\mu(A) = 0$ implies $P(\omega, A) = 0$ $\mu$-a.e.. If $\tau$ is a measurable transformation, then the stochastic kernel $P_\tau$ is null preserving iff $\tau$ is null preserving. In this special case the operators defined by (1.1) and (1.2) have the form $h \to h \circ \tau$, and $v \to v \circ \tau^{-1}$.

**Lemma 1.1.** *The kernel $P$ is null preserving iff $T$ maps $\tilde{L}_1$ into $\tilde{L}_1$.*

*Proof.* Assume $P$ null preserving, $v \in \tilde{L}_1$, and $\mu(A) = 0$. Then $Tv(A) = \int P(\omega, A)\, v(d\omega) = 0$. Hence $Tv \ll \mu$. Conversely, if $P$ is not null preserving, there exist $A, B$ with $\mu(A) = 0$, $0 < \mu(B) < \infty$ and $P(\omega, A) > 0$ for $\omega \in B$. Then $v$, defined by $v(C) = \mu(C \cap B)$, belongs to $\tilde{L}_1$, but $Tv$ not. □

For many purposes the assumption that $P$ is null preserving is not a restriction of generality. If $\mu$ is a given $\sigma$-finite measure and $\mu'$ a finite measure with $\mu' \sim \mu$, then $P$ is null preserving for the measure $\bar{\mu} = \sum_{i=0}^{\infty} 2^{-i} T^i \mu' \gg \mu$; (exercise). However, note that $\mu$-integrable functions need not be $\bar{\mu}$-integrable.

It will be convenient to work mainly with $L_1$ instead of $\tilde{L}_1$. By the Radon-Nikodym theorem there is a (linear, isometric, order preserving) isomorphism $\varphi_\mu$ between $\tilde{L}_1$ and $L_1$ associating with each $v \in \tilde{L}_1$ its Radon-Nikodym derivative $\varphi_\mu(v) = dv/d\mu$. If the kernel $P$ is null preserving, we may therefore regard $T$ as an operator in $L_1$ by identifying $\tilde{L}_1$ and $L_1$. More precisely: $T$ in $L_1$ is the operator

$Tf = \varphi_\mu T \varphi_\mu^{-1} f$. Because of the identification we use the same symbol for the operators in $\tilde{L}_1$ and $L_1$. $T$ is a positive contraction in $L_1$.

If the kernel $P$ is null preserving, (1.1) defines also an operator $T^*$ in $L_\infty(\mu) = L_\infty$: In this case $\mu(A) = 0$ implies $T^* 1_A = 0$ $\mu$-a.e. and, by approximation with step functions, $h = 0$ $\mu$-a.e. implies $T^* h = 0$ $\mu$-a.e.. $L_\infty$ is the dual space of $L_1$, with scalar product $\langle f, h \rangle = \int f \cdot h \, d\mu$, and, by the identification of $\tilde{L}_1$ and $L_1$, (1.3) translates into $\langle Tf, h \rangle = \langle f, T^* h \rangle$. $T^*$ is the adjoint of $T$, and the notation is justified.

One may ask if *each* positive contraction $T$ in $L_1$ is given by a null preserving substochastic kernel. This is in fact true, if $(\Omega, \mathscr{A})$ satisfies some regularity conditions, e.g., if $\Omega$ is a complete separable metric space and $\mathscr{A}$ the $\sigma$-algebra of Borel sets; see e.g. Neveu [1965: 192]. In the general case we may still find a "generalized" null preserving substochastic kernel: For fixed $A$, $P(\omega, A) := T^* 1_A(\omega)$ defines $P(\cdot, A)$ only up to nullsets. It is not hard to see that one may choose $P(\omega, A)$ in such a way that (k1) and the null preserving property hold, and (k2) is replaced by

(k2i) for all $\omega \in \Omega$, $A \in \mathscr{A}$, $0 \leqq P(\omega, A) \leqq 1$, and

(k2ii) for all disjoint $A_1, A_2, \ldots \in \mathscr{A}$, $P(\omega, \bigcup_1^\infty A_i) = \sum_{i=1}^\infty P(\omega, A_i)$ $\mu$-a.e..

If the equation in (k2ii) would hold *everywhere*, $P(\cdot, \cdot)$ would be a null preserving substochastic kernel.

We shall develop the theory for positive contractions in $L_1$ rather than for kernels, although the gain in generality is small. The advantage consists mainly in greater elegance and shortness of certain arguments.

The equations (1.1) and (1.2) still make sense for the generalized kernel $P(\cdot, \cdot)$, $h \in L_\infty$, and $v \in \tilde{L}_1$. But $P(\omega, \cdot)$ need not be $\sigma$-additive, and the integral in (1.1) must be defined as the limit of

$$\int h_n(\eta) \, P(\cdot, d\eta) = \sum_{i=1}^{k_n} a_{i,n} P(\cdot, A_{i,n})$$

where $h_n = \sum_i a_{i,n} 1_{A_{i,n}}$ is a sequence of step functions with $\|h_n - h\|_\infty \to 0$.

A positive contraction $S$ in $L_\infty$ is called a *sub-Markovian* operator if $Sh_n \to 0$ a.e. holds for any sequence $h_n$ decreasing to 0. If $T$ is a positive contraction in $L_1$, it is easy to see that $T^*$ is a sub-Markovian operator. Conversely, if $S$ is a sub-Markovian operator, we may define a positive contraction $T$ in $\tilde{L}_1$ with $T^* = S$ by $Tv(A) = \int S 1_A \, dv$. $T$ and $T^*$ are called *Markovian* if $T^* \mathbb{1} = \mathbb{1}$. An equivalent condition is $\int Tf \, d\mu = \int f \, d\mu$ for all $f \in L_1$. (Simply use $\int f \, d\mu = \langle f, \mathbb{1} \rangle$.)

Sometimes it will be useful to extend the range of definition of $T$: For measurable $f$ with $f^- \in L_1$ we may uniquely define $Tf = \lim Tf_n$ where $f_n$ is an increasing sequence of integrable functions tending to $f$; see § 1.6. Similarly, one may extend the range of definition of $T^*$.

We have set the stage and may start the investigation of positive contractions in

$L_1$, or – from the dual point of view – of sub-Markovian operators. We first give three different but equivalent descriptions of the important *Hopf decomposition* of $\Omega$ into two sets $C$ and $D = \Omega \setminus C$, called the *conservative* and the *dissipative part* of $\Omega$ (for $T$).

**Definition 1.2.** A function $h$ with $T^*h = h$ is called *harmonic*, a non negative function $h$ with $h \geqq T^*h$ is called *superharmonic*. A superharmonic $h$ is called *strictly superharmonic* on $A$ if $h > T^*h$ on $A$.

(We may delete "a.e." since we are dealing with equivalence classes of functions.)

**Theorem 1.3.** *If $T$ is a positive contraction in $L_1$, there exists a decomposition of $\Omega$ into disjoint sets $C$, $D$, determined uniquely* mod $\mu$ *by:*

*(C1) If $h$ is superharmonic, then $h = T^*h$ on $C$;*

*(D1) There exists a bounded superharmonic $h_0$, which is strictly superharmonic on $D$.*

*$h_0$ may be constructed with the additional properties $T^{*n}h_0 \to 0$ on $D$, and $h_0 = 0$ on $C$.*

*Proof.* Let us say that a set $A$ has the property (ss) if there exists a bounded superharmonic $g_A$ with $g_A > T^*g_A$ on $A$. By an exhaustion argument (see § 1.3) we may find countably many $D_1, D_2, \ldots$ with the property (ss), such that no subset of positive measure of $C = \Omega \setminus D$ with $D = \bigcup_{i=1}^{\infty} D_i$ has the property (ss). If $c_i > 0$ is a bound for $g_{D_i}$, then $h_1 = \sum_{i=1}^{\infty} c_i^{-1} 2^{-i} g_{D_i}$ is bounded, superharmonic, and strictly superharmonic on $D$. Assume that there exists a set $B \subset C$ with $\mu(B) > 0$ and an unbounded superharmonic $h_B$ with $h_B > T^*h_B$ on $B$. Choose some $\alpha$ for which $B' = B \cap \{h_B \geqq \alpha\} \cap \{T^*h_B < \alpha\}$ has positive measure. Then, because of $0 \leqq T^*\mathbb{1} \leqq \mathbb{1}$, the function $h = h_B \wedge \alpha$ is bounded, superharmonic, and strictly superharmonic on $B'$. This proves (C1), (D1) and the uniqueness.

By induction the sequence $h_n = T^{*(n-1)}h_1$ is decreasing and non negative, and each $h_n$ is superharmonic. Put $h_\infty = \lim h_n$. As $T^*$ is sub-Markovian $T^*h_\infty = \lim T^*h_n = \lim h_{n+1} = h_\infty$. We may take $h_0 = h_1 - h_\infty$. □

**Corollary 1.4.** $T^*\mathbb{1} = \mathbb{1}$ *on* $C$.

Recall the notation $S_n f = \sum_{i=0}^{n-1} T^i f$, $(n \in \mathbb{N} \cup \{\infty\})$.

**Theorem 1.5.** *The sets $C$, $D$ of the Hopf decomposition are also determined uniquely* mod $\mu$ *by:*

*(C2) For all $h \in L_\infty^+$, $S_\infty^* h = \infty$ on $C \cap \{S_\infty^* h > 0\}$ and*

*(D2) there exists an $h_D \in L_\infty^+$ with $\{h_D > 0\} = D$ and $S_\infty^* h_D \leqq \mathbb{1}$.*

*Proof.* Multiplying $h_0$ by a positive constant we may assume $h_0 \leqq \mathbb{1}$. $h_D = h_0 - T^* h_0$ is strictly positive on $D$ and vanishes on $C$, and $S_n^* h_D = (h_0 - T^* h_0) + (T^* h_0 - T^{*2} h_0) + \ldots + (T^{*(n-1)} h_0 - T^{*n} h_0) \leqq h_0$. This proves (D2). Now assume the existence of an $h \in L_\infty^+$, an $n \in \mathbb{N}$, and an $A \subset C$ with $\mu(A) > 0$, $T^{*n} h > 0$ on $A$ and $S_\infty^* h < \infty$ on $A$. Consider $\tilde{h} = \operatorname{Min}(\mathbb{1}, \sum_{i=n}^{\infty} T^{*i} h)$. $T^* \mathbb{1} \leqq \mathbb{1}$ and $T^* \tilde{h} \leqq \sum_{i=n+1}^{\infty} T^{*i} h$ imply $T^* \tilde{h} \leqq \tilde{h}$, and $T^{*(j+1)} \tilde{h} \leqq T^{*j} \tilde{h}$ for all $j \geqq 0$. Hence $T^{*j} \tilde{h} = \tilde{h}$ on $C$ for all $j$. But $T^{*j} \tilde{h} \leqq \sum_{i=n+j}^{\infty} T^{*i} h$ tends to 0 on $A$. This contradicts to $\tilde{h} > 0$ on $A$, and proves (C2). □

Using theorem 1.5 it is not hard to check that the conservative part studied in § 1.3 for a null preserving transformation $\tau$ coincides with the conservative part for the contraction $v \to Tv = v \circ \tau^{-1}$, or, equivalently, for $T^*$ given by $T^* h = h \circ \tau$. Extending the terminology used there we may call $T$ and $T^*$ *conservative* if $\Omega = C$ and *dissipative* if $\Omega = D$.

The Hopf decomposition generalizes also another well known decomposition. Consider a Markov chain on an at most countable state space $\Omega$ given by a matrix $P = (p_{ik})$ of transition probabilities $(i, k \in \Omega)$. Let $\mu$ be the counting measure: $\mu(A) = \operatorname{card}(A)$. Then $P(i, A) = \sum_{j \in A} p_{ij}$ is a stochastic kernel, which trivially is null preserving, because there are no non empty nullsets. If $P^n = (p_{ik}^n)$ is the matrix of the $n$-step transition probabilities, the contraction in $L_\infty$ belonging to the kernel may be described by $T^{*n} 1_{\{k\}}(i) = p_{ik}^n$. Recall that a state $i \in \Omega$ is called *recurrent* if $\sum_{n=1}^{\infty} p_{ii}^n = \infty$ and *transient* if $\sum_{n=1}^{\infty} p_{ii}^n < \infty$. Theorem 1.5 implies that $C$ is the set of recurrent states and $D$ is the set of transient states.

Finally, we describe the Hopf decomposition in terms of $T$:

**Theorem 1.6.** *The sets $C, D$ of the Hopf decomposition are also determined uniquely* mod $\mu$ *by:*

*(C3) For all $f \in L_1^+$, $S_\infty f = \infty$ on $C \cap \{S_\infty f > 0\}$, and*

*(D3) for all $f \in L_1^+$, $S_\infty f < \infty$ on $D$.*

*Proof.* $\langle S_\infty f, h_D \rangle = \langle f, S_\infty^* h_D \rangle \leqq \langle f, \mathbb{1} \rangle < \infty$ and $h_D > 0$ on $D$ imply (D3). Now assume $S_\infty f < \infty$ on a set $F \subset C \cap \{f > 0\}$ with $0 < \mu(F) < \infty$. $h = 1_F(\mathbb{1} + S_\infty f)^{-1}$ is strictly positive on $F$. By (C2) $S_\infty^* h = \infty$ on $F$. But

$$\langle f, S_\infty^* h \rangle = \langle S_\infty f, h \rangle = \int (S_\infty f) 1_F (\mathbb{1} + S_\infty f)^{-1} \mathrm{d}\mu \leqq \mu(F) < \infty,$$

which contradicts to $f > 0$ on $F$. Hence $S_\infty f = \infty$ on $C \cap \{f > 0\}$. Repeating the argument with $T^n f$ we find that $S_\infty f = \infty$ on $C \cap \{T^n f > 0\}$ for all $n \geqq 0$. □

We shall use the notation $L_p(B) = \{f \in L_p \colon \{f \neq 0\} \subset B\}$, $(1 \leqq p \leqq \infty)$.

**Definition 1.7.** A set $B \in \mathscr{A}$ is called *T-absorbing* if $Tf \in L_1(B)$ holds for all $f \in L_1(B)$.

Intuitively: if $f$ is a distribution of matter with support in $B$ then $T$ cannot carry any of the matter into $B^c$. The probability of a transition from a state in $B$ into $B^c$ is zero.

**Theorem 1.8.** *The conservative part $C$ for a positive contraction $T$ in $L_1$ is $T$-absorbing.*

*Proof.* Let $p$ be a strictly positive element of $L_1$. By (C3) and (D3), $C = \{S_\infty p = \infty\}$. For $f \in L_1^+(C)$ we have $nf \leqq S_\infty p$ for all $n$. Hence $nTf \leqq TS_\infty p = \sum_{i=1}^{\infty} T^i p$ for all $n$. In particular, $S_\infty p = \infty$ on $\{Tf > 0\}$. This proves $Tf \in L_1(C)$. □

Sometimes the following characterization of $T$-absorbing sets in terms of $T^*$ is useful:

**Theorem 1.9.** *$B \in \mathscr{A}$ is $T$-absorbing iff $T^* 1_{B^c} \leqq 1_{B^c}$.*

*Proof.* First let $B$ be $T$-absorbing and let $f$ be an element of $L_1^+$ with $\{f > 0\} = B$. Then $Tf \in L_1(B)$ implies $0 = \langle Tf, 1_{B^c} \rangle = \langle f, T^* 1_{B^c} \rangle$. Hence $T^* 1_{B^c} = 0$ on $B$. Using $T^* 1_{B^c} \leqq \mathbb{1}$ we obtain $T^* 1_{B^c} \leqq 1_{B^c}$. On the other hand, if this inequality holds and $g$ belongs to $L_1^+(B)$, then $0 \leqq \langle Tg, 1_{B^c} \rangle = \langle g, T^* 1_{B^c} \rangle \leqq \langle g, 1_{B^c} \rangle = 0$ and $Tg$ must belong to $L_1(B)$. For general $g \in L_1(B)$ use $g = g^+ - g^-$. □

## Notes

The foundations of the ergodic theory of positive contractions in $L_1$ were developed by E. Hopf [1954]. Foguel [1979] noticed that one does not need the maximal ergodic theorem. The theory of positive contractions in $L_1$ has been discussed in the books of Neveu [1965], Foguel [1969], Garsia [1970], and Revuz [1975].

Mutatis mutandis, most of the results of § 1.3 admit generalizations to positive contractions in $L_1$. For a positive contraction $T$ in $L_1$ the conservative part agrees with that of $T^k$ for all $k \in \mathbb{N}$. If $\{T_t, t > 0\}$ is a strongly continuous semigroup of positive contractions, all $T_t$ have the same conservative part, and it agrees with $\{\int_0^\infty T_s f ds = \infty\}$ for all strictly positive $f \in L_1$. If $\varphi(t)$ is decreasing, $\geqq 0$, with $\int_0^\infty \varphi(t) = 1$ and $\int_0^\infty \varphi(t) t dt < \infty$, also the operator $Q$ with $Qg = \int_0^\infty \varphi(t) T_t g dt$ has the same conservative part; see Horowitz [1974]. Hoover, Lambert and Quinn [1982] have determined the conservative part for operators $Tf = hf \circ \tau$.

An extensive bibliography on Markov chains with a general state space has been prepared by Šidak [1976].

*The conservative part of a Markov shift.* Let $(E, \mathscr{F}, \varrho)$ be a $\sigma$-finite measure space and $T$ a positive contraction in $L_1(\varrho)$ with conservative part $C$, and with $T^*\mathbb{1} = \mathbb{1}$. As in § 1.4 take $\Omega^+ = E^{\mathbb{Z}^+}$, $X_j = j$-th coordinate map. Generalizing 1.(4.6) put

$$\mu_\varrho(X_0 \in A_0, \ldots, X_n \in A_n) = \int 1_{A_0} T^* 1_{A_1} \ldots T^* 1_{A_n} d\varrho .$$

Under slight assumptions (e.g. $E$ is a Borel set in a Polish space, and $\mathscr{F}$ its Borel subsets), $\mu_\varrho$ extends to a measure in the product $\sigma$-algebra. The shift $\theta$ is null preserving in $(\Omega^+, \mu_\varrho)$. In the case $\varrho = T\varrho$ the shift is $\mu_\varrho$-preserving and we can consider the bilateral extension to $\Omega = E^{\mathbb{Z}}$.

**Theorem 1.10.** *The conservative part of $\theta$ in $\Omega^+$ is $C^{\mathbb{Z}^+}$, and the conservative part of the shift in $\Omega$ is $C^{\mathbb{Z}}$.*

This was proved by Harris and Robbins [1953] for $\Omega$ and countable $E$, and by Moy [1965] for Markov processes with transition probabilities having a density. The general case, due to Simons [1971], is similar.

When $X_0, X_1, \ldots$ is an aperiodic, transient, irreducible random walk on $E = \mathbb{Z}$, it may be seen that the remote $\sigma$-algebra $\mathscr{A}_\infty$ is trivial. Hence, $\theta$ is then an ergodic dissipative measure preserving transformation; see Krengel-Sucheston [1969]. [In $\Omega^+$ it is clear that the $\theta$-invariant sets belong to $\mathscr{A}_\infty$. Krengel-Sucheston showed that this remains so for the bilateral shift in the conservative case, but not in the dissipative case.]

Several authors have considered the problem of approximation of certain Markov operators by operators with additional properties or by convex combinations of such operators; see Brown [1966], Kim [1972], Iwanik [1980], Choksi and Prasad [1983].

An interesting class of Markov operators is the class of *convolutions* by a probability measure $\varrho$ on a locally compact group $G$. A stochastic kernel is given by $P(\omega, A) = \varrho(A\omega^{-1})$. Then $\varrho$ is the distribution of the size of the steps in a random walk on $G$.

For simplicity let us assume that $G$ is Abelian with Haar measure $\mu$. The operator $T$ is given by $Tv = \mu * v$. Its restriction to $L_1(\mu)$ has the form $Tf(\omega) = \mu * f(\omega) = \int f(\omega - \eta)\varrho(d\eta)$. $T^*$ is computed as $T^*g(\omega) = \int g(\omega + \eta)\varrho(d\eta)$.

We refer to the papers of Kerstan-Matthes [1965], Foguel [1975], Derriennic [1976], Glasner [1976a], Lin [1977], and Derrienic-Lin [1984], for the study of the asymptotic properties of these convolutions.

For the construction of conservative and dissipative Bernoulli-shifts with nonidentical factor measures, see Krengel [1970], Hamachi [1981], Grewe [1983].

## § 3.2 The Chacon-Ornstein theorem

Let $T$ be a positive contraction in $L_1$. We shall now study the a.e.-convergence of $S_n f / S_n g$ on $\{S_\infty g > 0\}$ for $f \in L_1$, $g \in L_1^+$. This will be independent of the Hopf decomposition.

It is of interest that part of the proof works for an arbitrary positive linear operator $T$ in a general vector lattice of measurable functions on $\Omega$ (or of equivalence classes of functions). For the heuristic interpretation we think of $L = L_1$ and of the transport of matter by $T$.

We write $f \xrightarrow{1} g$ if $f \in L^+$ and there exist $r, s \in L^+$ with $f = r + s$ and $g = r + Ts$. ($g$ results if the part $r$ of the matter $f$ remains fixed and the part $s$ is mapped with $T$). For $n \geqq 2$ we write $f \xrightarrow{n} g$ if there exists a $g_1$ with $f \xrightarrow{n-1} g_1$ and $g_1 \xrightarrow{1} g$. The *nonlinear* operator $U$ in $L$ given by $Uh = Th^+ - h^-$ is called the *filling operator*. If $h = f - g$ with $f, g \in L^+$ then $f$ may be considered as matter, $g$ as antimatter. $U(f-g)$ is obtained as follows: First the part $f \wedge g$ of $f$ and $g$ cancel each other, and then the remaining part $h^+$ of $f$ is mapped, while the remaining part $h^-$ of $g$ stays where it is. Frequently, $g$ is thought of as a hole. The part $f \wedge g$ of $f$ falls into the hole and partly *fills* it. $h^-$ is the remaining hole. The part $h^+$ of $f$ which did not drop into the hole is mapped. If we apply $U$ a second time part of this mapped matter $Th^+$ will drop into the remaining hole and again fill it partly, the rest will be mapped again. The repeated application of $U$ is therefore called the *filling scheme*.

**Lemma 2.1.** *For any $f, g \in L$ and $n \geqq 0$ there exists an $f_n \in L^+$ with $f \xrightarrow{n} f_n$ and $U^n(f-g) = f_n - g$.*

*Proof.* For $n = 0$ we may take $f_0 = f$. If $f_n$ has already been constructed, then $t_n = f_n - [U^n(f-g)]^+$ is non negative and $[U^n(f-g)]^- = -U^n(f-g) + [U^n(f-g)]^+ = g - t_n$. Hence $U^{n+1}(f-g) = T(f_n - t_n) - (g - t_n)$ and we may put $f_{n+1} = T(f_n - t_n) + t_n$. □

Intuitively $t_n$ is the part of the hole that has been filled up to time $n$ and $f_n$ is the sum of $t_n$ and the remaining matter. Recall the notation $M_n^S$, $M_n$ from §1.2. The next theorem asserts that within the set $\{M_n^S h > 0\}$ the hole will be filled completely by time $n-1$:

**Theorem 2.2.** *For $h \in L$, $n \geqq 1$, and $T, U$ as above we have $U^{n-1} h \geqq 0$ in $\{M_n^S h > 0\} = \{M_n h > 0\}$.*

*Proof.* By the definition of $U$ the sequence $(U^k h)^-$, $(k = 0, 1, 2, \ldots)$, is decreasing. It is therefore sufficient to show that for almost all $\omega \in \{M_n^S h > 0\}$ there exists some $m$ with $0 \leqq m \leqq n-1$ and $(U^m h)^+(\omega) > 0$. This is equivalent to showing that $\varphi_k(\omega) = \sum_{m=0}^{k} (U^m h)^+(\omega)$ is strictly positive for some $0 \leqq k \leqq n-1$. As $(U^m h)^+$ is that part of $T^m h^+$ which did not drop into the hole up to time $m$ one should have

$$(2.1) \qquad \varphi_k \geqq \sum_{m=0}^{k} T^m h = S_{k+1} h.$$

We can complete the proof by showing that (2.1) indeed holds. For $k = 0$ (2.1) is the obvious inequality $h^+ \geqq h$. If (2.1) has been verified for some $k$, we obtain

$$\begin{aligned} S_{k+2}f &= h + TS_{k+1}f \leqq h + T\varphi_k = h + \sum_{m=0}^{k} T(U^m h)^+ \\ &= h + \sum_{m=0}^{k} (U^{m+1}h - (U^m h)^-) \\ &= h + \sum_{m=0}^{k} [(U^{m+1}h)^+ - (U^{m+1}h)^- + (U^m h)^-] \\ &= h + (\varphi_{k+1} - h^+) + h^- - (U^{k+1}h)^- \leqq \varphi_{k+1}. \quad \square \end{aligned}$$

*From now on $T$ will be a positive contraction in $L = L_1$.*

Before we continue it may be useful to point out that theorem 2.2 yields a simple proof of Hopf's maximal ergodic theorem: For $h \in L_1$ put $f = h^+$, $g = h^-$ and find $f_{n-1}$ with $(f_{n-1} - g) = U^{n-1}h$ and $f \xrightarrow{n-1} f_{n-1}$ as in Lemma 2.1. By induction $f \xrightarrow{k} f'$ implies $\int f \geqq \int f'$. As $\{h^+ > 0\}$ is contained in $E := \{M_n h > 0\}$ we obtain

$$\int_E h = \int_E h^+ - \int_E h^- = \int_\Omega h^+ - \int_E g \geqq \int_\Omega f_{n-1} - \int_E g \geqq \int_E U^{n-1}h \geqq 0.$$

This is the maximal ergodic theorem with $\{M_n h \geqq 0\}$ replaced by $\{M_n h > 0\}$. The slightly sharper formulation in theorem 1.2.1 is obtained by approximation: If $h_k$ is a sequence in $L_1$ strictly decreasing to $h \in L_1$, then the sets $\{M_n h_k > 0\}$ decrease to $\{M_n h \geqq 0\}$ and therefore

$$\int_{\{M_n h \geqq 0\}} h\,d\mu = \lim_k \int_{\{M_n h_k > 0\}} h_k\,d\mu \geqq 0.$$

The following lemma will be rather crucial:

**Lemma 2.3** (Chacon-Ornstein). *If $T$ is a positive contraction in $L_1, f \in L_1$ and $g \in L_1^+$, then $\lim_{n\to\infty} T^n f / S_{n+1} g = 0$ a.e. on $\{S_\infty g > 0\}$.*

*Proof.* We may assume $f \geqq 0$. For $\varepsilon > 0$ set $r_n = T^n f - \varepsilon S_{n+1} g$ and $A_n = \{T^n f / S_{n+1} g > \varepsilon\} \cap \{g > 0\} = \{r_n > 0\} \cap \{g > 0\}$, $(n \geqq 0)$. We first give a heuristic argument: Imagine that we are colouring the amount $1_{A_n} \varepsilon g$ of the matter $T^n f$ at time $n$. As the matter already coloured at time $k < n$ has moved $n - k$ times, the amount of matter still available for the colouring at time $n$ is at least $(T^n f - \varepsilon T S_n g)^+ \geqq 1_{A_n} \varepsilon g$ and suffices. As the total amount of matter coloured in all steps cannot exceed $\int f d\mu$ we obtain

$$(2.2) \qquad \int \sum_{n=0}^{\infty} 1_{A_n} \varepsilon g\,d\mu \leqq \int_\Omega f\,d\mu < \infty.$$

Formally, $r_n = Tr_{n-1} - \varepsilon g \leqq Tr_{n-1}^+ - \varepsilon g$, $(n \geqq 1)$. From this we derive $r_n^+ \leqq Tr_{n-1}^+ - \varepsilon 1_{A_n} g$ by verifying this inequality on each of the sets $\{g = 0\}$, $A_n$, $\{g > 0\} \cap A_n^c$. Hence $\int \varepsilon 1_{A_n} g \leqq \int Tr_{n-1}^+ - \int r_n^+ \leqq \int r_{n-1}^+ - \int r_n^+$. Summing over n $\geqq 1$ and noticing $r_0^+ = f - \varepsilon 1_{A_0} g$ we indeed obtain (2.2). Thus, a.e. $\omega \in \{g > 0\}$ belongs to only finitely many $A_n$. As $\varepsilon > 0$ was arbitrary the convergence is proved on $\{g > 0\}$. For the convergence on $\{T^m g > 0\}$ repeat the argument with $T^m f$ and $T^m g$. □

**Lemma 2.4.** *Assume* $f, g \in L_1^+$, $n \geqq 1$, $g \xrightarrow{n} g_1$, *and* $\gamma > 1$, *then* $\{\limsup S_n(f-g) > 0\} \subset \{\limsup S_n(\gamma f - g_1) > 0\}$.

*Proof.* As $\gamma$ may be written as a product $\gamma_1 \cdot \gamma_2 \ldots \gamma_n$ with $\gamma_i > 1$ we may assume n = 1 (and then use induction). Now $g_1 = r + Ts$ with $r + s = g$ and $r, s \in L_1^+$. The result follows from $S_n(\gamma f - g_1) = \sum_{k=0}^{n-1} T^k(\gamma f - r - Ts) = S_n(\gamma f - g) + s - T^n s$ $\geqq S_n(f-g) + (\gamma - 1) S_n f - T^{n-1}(Ts)$, because by Lemma 2.3 $T^{n-1}(Ts)$ $\leqq (\gamma - 1) S_n f$ for large $n$ on $\{S_\infty f > 0\}$. □

For $H \in \mathscr{A}$, $f \in L_1^+$ and $n \geqq 0$ we define

$$\Psi_H^n f = \sup\{\int_H g\,d\mu : f \xrightarrow{n} g\} \quad \text{and} \quad \Psi_H f = \lim_{n\to\infty} \Psi_H^n f.$$

The sequence $\Psi_H^n f$ is increasing because $f \xrightarrow{n} g$ implies $f \xrightarrow{n+1} g$. Heuristically, $\Psi_H^n f$ is the maximum amount of matter we can bring into $H$ by mapping various parts of $f$ at most $n$ times. As $f \xrightarrow{1} g_1$ implies $\|g_1\|_1 \leqq \|f\|_1$, it is clear that $\Psi_H f \leqq \|f\|_1$. Obviously $\Psi_H^n$ and $\Psi_H$ are positively homogeneous: for $\alpha > 0$, $\Psi_H^n(\alpha f) = \alpha \Psi_H^n f$ and $\Psi_H(\alpha f) = \alpha \Psi_H f$. Let $E_\infty(h)$ denote the union of the sets $E_n(h) = \{M_n h > 0\}$.

**Lemma 2.5.** *For* $f, g \in L_1^+$, $h = f - g$ *and* $n \geqq 1$, $H \subset E_n(h)$ *implies* $\Psi_H^{n-1} f \geqq \int 1_H g\,d\mu$ *and* $H \subset E_\infty(h)$ *implies* $\Psi_H f \geqq \int 1_H g\,d\mu$.

*Proof.* By theorem 2.2, $U^{n-1}h$ is non negative on $E_n(h)$, and by Lemma 2.1, $U^{n-1}h = f_{n-1} - g$ with $f \xrightarrow{n-1} f_{n-1}$. If $H \subset E_n(h)$, then $0 \leqq \int 1_H U^{n-1} h$ $= \int 1_H(f_{n-1} - g)$ implies the first assertion, and the second follows by applying the first to $H_n = H \cap E_n(h)$ and letting $n$ tend to $\infty$. □

**Lemma 2.6.** $H \subset \{\limsup S_n(f-g) > 0\}$ *with* $f, g \in L_1^+$ *implies* $\Psi_H f \geqq \Psi_H g$.

*Proof.* If, for some $n \geqq 1$, $g \xrightarrow{n} g_1$, then by Lemma 2.4 lim sup $S_n(\gamma f - g_1)$ is strictly positive on $H$ for all $\gamma > 1$. Hence $H \subset E_\infty(\gamma f - g_1)$. By the previous lemma $\gamma \Psi_H f = \Psi_H(\gamma f) \geqq \int 1_H g_1$. As $\gamma > 1$ was arbitrary $\Psi_H f \geqq \int 1_H g_1$. As this holds for all $n$ and for all $g_1$ with $g \xrightarrow{n} g_1$ the assertion follows. □

With these tools it is easy to prove the theorem of Chacon-Ornstein [1960]:

**Theorem, 2.7.** *Let* $T$ *be a positive contraction in* $L_1$, $f \in L_1$ *and* $g \in L_1^+$, *then* $S_n f / S_n g$ *converges a.e. on* $\{S_\infty g > 0\}$ *to a finite limit.*

*Proof.* We may assume $f \geqq 0$. First notice that $\Psi_H g$ is strictly positive for any $H \subset \{S_\infty g > 0\}$ with $\mu(H) > 0$ because $g \xrightarrow{n} T^n g$. On $\{S_\infty g > 0\}$ we set $\bar{h} = \limsup (S_n f / S_n g)$ and $\underline{h} = \liminf (S_n f / S_n g)$. We need not define $\bar{h}$ and $\underline{h}$ elsewhere. On $\{\bar{h} > \alpha\}$ we have $\limsup (S_n(f - \alpha g)/S_n g) > 0$. As $S_n g$ increases this shows that $\{\bar{h} > \alpha\} \subset \{\limsup S_n(f - \alpha g) > 0\}$.

Assume that $H = \{\bar{h} = \infty\} \cap \{S_\infty g > 0\}$ has positive measure. Then, by lemma 2.6, we obtain for *all* $\alpha > 0$

$$\Psi_H f \geqq \Psi_H(\alpha g) = \alpha \Psi_H g \quad \text{with } \Psi_H g > 0.$$

This is a contradiction to $\Psi_H f \leqq \|f\|_1 < \infty$. Hence $\bar{h} < \infty$ on $\{S_\infty g > 0\}$. If the assertion of the theorem is wrong there exists a subset $G \subset \{S_\infty g > 0\}$ with $\mu(G) > 0$ and some rational numbers $\alpha$, $\beta$ with $G \subset \{\underline{h} < \alpha < \beta < \bar{h}\}$. Now $G$ is contained in $\{\limsup S_n(f - \beta g) > 0\}$ and in $\{\limsup S_n(\alpha g - f) > 0\}$. By lemma 2.6, $\Psi_G f \geqq \beta \Psi_G g$ and $\Psi_G f \leqq \alpha \Psi_G g$. This contradicts $0 < \Psi_G g < \infty$. □

### Notes

The Chacon-Ornstein theorem unifies many results. The special case where $T$ is given by a Markov chain with countable state space and $f$, $g$ are indicator functions of single states goes back to Doeblin [1938]. The special case where $T$ is given by a conservative nonsingular transformation is due to Hurewicz [1944]; see also Halmos [1946]. For a Markoff process with finite invariant measure early results were due to Doob [1938], [1948] and to Kakutani [1940]. Here we have followed the approach of Akcoglu-Chacon [1970a]. It is rather straight-forward to derive the continuous parameter variant of the Chacon-Ornstein theorem from the present discrete time version.

Fukushima [1974] developed a potential theory for "standard Markov processes", and he strengthened the Chacon-Ornstein theorem for this setting, showing convergence except on an almost polar set. For generalizations; see Sur [1977], [1978], Fukushima [1983]. Direev [1981] extended this to additive functionals.

## § 3.3 Brunel's lemma and the identification of the limit

We have not yet identified the limit of the sequence $S_n f / S_n g$ in the Chacon-Ornstein theorem. This is now usually done via a deep lemma of Brunel [1963], which has considerably influenced the whole subject including the previous section. The identification of the limit will lead to an ergodic theorem of Dunford-Schwartz concerning Cesàro averages, and of a generalization due to Chacon. We also discuss the question of convergence of $A_n f$ in $L_1$-norm.

**1. Equilibrium potentials.** Let $T$ be a positive contraction in $L_1$ and $\alpha$ a measurable function on $\Omega$ with $0 \leqq \alpha \leqq 1$. Then $f \to T_\alpha f := \alpha f + T((1 - \alpha) f)$ is a positive contraction in $L_1$, called a *partial application* of $T$.

The filling operator $U$ has the form $Uf = T_\alpha f$ with $\alpha = 1_{\{f<0\}}$. For $f, g \in L_1^+$ the relation $f \xrightarrow{1} g$ is equivalent to the existence of an $\alpha$ with $g = T_\alpha f$. Hence, for $E \in \mathscr{A}$,

$$\Psi_E^n f = \sup \{ \int_E T_{\alpha_n} T_{\alpha_{n-1}} \dots T_{\alpha_1} f d\mu \},$$

where the sup ranges over all sequences consisting of $n$ measurable functions $\alpha_i\colon \Omega \to [0,1]$.

We are particularly interested in the case $\alpha = 1_E$. If $I_E$ is the operator $f \to I_E f = 1_E f$, then $T_E := T_{1_E}$ may be written as $T_E = I_E + T I_{E^c}$. The next lemma confirms the intuitive feeling that this partial application of $T$ should be most efficient for bringing matter into $E$:

**Lemma 3.1.** *For all $f \in L_1^+$ and $n \geqq 1$, $\Psi_E^n f = \int_E T_E^n f d\mu$.*

*Proof.* The lemma follows by induction if we prove that for all $n \geqq 0$, $f \in L_1^+$, and measurable $\alpha\colon \Omega \to [0,1]$:

$$\int_E T_E^n T_\alpha f \leqq \int_E T_E^{n+1} f. \tag{3.1}$$

To this end set $\beta = \alpha 1_E$. We first show by induction that

$$T_E^n T_\beta f - T_E^n T_\alpha f = (T I_{E^c})^{n+1} (\alpha f) - I_{E^c} (T I_{E^c})^n (\alpha f). \tag{3.2}$$

Indeed, for $n = 0$,

$$T_\beta f - T_\alpha f = \beta f + T((\mathbb{1} - \beta) f) - \alpha f - T((\mathbb{1} - \alpha) f) = T(1_{E^c} \alpha f) - 1_{E^c} \alpha f,$$

and if (3.2) holds for some $n$, then

$$\begin{aligned} T_E^{n+1} T_\beta f - T_E^{n+1} T_\alpha f &= (I_E + T I_{E^c}) ((T I_{E^c})^{n+1} (\alpha f) - I_{E^c} (T I_{E^c})^n (\alpha f)) \\ &= (T I_{E^c})^{n+2} (\alpha f) + I_E (T I_{E^c})^{n+1} (\alpha f) - (\mathrm{T} I_{E^c}) I_{E^c}) (T I_{E^c})^n (\alpha f) \\ &= (T I_{E^c})^{n+2} (\alpha f) - I_{E^c} (T I_{E^c})^{n+1} (\alpha f). \end{aligned}$$

The identity (3.2) implies $1_E T_E^n T_\alpha f \leqq 1_E T_E^n T_\beta f$.

We may therefore assume $\alpha 1_{E^c} = 0$ in the proof of (3.1), and obtain successively

$$\begin{aligned} T_E f - T_\alpha f &= (1_E - \alpha) f - T((1_E - \alpha) f) \\ T_E^2 f - T_E T_\alpha f &= (1_E - \alpha) f - T_E T((1_E - \alpha) f) \\ T_E^{n+1} f - T_E^n T_\alpha f &= (1_E - \alpha) f - T_E^n T((1_E - \alpha) f). \end{aligned}$$

(3.1) now follows from

$$\int_E T_E^n T(f(1_E - \alpha)) \leqq \int_\Omega f(1_E - \alpha) = \int_E f(1_E - \alpha). \quad \square$$

By induction we obtain

$$(3.3) \qquad T_E^n = \sum_{\nu=0}^{n-1} I_E(TI_{E^c})^\nu + (TI_{E^c})^n, \quad (n \geqq 1).$$

Hence, for $f \in L_1^+$,

$$(3.4) \qquad \Psi_E^n f = \int_E \sum_{\nu=0}^{n} (TI_{E^c})^\nu f d\mu = \int f \psi_E^n d\mu$$

and

$$(3.5) \qquad \Psi_E f = \lim_n \Psi_E^n f = \int_E \sum_{\nu=0}^{\infty} (TI_{E^c})^\nu f d\mu = \int f \psi_E d\mu$$

with

$$(3.6) \qquad \psi_E^n := \sum_{\nu=0}^{n} (I_{E^c} T^*)^\nu 1_E \quad \text{and} \quad \psi_E := \lim_n \psi_E^n = \sum_{\nu=0}^{\infty} (I_{E^c} T^*)^\nu 1_E .$$

(We need not distinguish $I_E$ and $I_E^*$ because also $I_E^*$ has the form $h \to 1_E h$). Inductively, we can see that $\psi_E^n \leqq \mathbb{1}$, because then $T^* \psi_E^n \leqq \mathbb{1}$ implies $\mathbb{1} \geqq 1_E \vee T^* \psi_E^n = 1_E + 1_{E^c} T^* \psi_E^n = \psi_E^{n+1}$. Lemma 3.1 and lemma 2.6 now almost instantly yield

**Theorem 3.2** (Brunel's lemma). *Let $T$ be a positive contraction in $L_1, f \in L_1$, and $E \subset \{\omega\colon S_n f(\omega) > 0$ for infinitely many $n\}$, then $\int f \psi_E d\mu \geqq 0$.*

*Proof.* Let $p \in L_1^+$ be strictly positive in $\Omega$ and $f_\varepsilon = f + \varepsilon p$.

Then, for all $\varepsilon > 0$, $E$ is contained in $\{\limsup_{n\to\infty} S_n f_\varepsilon > 0\}$. Lemma 2.6 implies $\Psi_E f_\varepsilon^+ \geqq \Psi_E f_\varepsilon^-$. When $\varepsilon$ decreases to 0, $f_\varepsilon^+ \downarrow f^+$ and $f_\varepsilon^- \uparrow f^-$ a.e. and in $L_1$-norm. As $\psi_E$ is bounded,

$$\begin{aligned} \int f^+ \psi_E d\mu &= \lim_\varepsilon \int f_\varepsilon^+ \psi_E d\mu = \lim_\varepsilon \Psi_E f_\varepsilon^+ \geqq \lim_\varepsilon \Psi_E f_\varepsilon^- \\ &= \lim_\varepsilon \int f_\varepsilon^- \psi_E d\mu = \int f^- \psi_E d\mu . \quad \square \end{aligned}$$

Heuristically, Brunel's lemma says that if $E$ is contained in $\{\omega\colon S_n f(\omega) > 0$ for infinitely many $n\}$ then we can bring more mass into $E$ starting with a mass distribution $f^+$ than with $f^-$. The function $\psi_E$ has been called the *equilibrium potential* or the *Brunel function* of $E$ in the literature. It is the minimal superharmonic function dominating $1_E$. In fact, the recursive relation between $\psi_E^n$ and $\psi_E^{n+1}$ is a special case of a construction in potential theory: For any $g_0 \in L_\infty^+$ put $g_{n+1} = g_0 \vee T^* g_n$, $(n \geqq 0)$. By induction $(g_n)$ is monotonely increasing and $\|g_n\|_\infty \leqq \|g_0\|_\infty$. If $g_\infty := \lim g_n$, then $T^* g_\infty = \lim T^* g_n \leqq \lim g_{n+1} = g_\infty$, as $T^*$ is sub-Markovian. If $h$ is any superharmonic function with $h \geqq g$ then by induction $h \geqq g_n$ for all $n$. Thus, $g_\infty$ is the minimal superharmonic function $\geqq g_0$.

It may also be worth mentioning the identity

$$(3.7) \qquad \psi_E^n = T_E^{*n} 1_E$$

which is trivial for $n = 0$ and follows for $n \geqq 1$ from

$$\langle f, \psi_E^n \rangle = \Psi_E^n f = \int_E T_E^n f \, d\mu = \langle T_E^n f, 1_E \rangle = \langle f, T_E^{*n} 1_E \rangle .$$

If $T$ is given by transition probabilities $P(\omega, A)$ as in § 3.1, then the transition probabilities $P_E(\omega, A)$ for $T_E$ are $= P(\omega, A)$ for $\omega \in E^c$, and for $\omega \in E$, $P_E(\omega, \cdot)$ admits only the deterministic transition $\omega \to \omega$. $T_E^{*n} 1_E(\omega)$ is the probability of being in $E$ at time $n$ given a start in $\omega$ subject to the transition probabilities $P_E(\cdot, \cdot)$. This is equal to the probability of being in $E$ before time $n$ or at time $n$ subject to the original transition probabilities. We may therefore interpret $\psi_E(\omega)$ as the probability of ever being in $E$ given a start in $\omega$.

**2. Identification of the limit.** Our next aim is to determine the limit of $S_n f / S_n g$ in the Chacon-Ornstein theorem on the conservative part $C$.

Let $\mathscr{C}$ be the family of $T$-absorbing subsets of $C$. First assume $\Omega = C$. Then $T^* \mathbb{1} = \mathbb{1}$ is harmonic. $B \in \mathscr{A}$ is $T$-absorbing iff $1_{B^c}$ is superharmonic. As $T$ is conservative any superharmonic function is harmonic. But then $1_B = \mathbb{1} - 1_{B^c}$ must be harmonic, too, and we see that $B$ belongs to $\mathscr{C}$ iff $T^* 1_B = 1_B$. Hence $\mathscr{C}$ is an algebra, and even a $\sigma$-algebra because $T^*$ is sub-Markovian.

In the general case $\mathscr{C}$ is a $\sigma$-algebra in $C$ because the restriction of $T$ to $L_1(C, C \cap \mathscr{A}, \mu)$ is conservative and $A \subset C$ is $T$-absorbing iff it is absorbing under this restriction.

$T$ is called *ergodic if* $T^* 1_A = 1_A$ implies $A = \emptyset$ or $A = \Omega \bmod \mu$. In the conservative case this means that $\mathscr{C}$ is trivial.

**Lemma 3.3.** *Assume $\Omega = C$, then $h \in L_\infty$ is harmonic iff $h$ is $\mathscr{C}$-measurable.*

*Proof.* If $h = T^* h$, then $T^* h^+ \geqq h$ and $T^* h^+ \geqq 0$ imply $T^* h^+ \geqq h^+$. Hence $g = \|h\|_\infty \cdot \mathbb{1} - h^+$ is superharmonic and therefore harmonic. Hence $T^* h^+ = h^+$. If $B = \{h > 0\}$, then

$$T^* 1_B = T^* (\lim_n (n h^+ \wedge \mathbb{1})) = \lim_n T^* (n h^+ \wedge \mathbb{1}) \leqq \lim (n h^+ \wedge \mathbb{1}) = 1_B .$$

Now we can again infer that $T^* 1_B = 1_B$ and $B \in \mathscr{C}$. Applying this argument to $h - a$ we obtain $\{h > a\} = \{(h - a) > 0\} \in \mathscr{C}$ for all $a$.

For the converse conclusion approximate $h$ with $\mathscr{C}$-measurable step functions and use that $B \in \mathscr{C}$ implies $T^* 1_B = 1_B$. □

In the general non conservative case we shall have to study the influence of the dissipative part in $C$, the contribution of $1_D f$ to the limit in $C$. For this we shall need (for $E = C$) the operator

(3.8) $$H_E = I_E \sum_{\nu=0}^{\infty} (T I_{E^c})^\nu .$$

As $T_E^n$ is a positive contraction in $L_1$ for each $n$ and $E \in \mathscr{A}$, (3.3) implies $\|I_E \sum_{\nu=0}^{n} (TI_{E^c})^\nu f\|_1 \leqq \|f\|_1$ for all $f \in L_1^+$. Therefore, $H_E$ is a positive contraction in $L_1$. Using $H_E$, the identity (3.5) can be written as

$$\Psi_E f = \langle H_E f, 1_E \rangle = \langle H_E f, \mathbb{1} \rangle \tag{3.9}$$

and we see that $\psi_E = H_E^* \mathbb{1}$.

The operator $H_E$ maps the mass $f$ to the place where it first arrives in $E$ under iterated applications of $T$. Recall that $g\mu$ denotes the measure $\nu$ with $d\nu/d\mu = g$. $\nu|\mathscr{C}$ will be the restriction of $\nu$ to the $\sigma$-algebra $\mathscr{C}$.

**Theorem 3.4** (Neveu-Chacon). *If $T$ is a positive contraction in $L_1$, $f \in L_1$, and $g \in L_1^+$, then*

$$\lim_{n\to\infty} \frac{S_n f}{S_n g} = \frac{d((H_C f)\,\mu|\mathscr{C})}{d((H_C g)\,\mu|\mathscr{C})} \quad \text{a.e. } on \ \{S_\infty g > 0\} \cap C.$$

If $\mu$ is a probability measure, the Radon-Nikodym derivative on the right side is the ratio $E(H_C f|\mathscr{C})/E(H_C g|\mathscr{C})$ of the conditional expectations. This is the most common way of writing the limit, but it does not simplify matters because a conditional expectation *is* a Radon-Nikodym derivative.

*Proof.* (a) We at first assume $\Omega = C$. Then $H_C$ is the identity. If $A = \{S_\infty g > 0\}$, then $S_n g$ tends to $\infty$ on $A$. For all $h \in L_1^+(A)$, $h_n = h \wedge S_n g$ tends to $h$ and $Th_n \leqq S_{n+1} g$ vanishes in $A^c$. Hence, $Th = \lim Th_n$ belongs to $L_1(A)$. This implies $A \in \mathscr{C}$. For any pair $a < b$ we now consider the set $B = \{a < \lim S_n f/S_n g < b\}$. On $B$, $S_n(f - ag)$ is positive infinitely often. Brunel's lemma implies $\int (f - ag)\,\psi_B\, d\mu \geqq 0$.

Put $B' = \{\psi_B = 1\}$, then $B \subset B' \in \mathscr{C}$ and $1_B \leqq 1_{B'} = T^* 1_B$ imply $\psi_B \leqq 1_{B'}$. Hence $\psi_B = 1_{B'}$ and $\int (f - ag)\, 1_{B'}\, d\mu \geqq 0$. The same argument with $B$ replaced by $B \cap G$ and $G \in \mathscr{C}$ shows $\int (f - ag)\, 1_{G \cap B'}\, d\mu \geqq 0$. As $G$ was arbitrary, $a \leqq d(f\mu|\mathscr{C})/d(g\mu|\mathscr{C})$ on $B'$. In the same way one can show $d(f\mu|\mathscr{C})/d(g\mu|\mathscr{C}) \leqq b$ on $B'$. As $a < b$ were arbitrary this proves the theorem for $\Omega = C$.

(b) Let us write $L(f, g) := \lim S_n f/S_n g$. We now consider the case $\Omega \neq C$. We may assume $f \geqq 0$. Choose some fixed $p \in L_1^+$ with $\{p > 0\} = C$. If $f = r + s$ and $f' = r + Ts$, then $S_n f' = S_n f + T^n s - s$. Using lemma 2.3 we therefore see that $f \overset{1}{\rightarrow} f'$ implies $L(f', p) = L(f, p)$ on $C$. By induction this remains true for $f \overset{m}{\rightarrow} f'$. Set $f_m = 1_C T_C^m f$. Then $f_m \leqq T_C^m f$ and $f \overset{m}{\rightarrow} T_C^m f$ imply $L(f_m, p) \leqq L(f, p)$ on $C$. We may already apply the identification of the limit in the conservative case to $f_m$ because $T$ restricted to $L_1(C)$ is conservative and $f_m \in L_1(C)$. This has the following consequence: if $f_m$ is an increasing sequence in $L_1(C)$ tending to some $f_\infty \in L_1(C)$, then $\lim_m L(f_m, p) = L(f_\infty, p)$ in $C$. Therefore $f_m = 1_C T_C^m f \uparrow H_C f$ yields $L(H_C f, p) \leqq L(f, p)$ in $C$.

On the other hand,

$I_C T^k f = I_C T^k(I_C f) + I_C T^{k-1}(TI_{C^c})f = \ldots = I_C T^k(I_C f) + I_C T^{k-1}(I_C(TI_{C^c})^1 f)$
$+ I_C T^{k-2}(I_C(TI_{C^c})^2 f) + \ldots + I_C(TI_{C^c})^k f$ implies

$$\sum_{k=0}^{n-1} I_C T^k f = \sum_{k=0}^{n-1} \sum_{\nu=0}^{k} I_C T^\nu (I_C(TI_{C^c})^{k-\nu} f) \leqq \sum_{\nu=0}^{n-1} I_C T^\nu H_C f.$$

Hence $S_n H_C f \geqq S_n f$ for all $n$ on $C$. Together with the inequality proved above we obtain $L(H_C f, p) = L(f, p)$ in $C$. Similarly $L(H_C g, p) = L(g, p)$ in $C$.

On $C \cap \{S_\infty g > 0\}$ we may write $S_n f(\omega)/S_n g(\omega) = (S_n f(\omega)/S_n p(\omega))$. $(S_n g(\omega)/S_n p(\omega))^{-1}$ for $n$ large enough. Hence $L(f, g) = L(f, p)/L(g, p)$. Similarly, $L(H_C f, H_C g) = L(H_C f, p)/L(H_C g, p)$ on $C \cap \{S_\infty H_C g > 0\}$ $= C \cap \{S_\infty g > 0\}$. Combining our results we have proved $L(f, g)$ $= L(H_C f, H_C g)$ and may complete the proof by applying the conservative case to $H_C f, H_C g \in L_1(C)$. □

**Changing the reference measure.** If $\mu'$ is a $\sigma$-finite measure with $\mu' \sim \mu$ the results above can also be expressed in terms of $L_1(\mu')$. $\mu' \sim \mu$ is equivalent to the existence of a strictly positive measurable $p$ with $\mu'(A) = \int 1_A p \, d\mu$, $(A \in \mathscr{A})$. Clearly, $\tilde{L}_1(\mu) = \tilde{L}_1(\mu')$. Exploiting the isomorphisms $L_1(\mu) \leftrightarrow \tilde{L}_1(\mu)$ and $L_1(\mu') \leftrightarrow \tilde{L}_1(\mu')$ the operator $T'$ in $L_1(\mu')$ corresponding to $T$ in $L_1(\mu)$ has the form $T'f' = p^{-1} T(f' \cdot p)$. $(f' \in L_1(\mu') \Leftrightarrow f' \cdot p \in L_1(\mu))$. It is trivial to check that $T'$ has the same conservative part and the same system of absorbing sets. Clearly $(T')^n f' = p^{-1} T^n(f' \cdot p)$. This allows us to pass always to a probability measure $\mu' \sim \mu$.

**3. Convergence of Cesàro averages.** The reader may have wondered why we have studied the question of convergence of $S_n f/S_n g$ before thinking about the more obvious question of convergence of $A_n f = n^{-1} S_n f$. There are several reasons. If there exists an integrable strictly positive $p$ with $Tp = p$, we have $A_n f$ $= p S_n f/S_n p$ and we see that the two questions are equivalent. We shall even be able to derive the convergence a.e. of $A_n f$ if there exists only a strictly positive $p$ with $Tp \leqq p$, which need not be integrable. In this case, however, the limit of $A_n f$ frequently is 0 and the convergence of $S_n f/S_n g$ may contain more information. Finally: if no extra condition is added, the averages $A_n f$ need not converge a.e., (§ 3.6), and we obtain only stochastic convergence of $A_n f$, (§ 3.4).

Passing to the reference measure $\mu' = p \cdot \mu$ the a.e.-convergence in the following theorem may already be deduced from theorem 1.7.3 even under the weaker assumption $Tp \leqq p$. But the full assumption $Tp = p$ allows to give simple expressions for the limit.

**Theorem 3.5** (Hopf). *Let $T$ be a positive contraction in $L_1(\mu)$ for which there exists a strictly positive measurable $p \in L_1$ with $Tp = p$. Then*

*(i) for* $f \in L_1(\mu)$, $A_n f$ *converges a.e. to* $p \cdot d(f \cdot \mu|\mathscr{C})/d(p \cdot \mu|\mathscr{C})$;
*(ii) for* $g$ *with* $pg \in L_1(\mu)$, *in particular for* $g \in L_\infty$,

$$\lim_{n\to\infty} A_n^* g = d((p \cdot g)\,\mu|\mathscr{C})/d(p\mu|\mathscr{C}) \quad \text{a.e..}$$

*Proof.* (i) Apply theorem 3.4 to $g = p$ and observe $\Omega = C$.

(ii) Let $\mu' = p\mu$. Then $pg \in L_1(\mu)$ is equivalent to $g \in L_1(\mu')$. For $g \in L_\infty$, $\int (T^* g)\,d\mu' = \int p(T^* g)\,d\mu = \int (Tp)g\,d\mu = \int pg\,d\mu = \int g\,d\mu'$. Hence $T^*$ preserves $\mu'$-integrals and may be extended to a positive contraction in $L_1(\mu')$. In particular, $S_n^* g$ is well defined for all $g \in L_1(\mu')$. As $T$ is conservative $\mathbb{1} = T^*\mathbb{1}$. Hence $A_n^* g = S_n^* g / S_n^* \mathbb{1}$. As $\mathbb{1}$ is $T^*$-invariant and integrable, $T^*$ is a conservative contraction in $L_1(\mu')$. Check that is has the same absorbing sets as $T$ and apply theorem 3.4. □

We now come to a rather general ratio ergodic theorem.

**Definition 3.6.** A sequence $(p_k)$, $(k = 0, 1, 2, \ldots)$, of non negative measurable functions $p_k$ on $\Omega$ is called *admissible* (for a positive contraction $T$ in $L_1$) if $Tp_k \leqq p_{k+1}$ holds for all $k \geqq 0$.

The $p_k$ need not be integrable. If $p$ is any measurable and non negative function the sequence $p_k = T^k p$ is admissible. If $p$ is a non negative function with $Tp \leqq p$, also the sequence $p_k \equiv p$ is admissible.

**Theorem 3.7** (Chacon [1963]). *Let* $(p_k)$ *be admissible for a positive contraction* $T$ *in* $L_1$ *and* $f \in L_1$, *then*

$$D_n(f, (p_k)) := \frac{S_n f}{\sum_{k=0}^{n-1} p_k}$$

*converges a.e. on* $\{\sum_{k=0}^{\infty} p_k > 0\}$.

*Proof.* We may assume $f \geqq 0$. We may also assume $\Omega = C$, since $S_\infty f$ converges in $D$, $C$ is $T$-absorbing, and the sequence $(1_C p_k)$ is again admissible. Let $e$ be a strictly positive integrable function. Set $q_k = d(p_k\mu|\mathscr{C})/d(e\mu|\mathscr{C})$. (This Radon-Nikodym derivative is well defined on the largest set (mod $\mu$) where the restriction of $p_k\mu$ to $\mathscr{C}$ is $\sigma$-finite. We put $q_k = \infty$ on the complement of this set). For $E \in \mathscr{C}$, $\langle p_k, 1_E\rangle = \langle p_k, T^* 1_E\rangle = \langle Tp_k, 1_E\rangle \leqq \langle p_{k+1}, 1_E\rangle$. It follows that the sequence $q_k$ is increasing. If $i \geqq 0$ is a fixed integer consider a sequence $g_\nu \in L_1$, with $\{g_\nu > 0\} = \{p_i > 0\}$, increasing to $p_i$.

For any $\nu$ we have on $\{p_i > 0\}$

$$\limsup_{n\to\infty} D_n(f,(p_k)) \leqq \limsup_{n\to\infty} \frac{\sum_{k=i}^{n} T^k f}{\sum_{k=i}^{n} p_k} \leqq \limsup_{j\to\infty} \frac{S_j T^i f}{S_j g_v}$$

$$= \frac{d(f\cdot\mu|\mathscr{C})}{d(g_v\cdot\mu|\mathscr{C})} = \frac{d(f\cdot\mu|\mathscr{C})}{d(e\mu|\mathscr{C})}\cdot\frac{d(e\mu|\mathscr{C})}{d(g_v\mu|\mathscr{C})}.$$

If $v$ tends to infinity $d(g_v\mu|\mathscr{C})/d(e\mu|\mathscr{C})$ tends to $q_i$ and we obtain

$$(3.10)\quad \limsup_{n} D_n(f,(p_k)) \leqq \frac{d(f\mu|\mathscr{C})}{d(e\mu|\mathscr{S})}\cdot\frac{1}{q_i}.$$

In particular, the limit of $D_n(f,(p_k))$ exists and is 0 on $\{q_\infty=\infty\}$, where $q_\infty=\lim q_i$. It remains to prove the convergence on the sets $A_K=\{q_\infty\leqq K\}$ for $K>0$. As these sets belong to $\mathscr{C}$ we may assume $\Omega=A_K$. Put

$$u_0=p_0,\ u_i=p_i-Tp_{i-1},\ w_n=\sum_{i=0}^{n} u_i,\ w=\sum_{i=0}^{\infty} u_i.$$

The $u_i$ are non negative. It follows from $\int Tp_{i-1}=\int p_{i-1}$ that $\|w_n\|_1=\|p_n\|_1$. As $q_\infty$ is bounded we have $\|p_n\|_1=\|q_n\|_1\leqq\|q_\infty e\|_1\leqq K\|e\|_1<\infty$, and $w$ is integrable. For all $E\in\mathscr{C}$,

$$\int_E u_i=\int_E p_i-\int_E Tp_{i-1}=\int_E p_i-\int_E p_{i-1}.$$

Hence $d(w_n\mu|\mathscr{C})/d(e\mu|\mathscr{C})=q_n$, and $d(w\mu|\mathscr{C})/d(e\mu|\mathscr{C})=q_\infty$.

By induction $p_k=Tp_{k-1}+(p_k-Tp_{k-1})$ yields

$$p_k=T^k p_0+\sum_{v=0}^{k-1} T^v(p_{k-v}-Tp_{k-1-v}),\quad (k\geqq 1).$$

Summing from 0 to $n$ we find

$$\sum_{k=0}^{n} p_k=\sum_{k=0}^{n} T^k p_0+\sum_{k=0}^{n-1} T^k(p_1-Tp_0)+\sum_{k=0}^{n-2} T^k(p_2-Tp_1)$$

$$+\ldots+(p_n-Tp_{n-1})\leqq\sum_{k=0}^{n} T^k w_n\leqq\sum_{k=0}^{n} T^k w.$$

Appealing to theorem 3.4 again we obtain

$$\liminf_{n\to\infty} D_n(f,(p_k))\geqq\lim S_n f/S_n w=(d(f\mu|\mathscr{C})/d(e\mu|\mathscr{C})\cdot q_\infty^{-1}.\quad\square$$

It is clear from the proof of this theorem that

$$(3.11)\quad \lim_{n\to\infty} D_n(f,(p_k))=\frac{d((H_C f)\mu|\mathscr{C})}{d(e\mu|\mathscr{C})}\cdot\frac{1}{q_\infty}\quad \text{on } C$$

where $q_\infty=\lim_k q_k$ and $q_k=d((1_C p_k)\mu|\mathscr{C})/d(e\mu|\mathscr{C})$.

**Corollary 3.8** (Dunford-Schwartz [1956]). *If $T$ is a positive $L_1 - L_\infty$-contraction $A_n f$ converges a.e. for all $f \in L_1$.*

*Proof.* Take $p_k \equiv \mathbb{1}$. □

**Definition 3.9.** A non negative real valued measurable (not necessarily integrable) function $p$ is called *subinvariant* (or $T$-subinvariant) if it satisfies $Tp \leqq p$. If $Tp = p$, it is called *invariant*. Similarly, a measure $\nu = p \cdot \mu$ is called subinvariant (invariant) if $p$ is subinvariant (invariant).

If a subinvariant $\sigma$-finite measure $p\mu = \nu \sim \mu$ exists, then we may use $\mu' = \nu$ as a new reference measure and $T'g = p^{-1} T(gp)$ is an $L_1 - L_\infty$-contraction in $L_1(\mu')$, because $T'\mathbb{1} = p^{-1} Tp \leqq \mathbb{1}$. An application of Corollary 3.8 to $T'$ shows that $A_n f$ converges $\mu$-a.e. for all $f \in L_1(\mu)$.

It follows from $\int T^* h \, d\mu' = \int (T^* h) p \, d\mu = \int h (Tp) \, d\mu \leqq \int hp \, d\mu = \int h \, d\mu'$, $(h \geqq 0)$, that $T^*$ is an $L_1 - L_\infty$-contraction in $L_1(\mu')$. Consequently also $A_n^* h$ converges a.e. for all $h \in L_1(\mu')$. $T^*$ in $L_1(\mu')$ is called the *dual Markov operator to $T$.*

**Lemma 3.10.** *If a measurable real valued $p \geqq 0$ is $T$-subinvariant, then $p = Tp$ holds on $C$. In this case $p_C = 1_C p$ is invariant. If a $T$-subinvariant, strictly positive real valued $p$ exists $D$ is $T$-absorbing.*

*Proof.* Choose some $h \in L_\infty^+$ with $\langle p, h \rangle < \infty$ and with $\{h > 0\} = C$. For all $n$,

$$\langle p - Tp, \sum_{i=0}^{n-1} T^{*i} h \rangle = \langle p, h \rangle - \langle p, T^{*n} h \rangle \leqq \langle p, h \rangle < \infty .$$

As $S_n^* h$ diverges on $C$ we obtain $p = Tp$ on $C$. As $C$ is $T$-absorbing and $Tp_C \leqq p$ the function $p_C$ is subinvariant and the second assertion follows from the first. But then $T(1_D p)$ must vanish on $C$, and thus, for all $f \in L_1^+(D)$, $Tf = \lim_n (T(f \wedge n 1_D p))$ must vanish on $C$. □

Lemma 3.10 shows that the assumption of existence of a $T$-invariant $\nu \sim \mu$ is fairly strong because it excludes any "influence" of the dissipative part in $C$. We therefore mention that the a.e.-convergence of $A_n f$ for all $f \in L_1$ follows also from the weaker assumption of existence of a subinvariant $p$ with $\{p > 0\} = C$. To prove this we may assume $f \geqq 0$. Now the convergence of $A_n f$ to 0 is trivial on $D \cup \{S_\infty f = 0\}$. On $C \cap \{S_\infty f > 0\}$ theorem 3.4 implies $\lim_{n \to \infty} S_n(H_C f)/S_n f = 1$. We can therefore replace $f$ by $H_C f$ and apply the result proved above to the restriction of $T$ to $L_1(C)$ and to $H_C f$.

The limit is determined by (3.11) with $p_k = p$. But there is a more explicit description: Let us say that a finite invariant measure exists on a set $A$ if there is some invariant $q \in L_1^+$ with $\{q > 0\} \supset A$. By an exhaustion argument we can find a sequence of sets $A_1, A_2, \ldots$ and invariant functions $\tilde{q}_1, \tilde{q}_2, \ldots \in L_1^+$ with $A_i = \{\tilde{q}_i > 0\}$ such that no finite invariant measure exists on any $B$ of positive measure in the complement of the union $\tilde{C}$ of $A_1, A_2, \ldots$. The function $\tilde{p} = \sum_{i=1}^{\infty} 2^{-i} \|\tilde{q}_i\|_1^{-1} \tilde{q}_i$ is $T$-invariant and integrable and has support $\{\tilde{p} > 0\} = \tilde{C}$. $\tilde{C}$ is the maximal set on which there exists a finite invariant measure. Clearly $\tilde{C} \subset C$. $\tilde{C}$ will be studied more carefully in the next section.

**Lemma 3.11.** *If a non negative measurable $p$ is $T$-subinvariant, its support $\{p > 0\}$ is $T$-absorbing.*

*Proof.* For $f \in L_1^+$ with $\{f > 0\} \subset \{p > 0\}$ the sequence $f \wedge np$ increases to $f$ and $Tf = \lim_n T(f \wedge np)$. □

**Theorem 3.12.** *Let $T$ be a positive contraction in $L_1$. Assume that there exists a real valued subinvariant $p \geqq 0$ with $\{p > 0\} = C$. Then $A_n f$ converges a.e. for all integrable $f$. The limit vanishes in the complement of $\tilde{C}$.* Set $\tilde{f} = 1_{\tilde{C}} H_{\tilde{C}} f = H_C f$. *Then the limit equals $\tilde{p} \cdot d(\tilde{f}\mu|\mathscr{C})/d(\tilde{p}\mu|\mathscr{C})$ on $\tilde{C}$.*

*Proof.* As $\tilde{C}$ is contained in $C$ and $T$-absorbing, also $C \setminus \tilde{C}$ is $T$-absorbing. Using this fact it is not hard to verify the heuristically obvious identity $1_{\tilde{C}} H_{\tilde{C}} f = H_C f$. The formula for the limit on $C$ therefore follows from theorem 3.5 and it remains to show that the limit vanishes on $C \setminus \tilde{C}$. We may assume $\Omega = C \setminus \tilde{C}$. Set $p_k = p$ again. The restriction of $p_k \mu$ to $\mathscr{C}$ has no sets of positive finite measure, because for any such set $A$ the function $p_k 1_A$ would be integrable and $T$-invariant. It follows that the $q_k$ in the proof to theorem 3.7 are $\equiv \infty$ and hence $q_\infty \equiv \infty$. By (3.11) $\lim_{n\to\infty} A_n f = \lim_{n\to\infty} D_n(f, (p_k)) \cdot p \equiv 0$. □

**Remarks on $L_1$-convergence.** In the dissipative case the sequence $A_n f$ for $f \in L_1$ may or may not converge in $L_1$-norm; see § 4.3.

In general, if $A_n f$ converges in norm to some $\bar{f}$ then $\bar{f}$ is $T$-invariant. If $T$ is conservative $T^* \mathbb{1} = \mathbb{1}$ implies $\int A_n f = \int f$. Hence, for $f \in L_1^+(C)$ the set $\{\bar{f} > 0\}$ is contained in $\tilde{C}$. As $C \setminus \tilde{C}$ is $T$-absorbing, norm convergence of $A_n f$ cannot hold for any $f \in L_1^+(C \setminus \tilde{C})$ with $f \neq 0$.

On the other hand $L_1$-convergence of $A_n f$ for $f \in L_1(\tilde{C})$ is a rather simple result; see § 2.1. It follows also from a.e.-convergence together with uniform integrability.

**Definition 3.13.** A sequence $(f_n)$ in $L_1$ is called *uniformly integrable* if, for each $\varepsilon > 0$, there exists a $g_\varepsilon \in L_1^+$ with $\int (|f_n| - g_\varepsilon)^+ d\mu < \varepsilon$ for all $n$.

To prove the uniform integrability of $(A_n f)_{n \geqq 1}$ for $f \in L_1(\tilde{C})$ one can take $g_\varepsilon = k_\varepsilon \tilde{p}$ with $k$ large enough since $(A_n |f| - k\tilde{p})^+ = (A_n(|f| - k\tilde{p}))^+ \leqq A_n(|f| - k\tilde{p})^+$.

We also prove a more general lemma valid even for non integrable $p$:

**Lemma 3.14.** *Let $p \geqq 0$ be $T$-subinvariant for the positive contraction $T$ in $L_1$, $h \in L_\infty^+(\{p>0\})$ and $\int hp\,d\mu < \infty$. Then, for any $f \in L_1$, the sequences $(T^n f)h$ and $(A_n f)h$ are uniformly integrable.*

*Proof.* We may assume $f \geqq 0$ and $h \leqq 1$. Set $E = \{p > 0\}$. For $f \in L_1(E)$ we may argue as above: The uniform integrability of $(hT^n f)$ follows from $(hT^n f - khp)^+ \leqq h(T^n f - kp)^+ \leqq h(T^n f - kT^n p)^+ \leqq hT^n(f-kp)^+$. For general $f$ put $f_\nu = I_E(TI_{E^c})^\nu f$. Then $\sum_{\nu=0}^{\infty} f_\nu = H_E f \in L_1$ and $1_E T^n f = \sum_{\nu=0}^{n} T^{n-\nu} f_\nu$.

As the $f_\nu$ belong to $L_1(E)$ each sequence $(T^n f_\nu)_{n \geqq 0}$ is uniformly integrable, and hence also each sequence $(g_{n,\nu})_{n\geqq 0}$ with $g_{n,\nu} = 0$ for $\nu > n$ and $g_{n,\nu} = T^{n-\nu} f_\nu$ for $\nu \leqq n$. Now, for all $K \geqq 1$,

$$hT^n f = h \sum_{\nu=0}^{K} g_{n,\nu} + h \sum_{\nu=K+1}^{n} T^{n-\nu} f_\nu .$$

As the integral of the second summand can be made arbitrarily small uniformly in $n$ by a large choice of $K$ and the first summand is uniformly integrable as a finite sum of uniformly integrable sequences the sequence $hT^n f$ must be uniformly integrable. But then this must be true also for $(hA_n f)$. □

If $f$ is an initial distribution of a Markov process for which the transition probabilities determine $T$, then $\langle T^k f, 1_A \rangle$ is the probability of a visit in $A$ at time $k$. The lemma can be combined with the Dunford-Schwartz ergodic theorem above to deduce the convergence of Cesàro averages of $\langle T^k f, 1_A \rangle$, when $A$ has finite subinvariant measure.

For functions $f$ with $\int f\,d\mu = 0$ there is $L_1$-convergence in the conservative ergodic case:

**Theorem 3.15.** *If $T$ is a positive contraction in $L_1$ and $f \in L_1(C)$ then $\| A_n f \|_1 \to 0$ holds iff $\int 1_E f\,d\mu = 0$ for all $E \in \mathscr{C}$.*

*Proof.* We may assume $\Omega = C$. The only if part is obvious because $\langle T^k f, 1_E \rangle = \langle f, T^{*k} 1_E \rangle = \langle f, 1_E \rangle$ holds for all $E \in \mathscr{C}$. Now assume $\langle f, 1_E \rangle = 0$ for all $E \in \mathscr{C}$. Then $\langle f, h \rangle = 0$ for all $\mathscr{C}$-measurable $h \in L_\infty$. By lemma 3.3 $h \in L_\infty$ is $\mathscr{C}$-measurable iff it is harmonic. Using $T^* h = h \Leftrightarrow \langle g, T^* h \rangle = \langle g, h \rangle\ \forall g \in L_1 \Leftrightarrow \langle Tg - g, h \rangle = 0\ \forall g \in L_1$ we see that $h$ is harmonic iff $h$ is orthogonal to $(T-I)L_1$. Hence, by Hahn-Banach, $f$ belongs to the closure of $(T-I)L_1$, and for such $f$ the convergence $\| A_n f \|_1 \to 0$ is simple to prove. □

## Notes

Lemma 3.1 is due to Akcoglu-Chacon [1970a]. The present simple as yet unpublished proof has been given by Giroux. In the original formulation of Brunel's lemma the set $\{\omega: S_n f(\omega) > 0$ for infinitely many $n\}$ was replaced by the smaller set $\{\omega: \sup_n T^k S_n f(\omega) > 0$ for all $k \geqq 0\}$. The sharper formulation was obtained by Akcoglu [1965]. Brunel [1966] has pointed out that it can also be derived from his formulation. In spite of alternative proofs by Meyer [1965], Garsia [1967] and Revuz [1975], Brunel's original proof may still be the shortest. A continuous time version of Brunel's lemma has been given by Akcoglu-Cunsolo [1970]. Neveu [1961] proved the identification of the limit in the case $\Omega = C$ and stated the general case. Independently, Chacon [1962a] proved the full result. The proof with Brunel's lemma is due to Meyer [1965a].

Akcoglu-Sharpe [1968] and Akcoglu-Chacon [1970] have developed an identification of the limit also on $D$. They study the partial order in $L_1^+$ defined by $f \prec g$ iff there exists an $n \geqq 0$ with $f \xrightarrow{n} g$. $A \in \mathscr{A}$ is called *asymptotically invariant* if, for all $f \in L_1^+$, $\lambda_f(A) = \lim (g\mu)(A)$ exists when $g$ ranges through $P_f := \{g \in L_1^+ : f \prec g\}$. It is shown that the family $\Sigma$ of asymptotically invariant sets is an algebra and that $\lambda_f$ is additive. For $g \in L_1^+$ the set $G := \{S_\infty g > 0\}$ belongs to $\Sigma$. $h = \lim_n S_n f / S_n g$ exists a.e. on $G$. The sets $\{h \leqq a\}$ belong to $\Sigma$ except for countably many a, and, for $A \in \Sigma$ with $A \subset G$, one has $\int_A h \, d\lambda_g = \lambda_f(A)$. On $C$ the algebra $\Sigma$ agrees with $\mathscr{C}$ and $\lambda_g(A) = (H_C g)\mu(A)$. $\Sigma$ may be fairly rich even in the case $\Omega = D$.

Berk [1968] has studied the identification of the limit on $C$, and Akcoglu-Cunsolo [1970a] on all of $\Omega$ for the continuous parameter situation. For a general continuous parameter ratio theorem see Terrell [1972a]. Tsurumi [1954], [1958a] proved Hopf type and ratio theorems under weakened stationarity assumptions.

Lemma 3.10 appears in the systematic study of subinvariant measures by Feldman [1962]. Theorem 3.15, observed independently by Sucheston [1967a] and Krengel [1966], is essentially equivalent to a factorization announced by Chacon [1962b]: If $T$ is conservative and $p \in L_1$ strictly positive, any integrable $f$ splits into a function $hp$ with $T(hp) = \mathrm{h}(Tp)$ and an element of $cl(T-I)L_1$; see also theorem 7.5.

Theorem 3.12 is a special case of results in Deriennic-Lin [1973]. Akcoglu-Chacon [1965] showed that conservative positive contractions in $L_1$ which are contractions in some $L_p$, $(1 < p < \infty)$, are contractions in all $L_p$, $(1 \leqq p \leqq \infty)$; M.M. Rao [1966] has related interpolation theorems in Orlicz spaces.

It is easy to see that the dual Markov operator constructed for a positive contraction $T$ admitting a strictly positive subinvariant function $p$, has the same conservative part as $T$ and the same $\sigma$-algebra $\mathscr{C}$.

**Induced operators.** For a positive contraction $T$ in $L_1$

$$T_{\to E} = (I_E T) \sum_{k=0}^{\infty} (I_{E^c} T)^k$$

is called the operator induced by $T$ on $E \in \mathscr{A}$. If $T$ is the contraction in $L_1$ corresponding to a null preserving $\tau$, $T_{\to E}$ corresponds to the induced transformation $\tau_E$. It is simple to check that $T_{\to E}$ is a positive contraction in $L_1(E)$ with conservative part $E \cap C$, and that the $T_{\to E}$-absorbing subsets of $E$ are just the restrictions to $E$ of the $T$-absorbing sets. One has $H_E = I_E + T_{\to E} I_{E^c}$.

For a more thorough discussion we refer to Horowitz [1968a] and Brunel [1971], and for a systematic study of related operators to Neveu [1972]. Simons and Overdijk [1979] investigate induced processes and the existence of "sweep-out-sets".

Scheller [1965] proved a generalization of the Kac recurrence theorem: For

$\omega \in E$, $T^*(I_{E^c}T^*)^\nu 1_E(\omega)$ is the probability that the first visit to $E$, given a start in $\omega$, happens at time $(\nu+1)$. Thus, $r_E(\omega) = \sum_{\nu=0}^{\infty} (\nu+1) I_E T^*(I_{E^c}T^*)^\nu 1_E(\omega)$ is the expected first return time. Scheller showed (both directly and via the Markov shift in $\Omega^+$): If $\mu = T\mu$ is a finite invariant measure, $\int_E r_E(\omega)\,d\mu = \mu(E^*)$, where $E^* = \{\psi_E = 1\}$ is the smallest $T$-absorbing set containing $E$.

**Supplement to § 1.6.** An application of the Brunel function enables us to complete the proof of proposition 1.6.8 also for non invertible $\tau$:

*Proof of Proposition 1.6.8.* Let $T$ be the contraction in $L_1$ given by $Tf = f \circ \tau$. As $\tau$ is conservative $T$ must be conservative. $\mathscr{C}$ is just the family of $\tau$-invariant sets. As $A^* = \bigcup_{i=0}^{\infty} \tau^{-i} A$ is $\tau$-invariant we may assume $\Omega = A^*$. As $T$ is conservative $T^*\psi_A = \psi_A$ and $\psi_A$ is $\mathscr{C}$-measurable. Hence $\psi_A \equiv 1$. The sets $\{r_A = n+1\} = A \cap \tau^{-1}A^c \cap \ldots \cap \tau^{-n}A^c \cap \tau^{-(n+1)}A$ with $n \geqq k$ form a disjoint decomposition of $A_k := A \cap \tau^{-1}A^c \cap \ldots \cap \tau^{-k}A^c$ when $k \geqq 1$.

The proof can therefore be completed by the following computation:

$$\begin{aligned}
\int_{A^*} g &= \langle \mathbb{1}, g\rangle = \langle \psi_A, g\rangle = \Big\langle \sum_{k=0}^{\infty} (I_{A^c}T^*)^k 1_A, g\Big\rangle = \Big\langle 1_A, \sum_{k=0}^{\infty} (TI_{A^c})^k g\Big\rangle \\
&= \int_A g + \sum_{k=1}^{\infty} \int_A (TI_{A^c})^k g = \int_A g + \sum_{k=1}^{\infty} \int_{A_k} g \circ \tau^k \\
&= \int_A g + \sum_{k=1}^{\infty} \sum_{n=k}^{\infty} \int_{\{r_A = n+1\}} g \circ \tau^k = \int_A g + \sum_{n=1}^{\infty} \sum_{k=1}^{n} \int_{\{r_A = n+1\}} g \circ \tau^k \\
&= \int_A g + \sum_{m=2}^{\infty} \int_{\{r_A = m\}} \sum_{k=1}^{m-1} g \circ \tau^k = \int_A \sum_{k=0}^{r_A - 1} g \circ \tau^k. \quad \square
\end{aligned}$$

## § 3.4 Existence of finite invariant measures

In this section we discuss necessary and sufficient conditions for the existence of finite invariant measures for positive contractions in $L_1$ and for nonsingular transformations of a measure space. An application is the proof of the stochastic ergodic theorem.

**1. Banach limits.** Recall from § 1.1 that a *Banach limit* $L$ is a positive linear functional $L$ on $\ell_\infty$, invariant under the shift $\theta: (x_0, x_1, \ldots) = x \to \theta x = (x_1, x_2, \ldots)$, and satisfying $L(\mathbf{1}) = 1$, where $\mathbf{1} = (1, 1, 1, \ldots)$.

The properties of Banach limits readily imply $\|L\| = 1$ and

(4.1) $\quad \liminf x_n \leqq L(x) \leqq \limsup x_n.$

**Theorem 4.1.** *(a) Banach limits exist.*

*(b) For fixed x the maximal value of Banach limits* $\sup\{L(x)$: *L is a Banach*

*limit} is given by*

(4.2) $$M(x) := \lim_{n\to\infty} \left(\sup_j n^{-1} \sum_{i=0}^{n-1} x_{i+j}\right).$$

*(c) For fixed x the values L(x) of all Banach limits agree and are equal to s iff*

$$\lim_{n\to\infty} n^{-1} \sum_{i=0}^{n-1} x_{i+j} = s \quad \textit{uniformly in } j.$$

*Proof.* Set $c_n = \sup_j n^{-1} \sum_{i=0}^{n-1} x_{i+j}$. For each $k$, $m$ one has $c_{km} \leqq c_m$. Thus

$$(r+km)\, c_{r+km} \leqq r\, c_r + km\, c_{km} \leqq r\, c_r + km\, c_m.$$

Dividing by $r+km$ and letting $k$ tend to $\infty$ we obtain $\limsup_{k\to\infty} c_{r+km} \leqq c_m$. Since this holds for $r = 0, 1, \ldots, m-1$, $\limsup c_n \leqq c_m$ for each $m$, and hence $\limsup c_n \leqq \liminf c_m$. Therefore the limit in (4.2) exists.

Let $\ell_\infty^0$ be the subspace of $\ell_\infty$ consisting of the convergent sequences. $L_0(x) = \lim_n x_n$ defines a linear functional on $\ell_\infty^0$. $M$ is a sublinear functional on $\ell_\infty$, i.e., $M(x+y) \leqq M(x) + M(y)$ and if $\alpha \geqq 0$ then $M(\alpha x) = \alpha M(x)$. As $L_0(x) = M(x)$ holds for $x \in \ell_\infty^0$, the Hahn-Banach theorem implies the existence of a linear functional $L$ on $\ell_\infty$ which agrees with $L_0$ on $l_\infty^0$ and for which $L(x) \leqq M(x)$ holds for all $x \in \ell_\infty$. Obviously, $L(\mathbf{1}) = L_0(\mathbf{1}) = 1$. For $x = (x_n)$ with $x_n \geqq 0$ for all $n$, $L(x) = -L(-x) \geqq -M(-x) \geqq 0$. Finally

$$|M(x - \theta x)| = |\lim_n \sup_j n^{-1}(x_j - x_{j+n})| \leqq \lim_n n^{-1}\, 2 \sup_k |x_k| = 0$$

together with the symmetric equation $M(\theta x - x) = 0$ imply $L(x) - L(\theta x) = L(x - \theta x) \leqq M(x - \theta x) = 0$ and $L(\theta x) - L(x) \leqq 0$. Thus, $L$ is a Banach limit.

To prove (b) we make use of the fact that, for fixed $x \in \ell_\infty \setminus \ell_\infty^0$, the extension procedure in the Hahn-Banach theorem may be started by assigning to $L(x)$ any value in the interval

$$[\sup\{(-M(-x-y) - L_0(y)) \colon y \in \ell_\infty^0\},\ \inf\{(M(x+y) - L_0(y)) \colon y \in \ell_\infty^0\}].$$

To show that, for any $x \in \ell_\infty$, there is a Banach limit $L^*$ with $L^*(x) = M(x)$, we may assume $x \in \ell_\infty \setminus \ell_\infty^0$. If $L^*$ is an extension of $L_0$ with $L^*(x) = \inf\{(M(x+y) - L_0(y)) \colon y \in \ell_\infty^0\}$, then $L^*(x) = M(x)$ since $M(x+y) = M(x) + L_0(y)$ for $y \in \ell_\infty^0$. Conversely, let $L$ be any Banach limit. Set $z_n^{(m)} = m^{-1} \sum_{i=0}^{m-1} x_{i+n}$, $(n = 0, 1, 2, \ldots)$. Then $L(x) = L(z^{(m)}) \leqq \sup_n z_n^{(m)}$. With $\lim_m \sup_n z_n^{(m)} = M(x)$ this yields $L(x) \leqq M(x)$ and completes the proof of (b).

(c) follows from (b) since now $-M(-x) = \lim_n \left(\inf_j n^{-1} \sum_{i=0}^{n-1} x_{i+j}\right)$ is the minimal value of Banach limits. □

**2. Existence of finite invariant measures.** Now let $T$ be a positive contraction in $L_1 = L_1(\Omega, \mathscr{A}, \mu)$, where $\mu$ is a $\sigma$-finite measure on $(\Omega, \mathscr{A})$. Recall that $T$ induces a transformation $v \to Tv$ in the space $\tilde{L}_1$ of finite signed measures $v \ll \mu$: If $v = f \cdot \mu$ (i.e., $f = dv/d\mu$), then $Tv = (Tf) \cdot \mu$. The question of existence of a finite invariant $v \in \tilde{L}_1^+$ is equivalent to the question of existence of a $T$-invariant $f \in L_1^+$. If $\mu'$ is a probability measure with $\mu' \sim \mu$, then $\tilde{L}_1(\mu') = \tilde{L}_1(\mu)$. Passing to a new reference measure we may and shall therefore always assume $\mu(\Omega) = 1$ in this section. Recall the notation $\langle v, h \rangle = \int h\, dv$, $L_p(B) = \{h \in L_p: \{h \neq 0\} \subset B\}$. The starting point of the investigation is

**Theorem 4.2.** *Let $B \in \mathscr{A}$. There exists an $f \in L_1^+$ with $B \subset \{f > 0\}$ and $Tf = f$ iff*

$$\inf_{n \geqq 0} \langle \mu, T^{*n} h \rangle > 0 \quad \textit{for all } h \textit{ with } 0 \neq h \in L_\infty^+(B). \tag{4.3}$$

*Proof.* The easy half is to show that (4.3) is necessary: We may assume $\mu(B) > 0$. If there exists an $f \in L_1^+$ with $\{f > 0\} \supset B$ and $Tf = f$, then $v = f \cdot \mu$ satisfies $T^n v = v$ for all $n$. For any $h \in L_\infty^+(B)$ with $h \neq 0$ there exists an $\varepsilon > 0$ with $\langle v, h \wedge \mathbb{1} \rangle > 2\varepsilon \|f\|_1$. Now $\langle v, T^{*n}(h \wedge \mathbb{1}) \rangle = \langle T^n v, h \wedge \mathbb{1} \rangle > 2\varepsilon \|f\|_1$ for all $n$. Set $A_n = \{T^{*n}(h \wedge \mathbb{1}) \geqq \varepsilon\}$, then $T^{*n}(h \wedge \mathbb{1}) \leqq 1$ implies $v(A_n) \cdot 1 + \varepsilon v(A_n^c) \geqq \langle v, T^{*n}(h \wedge \mathbb{1}) \rangle > 2\varepsilon \|f\|_1$, and hence $v(A_n) \geqq \varepsilon \|f\|_1$ for all $n$. By $v \ll \mu$ there is a $\delta > 0$ with $\mu(A_n) \geqq \delta$ for all $n$. Consequently, $\inf_n \langle \mu, T^{*n} h \rangle \geqq \inf_n \langle \mu, T^{*n}(h \wedge 1) \rangle \geqq \varepsilon\delta > 0$. Now assume (4.3). Take any Banach limit $L$ and define a linear functional $\varrho$ on $L_\infty$ by $\varrho(h) := L((\langle T^n \mu, h \rangle))$. Observe that $\varrho(h) \geqq 0$ for $h \in L_\infty^+$, and that $\varrho(T^* h) = \varrho(h)$. Define, for $h \in L_\infty^+$,

$$\tilde{v}(h) := \inf\{\sum_{n=1}^{\infty} \varrho(h_n): h = \sum_{n=1}^{\infty} h_n,\ h_n \in L_\infty^+\}.$$

For $h, h', h''$, and $(h_n) \in L_\infty^+$ with $h = h' + h''$ and $h = \sum h_n$ there exist sequences $h'_n, h''_n \in L_\infty^+$ with $h' = \sum h'_n$ and $h'' = \sum h''_n$: to see this start with $h'_1 = h' \wedge h_1$, $h''_1 = h_1 - h'_1$ and continue with $h - h_1$, $h' - h'_1$ and $h'' - h''_1$, etc. Hence $\tilde{v}(h) \geqq \tilde{v}(h') + \tilde{v}(h'')$.

To show that $\tilde{v}$ is "$\sigma$-additive" on $L_\infty^+$ it now suffices to verify that $h = \sum_{m=1}^{\infty} g_m$, ($g_m, h \in L_\infty^+$), implies $\tilde{v}(h) \leqq \sum_{m=1}^{\infty} \tilde{v}(g_m)$. By the definition of $\tilde{v}$ there are sequences $h_{n,m} \in L_\infty^+$ with $g_m = \sum_{n=1}^{\infty} h_{n,m}$ and $\sum_n \varrho(h_{n,m}) < \tilde{v}(g_m) + 2^{-(n+4)}\varepsilon$. Now $\tilde{v}(h) \leqq \sum_{n,m} \varrho(h_{n,m}) \leqq \sum_{m=1}^{\infty} \tilde{v}(g_m) + \varepsilon$ because $h = \sum_{n,m} h_{n,m}$. As $\varepsilon > 0$ was arbitrary $\tilde{v}$ is $\sigma$-additive and defines a finite measure $\tilde{v} \ll \mu$ by $\tilde{v}(A) = \tilde{v}(1_A)$. Set $\tilde{f} = d\tilde{v}/d\mu$.

We now want to check $B \subset \{\tilde{f} > 0\}$. Otherwise there is some $A \subset B$ with $\tilde{v}(A) = 0$ and $\mu(A) > 0$, and then we can find sequences $(h_{m,n}) \in L_\infty^+$ with $1_A = \sum_{n=1}^{\infty} h_{m,n}$

and $\sum_{n=1}^{\infty} \varrho(h_{m,n}) < m^{-1}$. If $k(m)$ is so large that $\sum_{m=1}^{\infty} \sum_{n=k(m)+1}^{\infty} \langle \mu, h_{m,n} \rangle < \langle \mu, h \rangle$, then $0 \leqq h^* := \inf_m \{ \sum_{n=1}^{k(m)} h_{m,n} \} \neq 0$. For all $m$, $\varrho(h^*) \leqq \varrho(\sum_{n=1}^{k(m)} h_{m,n}) = \sum_{n=1}^{k(m)} \varrho(h_{m,n}) < m^{-1}$, and hence $\varrho(h^*) = 0$. The function $h^*$ belongs to $L_\infty^+(B)$ since $h^* \leqq 1_A$ and $A \subset B$. Now $\varrho(h^*) = 0$ contradicts (4.3).

The $\theta$-invariance of $L$ implies $\varrho(T^* h) = \varrho(h)$. For each sequence $(h_n)$ with $h = \sum_{n=1}^{\infty} h_n$ we have $\tilde{v}(T^* h) \leqq \sum_{n=1}^{\infty} \varrho(T^* h_n) = \sum \varrho(h_n)$; we therefore obtain $\tilde{v}(T^* h) \leqq \tilde{v}(h)$; $\tilde{v}$ is sub-invariant. Next, observe that $B$ is contained in $C$: If $B$ has a non empty intersection with the dissipative part $D = \{ \sum_{n=0}^{\infty} T^n \mathbb{1} < \infty \}$ there exists an $h \in L_\infty^+$ with $\{h > 0\} = B \cap D \neq \emptyset$ for which $\int (\sum_{n=0}^{\infty} T^n \mathbb{1}) \cdot h \, d\mu = \sum_{n=0}^{\infty} \langle \mu, T^{*n} h \rangle$ is finite. This is impossible by (4.3).

Put $v(A) := \tilde{v}(A \cap C)$. The density $f = dv/d\mu = 1_C \tilde{f}$ is strictly positive on $B$. As $Tv$ has support in $C$ by the $T$-invariance of $C$, and $Tv \leqq T\tilde{v} \leqq \tilde{v}$, we obtain $Tv \leqq v$. But then $v$ is invariant since the support of $v$ is contained in $C$. $v$ is the desired invariant measure. □

We now define weakly wandering functions which will help us to deduce stronger and more useful conditions.

**Definition 4.3.** A *weakly wandering function h* is an element of $L_\infty^+$ for which there exists a strictly increasing sequence $0 = k_0 < k_1 < \ldots$ of integers with

(4.4) $$\| \sum_{\nu=0}^{\infty} T^{*k_\nu} h \|_\infty < \infty .$$

If even the stronger condition

(4.5) $$\sup_{j \geqq 0} \| \sum_{k_\nu \geqq j} T^{*k_\nu - j} h \|_\infty < \infty$$

holds, $h$ is called *m-weakly wandering*.

**Lemma 4.4.** *If $g \in L_\infty^+$ has the property*

(4.6) $$\liminf_{n \to \infty} \langle T^n \mathbb{1}, g \rangle = 0$$

*then for any $\varepsilon > 0$ there exists an m-weakly wandering function $h \leqq g$ with*

(i) $\int (g - h) \, d\mu < \varepsilon$ *and* (ii) $\{h > 0\} = \{g > 0\}$.

*Proof.* We may assume $\|g\|_1 \leqq \|g\|_\infty \leqq 1$. (4.6) is equivalent to

(4.7) $$\liminf_{n \to \infty} \langle T^n f, g \rangle = 0 \quad \text{for all } f \in L_1^+ .$$

For a number $b$ with $0 < b < 1$, which will be specified later, there exist $\varepsilon_1 > \varepsilon_2 > \varepsilon_3 > \ldots > 0$ with

$$1 + b = \prod_{\nu=1}^{\infty} (1 + \varepsilon_\nu).$$

Using (4.6) we find $i_1 > 0$ with $\|(I + T^{*i_1})g\|_1 = \langle \mathbb{1}, g\rangle + \langle T^{i_1}\mathbb{1}, g\rangle \leqq 1 + \varepsilon_1$. Applying (4.7) with $f = (I + T^{i_1})\mathbb{1}$ we find $i_2 > 0$ with $\|(I + T^{*i_1})(I + T^{*i_2})g\|_1 \leqq (1 + \varepsilon_1) \cdot (1 + \varepsilon_2)$. Continuing in this way we inductively define a sequence $i_1$, $i_2$, $i_3$, ... of positive integers with

$$\|(I + T^{*i_1})(I + T^{*i_2}) \ldots (I + T^{*i_n})g\|_1 \leqq \prod_{\nu=1}^{n} (1 + \varepsilon_\nu).$$

The sequence $g^{(n)} = (I + T^{*i_1})(I + T^{*i_2}) \ldots (I + T^{*i_n})g$ increases to a limit $g^{(\infty)}$ with $\|g^{(\infty)}\|_1 \leqq 1 + b$. Our first aim is to show that the function $\varphi = (2g - g^{(\infty)})^+$ is $m$-weakly wandering for the sequence $k_n = \sum_{\nu=1}^{n} i_\nu$, $(n \geqq 0)$. From $\varphi \leqq (2g - (I + T^{*i_n})g)^+ = (g - T^{*i_n}g)^+$ it follows that $\varphi \leqq g - T^{*i_n}g$ holds on $\{\varphi > 0\}$, and hence that $\varphi + T^{*i_n}\varphi \leqq \varphi + T^{*i_n}g \leqq g \vee T^{*i_n}g \leqq \|g\|_\infty \leqq \mathbb{1}$. The inequality

$$\text{(4.8)} \qquad \varphi^{(n,p)} := \varphi + T^{*i_p}\varphi + T^{*(i_p + i_{p+1})}\varphi + \ldots + T^{*(i_p + i_{p+1} + \ldots + i_n)}\varphi \leqq \mathbb{1}$$

is therefore valid for $1 \leqq p = n$. Once it is proved for $2 < p \leqq n$, we can derive $\varphi^{(n,p-1)} \leqq \mathbb{1}$: Note that

$$\varphi \leqq (2g - (I + T^{*i_{p-1}})(I + T^{*i_p}) \ldots (I + T^{*i_n})g)^+.$$

This implies that, on $\{\varphi > 0\}$, the inequalities $\varphi \leqq 2g - (I + T^{*i_{p-1}})\varphi^{(n,p)} \leqq 2g - g - T^{*i_{p-1}}\varphi^{(n,p)} = g - T^{*i_{p-1}}\varphi^{(n,p)}$ hold true. Consequently, $\varphi^{(n,p-1)} = \varphi + T^{*i_{p-1}}\varphi^{(n,p)} \leqq \mathbb{1}$. Using backward induction we have verified (4.8) for $1 \leqq p \leqq n$.

If $j \geqq 0$ is fixed let $t$ be the first index with $k_t \geqq j$. Then, by (4.8),

$$\sum_{k_\nu \geqq j} T^{*(k_\nu - j)}\varphi = T^{*(k_t - j)}(\varphi + \sum_{\nu=t+1}^{\infty} T^{*(i_{t+1} + i_{t+2} + \ldots + i_\nu)}\varphi \leqq \mathbb{1}.$$

Thus, $\varphi$ is $m$-weakly wandering. It is clear that $0 \leqq \varphi \leqq g$. For $b < \varepsilon$ the inequalities

$$\int (g - \varphi)\, d\mu \leqq \int (g - (2g - g^{(\infty)}))\, d\mu = \int (g^{(\infty)} - g)\, d\mu \leqq b$$

show that $\varphi$ has all the properties requested for $h$ except (ii).

To find $h$ we have to modify the above construction. Let $1 > b_1 \geqq b_2 \geqq b_3 \geqq \ldots > 0$ be reals to be chosen later. There exist $\varepsilon_1, \varepsilon_2, \ldots > 0$ and $i_1, i_2, i_3, \ldots \geqq 1$ with

(4.9) $$\prod_{\nu=n}^{\infty}(1+\varepsilon_\nu) \leqq (1+b_n), \quad (n \geqq 1),$$

and

$$\|(I+T^{*i_n})(I+T^{*i_{n+1}})\dots(I+T^{*i_{n+m}})g\|_1 \leqq \prod_{\nu=n}^{n+m}(1+\varepsilon_\nu), \qquad (n \geqq 1, m \geqq 0).$$

As above $g_n^{(\infty)} = \lim\limits_{m\to\infty} \prod\limits_{\nu=n}^{n+m}(I+T^{*i_\nu})g$ satisfies $\|g_n^{(\infty)}\|_1 \leqq (1+b_n)$ and $g_n^{(\infty)} \geqq g$. The function $\varphi_n = (2g - g_n^{(\infty)})^+ \leqq g$ again has the analogous properties as above: $\|g-\varphi_n\|_1 \leqq b_n$ and

$$\varphi_n + \sum_{\nu=n}^{\infty} T^{*(i_n+i_{n+1}+\dots+i_\nu)}\varphi_n \leqq \mathbb{1}.$$

If $t$ is the first index with $k_t \geqq j$, and $t \geqq n-1$, then $\sum\limits_{k_\nu \geqq j} T^{*(k_\nu - j)}\varphi_n \leqq \mathbb{1}$ as in the above argument for $\varphi$. In the case $t \leqq n-2$,

$$\sum_{k_\nu \geqq j} T^{*(k_\nu-j)}\varphi_n = \sum_{\nu=t}^{n-1} T^{*(k_\nu-j)}\varphi_n + \sum_{\nu \geqq n} T^{*(k_\nu-j)}\varphi_n \leqq n+1.$$

Now

$$h = \left(1-\frac{\varepsilon}{2}\right)\varphi_1 + \sum_{n=2}^{\infty} \varepsilon 2^{-n}(n+1)^{-1}\varphi_n$$

is $m$-weakly wandering because

$$\sum_{k_\nu \geqq j} T^{*k_\nu - j}h \leqq \mathbb{1}$$

holds for all $j \geqq 0$. $h \leqq g$ follows from $\varphi_n \leqq g$, $(n \geqq 1)$. For $b_1 < \varepsilon/2$ we have $\|g-h\|_1 \leqq \|g-\varphi_1\|_1 + 2^{-1}\varepsilon < \varepsilon$. It remains to show that $\{h>0\} = \{g>0\}$ holds if the $b_n$'s are sufficiently small:

Because of $\mu(g-\varphi_n \geqq n^{-1}) \leqq n\|g-\varphi_n\| \leqq nb_n$ we can obtain $\sum\limits_{n=1}^{\infty}\mu(g-\varphi_n \geqq n^{-1}) < \infty$ by choosing the $b_n$'s so that $b_n < n^{-3}$. By the Borel-Cantelli lemma then almost every $\omega$ belongs to only finitely many sets $\{g-\varphi_n \geqq n^{-1}\}$.

Consequently, a.e. $\omega \in \{g>0\}$ belongs to infinitely many of the sets $\{\varphi_n > 0\} \supset \{g > n^{-1}\} \setminus \{g-\varphi_n \geqq n^{-1}\}$. Hence $\{h>0\} = \{g>0\}$. □

**Lemma 4.5.** *If* $0 = k_0 < k_1 < \dots < k_n$ *are integers, and, for some* $h \in L_\infty^+$ *and* $c \in \mathbb{R}^+$, *the inequality* $\sum\limits_{i=0}^{n} T^{*k_i}h \leqq c$ *holds* a.e., *then*

$$\limsup_{m\to\infty} \|A_m^* h\|_\infty \leqq (n+1)^{-1}c.$$

*In particular, if h is weakly wandering,* $\lim\limits_{m\to\infty}\|A_m^* h\|_\infty = 0$.

*Proof.* Simply observe that

$$\begin{aligned}(n+1)\sum_{i=0}^{m-1} T^{*i}h &\leqq \sum_{i=k_0}^{k_0+m-1} T^{*i}h + \sum_{i=k_1}^{k_1+m-1} T^{*i}h + \dots \\ &\quad + \sum_{i=k_n}^{k_n+m-1} T^{*i}h + (n+1)k_n\|h\|_\infty \\ &= \sum_{i=0}^{m-1} T^{*i}\Big(\sum_{\nu=0}^{n} T^{*k_\nu}h\Big) + (n+1)k_n\|h\|_\infty \\ &\leqq mc + (n+1)k_n\|h\|_\infty. \quad \square\end{aligned}$$

We are now in a position to prove a theorem which unifies many of the known necessary and sufficient criteria for existence of a finite invariant measure.

**Theorem 4.6.** *If $T$ is a positive contraction in $L_1(\Omega, \mathscr{A}, \mu)$, there exists a decomposition of $\Omega$ into two disjoint sets $\tilde{C}$, $\tilde{D}$, uniquely determined up to nullsets by the properties*
*(i) there exists a $p_0 \in L_1^+$ with $Tp_0 = p_0$ and $\{p_0 > 0\} = \tilde{C}$, and*
*(ii) there exists an m-weakly wandering $h_0 \in L_\infty^+$ with $\{h_0 > 0\} = \tilde{D}$.*
*The set $\tilde{C}$ belongs to $\mathscr{C}$.*

We still assume $\mu(\Omega) = 1$, but as both statements (i) and (ii) are invariant under a passage to an equivalent measure $\mu' \sim \mu$ (with a different $p_0'$), the theorem is equally valid for $\sigma$-finite measures $\mu$. In the special case of a Markov chain with countable state space, $\tilde{C}$ is the set of positively recurrent states and $\tilde{D}$ the set of null states. $\tilde{C}$ is called the *strongly conservative part* of $\Omega$ for $T$, or the *positive part*. $\tilde{D}$ is called the *null part*.

*Proof.* Set $\mathscr{N} = \{g \in L_\infty^+ : \lim \|A_m^* g\|_\infty = 0\}$. Let $g_n$, $(n \geqq 1)$, be a sequence with $g_n \in \mathscr{N}$ and $\mu(g_n > 0) \to \eta := \sup\{\mu(g > 0) : g \in \mathscr{N}\}$. Then $g_0 = \sum_{n=1}^{\infty} (g_n \wedge \mathbb{1}) 2^{-n}$ belongs to $\mathscr{N}$ and we have

$$\liminf_{m\to\infty} \langle T^m \mathbb{1}, g_0\rangle = \liminf_{m\to\infty} \int T^{*m} g_0 \, d\mu = 0.$$

By lemma 4.4 there exists an $m$-weakly wandering $h_0 \leqq g_0$ with $\{h_0 > 0\} = \{g_0 > 0\}$ and we obtain (ii) by setting $\tilde{D} = \{h_0 > 0\}$. If there exists a non zero $h \in L_\infty^+$ with support in $\tilde{C} := \Omega \setminus \tilde{D}$ and $\liminf_{n\to\infty} \langle T^n \mathbb{1}, h\rangle = 0$, there exists by lemma 4.4 a weakly wandering $h' \neq 0$ with $\{h' > 0\} \subset \tilde{C}$. By lemma 4.5, $h'$ belongs to $\mathscr{N}$, and in this case $g_0 + h'$ would be an element of $\mathscr{N}$ with $\mu(g_0 + h' > 0) = \eta + \mu(h' > 0) > \eta$ contrary to the definition of $\eta$. It follows that $\liminf_n \langle T^n \mathbb{1}, h\rangle > 0$ for $0 \neq h \in L_\infty^+(\tilde{C})$, and therefore also that $\inf_n \langle \mu, T^{*n} h\rangle > 0$ for $0 \neq h \in L_\infty^+(\tilde{C})$. By theorem 4.2 there exists a $p_0 \in L_1^+$ with $Tp_0 = p_0$ and $\tilde{C} \subset \{p_0 > 0\}$.

If $p \in L_1^+$ is $T$-invariant and $h \in L_\infty^+$ is weakly wandering for the sequence $k_0 < k_1 < k_2 < \ldots$, the finiteness of $\langle p, \sum_{i=0}^{\infty} T^{*k_i} h\rangle = \langle \sum_{i=0}^{\infty} T^{k_i} p, h\rangle = \langle \infty \cdot p, h\rangle$ implies $\{p > 0\} \cap \{h > 0\} = \emptyset$. Hence $\tilde{C} = \{p_0 > 0\}$, and the same argument implies the uniqueness mod nullsets of the decomposition $\Omega = \tilde{C} \cup \tilde{D}$. By $\{\sum_{i=0}^{\infty} T^i p_0 = \infty\} = \{\infty \cdot p_0 = \infty\} = \{p_0 > 0\}$ the set $\tilde{C}$ is a subset of $C$. $\tilde{C} \in \mathscr{C}$ follows from lemma 3.10. □

Let us now show how theorem 4.6 yields some known criteria:

**Corollary 4.7.** *Each of the following conditions is equivalent to the condition of existence of a finite T-invariant* $\nu \sim \mu$:

*(F) There is no A with* $\mu(A) > 0$ *and* $\lim_{n\to\infty} \|A_n^* 1_A\|_\infty = 0$;

*(DSN) For all A with* $\mu(A) > 0$, $\limsup_{n\to\infty} \sup_j n^{-1} \sum_{i=j}^{j+n-1} \langle \mu, T^{*i} 1_A\rangle > 0$;

*($I_1$) For all A with* $\mu(A) > 0$; $\limsup n^{-1} \sum_{i=0}^{n-1} \langle \mu, T^{*i} 1_A\rangle > 0$;

*($I_2$) For all A with* $\mu(A) > 0$, $\inf_{n \geq 0} \langle \mu, T^{*n} 1_A\rangle > 0$;

*(B)* $cl((I - T^*)L_\infty) \cap L_\infty^+ = \{0\}$, *where cl is the norm closure.*

*Proof.* The existence of $\nu$ implies ($I_2$) by theorem 4.2, and the implications ($I_2$) ⇒ ($I_1$) ⇒ (DSN) ⇒ (F) are obvious. Now assume (F). If no finite invariant $\nu \sim \mu$ exists, $\tilde{D}$ is non empty and there is a weakly wandering $h_0 \neq 0$. For small enough $\beta > 0$, $A := \{h_0 \geqq \beta\}$ has positive measure and $1_A \leqq \beta^{-1} h_0$ is weakly wandering. By lemma 4.5 this contradicts (F).

It remains to prove the equivalence of $(B)$: If there exists a strictly positive $f \in L_1^+$ with $Tf = f$, then $\langle f, (I - T^*)g\rangle = 0$ for all $g \in L_\infty$, and hence $\langle f, h\rangle = 0$ for $h \in H_0 := cl((I - T^*)L_\infty)$, while $\langle f, h\rangle > 0$ for $0 \neq h \in L_\infty^+$. Consequently, in this case $(B)$ holds. If such an $f$ does not exist, consider $A = \{h_0 \geqq \beta\}$ as above, and $H_1 = \{g \in L_\infty^*: T^{**} g = g\}$. From $\sum_{i=0}^{\infty} T^{*k_i} 1_A \in L_\infty$ it follows that $\langle 1_A, g\rangle = 0$ for all $f \in H_1 = H_0^\perp$. By the Hahn-Banach theorem $\{g \in L_\infty: \langle g, f\rangle = 0$ for $f \in H_1\} = H_0$. Hence $1_A \in H_0 \cap L_\infty^+$ and in this case $(B)$ does not hold. □

**3. The stochastic ergodic theorem.** Now we show that theorem 4.6 can be used to derive stochastic convergence for Cesàro averages if $T$ is a contraction in $L_1$ which need not contract the $L_\infty$-norm. A.e.-convergence and $L_1$-convergence fail to hold in this generality.

**Definition 4.8.** A sequence $\{f_k\}$ of measurable functions on a $\sigma$-finite measure space $(\Omega, \mathscr{A}, \mu)$ is said to *converge stochastically* to $f_\infty$ if $\mu(A \cap \{|f_k - f_\infty| > \varepsilon\})$ converges to 0 for all $\varepsilon > 0$ and all $A$ with $\mu(A) < \infty$.

**Theorem 4.9** (Stochastic ergodic theorem; Krengel [1966]). *If $T$ is a positive contraction in $L_1$ of a $\sigma$-finite measure space $(\Omega, \mathscr{A}, \mu)$ then, for any $f \in L_1$, the averages $A_n f$ converge stochastically. The limit is $T$-invariant and vanishes in $\tilde{D}$. For $f \in L_1^+$ it agrees a.e. with the pointwise* $\liminf A_n f$.

(All assertions but the last in this theorem remain true for non positive contractions $T$ in $L_1$ if $\tilde{D}$ is the complement of the strongly conservative part of $\Omega$ for the linear modulus $|T|$; see § 4.1).

*Proof.* By theorem 4.6 there exist a weakly wandering $h_0 \in L_\infty^+$ and a $T$-invariant $p_0 \in L_1^+$ with $\{h_0 > 0\} = \tilde{D}$ and $\{p_0 > 0\} = \tilde{C}$. By the Chacon-Ornstein theorem the ratios $S_n f / S_n p_0 = p_0^{-1} A_n f$ converge a.e. in $\tilde{C}$, and hence $A_n f$ converges a.e. in $\tilde{C}$. Because of $H_C f = H_C H_C f$ theorem 3.4 implies that the limit on $\tilde{C}$ is the same as that of $A_n(H_C f)$. As $\tilde{C}$ is $T$-absorbing it is also the same as that of $A_n(1_{\tilde{C}} H_{\tilde{C}} f)$. In view of the $L_1$-convergence in $L_1(\tilde{C})$ the limit must be $T$-invariant. Let us now show stochastic convergence to 0 $\tilde{D}$. We may assume $f \geqq 0$. Let $A$ be a set of finite measure and $A(\beta) = A \cap \{h_0 \geqq \beta\}$. For given $\eta > 0$ there exists $\beta > 0$ with $\mu(A \setminus A(\beta)) < \eta/2$. As $1_{A(\beta)} \leqq \beta^{-1} h_0$ is weakly wandering, lemma 4.5 implies $\|A_n^* 1_{A(\beta)}\|_\infty \to 0$ and hence $\lim \langle A_n f, 1_{A(\beta)} \rangle = 0$. For $n \geqq N_\eta$ and $N_\eta$ large enough we have $\langle A_n f, 1_{A(\beta)} \rangle < \eta\varepsilon/2$, and then

$$\begin{aligned}\mu(A \cap \{A_n f > \varepsilon\}) &\leqq \mu(A \setminus A(\beta)) + \mu(A(\beta) \cap \{A_n f > \varepsilon\}) \\ &\leqq \eta/2 + \varepsilon^{-1} \langle A_n f, 1_{A(\beta)} \rangle < \eta\end{aligned}$$

completes the proof of stochastic convergence to 0 on $\tilde{D}$.

As there is even a.e.-convergence on $\tilde{C}$ the last assertion follows, too. □

**4. Null preserving transformations.** Many of the results on existence of finite invariant measures have first been studied for contractions $T$ given by a null preserving $\tau: \Omega \to \Omega$ in a probability space $(\Omega, \mathscr{A}, \mu)$. Then $T\nu(A) = \nu(\tau^{-1} A)$ and $T^* h = h \circ \tau$. In this case the concept "weakly wandering" has emerged in a stronger and more intuitive form. A set $W \in \mathscr{A}$ is called *weakly wandering* if there exists a sequence $0 = k_0 < k_1 < k_2 < \dots$ of integers such that the sets $\tau^{-k_i} W$ are disjoint. Theorem 4.2 implies that a finite invariant $\nu \sim \mu$ exists iff there exists no $A$ with $\mu(A) > 0$ and

(4.10) $\inf \{\mu(\tau^{-k} A) \colon 0 \leqq k < \infty\} = 0.$

If we start with such a set, repeated applications of the following lemma (with sufficiently small $\varepsilon_r$'s) produce in the limit a weakly wandering set of positive measure:

**Lemma 4.10.** *If $A \in \mathscr{A}$ is such that for a finite sequence $0 = k_0 < k_1 < \ldots < k_r$, the sets $\tau^{-k_i}A$, $(i = 0, \ldots, r)$, are disjoint, and* (4.10) *holds, then for each $\varepsilon_r > 0$ there exists a set $A' \subset A$ with $\mu(A \setminus A') < \varepsilon_r$ and an integer $k_{r+1} > k_r$ such that the sets $\tau^{-k_i}A'$, $(i = 0, \ldots, r+1)$, are disjoint.*

*Proof.* Let $\delta_r > 0$ be such that $\mu(B) < \delta_r$ implies $\mu(\tau^{-i}B) < (r+1)^{-1}\varepsilon_r$ for $i = k_r - k_j$, $(j = 0, \ldots, r)$. Choose $k_{r+1} > k_r$ such that $\mu(\tau^{-(k_{r+1}-k_r)}A) < \delta_r$. Take $A' = A \setminus \bigcup_{i=0}^{r} \tau^{-(k_{r+1}-k_i)}A$. □

We have proved:

**Theorem 4.11** (Hajian-Kakutani). *A null preserving $\tau$ in $(\Omega, \mathscr{A}, \mu)$ admits a finite $\tau$-invariant measure $\nu \sim \mu$ iff there exists no weakly wandering set of positive measure.*

## Notes

The existence of Banach limits is due to Banach [1932], Theorem 4.1 (b) to Sucheston [1964], (c) to Lorentz [1948]. The first necessary and sufficient conditions for existence of finite invariant measures, now contained in Hajian's and Kakutani's theorem, were derived by E. Hopf [1932]. Calderon [1955] and Y.N. Dowker [1955], [1956] gave progressively stronger results, before Hajian and Kakutani [1964] introduced the simple but powerful tool of weakly wandering sets. For an approach which avoids Banach limits see Hajian-Ito [1969]. A streamlined introduction to the special case of nonsingular point mappings $\tau$ can be found in the paper of Jones and Krengel [1974]. They have also strengthened the result of Hajian-Kakutani for invertible $\tau$:

**Theorem 4.12** (Jones-Krengel). *If $\tau$ is invertible and nonsingular in $(\Omega, \mathscr{A}, \mu)$ there exists a set $W \in \mathscr{A}$ and a sequence $0 = k_0 < k_1 < \ldots$ of integers such that the sets $\tau^{-k_i}W$ are disjoint and their union is $\tilde{D}$.*

Note that the existence of such an "exhaustive weakly wandering set" is not even simple if $\tau$ is the translation $\tau: x \to x+1$ in $\mathbb{Z}$.

The same paper also contains other conditions; e.g., if $\mathscr{A}$ is countably generated, we have $\tilde{C} = \emptyset$ iff there exists an $A$ with an orbit $\{\tau^k A, k \geqq 0\}$ which is dense in $\mathscr{A}$ with respect to the metric $d(A, B) := \mu(A \Delta B)$. (We assume $\mu(\Omega) = 1$.) Kamae [1983] has characterized the set of weakly wandering sequences; see also Ellis-Friedman [1978], [1978a].

By Corollary 4.7 there exists an invariant $\nu \sim \mu$ iff $\lim_n \sup_j n^{-1} \sum_{i=j}^{n+j-1} \mu(\tau^{-i}A) > 0$ for all $A$ with $\mu(A) > 0$, as was shown by Sucheston [1964]. It therefore is natural to ask if even the condition

$$(4.11) \qquad \limsup_i \mu(\tau^{-i}A) > 0 \quad \text{for all } A \text{ with } \mu(A) > 0$$

is sufficient for the existence of finite invariant measures. A transformation with this property is called of *positive type*. Hajian and Kakutani [1964] observed that (4.11) is not sufficient; a construction of ergodic transformations of positive type with an infinite and hence no finite invariant measure can be found in the paper of Osikawa and Hamachi [1971]. Hajian and Ito [1978] have prepared a survey on transformations without finite invariant measure.

Conditions for the existence of finite invariant measures for positive contractions in $L_1$ (in the form of stochastic kernels) were first studied by Y. Ito [1964]. This paper contained the conditions ($I_1$) and ($I_2$) in Corollary 4.7. Condition (DSN) has been obtained independently by Dean-Sucheston [1966], and Neveu [1967]. Neveu also constructed weakly wandering functions $h$ with $\mu(\tilde{D} \setminus \{h > 0\}) < \varepsilon$. Foguel [1969] showed, by a much longer argument, that (F) holds for weakly wandering functions $1_A$. Takahashi [1971] constructed a strictly positive weakly wandering $h$ in the case $\Omega = \tilde{D}$. Condition $(B)$ and the concept of $m$-weakly wandering functions as well as the first half of lemma 4.4 are taken from the thesis of Brunel [1966]. Theorem 4.6, obtained jointly with Brunel, clearly owes a lot to the above papers, primarily to Neveu, who stated the decomposition $\Omega = \tilde{C} \cup \tilde{D}$ with a different characterization.

Its purpose is to unify and strengthen the above conditions and various other conditions of Neveu, Y. Ito [1964], Hajian and Ito [1965], [1967], Krengel [1967], and S. Horowitz [1968]; see also Helmberg [1966a]. [Apparently, Foguel [1969] overlooked that the decomposition, even in Neveu's form, requires the present generality of theorem 4.2 and not just the case $\Omega = B$.]

By theorem 4.6 $\tilde{D}$ can we written as a countable disjoint union of sets $X(i)$ with $\lim \langle \mu, A_n^* 1_{X(i)} \rangle = 0$. One can take $X(1) = \{h_0 \geqq 1\}$, $X(n) = \{n^{-1} \leqq h_0 < (n-1)^{-1}\}$, $(n \geqq 2)$. This can be used to derive also conditions of a pointwise nature. We leave the proof of the following elementary lemma as an exercise: If, for each $i \geqq 0$, $c_{in}$, $(n = 1, 2, \dots)$, is a non negative null sequence and the convergent sums $\sum_{i=0}^{\infty} c_{in}$ tend to some $c > 0$ for $n \to \infty$, there exists some $V \subset \{0, 1, \dots\}$ for which the sequence $\sum_{i \in V} c_{in}$ diverges. Assume that the limit of the decreasing sequence $\langle \mu, T^{*n} 1_D \rangle$ is positive; (e.g., $T$ is given by a nonsingular $\tau$ with $\Omega \neq \tilde{C}$ or $C \setminus \tilde{C}$ has positive measure). Apply the lemma with $c_{in} = \langle \mu, A_n^* 1_{X(i)} \rangle$. Then $A = \bigcup_{i \in V} X(i)$ has the property that the sequence $\langle \mu, A_n^* 1_A \rangle$ diverges.

In particular, $A_n^* 1_A$ cannot converge a.e. (and not even stochastically). On the other hand, we know from the previous section that in the case of existence of a finite invariant measure $\mu_0 \sim \mu$ the sequence $A_n^* 1_A$ converges a.e. for all $A$. We find that for conservative $T$ and for $T$ given by a nonsingular point mapping $\tau$ the a.e.-convergence of $A_n^* 1_A$ for all $A \in \mathscr{A}$ is necessary and sufficient for the existence of a finite invariant $\mu_0 \sim \mu$. Such conditions are due to Mrş. Dowker [1951], [1955] for point mappings and to Feldman [1962] for operators, see also § 4.3, and Chersi [1982].

In the case $\Omega = \tilde{C}$ the sequence $A_n \mathbb{1}$ is uniformly integrable by the remark after definition 3.13. On the other hand, if $A_n \mathbb{1}$ is uniformly integrable, then the stochastic convergence of $A_n \mathbb{1}$ implies that $\langle A_n \mathbb{1}, 1_A \rangle = \langle \mu, A_n^* 1_A \rangle$ converges for all $A$. Hence a conservative $T$ admits a finite invariant measure $\mu_0 \sim \mu$ iff the sequence $A_n \mathbb{1}$ is uniformly integrable; see Y. Ito [1965], and Kim [1968].

An interesting condition has been proposed by Brunel [1970]:

**Theorem 4.13.** *A positive contraction $T$ in $L_1$ possesses a finite invariant measure $\mu_0 \sim \mu$ iff for all $a = (a_i)_{i \geqq 0}$ with $a_i \geqq 0$ and $\sum_{i=0}^{\infty} a_i = 1$ the operator $T_a = \sum_{i=0}^{\infty} a_i T^i$ is conservative.*

An application showing that all random walks on a locally compact metrizable group $G$ are recurrent iff $G$ is compact has been given by Brunel and Revuz [1974]. For generalizations and related work see Horowitz [1972], Falkowitz [1973], Foguel-Weiss [1973].

Krengel [1967] has given a different condition involving weights. He shows that $\tilde{C}$ is a special *weighted conservative part*. If $w = (w_i)_{i \geqq 0}$ is a decreasing sequence of positive numbers with divergent sum, $C_w := \{\sum_{k=0}^{\infty} w_k T^k f_0 = \infty\}$ is independent of the choice of a strictly positive integrable $f_0$. $\tilde{C}$ is contained in all sets $C_w$ and, for some $w$, $\tilde{C} = C_w$. Aaronson [1983] used this to show that the larger the eigenvalue group of a nonsingular conservative $\tau$ is, the less recurrent it is.

D. Maharam [1969] derived rather different necessary and sufficient conditions for the existence of an equivalent finite invariant measure. They involve the pointwise relative frequencies of the sequence of $n$'s for which $T^n \mathbb{1}(\omega) \geqq \alpha$ holds.

The conditions for the existence of a finite invariant measure for a strongly continuous semigroup $\{T_t, t \geqq 0\}$ of positive contractions in $L_1$ are quite analogous; see Lin [1972]. If $p_0 \in L_1^+$ is $T_1$-invariant $S_{0,1} p_0 = \int_0^1 T_s p_0 \, ds$ is $T_t$-invariant for all $t \geqq 0$. If $\tilde{C}$ is the strongly conservative part for $T_1$ and $p_0$ is such that $\{p_0 > 0\} = \tilde{C}$ holds, then we must also have $\{S_{0,1} p_0 > 0\} = \tilde{C}$, because $\supset$ holds by the strong continuity at $t = 1$ and $\subset$ because $S_{0,1} p_0$ is $T_1$-invariant. Thus, the strongly conservative parts agree for all $T_t$, $(t > 0)$.

The general conditions derived in this section have been used in proofs of some general theoretical results. They are frequently not very helpful for specific examples which require methods tailored to their special structure; see, e.g., Lasota-Yorke [1973], Veech [1982], Cornfeld-Fomin-Sinai [1982].

The existence of finite or $\sigma$-finite invariant measures for nonsingular transformations is closely related to the theory of orbit equivalence developed by Dye, Krieger, and others. We refer to Weiss [1981] for an introduction, and to Hamachi-Osikawa [1981], Hajian-Ito-Kakutani [1972], [1974].

For abstract generalizations in Banach lattices of several results of this section, see Shields [1967], Kühne [1982a], Akcoglu-Sucheston [1984], and Brunel-Sucheston [1984/85].

More on invariant measures can be found in § 4.3 (Power bounded $T$), § 6.3 (semigroups), and in the supplement ($\sigma$-finite measures).

## § 3.5 The subadditive ergodic theorem for positive contractions in $L_1$

We now prove the extension of Kingman's subadditive ergodic theorem to positive contractions in $L_1$, due to Akcoglu-Sucheston [1978]. The key step is a generalization of Kingman's decomposition of a subadditive process into an additive process and a non negative subadditive process with time constant 0.

A family $F = \{F_n, n \geqq 1\}$ of integrable functions is called a *subadditive process* (for a positive contraction $T$) if the conditions

(i) $\quad F_{m+n} \leqq F_m + T^m F_n, \quad (m, n \geqq 1),$

(ii) $\quad \sup_n \|n^{-1} F_n\|_1 < \infty$

are satisfied. If $-F$ is subadditive $F$ is called *superadditive*, and if $F$ is subadditive

and superadditive $F$ is called *additive*. If $T$ is the composition with an endomorphism $\tau$ of $(\Omega, \mathscr{A}, \mu)$ this agrees with the definition in § 1.5:
If $\{F_{i,k}: (i,k) \in Q\}$ is given as in § 1.5 we may put $F_n = F_{0,n}$ and obtain the formulation used here. Conversely, if $\{F_n: n \geqq 1\}$ is given, we may put $F_{i,k} = T^i F_{k-i}$ to obtain the formulation with the double index.

For the operator $T = 0$ every decreasing sequence $F_n$ with the property (ii) is a subadditive process and simple examples show that $n^{-1} F_n$ need not converge in any sense.

Recall that $T$ had been called Markovian if $T^* \mathbb{1} = \mathbb{1}$ holds, or – equivalently – if $\int Tf d\mu = \int f d\mu$ holds for all $f \in L_1$. This is certainly the case for the point mapping case $Tf = f \circ \tau$. If $T$ is Markovian, condition (i) yields $\int F_{m+n} \leqq \int F_n + \int F_m$ and lemma 1.5.1 shows that the sequence $n^{-1} \int F_n$ converges. We may then define the time constant as before by

$$\gamma(F) = \lim n^{-1} \int F_n d\mu.$$

It is convenient to work mainly with superadditive processes $(F_n)$. Then the process $F_n' := F_n - \sum_{i=0}^{n-1} T^i F_1$ is non negative and superadditive and differs from the original process only by an additive process.

An additive process $G = (G_n)$ necessarily is of the form $G_n = \sum_{i=0}^{n-1} T^i G_1$ for some $G_1 \in L_1$. It is called an *exact dominant* for a superadditive process $(F_n)$ if $G_n \geqq F_n$ holds for all $n \geqq 1$ and the time constants agree: $\gamma(G) = \gamma(F)$. Sometimes $G_1$ is called the exact dominant because it determines $(G_n)$. As $T$ is Markovian we have $\gamma(G) = \int G_1 d\mu$. By $(-F_n) = (-G_n) + (G_n - F_n)$ the existence of an exact dominant for a superadditive process $(F_n)$ is equivalent to the existence of a Kingman decomposition of a subadditive process $(-F_n)$ into an additive process and a non negative subadditive process with time constant 0. Therefore the following result of Akcoglu-Sucheston is a precise generalisation of Kingman's decomposition theorem to the case of Markovian $T$:

**Theorem 5.1.** *Let $T$ be a Markovian positive contraction in $L_1$, then each superadditive process $(F_n)$ for $T$ has an exact dominant.*

The following lemma is due to Kingman in the case of measure preserving transformations:

**Lemma 5.2.** *Let $(F_m)$ be a non negative superadditive process for a Markovian $T$. The sequence*

$$\varphi_m = m^{-1} \sum_{i=1}^{m} (F_i - TF_{i-1}), \quad (\text{with } F_0 = 0)$$

*of non negative integrable functions has the properties*

(α) $\int \varphi_m d\mu \leqq \gamma(F), \quad (m \geqq 1)$,

(β) $\sum_{j=0}^{n-1} T^j \varphi_m \geqq \left(1 - \frac{n-1}{m}\right) F_n, \quad (m \geqq 1, 1 \leqq n < m)$.

*Proof.* $F_i \geqq F_1 + TF_{i-1}$ implies $\varphi_m \geqq 0$. For superadditive $(F_m)$, $\gamma(F)$ is equal to $\sup m^{-1} \int F_m d\mu$. We therefore obtain

$$\int \varphi_m d\mu = m^{-1} \sum_{i=1}^{m} \int (F_i - TF_{i-1}) d\mu = m^{-1} \sum_{i=1}^{m} \int (F_i - F_{i-1}) d\mu$$
$$= m^{-1} \int F_m d\mu \leqq \gamma(F).$$

The inequality (β) is a consequence of the estimates

$$m \sum_{j=0}^{n-1} T^j \varphi_m = (I - T^n) \sum_{i=1}^{m-1} F_i + \sum_{j=0}^{n-1} T^j F_m$$
$$= \sum_{i=1}^{n-1} F_i + F_n + \sum_{i=1}^{m-n} (F_{n+i} - T^n F_i)$$
$$+ \sum_{j=1}^{n-1} (T^j F_m - T^n F_{j+m-n}) \geqq (1 + (m-n)) F_n,$$

where the last inequality follows from $F_i \geqq 0$, $F_{n+i} - T^n F_i \geqq F_n$, and $T^j(F_m + T^{n-j} F_{j+m-n}) \geqq T^j F_{n-j} \geqq 0$. □

If the sequence $\varphi_m$ would converge in $L_1$ the proof of theorem 5.1 would already be complete because we could take $G_1 = \lim \varphi_m$. Unfortunately, the sequence $\varphi_m$ need not converge and a limit point has to be determined by some sort of compactness or subsequence argument.

*Proof of theorem 5.1.* Let us first assume that $\mathscr{A}$ is generated by a countable algebra $\mathscr{A}_0 \subset \mathscr{A}$. Let $h$ be a strictly positive integrable function. $\mathscr{A}_0$ is dense in $\mathscr{A}$ in the metric $d(A, B) = (h\mu)(A \Delta B)$. For each $i \geqq 0$, $j \geqq 1$, and $A \in \mathscr{A}_0$ the sequence $\langle T^i \varphi_m \wedge (jh), 1_A \rangle$ is bounded. By the diagonal procedure we may find a subsequence $M \subset \mathbb{N}$ such that

$$\lim_{m \in M} \langle T^i \varphi_m \wedge (jh), 1_A \rangle$$

exists for all $i \geqq 0, j \geqq 1$ and $A \in \mathscr{A}_0$ and hence also for all $A \in \mathscr{A}$. Write $w\text{-}\lim_M$ for the weak limit in $L_1$ along the subsequence. We have obtained the existence of elements $\lambda_{ij} \in L_1^+$ with $\lambda_{ij} = w\text{-}\lim_M (T^i \varphi_m \wedge (jh))$. For fixed $i$ the sequence $\lambda_{ij}$ is non decreasing in $j$ and norm bounded by $\sup \|\varphi_m\|_1 \leqq \gamma(F) < \infty$. Hence $\lim_{j \to \infty} \lambda_{ij} = \lambda_i$ exists a.e. and in $L_1$, and $\int \lambda_i d\mu \leqq \gamma(F)$. The inequality (β) above implies that if $n < m$, then

$$\sum_{i=0}^{n-1} [(T^i \varphi_m) \wedge (jh)] \geqq [\sum_{i=0}^{n-1} T^i \varphi_m] \wedge (jh) \geqq \left(1 - \frac{n}{m}\right) F_n \wedge (jh).$$

Taking the $w$-$\lim_M$ and then passing with $j$ to $\infty$ we obtain

$$(5.1) \qquad \sum_{i=0}^{n-1} \lambda_i \geqq F_n, \quad (n \geqq 1).$$

For each $j \geqq 1$ and $\varepsilon > 0$ we can find an integer $k$ and a function $g \in L_1^+$ with $\|g\|_1 < \varepsilon$ and $T(jh) \leqq k \cdot h + g$. As the continuity of $T$ in the strong topology of $L_1$ implies also the continuity in the weak topology we have

$$\begin{aligned} T\lambda_{ij} &= T\{w\text{-}\lim_M (T^i \varphi_m \wedge (jh)\} = w\text{-}\lim_M T(T^i \varphi_m \wedge (jh)) \\ &\leqq w\text{-}\lim_M (T^{i+1} \varphi_m \wedge (kh)) + g \leqq \lambda_{i+1,k} + g \leqq \lambda_{i+1} + g\,. \end{aligned}$$

As $j$ was arbitrary and $\varepsilon > 0$ arbitrarily small, $T\lambda_i \leqq \lambda_{i+1}$ follows. Let us show that

$$G_1 = \lambda_0 + \sum_{i=0}^{\infty} (\lambda_{i+1} - T\lambda_i)$$

has the required properties. All summands are non negative.

Using $\int T\lambda_i = \int \lambda_i$ and $\int \lambda_i \leqq \gamma(F)$ we can show

$$\int G_1 = \lim_{n\to\infty} \int (\lambda_0 + \sum_{i=0}^{n} (\lambda_{i+1} - T\lambda_i)) = \lim_{n\to\infty} \int \lambda_{n+1} \leqq \gamma(F).$$

We can write

$$\lambda_i = (\lambda_i - T\lambda_{i-1}) + T(\lambda_{i-1} - T\lambda_{i-2}) + \ldots + T^{i-1}(\lambda_1 - T\lambda_0) + T^i \lambda_0\,.$$

With (5.1) this yields

$$F_n \leqq \sum_{i=0}^{n-1} \lambda_i \leqq \sum_{\nu=0}^{n-1} T^\nu G_1 =: G_n\,.$$

The additive process $(G_n)$ dominates $(F_n)$. Therefore we also have $\int G_1 = \gamma(G) \geqq \gamma(F)$.

It remains to eliminate the assumption that $\mathscr{A}$ is countably generated. This is done with a standard argument: If $\mathscr{A}_1$ is the smallest $\sigma$-algebra in which $F_1, F_2, \ldots$ are measurable then $\mathscr{A}_1$ is countably generated. Once a countably generated $\mathscr{A}_k$ has been constructed we know that $L_1(\Omega, \mathscr{A}_k, \mu)$ is separable. Let $f_{k1}, f_{k2}, \ldots$ be dense in this space and let $\mathscr{A}_{k+1}$ be the $\sigma$-algebra generated by all $T^i f_{kj}$, $(i \geqq 0, j \geqq 1)$. The $\sigma$-algebra $\mathscr{A}_\infty$ generated by the sequence $\mathscr{A}_1, \mathscr{A}_2, \ldots$ is again countably generated, all $F_n$ are $\mathscr{A}_\infty$-measurable, and $T$ maps $L_1(\Omega, \mathscr{A}_\infty, \mu)$ into itself. We can therefore apply the above argument in this space. □

It is now not hard to prove a convergence theorem for superadditive processes:

**Theorem 5.3** (Akcoglu-Sucheston). *Let $T$ be a conservative positive contraction in $L_1$ and let $(F_n)$ be a non negative superadditive process with exact dominant $(G_n)$. Then* $\lim_{n\to\infty} F_n/G_n = 1$ *a.e. on* $\{S_\infty G_1 > 0\}$.

*Proof*. We may assume $\Omega = \{S_\infty G_1 > 0\}$. We shall use the following notation: $\underline{h} = \liminf_{n-1} F_n/G_n$, $\bar{h} = \limsup_{n-1} F_n/G_n$, $L(f, g) = \lim_n S_n f/S_n g$, $F_n^k = \sum_{i=0}^{n-1} T^{ki} F_k$. As $(G_n)$ is a dominant, $\bar{h} \leqq 1$. As the sequences $(F_n)$ and $(G_n)$ are increasing in $n$ and $\lim G_{nk}/G_{(n+1)k} = 1$ by Lemma 2.3 we see that $\underline{h} = \liminf_n F_{(n+1)k}/G_{nk}$. The superadditivity implies $T^j F_{nk} \leqq F_{(n+1)k}$ for $j = 0, \ldots, k-1$ and $F_{nk} \geqq F_n^k$. It follows that $k^{-1} S_{nk} F_k = k^{-1} \sum_{j=0}^{k-1} \sum_{i=0}^{n-1} T^{ki+j} F_k \leqq F_{(n+1)k}$ and hence also that

$$\int \underline{h}\, d(G_1 \cdot \mu) \geqq \int L(k^{-1} F_k, G_1)\, d(G_1 \cdot \mu) = k^{-1} \int F_k\, d\mu.$$

As $k^{-1} \int F_k\, d\mu$ tends to $\gamma(F) = \int G_1\, d\mu \geqq \int \bar{h}\, d(G_1 \cdot \mu)$ we have proved $\underline{h} = \bar{h}$ on $\{G_1 > 0\}$.

The assumption $\{G_1 > 0\} = \Omega$ is no restriction of generality: Simply let $G_1' = \sum_{\nu=0}^{\infty} 2^{-(\nu+1)} T^\nu G_1$, $G_n'' = G_n + S_n G_1'$, $F_n'' = F_n + S_n G_1'$. Then the result just proved yields $\lim F_n''/G_n'' = 1$. As $\int 1_E G_1' = \int 1_E G_1$ holds for all $E \in \mathscr{C}$, we have $S_n G_1'/G_n'' \to 1/2$ and $G_n''/G_n \to 2$ on $\{S_\infty G_1 > 0\} = \Omega$, and hence $F_n/G_n \to 1$. □

One can also obtain convergence a.e. on $C$ when $T$ is not conservative by observing that for non negative superadditive $(F_n)$ the restriction $(1_C F_n)$ is superadditive. For non negative *subadditive* $(F_n)$ the sequence $F_n$ converges a.e. in $D$ because of $\sum_{n=1}^{\infty} (F_{n+1} - F_n)^+ \leqq \sum_{n=1}^{\infty} T^n F_1 < \infty$. One can also study ratios $F_n/H_n$ where $H_n$ is non negative and satisfies $H_{m+n} \geqq H_m + T^m H_n$, but the boundedness assumption $\sup \|m^{-1} H_m\|_1 < \infty$ fails. This is analogous to Chacon's admissible sequences in the previous section and generalizes them.

## Notes

It seems that at present all proofs of the existence of the Kingman decomposition depend on variants of lemma 5.2 and differ only in the limiting argument. Kingman used $w^*$-compactness. Burkholder [1973] gave a short proof, which – however – depends on the deep theorem of Komlos [1967] that each $L_1$-bounded sequence admits a subsequence for which the Cesàro averages converge a.e.. Del Junco [1977] gave a martingale argument. The present argument of Akcoglu-Sucheston has the advantage of being both short and elementary. Brunel and Sucheston [1979] have studied the existence of dominants for non negative processes with $F_{n+m} \geqq F_n + T^n F_m$, $(m, n \geqq 1)$, when the positive contraction $T$ need not be Markovian. Call an additive process $(G_n)$ a dominant if $G_n \geqq F_n$ $(n \geqq 1)$ and exact if $\int G_n$ is minimal among all dominants. They show that the sequence $\int \varphi_m$ converges to some limit $\alpha$ with $0 \leqq \alpha \leqq \infty$ and an exact dominant exists iff $\alpha$ is finite. Then $\alpha = \lim \int \varphi_m$. Akcoglu and Sucheston [1982] have a superadditive version of the lemma of Brunel for $E \subset C$. Fong [1979] has studied ratio and stochastic superadditive ergodic theorems for positive Cesàro bounded linear operators $T$ in $L_1$.

Brooks and Chacon [1980] derived the existence of exact dominants from a general lemma on additive set functions. Akcoglu-Sucheston [1984b] considered processes $F_n = \psi(S_n f)$ with $f \geqq 0$ and subadditive or superadditive functions $\psi$ on $\mathbb{R}^+$.

## § 3.6 An example with divergence of Cesàro averages

We shall now consider Chacon's [1964a] example of a positive conservative and ergodic contraction $T$ in $L_1$, for which the averages $A_n f$ diverge a.e. for all non zero $f \in L_1^+$. $T$ will be induced by an invertible nonsingular point mapping $\tau$: for $v \in \tilde{L}_1$ we shall have $Tv = v \circ \tau^{-1}$. $T^*$ has the form $T^* h = h \circ \tau$. As the averages $A_n f$ converge a.e. in the case of existence of a $\sigma$-finite invariant measure $\mu_0 \sim \mu$, the example also shows that such a $\mu_0$ need not exist in general.

It will be enough to show the existence of one element $f \in L_1^+$ for which $A_n f$ diverges a.e., because then the divergence for all other $0 \neq f' \in L_1^+$ follows from the Chacon-Ornstein theorem. The $f$ in the construction shall even satisfy

(6.1) $\liminf A_n f = 0$ a.e., and

(6.2) $\limsup n^{-1} T^{n-1} f = \infty$ a.e..

$\Omega$ shall be (mod nullsets) a subinterval $[0, a]$ of $\mathbb{R}^+$ endowed with the Lebesgue measure $\mu$. The value of $a > 1$ shall be determined in the construction. If $I = [\alpha, \beta[$ and $J = [\gamma, \delta[$ with $\alpha < \beta$ and $\gamma < \delta$ are two intervals the affine transformation from $I$ to $J$ shall be the map

$$\tau(\omega) = \gamma + \frac{\delta - \gamma}{\beta - \alpha}(\omega - \alpha).$$

$\tau$ induces a positive contraction $T$ from $L_1(I)$ to $L_1(J)$: For any $g \in L_1(I)$ we have $(Tg)(\eta) = (\beta - \alpha)(\delta - \gamma)^{-1} f(\tau^{-1}\eta)$ for $\eta \in J$. This is just the description of the map $v \to Tv = v \circ \tau^{-1}$ from $\tilde{L}_1(I)$ to $\tilde{L}_1(J)$ in terms of $L_1(I)$ and $L_1(J)$. (We shall use the letters $\tau$, $T$ also for the restrictions of the ultimate transformations to subsets).

$\tau$ on $\Omega$ shall be defined by an inductive procedure. We start with $I_1^1 = [0, 1/2[$, $I_2^1 = [1/2, 1[$, $N_1 = 2$, and $\tau$ on $I_1^1$ shall be the affine map $I_1^1 \to I_2^1$. At the $n$-th stage of the construction we have defined a finite number $N_n$ of left-closed, right-open disjoint intervals $I_1^n, \ldots, I_{N_n}^n$ such that their union $\Omega_n$ is an interval $[0, a_n[$. The sequence $\Omega_n$ shall be increasing with union $\Omega$. $\tau$ on $I_\nu^n$ is the affine map to $I_{\nu+1}^n$, $(\nu = 1, \ldots, N_n - 1)$. Set $f = 1_{[0,1/2[}$. To simplify the notation we write $N$ for $N_n$, $N'$ for $N_{n+1}$, $I_\nu$ for $I_\nu^n$ and $I'_\nu$ for $I_\nu^{n+1}$ in the inductive step.

*Passage $n \to n+1$ for $n$ odd.* We take $I'_\nu = I_\nu$ for $\nu = 1, \ldots, N$ and we add new disjoint intervals $I'_\nu$, $(\nu = N+1, \ldots, N')$. The number $N'$ will be specified in a moment. The new intervals shall form a subdivision of $[a_n, a_{n+1}[$ with $a_{n+1} = a_n$

$+ \mu(I_1)/2$. If $M_n$ is any integer with $M_n \geqq N+1$ and $N' = 2M_n$ we can be sure that $(A_k f)(\omega)$ is well defined at this stage of the construction for all $k \leqq M_n$ and $\omega \in \Omega_n$: If $\omega$ belongs to $I_j$, $(1 \leqq j \leqq N)$, then $(T^i f)(\omega)$ is determined for $0 \leqq i \leqq j-1$ by the sizes of the intervals $I_j, I_{j-1}, \ldots, I_{j-i}$ and by $f(\tau^{-i}\omega)$. For $j \leqq i < M_n$ we can already be sure that $\tau^{-i}\omega$ shall not belong to $[0, 1/2[$, no matter how we continue the construction. For these $i$ we shall therefore necessarily have $(T^i f)(\omega) = 0$. This shows that by a sufficiently large choice of $M_n$ we can achieve

(6.3) $$0 \leqq A_{M_n} f(\omega) < 1/n, \quad (\omega \in \Omega_n).$$

We complete this step of the construction by fixing such an $M_n$. $\tau$ is defined on $\Omega_{n+1} \setminus I'_{N'}$ by the requirement that $\tau$ maps $I'_\nu \to I'_{\nu+1}$, $(\nu = 1, \ldots, N'-1)$ in an affine way.

*Passage $n \to n+1$ for $n$ even.* Now the construction depends on an integer $K = K_n \geqq 2$ and positive numbers $\alpha_{nk}$, $(0 \leqq k < K)$, with $\sum_{k=0}^{K-1} \alpha_{nk} = 1$, to be determined in a moment. Each of the intervals $I_i$ is cut into $K$ disjoint subintervals $I_{i,k}$ of length $\alpha_{nk}\,\mu(I_i)$ such that, for $k = 0, \ldots, K-2$, the interval $I_{i,k+1}$ lies to the right of $I_{i,k}$. The restriction of $\tau$ to $I_{i,k}$ automatically is the affine map to $I_{i+1,k}$.

Set $N' = N \cdot K$. Each $j$ with $1 \leqq j \leqq N'$ has a unique representation of the form $j = kN + i$ with $0 \leqq k < K$ and $1 \leqq i \leqq N$. Define $I'_j = I_{i,k}$. On the set $\Omega_n \setminus I_N$ where $\tau$ has already been defined previously we just observed that $\tau$ restricted to $I'_j$ is the affine map to $I'_{j+1}$. In order to achieve this also for all other $j < N'$ we define $\tau$ on $I'_{mN} = I_{N,m-1}$ as the affine map to $I'_{mN+1} = I_{1,m}$, $(m = 1, \ldots, K-1)$. We arrive at the desired situation $\tau$: $I'_1 \to I'_2 \to \ldots \to I'_{N'}$.

We now determine the parameters $K = K_n$ and $\alpha_{nk}$: Set $\varepsilon_n = 2^{-n}$. There exists a constant $c_n = c > 0$ and an integer $K \geqq 2$ with

(6.4) $$\varepsilon_n > cnN \quad \text{and} \quad 0 < 1 - \left(\varepsilon_n + c \sum_{k=1}^{K-2} (k+1)^{-1}\right) < \varepsilon_n .$$

Define $\alpha_{n0} = \varepsilon_n$, $\alpha_{nk} = c(k+1)^{-1}$, $(k = 1, \ldots, K-2)$

$$\alpha_{n,K-1} = 1 - \sum_{k=0}^{K-2} \alpha_{nk} .$$

By (6.4) we have $\alpha_{nk} \leqq \alpha_{n0}$, $(k = 1, \ldots, K-1)$. As this is true for each stage of the construction and the added intervals for the steps with $n$ odd were shorter than the first interval $I_1$ of that step, we can be certain that in each step the first interval in the chain $I_1, \ldots, I_N$ is the longest. It is also always a subinterval of $[0, 1/2[$.

The construction is complete. As the measure of the set $I^n_{N_n} \cup (\Omega \setminus \Omega_n)$ where $\tau$ is not yet defined at stage $n$ tends to 0, $\tau$ shall ultimately be defined on almost all of $\Omega$, and similarly also $\tau^{-1}$. After the elimination of a nullset, $\tau$ and $\tau^{-1}$ are defined everywhere. (6.1) is a consequence of (6.3). Let us now check (6.2):

For $n$ even, let $B_n$ be the set of points $\omega$ belonging to some $I'_j = I_j^{n+1}$ with $N+1 \leqq j \leqq (K-1)N$. We may write $j = kN + i$ with $0 < k \leqq K-2$ and $1 \leqq i \leqq N$. $\tau^{j-1}$ maps $I'_1$ onto $I'_j$. We therefore have

$$(T^{j-1}f)(\omega) = \frac{\mu(I'_1)}{\mu(I'_j)} \cdot f(\tau^{j-1}\omega) = \frac{\mu(I'_1)}{\mu(I'_j)}$$

$$\geqq \frac{\mu(I'_i)}{\mu(I'_j)} = \frac{\varepsilon_n}{c(k+1)^{-1}} > nN(k+1) \geqq n \cdot j.$$

Using $\mu(\Omega_n \setminus B_n) \leqq 2\varepsilon_n \mu(\Omega_n)$, we see that almost all $\omega$ belong for infinitely many even $n$ to $B_n$. This implies (6.2). In particular, $T$ must be conservative.

It remains to show that $T$ is ergodic. Let $E$ be a $\tau$-invariant set of positive measure. Let $\mathscr{A}_n$ be the algebra generated by $I_1^n, \ldots, I_{N_n}^n$, and $\mathscr{A}_\infty = \cup \mathscr{A}_n$. As the length of the longest interval $I_1^n$ of step $n$ tends to 0, $\mathscr{A}_\infty$ generates the Borel-$\sigma$-algebra $\mathscr{A}$ in $\Omega$. Hence, for any $\eta > 0$, we can find an $n$ and a set $F_n \in \mathscr{A}_n$ with $F_n \subset \Omega_n$ and $\mu(F_n \cap E) \geqq (1-\eta)\,\mu(E)$.

As $F_n$ can be chosen arbitrarily close to $E$ we can also attain $\mu(F_n \cap E) \geqq (1-2\eta)\,\mu(F_n)$. But then there must be at least one of the intervals $I_\nu^n$ with $\mu(I_\nu^n \cap E) \geqq (1-2\eta)\,\mu(I_\nu^n)$.

As $E$ is invariant and $\tau$ affine we must then have $\mu(I_\nu^n \cap E) \geqq (1-2\eta)\,\mu(I_\nu^n)$ for all $\nu$ with $1 \leqq \nu \leqq N_n$. This implies $\mu(E) \geqq (1-2\eta)\,\mu(\Omega_n)$. As $\eta$ was arbitrarily small and $n$ arbitrarily large we obtain $\mu(E) = \mu(\Omega)$. □

## Notes

The first example of a nonsingular transformation $\tau$ admitting no $\sigma$-finite invariant $\mu_0 \sim \mu$ was given by Ornstein [1960]. The construction of Chacon is partly similar to Ornstein's example. Such constructions are now called stacking constructions, and they are discussed in some detail in the book of Friedman [1970]. Also the simple example of Brunel [1966a] belongs in this class. Hamachi [1981] gave a very different example: a Bernoulli shift with nonidentical factor measures.

It is simple to see that there exist positive linear operators in $L_1$ with $\|T\| \leqq 1 + \varepsilon$ and $\|T\|_{1,\infty} \leqq 1$ for which $A_n f$ need not converge a.e..

Modifying the example above, Friedman [1966] has proved that also $\|T\| \leqq 1$, $\|T\|_{1,\infty} \leqq 1 + \varepsilon$ is not sufficient. A. Ionescu-Tulcea [1965] has proved a related category theorem: In the class of positive invertible isometries of $L_1([0,1])$ the set of $T$ for which $A_n f$ converges a.e. for all $f \in L_1$ is a set of first category in the strong operator topology.

Weiner [1968] showed that there exists a nonsingular $\tau$ without $\sigma$-finite invariant measure equivalent to $\mu$, such that $A_n(T)f$ converges a.e. for all $f$, where $T$ is the corresponding contraction in $L_1$. Y. Ito [1981] has simplified and generalized the construction.

Positive results can be proved using stronger summation methods; see § 8.2.

## § 3.7 More on the filling scheme

As the filling scheme has been the key for the proof of the Chacon-Ornstein theorem we explore it and its consequences a bit further. We also report briefly on abstract generalizations of the ratio theorem. Except for the first lemma below which will be applied in § 7.1, the results of this section will not be needed in the sequel.

We again start with the general situation, where $T$ is an arbitrary positive linear operator in a general vector lattice $L$ of measurable functions on $\Omega$. Again $Uh := Th^+ - h^-$. Given $f, g \in L^+$ we shall be interested in the sequences $h_n = U^n h$, $(n \geqq 0)$, with $h = f - g$. $f_n := h_n^+$ is the matter and $g_n := h_n^-$ the antimatter (or hole) remaining at time n. The portion $d_n$ of the hole filled exactly at time $n$ is given by $d_0 = f \wedge g = g - g_0$, and $d_n = g_{n-1} - g_n$, $(n \geqq 1)$. The following lemma essentially is the original formulation of the filling scheme:

**Lemma 7.1.** *For all* $n \geqq 0$, $T^n f = f_n + \sum_{k=0}^{n} T^{n-k} d_k$. *On the set* $B = \{M_{n+1} h > 0\}$ *one has* $g_n = 0$, *or (equivalently)* $g = \sum_{k=0}^{n} d_k$.

*Proof.* For $n = 0$ the first assertion follows from $f = (f-g)^+ + (f \wedge g)$ and the induction hypothesis yields

$$\begin{aligned} T^{n+1} f &= Tf_n + \sum_{k=0}^{n} T^{(n+1)-k} d_k \\ &= (Tf_n - g_n)^+ - (Tf_n - g_n)^- + g_n + \sum_{k=0}^{n} T^{(n+1)-k} d_k \\ &= f_{n+1} - g_{n+1} + g_n + \sum_{k=0}^{n} T^{(n+1)-k} d_k \\ &= f_{n+1} + \sum_{k=0}^{n+1} T^{(n+1)-k} d_k . \end{aligned}$$

The second assertion is equivalent to theorem 2.2. □

The effect of the filling scheme may be more visible if one considers the sums of $T^n f$, $(n = 0, \ldots, N)$ and of $T^j g = T^j(g_{N-j} + d_0 + \ldots + d_{N-j})$, $(j \leqq N)$. Then

$$(7.1) \qquad S_{N+1} f = \sum_{n=0}^{N} f_n + u_{N+1} \quad \text{and} \quad S_{N+1} g = \sum_{j=0}^{N} T^j g_{N-j} + u_{N+1}$$

with

$$u_{N+1} = (d_0 + \ldots + d_N) + T(d_0 + \ldots + d_{N-1}) + \ldots + T^N d_0 \geqq 0 .$$

In this form the scheme has been used by Neveu [1979] to give a transparent alternative proof of the Chacon-Ornstein theorem including the identification of the limit in the conservative ergodic case. (It seems simpler to derive the general case directly than to deduce it from the ergodic case because the ergodic decomposition of $T$, due to Jacobs [1962/63], (see also Krengel [1963]), is non trivial).

In the case considered by Neveu, the assumption $g_\infty := \lim g_n \not\equiv 0$ (that the hole is not filled completely) implies $\psi_A \equiv \mathbb{1}$ in the following lemma:

**Lemma 7.2.** *Let $T$ be a positive contraction in $L_1$. Set $A = \{g_\infty > 0\}$, then*

$$(7.2) \qquad \sum_{n=0}^{\infty} f_n < \infty \quad \text{on } \{\psi_A > 0\}$$

*and*

$$(7.3) \qquad \lim_n \int f_n \psi_A = 0.$$

*Proof.* Set $v_p = (I_{A^c} T^*)^p 1_A$. Then $\psi_A = \sum_{p=0}^{\infty} v_p$. Because of $f_n \wedge g_n = h_n^+ \wedge h_n^- = 0$ and $g_n \geqq g_\infty$ the functions $f_n$ must vanish on $A$. Hence $f_{n+1} \leqq Tf_n = TI_{A^c} f_n$, and, by induction, $f_n \leqq (TI_{A^c})^n f_0$. This yields

$$(7.4) \qquad \langle f_n, v_p \rangle \leqq \langle (\mathrm{TI}_{A^c})^n f_0, (I_{A^c} T^*)^p 1_A \rangle = \langle f_0, v_{p+n} \rangle .$$

Summing over $p$ gives $\langle f_n, \psi_A \rangle \leqq \langle f_0, \sum_{j=n}^{\infty} v_j \rangle \to 0$, i.e., (7.3). Summing over $n$ gives $\langle \sum_n f_n, v_p \rangle \leqq \langle f_0, \sum_n v_{p+n} \rangle \leqq \langle f_0, \mathbb{1} \rangle < \infty$, so that $\sum f_n$ converges on $\{v_p > 0\}$ for each $p \geqq 0$. □

Rost [1971] has studied the following two questions on the filling scheme for a positive contraction $T$ in $L_1$:

($\alpha$) When will the hole be filled completely?

($\beta$) When will all the matter fall into the hole or disappear?

In other words: When do we have $g_\infty = 0$, resp. $\lim \| f_n \|_1 = 0$? We shall need the partial order in $L_1$ defined by $h' \vdash h''$ iff $\langle h', u \rangle \geqq \langle h'', u \rangle$ holds for all superharmonic $u \in L_\infty$.

For $f = r + s$, $(r, s \in L_1^+)$, and $f' = Tr + s$, we clearly have $\langle f, u \rangle = \langle r, u \rangle + \langle s, u \rangle \geqq \langle r, T^* u \rangle + \langle s, u \rangle = \langle f', u \rangle$. We therefore see that $f \xrightarrow{n} \bar{f}$ implies $f \vdash \bar{f}$. Similarly, for any $d \in L_1$ the inequalities $\langle d^+ - d^-, u \rangle \geqq \langle d^+, T^* u \rangle - \langle d^-, u \rangle = \langle Td^+ - d^-, u \rangle = \langle Ud, u \rangle$ give $d \vdash Ud$. As $\langle U^n(f-g), u \rangle$ decreases, we see $\langle f-g, u \rangle \geqq \lim \langle U^n(f-g), u \rangle = \lim \langle f_n - g_n, u \rangle \geqq \lim \langle -g_n, u \rangle = -\langle g_\infty, u \rangle$, i.e.,

$$(7.5) \qquad f - g \vdash -g_\infty .$$

In particular, $f \vdash g$ is obviously necessary for $g_\infty = 0$.

**Theorem 7.3** (Rost). *In the filling scheme with $f, g \in L_1^+$ the relation $f \vdash g$ is necessary and sufficient for $g_\infty = 0$.*

*Proof.* Let us first show, for $d \in L_1$, that $d \vdash 0$ implies $Ud \vdash 0$. Set $B = \{d^- > 0\}$. For superharmonic $u$ define $v = u 1_B + (T^* u) 1_{B^c}$. Then $T^* u \leqq v \leqq u$ shows $T^* v \leqq T^* u \leqq v$ so that $v$ is superharmonic. It follows that $\langle d, v\rangle \geqq 0$. Using the definition of $v$ and $\{d^+ > 0\} \subset B^c$ we find $\langle Ud, u\rangle = \langle Td^+ - d^-, u\rangle = \langle d^+, T^* u\rangle - \langle d^-, u\rangle = \langle d^+, v\rangle - \langle d^-, v\rangle \geqq 0$. As the superharmonic $u$ was arbitrary we have proved $Ud \vdash 0$. By induction $U^n d \vdash 0$ for all $n$.

Now assume $f \vdash g$ and set $A = \{g_\infty > 0\}$. We have just proved $\langle g_n, u\rangle \leqq \langle f_n, u\rangle$ for all superharmonic $u$. As $\psi_A$ is superharmonic, $\langle g_n, \psi_A\rangle \leqq \langle f_n, \psi_A\rangle$. By (7.3) the right hand side tends to 0 and we obtain $\langle g_\infty, \psi_A\rangle = 0$. Now $\psi_A \geqq 1_A$ yields $g_\infty = 0$. □

As each step in the filling procedure is a partial application of the remaining matter the theorem tells us that, for $f, g \in L_1^+$, the relation $f \vdash g$ is equivalent to the existence of a sequence of measurable functions $\alpha_i$: $\Omega \to [0, 1]$ with $g = \alpha_1 f + \alpha_2(T(\mathbb{1} - \alpha_1)f) + \alpha_3(T(\mathbb{1} - \alpha_2)(T(\mathbb{1} - \alpha_1)f)) + \ldots$.

The answer to question ($\beta$) is quite symmetric: As in Meyer [1969/70] we write $h' \dashv h''$ iff $\langle h', u\rangle \leqq \langle h'', u\rangle$ holds for all subharmonic $u$.

**Theorem 7.4** (Rost). *In the filling scheme with $f, g \in L_1^+$, $f \dashv g$ is necessary and sufficient for $\|f_n\|_1 \to 0$.*

*Proof.* For subharmonic $u$ the inequalities $\langle Ud, u\rangle = \langle Td^+, u\rangle - \langle d^-, u\rangle = \langle d^+, T^* u\rangle - \langle d^-, u\rangle \geqq \langle d, u\rangle$, valid for any $d \in \mathrm{L}_1$, show that the sequence $\langle f_n - g_n, u\rangle = \langle U^n(f - g), u\rangle$ increases. If $\|f_n\|_1$ tends to 0 we obtain $0 \geqq \lim \langle -g_n, u\rangle = \lim \langle f_n - g_n, u\rangle \geqq \langle f - g, u\rangle$ for all such $u$ and hence $f \dashv g$.

Now assume $f \dashv g$. Exactly the same argument as in the previous proof shows that $d \dashv 0$ implies $Ud \dashv 0$. Hence $f_n \dashv g_n$ holds for all $n$. Let $e$ denote the (decreasing) limit of $T^{*n}\mathbb{1}$. Then we have $T^* e = e$, and $\langle T^n f, \mathbb{1} - e\rangle = \langle f, T^{*n}\mathbb{1} - e\rangle$ tends to 0. Because of $f_n \leqq T^n f$ we obtain $\langle f_n, \mathbb{1} - e\rangle \to 0$. It remains to show $\langle f_n, e\rangle \to 0$. Set $A = \{g_\infty > 0\}$. As $\psi_A$ is superharmonic $T^*(e - \psi_A)^+ \geqq T^*(e - \psi_A) \geqq e - \psi_A$ shows that $(e - \psi_A)^+$ is subharmonic. Now $f_n \dashv g_n$ yields $\langle f_n, e\rangle \leqq \langle f_n, \psi_A\rangle + \langle f_n, (e - \psi_A)^+\rangle \leqq \langle f_n, \psi_A\rangle + \langle g_n, (e - \psi_A)^+\rangle$. The first term tends to 0 by (7.3) and the second because $g_n$ decreases to 0 in $A^c$ and $(e - \psi_A)^+$ vanishes in $A$. □

For further related results and the interpretation of $g$ (in the case $g_\infty = 0$) as the distribution of a stopped Markov process the reader should consult the papers of Rost and Meyer.

## Notes

Let $T$ be a positive linear operator in a vector lattice $L$ of measurable functions on $(\Omega, \mathcal{A}, \mu)$. Ornstein [1970] calls $\bar{f} \in L^+$ a *modification* of $f \in L^+$ if there exists some $u \in L^+$

with $\bar{f} = f - u + Tu$. E.g., for $f = r + s$, $(r, s \in L^+)$, the function $\bar{f} = Tr + s$ is a modification of $f$ because we may take $u = r$. If $f' = f - u + Tu$ is a modification of $f$ and $f'' = f' - u' + Tu'$ a modification of $f'$, then $f'' = f - (u + u') + T(u + u')$ is also a modification of $f$. We thus see that $f \overset{n}{\to} \bar{f}$ implies that $\bar{f}$ is a modification of $f$. (However, a modification $\bar{f}$ of $f$ need not satisfy $f \overset{n}{\to} \bar{f}$ for any $n$. E.g., we may have $\bar{f} = \sum_{n=0}^{\infty} 2^{-(n+1)} T^n f$.)

Before we continue let us give an application of modifications:

**Theorem 7.5** (Fong-Sucheston [1973]). *If $T$ is a conservative positive contraction in $L_1$ the subspace $L_0 = \{h \in L_1 \colon \langle h, 1_E \rangle = 0$ for all $E \in \mathscr{C}\}$ is the closure of $\{u - Tu \colon u \in L_1^+\}$.*

*Proof.* For any $u \in L_1^+$, we have $u - Tu \in L_0$ since $E \in \mathscr{C}$ is equivalent to $T^* 1_E = 1_E$. For $h \in L_0$ consider the filling scheme for $f = h^+$, $g = h^-$. As $T$ is conservative, the superharmonic functions are harmonic and $\mathscr{C}$-measurable. Hence $f \vdash g$. For $\varepsilon > 0$ we may find an $n$ with $\|g_n\|_1 < \varepsilon$. Recall $g_n = g - (d_0 + \ldots + d_n)$. As $T$ is Markovian we have $\|f_i\|_1 = \|f_{i+1}\|_1 + \|d_i\|_1$. We obtain $\|f_n\|_1 = \|f\|_1 - \sum_{i=0}^{n} \|d_i\|_1 = \|g\|_1 - \sum_{i=0}^{n} \|d_i\|_1 = \|g_n\|_1 < \varepsilon$. $\bar{f} = \sum_{i=0}^{n} d_i + f_n$ is a modification $f - u + Tu$ of $f$. $h = f - g = u - Tu - g_n + f_n$ now gives $\|h - (u - Tu)\|_1 < 2\varepsilon$. □

If $T$ is a positive contraction in $L_1$ a modification $\bar{f}$ satisfies $\|\bar{f}\|_1 = \|f - u + Tu\|_1 = \|f\|_1 + \|Tu\|_1 - \|u\|_1 \leqq \|f\|_1$. Therefore the following theorem of Ornstein contains the Chacon-Ornstein theorem:

**Theorem 7.6.** *Let $T$ be a positive linear operator in a vector lattice $L$ of measurable functions on $(\Omega, \mathscr{A}, \mu)$, and $f, g \in L^+$. If $S_n f / S_n g$ fails to converge on a subset $E$ of $\{S_\infty g > 0\}$ there exists a sequence $e_n$ of modifications of $f + g$ with* $\lim e_n(\omega) = \infty$ a.e. *on $E$.*

Ornstein also gives a partial converse. Under suitable assumptions, satisfied e.g. for positive contractions in $L_2$, the set where the ratio convergence fails for some $f, g$ agrees mod $\mu$ with the set where there exist arbitrarily large modifications. These results and those in the work of Fong and Lin [1976] show that under fairly general conditions the ratios $S_n f / S_n g$ converge a.e. for all $f, g \in L^+$ *only* if $T$ is a positive contraction in $L_1$ after a suitable change of the underlying measure.

Foguel [1980] has a Hopf decomposition in vector lattices.

Brunel [1976] has extended many of the results of § 3 of this chapter to the abstract situation studied by Ornstein. He calls $E$ a *good set* if there exists a non negative $\chi \geqq T^* \chi$ in the space $L^*$ of bounded linear forms on $L$ which is strictly positive on $E$. (This means that there is a $\varphi \in L^+$ with $E \subset \{\varphi > 0\}$ and $\chi(f) \geqq \int \varphi f \, d\mu \; \forall f \in L^+$.)

The condition that there exists no sequence $e_n$ of modifications of $u \in L^+$ with $e_n \to \infty$ a.e. on $E$ is shown to be equivalent to the condition that $E$ is good when $L$ is replaced by the space $L_u$ of all functions bounded by finite linear combinations of functions $T^i u$, $(i \geqq 0)$. There is also an abstract form of Chacon's convergence theorem for ratios $S_n f / \sum_{i=0}^{n} p_i$ with $(p_i)$ admissible (asserting convergence on good sets in $\{\sum_{i=0}^{\infty} p_i > 0\}$.), and of Brunel's lemma. Gologan [1979] has dropped the condition of positivity of $T$ for suitable spaces. In these generalizations the basic setup is still defined in terms of a linear operator. As the convergence of ratios $q_n = (f_0 + f_1 + \ldots + f_{n-1}) / (p_0 + \ldots + p_{n-1})$ depends only on the joint distribution of $(f_0, p_0, f_1, p_1, \ldots)$ Dinges [1971] has proposed to investigate conditions for convergence which should be stated in terms of these distributions. Let $P_n$ denote the joint distribution of $(f_n, p_n, f_{n+1}, p_{n+1}, \ldots)$. It was shown that the condition $\ll \int h\,dP_n \geqq \int h\,dP_{n+1}$ for all non negative, positively homogeneous convex functions $h$ and all $n \geqq 0 \gg$ could be used to prove convergence a.e. of $q_n$. However, it was shown by Rost [1971a] and Engmann [1976] that under such a condition $q_n$ did arise from a positive contraction. Engmann also discussed the ergodic theorem of Cuculescu and Foias [1966], which deals with a sequence $T_1 \geqq T_2 \geqq \ldots$ of positive contractions and functions $f_n, p_n$ with $T_{n+1} f_n \geqq f_{n+1}$, $T_{n+1} |p_n| \leqq p_{n+1}$. (See also Gologan [1977]). In the course of his program Dinges [1970] systematically studied inequalities and identities between partial sums $s_k^i = \sum_{j=i}^{k-1} f_j$ of functions and various maxima and minima defined in terms of some $s_k^i$.

The book by Dellacherie-Meyer [1983] and the articles by Dinges [1974] and Revuz [1978] are further references related to the filling scheme.

# Chapter 4: Extensions of the $L_1$-theory

Several results from the previous chapter are now generalized in three different directions. First we drop the positivity assumption. Then we allow that the functions take values in $\mathbb{R}^n$ or even in a Banach space. Finally, we show that the case of power bounded or Cesàro bounded operators can often be reduced to the contraction case.

## § 4.1 Non positive contractions in $L_1$

**1. The linear modulus of $T$.** Quite a few ergodic theorems have first been derived for positive operators in $L_1$, but can be extended to the case where $T$ need not be positive with the help of the "linear modulus" of $T$. We now discuss this device and some applications.

For positive operators it has not been a restriction of generality to work always with real $L_1$-spaces, since a complex $f$ can be decomposed into its real and imaginary parts. In this section we shall always work with complex $L_1$-spaces. Some of the interesting non positive operators like $Tf(\omega) = e^{i\alpha(\omega)} f(\tau\omega)$, (for $\alpha \geqq 0$ and measure preserving $\tau$), make sense only in the complex $L_1$. The results will remain true for operators in the real $L_1$. In fact, in that case some statements can be strengthened or are simpler due to the lattice structure of the real $L_1$. $\mathbf{T}$ in the following theorem is called the *linear modulus* of $T$. It is frequently denoted by $|T|$.

**Theorem 1.1.** *For every bounded linear operator $T$: $L_1 \to L_1$ there exists a unique bounded linear positive operator $\mathbf{T}$: $L_1 \to L_1$ such that*
*(i)* $\|\mathbf{T}\| = \|T\|$,
*(ii) for all* $f \in L_1$, $|Tf| \leqq \mathbf{T}|f|$,
*(iii) for all* $f \in L_1^+$, $\mathbf{T}f = \sup\{|Tg| : g \in L_1, |g| \leqq f\}$.

*Proof.* Let $\mathscr{P}$ denote the family of all finite measurable partitions $\pi = \{B_1, B_2, \ldots, B_m\}$ of $\Omega$. $\mathscr{P}$ is partially ordered in the usual way: $\pi \leqq \pi'$ means that $\pi'$ is a refinement of $\pi$, i.e., the sets $B_i$ are unions of sets of $\pi'$. For $f \in L_1^+$ define $\mathbf{T}_\pi f := \sum_{i=1}^{m} |T(1_{B_i} f)|$. Clearly $\pi \leqq \pi'$ implies $\mathbf{T}_\pi f \leqq \mathbf{T}_{\pi'} f$. From $\|f\|_1 = \sum_{i=1}^{m} \|1_{B_i} f\|_1$ we obtain $\|\mathbf{T}_\pi f\|_1 \leqq \|\mathbf{T}\| \, \|f\|_1$. As $\{\mathbf{T}_\pi f : \pi \in \mathscr{P}\}$ is increasing on $\mathscr{P}$

and norm bounded we may define

$$(1.1)\qquad \mathbf{T}f := \lim_{\pi\in\mathscr{P}} \mathbf{T}_\pi f, \quad (f\in L_1^+).$$

(The monotone convergence theorem for families $\{f_\pi\}\subset L_1^+$ indexed by elements a directed set $\mathscr{P}$ follows easily from its sequence version. If $\pi_1 \leqq \pi_2 \leqq \ldots$ is an increasing sequence with $\|f_{\pi_n}\|_1 \to \alpha := \sup\{\|f_\pi\|_1 : \pi\in\mathscr{P}\} < \infty$, then $f_{\pi_1} \leqq f_{\pi_2} \leqq \ldots$ converges to some $f_\infty \in L_1^+$ with norm $\alpha$, and it is easy to check that $f_\infty = \lim f_\pi$ does not depend on the sequence $(\pi_n)$.)

We clearly have

$$(1.2)\qquad \|\mathbf{T}f\|_1 \leqq \|T\|\,\|f\|_1, \quad (f\in L_1^+)$$

and $\mathbf{T}(f+g) \leqq \mathbf{T}f + \mathbf{T}g$, $(f, g\in L_1^+)$. If $f$ and $g$ are simple functions, (i.e., take only finitely many values), and $\pi$ is at least as fine as the partition generated by the sets of constancy of $f$ and $g$, we have $\mathbf{T}_\pi(f+g) = \mathbf{T}_\pi f + \mathbf{T}_\pi g$, and, in the limit, $\mathbf{T}(f+g) = \mathbf{T}f + \mathbf{T}g$. By approximation this remains true for all $f, g\in L_1^+$ and $\mathbf{T}$ acts as a bounded positive linear operator on $L_1^+$. Now $\mathbf{T}$ can be extended by linearity to all of $L_1$. This extension is again denoted by $\mathbf{T}$.

For $f\in L_1^+$ and $|g|\leqq f$ we obtain $\mathbf{T}f \geqq |Tg|$ by approximation with simple functions. This yields (ii) and the inequality $\geqq$ in (iii). If $\leqq$ fails in (iii) there exists an $f\in L_1^+$, a partition $\pi = \{B_1, \ldots, B_m\}$, a set $A_0$ with $\mu(A_0) > 0$ and an $\varepsilon > 0$ such that

$$(1.3)\qquad \mathbf{T}_\pi f \geqq \sup_{|g|\leqq f} |Tg| + \varepsilon \quad \text{on} \quad A_0.$$

Now there exists a complex number $a_1$ with $|a_1| = 1$ and a set $A_1 \subset A_0$ with $\mu(A_1) > 0$ such that $|a_1 T(1_{B_1} f) - |T(1_{B_1} f)|| \leqq \varepsilon/2m$ on $A_1$. Continuing in this way we find $a_i$ with $|a_i| = 1$, $(i = 1, \ldots, m)$, and $A_0 \supset A_1 \supset A_2 \ldots \supset A_m$ with $\mu(A_m) > 0$ such that $|a_i T(1_{B_i} f) - |T(1_{B_i} f)|| \leqq \varepsilon/2m$ on $A_i$. Using the function $g = \sum_{i=1}^{m} a_i f 1_{B_i}$, a contradiction to (1.3) results. The identity (i) follows from (1.2) and (ii). □

The proof of the following proposition is straightforward and is left as an exercise:

**Proposition 1.2.** *The mapping $T \to |T|$ in the previous theorem has the following properties:*

*(a)* $T \geqq 0$ *implies* $T = |T|$;

*(b)* $|aT| = |a|\,|T|$; $|T_1 + T_2| \leqq |T_1| + |T_2|$;

*(c)* $|T_1 T_2| \leqq |T_1|\,|T_2|$, ($<$ *occurs*);

*(d)* $|T|^n |f| \geqq |T^n f|$ *for all* $f\in L_1$, $n \geqq 0$;

*(e)* $\|T\|_{1,\infty} = \||T|\|_{1,\infty}$; (see § 1.6);

*(f) The map $T \to |T|$ is norm continuous. (In general it is not continuous in the strong operator topology.)*

The arguments above can be modified to show that the space of bounded linear operators in a *real* $L_1$ is a lattice. E.g. $T^+ = T \vee 0$ and $T^- = (-T) \vee 0$ may be constructed by setting (for $f \geqq 0$)

$$\mathbf{T}_\pi^+ f = \sum_{i=1}^m (T(1_{B_i} f))^+, \ \mathbf{T}_\pi^- f = \sum_{i=1}^m (T(1_{B_i} f))^-;$$

$$T^+ f = \lim \mathbf{T}_\pi^+ f \quad \text{and} \quad T^- f = \lim \mathbf{T}_\pi^- f.$$

It is then simple to check $|T| = T^+ - T^-$.

The results above are due to Chacon and Krengel [1964]. Dunford-Schwartz [1956] gave a proof of the special case of theorem 1.1 in which $T$ satisfies the additional assumption $\|T\|_{1,\infty} < \infty$. However, in the real case Kantorovič [1940] has already given a (different) construction of the linear modulus in a lattice theoretic context.

**Remark.** Call a linear operator $T$ in $L_p$ with $1 \leqq p \leqq \infty$ *majorizable* or *regular* if there exists a bounded linear operator $T_0$ in $L_p$ with $|Tf| \leqq T_0 |f|$ for all $f \in L_p$.

If $T$ is majorizable essentially the same proof yields the existence of a linear modulus $\mathbf{T}$ satisfying (ii) and (iii); while (i) is replaced by $\|\mathbf{T}\| \leqq \|T_0\|$.

**Theorem 1.3.** *All bounded linear operators $R$ in $L_\infty$ of a $\sigma$-finite space admit a linear modulus $\mathbf{R}$. It satisfies $\|\mathbf{R}\| = \|R\|$ and*

(1.2) $\quad \mathbf{R}f = \sup\{|Rg| : |g| \leqq f\}$ *for* $f \in L_\infty^+$.

*If $T$ is a bounded linear operator in $L_1$ and $R = T^*$, then $\mathbf{R} = \mathbf{T}^*$.*

(Except for the easy inequality $|Rg| \leqq \mathbf{T}^* |g|$, this theorem, due to Krengel [1963], will be used only in the supplement.)

*Proof.* Define $\mathbf{R}_\pi$ in analogy to $\mathbf{T}_\pi$. A priori the almost sure limit $\mathbf{R}f = \lim \mathbf{R}_\pi f$ might assume the value $\infty$, but the second part of the proof of theorem 1.1 remains valid and shows (1.4). Hence the existence of the linear modulus and $\|\mathbf{R}\| = \|R\|$ follows.

For any $h \in L_1^+$ we have

$$|\langle h, Rg \rangle| = |\langle Th, g \rangle| \leqq \langle \mathbf{T}h, |g| \rangle = \langle h, \mathbf{T}^* |g| \rangle.$$

Hence $|Rg| \leqq \mathbf{T}^* |g|$, and $\mathbf{R} \leqq \mathbf{T}^*$.

To prove $\geqq$ we use the fact that the natural embedding $\kappa$: $L_1 \to L_1^{**}$ satisfies $\kappa(h_1 \vee h_2) = (\kappa h_1) \vee (\kappa h_2)$; see Dunford-Schwartz [1958: 303]. Here $h_1, h_2$ are real valued, but $|\kappa h| = \kappa |h|$ holds for complex valued $h \in L_1$ by approximation with simple functions.

For $f \in L_\infty^+$, $h \in L_1^+$ we obtain

$$|\langle f, \kappa Th\rangle| = |\langle f, Th\rangle| = |\langle Rf, h\rangle| \leqq \langle |Rf|, h\rangle \leqq \langle \mathbf{R}f, h\rangle = \langle f, \mathbf{R}^* \kappa h\rangle.$$

As $f \in L_\infty^+$ was arbitrary, $\kappa|Th| = |\kappa Th| \leqq \mathbf{R}^* \kappa h$. For $\pi = \{B_1, \ldots, B_m\}$ put $h_i = h 1_{B_i}$. Then

$$\kappa|Th_i| \leqq \mathbf{R}^* \kappa h_i \leqq \mathbf{T}^{**} \kappa h_i = \kappa \mathbf{T} h_i.$$

Summation yields $\kappa \mathbf{T}_\pi h = \mathbf{R}^* \kappa h \leqq \mathbf{T}^{**} \kappa h = \kappa \mathbf{T} h$, and a passage to the limit implies $\kappa \mathbf{T} h \leqq \mathbf{R}^* \kappa h \leqq \mathbf{T}^{**} \kappa h = \kappa \mathbf{T} h$. Hence equality holds. Now

$$\langle h, \mathbf{T}^* f\rangle = \langle \mathbf{T}^{**} \kappa h, f\rangle = \langle \mathbf{R}^* \kappa h, f\rangle = \langle \kappa h, \mathbf{R} f\rangle = \langle h, \mathbf{R} f\rangle$$

shows $\mathbf{R} = \mathbf{T}^*$. □

Theorem 1.1 can be used to prove the ergodic theorem of Dunford-Schwartz [1956] which asserts that $A_n f$ converges a.e. for all $f \in L_1$ if $T$ is an arbitrary $L_1 - L_\infty$-contraction which need not be positive.

For economical reasons we proceed right away to the proof of the deeper ratio theorem of Chacon via a representation theorem of Akcoglu and Brunel [1971].

**2. The structure of $T$ on the conservative part of $|T|$.** In the sequel $T$ will be a fixed positive contraction in $L_1$, $\mathbf{T}$ its linear modulus, $D$ and $C$ the dissipative and conservative part of $\Omega$ for the operator $\mathbf{T}$ and $\mathscr{C}$ the $\sigma$-algebra (in $C$) of $\mathbf{T}$-absorbing subsets of $C$. It is not difficult to check that a set $A \in \mathscr{A}$ is $\mathbf{T}$-absorbing iff $T$ maps $L_1(A) = \{f \in L_1 \colon \{f \neq 0\} \subset A\}$ into $L_1(A)$.

**Lemma 1.4.** *If $h \in L_\infty$ and $T^* h = h$ then $\mathbf{T}^*|h| = |h|$ on $C$.*

*Proof.* $|T^*| = \mathbf{T}^*$ yields $|h| = |T^* h| \leqq \mathbf{T}^*|h|$. Apply theorem 3.1.3. □

**Lemma 1.5.** *Assume $A \in \mathscr{C}$* and $h \in \mathrm{L}_\infty$. *Then $\mathbf{T}^* h = h$ holds on $A$ iff $1_A h$ has a $\mathscr{C}$-measurable restriction to $C$.*

*Proof.* This is essentially a rewording of lemma 3.3.3. Because of $\mathbf{T}^* 1_{A^c} \leqq 1_{A^c}$, $\mathbf{T}^* h$ in $A$ does not depend on the values $h$ takes in $A^c$ and we may assume $\Omega = A = C$. Now it does not matter that $h \geqq 0$ has not been assumed, because we may add a multiple of $\mathbb{1} = \mathbf{T}^* \mathbb{1}$ to $h$ to achieve a reduction to the case $h \geqq 0$. □

**Lemma 1.6.** *Let $A \in \mathscr{C}$ and $h \in L_\infty$. Then* (i): $\mathbf{T}(hf) = h \cdot \mathbf{T} f$ *holds for all $f \in L_1(A)$, is equivalent to* (ii): $\mathbf{T}^* h = h$ *on $A$.*

*Proof.* $\mathbf{T}$ is Markovian on $A \subset C$. Hence $hf \in L_1(A)$ and (i) yield $\langle f, h\rangle = \int hf = \int \mathbf{T}(fh) = \int (h \cdot \mathbf{T} f) = \langle \mathbf{T} f, h\rangle = \langle f, \mathbf{T}^* h\rangle$. As $f \in L_1(A)$ was arbitrary (ii) follows.

Now let $\mathscr{H}$ be the class of all $h \in L_\infty$ satisfying (i). If $B \in \mathscr{C}$ is contained in $A$ and $f \in L_1(A)$, then $\mathbf{T}(1_B f) = 1_B \mathbf{T}(1_B f) = 1_B \mathbf{T}(f - (1_{A \setminus B} f)) = 1_B \mathbf{T} f$, which means that $1_B$ belongs to $\mathscr{H}$. Since $\mathscr{H}$ is a closed linear manifold in $L_\infty$ containing all $h$ with $h 1_A = 0$ we see that $\mathscr{H}$ contains all $h$ satisfying (ii). □

**Lemma 1.7.** *If $f, g$ are integrable, $|f| \leqq g$ and $\int (g-f)\,d\mu = 0$, then $f = g$ a.e..*

*Proof.* Since $Re(g-f) = g - Re(f) \geqq 0$ and $Re(\int (g-f)) = 0$, it follows that $g = Re(f)$ and hence that $g = f$. □

The next lemma will give us the key for the representation.

**Lemma 1.8.** *Assume that* $\mathbf{T}$ *is conservative and that there exists an $h \in L_\infty$ with $h \neq 0$* a.e. *and $T^* h = h$. Let $s = h/|h|$. Then $\mathbf{T} f = sT(\bar{s} f)$ for any $f \in L_1$, where $\bar{s} = 1/s$ is the complex conjugate of $s$.*

*Proof.* It is enough to prove the lemma for $f \geqq 0$. Using lemma 1.4 we see that $\mathbf{T}^*|h| = |h|$. Now $\int (\mathbf{T}f) \cdot |h| = \langle f, \mathbf{T}^*|h| \rangle = \langle f, |h| \rangle = \int (\bar{s} f) h = \int (\bar{s} f) T^* h = \int (T(\bar{s} f)) h = \int (sT(\bar{s} f)) |h|$ and $|sT(\bar{s} f)|\,|h| = |T(\bar{s} f)|\,|h| \leqq (\mathbf{T}|\bar{s} f|)\,|h| = (\mathbf{T} f)|h|$, together with the previous lemma yield $(sT(\bar{s} f))|h| = (\mathbf{T} f)|h|$. Because of $h \neq 0$ a.e. the proof is complete. □

Applying this lemma to the restriction of $T$ to absorbing subsets of $C$ we obtain:

**Lemma 1.9.** *Let $G \in \mathscr{C}$ and assume that there exists an $h \in L_\infty$ with $h \neq 0$ a.e. on $G$ and $T^* h = h$ on $G$. Let $s$ be any function such that $s = h/|h|$ on $G$. Then $\mathbf{T} f = sT(\bar{s} f)$ for all $f \in L_1(G)$.*

*Proof.* $\mathbf{T} L_1(G) \subset L_1(G)$ and $|Tg| \leqq \mathbf{T}|g|$ shows $T L_1(G) \subset L_1(G)$. If $\mathbf{T}'$, $T'$, and $h'$ are the restrictions of $\mathbf{T}$, $T$, $h$ to $G$, then it is simple to check that $\mathbf{T}'$ is the linear modulus of $T'$ and $\mathbf{T}'$, $T'$, $h'$ satisfy the assumptions of the previous lemma. □

We are finally ready to prove

**Theorem 1.10** (Akcoglu-Brunel). *Let $T$ be a contraction in $L_1$,* $\mathbf{T}$ *its linear modulus, $C$ the conservative part of $\Omega$ for* $\mathbf{T}$ *and $\mathscr{C}$ the system of* $\mathbf{T}$*-absorbing subsets of $C$. There exists a set $\Gamma \in \mathscr{C}$ and a function $s \in L_\infty(\Gamma)$ satisfying*

*(i) $|s| = 1$ on $\Gamma$ and $Tf = \bar{s}\mathbf{T}(sf)$ for any $f \in L_1(\Gamma)$, and*

*(ii) if $\Delta = C \setminus \Gamma$, then $(I - T) L_1(\Delta)$ is dense in $L_1(\Delta)$ in the norm topology.*

*The partition of $C$ into $\Gamma$ and $\Delta$ is determined uniquely* mod $\mu$ *by* (i) and (ii). *A function $r$ has the properties stated for $s$ iff there exists a function $l \in L_\infty(\Gamma)$ with $|l| = 1$ on $\Gamma$, $\mathbf{T}^* l = l$ on $\Gamma$, and $r = s \cdot l$.*

*Proof.* $\Gamma$ can be constructed by a simple exhaustion argument: Let $\mathscr{G}$ be the class of all sets $G \in \mathscr{C}$ for which there exists an $h \in L_\infty$ with $T^* h = h$ on $G$ and $h \neq 0$ on $G$, and let $\nu$ be a finite measure equivalent to $\mu$. Set $\eta = \sup\{\nu(G): G \in \mathscr{G}\}$.

Let $G'_n$ be a sequence in $\mathscr{G}$ with $\nu(G'_n) \to \eta$. $\Gamma$ shall be the (disjoint) union of the sets $G_1 = G'_1$, $G_2 = G'_2 \setminus G_1$, $G_3 = G'_3 \setminus (G_1 \cup G_2)$, ... By the construction there exist functions $h_n \in L_\infty$ with $T^* h_n = h_n$ on $G'_n$ and $h_n \neq 0$ on $G'_n$. The function $s \in L_\infty(\Gamma)$ is defined as $s = h_n / |h_n|$ on $G_n$. The desired property of $s$ follows from lemma 1.9.

If $h \in L_\infty$ has the property that $\langle (I-T)f, h\rangle = 0$ holds for all $f \in L_1(\Delta)$, then we have $\langle f, T^* h - h\rangle$ for all $f \in L_1(\Delta)$ and hence $T^* h = h$ on $\Delta$. By the definition of $\Gamma$ we must have $h = 0$ on $\Delta$. Therefore any bounded linear functional vanishing in $(I-T)L_1(\Delta)$ vanishes in $L_1(\Delta)$ and (ii) is proved.

If $\Delta' \cup \Gamma'$ is a different decomposition of $C$ with the same properties and $B \in \mathscr{C}$ contained in $\Delta'$, then, for any $h \in L_1(B)$ with $T^* h = h$, we have $\langle f, T^* h - h\rangle = 0 \;\forall f \in L_1(B)$ and hence $\langle (I-T)f, h\rangle = 0$ for all $f \in L_1(B)$. As $(I-T)L_1(B)$ must be dense in $L_1(B)$ by the properties of $\Delta'$, we see $\langle f, h\rangle = 0$ for all $f \in L_1(B)$, i.e., $h = 0$. This proves $\Delta' \subset \Delta$. $\Gamma' \subset \Gamma$ follows from the maximality of $\Gamma$, and the uniqueness is proved.

Now let $r \in L_\infty(\Gamma)$ with $|r| = 1$ on $\Gamma$ be given. There exists an $l \in L_\infty(\Gamma)$ with $r = sl$ and $|l| = 1$ on $\Gamma$. We have $Tf = \bar{r}\mathbf{T}(rf)\;\forall f \in L_1(\Gamma) \Leftrightarrow \bar{s}\mathbf{T}(sf) = \bar{s}\bar{l}\mathbf{T}(slf)$ $\forall f \in L_1(\Gamma) \Leftrightarrow \mathbf{T}g = \bar{l}\mathbf{T}(lg)\;\forall g \in L_1(\Gamma)$. As the last condition is equivalent to $\mathbf{T}^* l = l$ on $\Gamma$ by lemma 1.6., the last assertion in the theorem follows. □

**3. Chacon's general ratio theorem.** We proceed to apply theorem 1.10 to the proof of a very general ratio ergodic theorem. A sequence $(p_k)_{k \geqq 0}$ of measurable functions $p_k: \Omega \to \mathbb{R}^+$ is called *T-admissible* if, for all $k \geqq 0$, $|Tf| \leqq p_{k+1}$ holds for all $f \in L_1$ with $|f| \leqq p_k$. It follows immediately from theorem 1.1 that $(p_k)$ is admissible for a contraction $T$ in $L_1$ iff it is admissible in the sense of § 3.3 for the linear modulus $\mathbf{T}$.

**Theorem 1.11** (Chacon). *If $T$ is an arbitrary contraction in $L_1$, $f \in L_1$ and $(p_k)$ T-admissible, the ratios $S_n f / \sum_{i=0}^{n-1} p_i$ converge to a finite limit a.e. in $\{\sum_{i=0}^{\infty} p_i > 0\}$.*

*Proof.* Let $p$ be a strictly positive integrable function. Let us write $\mathbf{S}_n g = \sum_{i=0}^{n-1} \mathbf{T}^i g$. Applying theorem 3.3.7 to $\mathbf{T}$ we see that it is enough to prove the a.e.-convergence of $S_n f / \mathbf{S}_n p$. Let us first look at the case $\Omega = C$. For $f \in L_1(\Gamma)$ the identity $Tf = \bar{s}\mathbf{T}(sf)$ implies $T^k f = \bar{s}\mathbf{T}^k(sf)$. We see that $S_n f / \mathbf{S}_n p = \bar{s}\mathbf{S}_n(sf)/\mathbf{S}_n p$ converges a.e. by the Chacon-Ornstein theorem.

For $f \in L_1(\Delta)$ and $\varepsilon > 0$ we may find functions $f_\varepsilon, g_\varepsilon \in L_1(\Delta)$ with $f = (I-T)f_\varepsilon + g_\varepsilon$ and $\|g_\varepsilon\|_1 < \varepsilon$. The estimate $|S_n(I-T)f_\varepsilon| \leqq |f_\varepsilon| + |T^n f_\varepsilon| \leqq |f_\varepsilon| + \mathbf{T}^n |f_\varepsilon|$ and lemma 3.2.3 show that $S_n(I-T)f_\varepsilon / \mathbf{S}_n p$ tends to 0. Using $|T^i g_\varepsilon| \leqq \mathbf{T}^i |g_\varepsilon|$ we obtain $|\limsup S_n f/\mathbf{S}_n p| \leqq \lim \mathbf{S}_n |g_\varepsilon|/\mathbf{S}_n p = d(|g_\varepsilon|\mu|\mathscr{C})/d(p\mu|\mathscr{C})$. As $\varepsilon > 0$ was arbitrarily small we see that the desired limit exists and is 0 for all $f \in L_1(\Delta)$.

To prove the theorem also in the case $\Omega \neq C$ first note that, due to $|T^i f| \leqq \mathbf{T}^i |f|$, the sum $\sum_{i=0}^{\infty} T^i f$ converges absolutely a.e. in $D$. It therefore remains to study the influence of $D$ in $C$. Recall $T_E = I_E + TI_{E^c}$. From § 3.3 we know that

$$\mathbf{H}_C = I_C \sum_{\nu=0}^{\infty} (\mathbf{T}I_D)^\nu \text{ and hence also } H_C = I_C \sum_{\nu=0}^{\infty} (TI_D)^\nu$$

is a contraction in $L_1$. We want to show

$$\text{(1.4)} \qquad \lim_{n\to\infty} S_n f/\mathbf{S}_n p = \lim_{n\to\infty} S_n H_C f/\mathbf{S}_n p \text{ in } C.$$

This will complete the proof and provide an identification of the limit. First note that $|S_n(f - T_C f)| \leqq |f 1_D| + \mathbf{T}^n |f 1_D|$ implies that

$$\text{(1.5)} \qquad \lim_{n\to\infty} S_n(f - T_C^k f)/\mathbf{S}_n p = 0 \text{ on } C$$

holds for $k = 1$, and, by induction, for all $k$. The identity (3.3.3) remains true for non positive operators and gives

$$\text{(1.6)} \qquad T_C^k = \sum_{\nu=0}^{k-1} I_C (TI_D)^\nu + (TI_D)^k, \quad (k \geqq 1).$$

To estimate the influence of the last summand observe that the identities $T^i(TI_D)^k = T^i I_C (TI_D)^k + T^{i-1}(TI_D)^{k+1} = \dots$
$= T^i I_C (TI_D)^k + T^{i-1} I_C (TI_D)^{k+1} + \dots + TI_C (TI_D)^{k+i-1} + (TI_D)^{k+i}$ yield

$$\text{(1.7)} \qquad |I_C S_n (TI_D)^k f| \leqq \mathbf{S}_n \sum_{\nu=k}^{\infty} I_C (\mathbf{T}I_D)^\nu |f|.$$

The functions

$$g_k = \sum_{\nu=k}^{\infty} I_C (TI_D)^\nu f \quad \text{and} \quad h_k = \sum_{\nu=k}^{\infty} I_C (\mathbf{T}I_D)^\nu |f|$$

belong to $L_1(C)$. Using (1.6) we have

$$T_C^k f - H_C f = (TI_D)^k f - g_k.$$

By (1.7) we obtain $|S_n(T_C^k f - H_C f)| \leqq \mathbf{S}_n h_k + \mathbf{S}_n |g_k|$ on $C$. Choosing $k$ large the norm of $h_k$ and $g_k$ can be made as small as we want, and therefore

$$\lim_{n\to\infty} \mathbf{S}_n (h_k + |g_k|)/\mathbf{S}_n p = d((h_k + |g_k|) \cdot \mu|\mathscr{C})/d(p\mu|\mathscr{C})$$

can be made arbitrarily small. Together with (1.5) and the existence of $\lim_{n\to\infty} S_n H_C f/\mathbf{S}_n p$ this implies (1.4). □

## Notes

The linear modulus of $T$ is discussed in the books of Peressini [1967] and Schaefer [1974]), who treats also the complexification of Banach lattices. Krengel [1963] showed $|T| = \sup\{Re(aT): |a| = 1\}$, where sup is taken in the order complete lattice of real regular operators in $L_p$. (This was also proved by Schaefer [1974]). Krengel [1963] showed that the bounded linear operators in $L_2$ need not be majorizable (and form a lattice iff $L_2$ is finite dimensional). Peressini and Sherbert [1966] noticed that the Hilbert transform can serve as an example in $L_p$ for all $1 < p < \infty$; see also Starr [1971]. An application of $|T|$ in $L_1$ is the construction of the *ergodic decomposition* of $T$, when $|T|$ is conservative.

For further related results, see Feyel [1978], and for a semigroup extension of the notion of linear modulus see § 7.2.

The original proof of Chacon's theorem 1.11 was complicated. Simpler proofs were given by Akcoglu [1966] and by Akcoglu and Chacon [1970a]. The argument here with (1.4) seems new. The identification of the limit determined by (1.4), which is the same as for positive $T$, seems both simpler and more explicit than that given by Akcoglu and Brunel.

For a rather short proof of theorem 1.11 in the real case see Gologan [1976].

An alternative proof of Chacon's theorem has been given by Bishop [1968]. He derives an upcrossing type inequality which is also of independent interest: Let $p_0, p_1, p_2, \ldots$ and $q_0, q_1, q_2, \ldots$ be admissible sequences for a contraction $T$ in $L_1$, and let $g_0, g_1, \ldots$ be a sequence of integrable functions with

$$|T|\,|\sum_{j=k}^{i} g_j| \geqq |\sum_{j=k+1}^{i+1} g_j|, \quad (0 \leqq k \leqq i).$$

For each $i \geqq 0$ write $f_i = g_i - q_i$. For any $n \in \mathbb{N}$ and $\omega \in \Omega$ let $U_n(\omega)$ be the maximum integer $N$ such that there exist integers

$$-1 \leqq s_1(\omega) < t_1(\omega) < s_2(\omega) < t_2(\omega) \ldots < s_N(\omega) < t_N(\omega) \leqq n$$

with

$$\sum\{f_k(\omega): 0 \leqq k \leqq s_i\} \leqq \sum\{f_k(\omega) - p_k(\omega): 0 \leqq k \leqq t_i\}, \quad (1 \leqq i \leqq N)$$

and

$$\sum\{f_k(\omega): 0 \leqq k \leqq s_i\} \leqq \sum\{f_k(\omega) - p_k(\omega): 0 \leqq k \leqq t_{i-1}\}, \quad (2 \leqq i \leqq N).$$

(A sum over a void set has the value 0. If no such $N$ exists, then $U_n(\omega) = 0$). The result of Bishop is:

**Theorem 1.12.** *The function $p_0 U_n$ is integrable, and*

(1.8) $$\int p_0 U_n \, d\mu \leqq \int \{f_0^+ + (-g_0 - p_0)^+\} \, d\mu.$$

Bishop mentions that the integral on the right side need only be taken over $\{U_N \geqq 1\}$, and that for positive $T$ the right side may be replaced by $\int f_0^+ \, d\mu$. If, in addition, $s_1(\omega)$ is required to be $-1$, then the right side may be replaced by $\int f_0 \, d\mu$. The inequality (1.8) does not give the identification of the limit.

## § 4.2 Vector valued ergodic theorems

We shall now consider operators $T$ acting in the space $L_1^{\mathfrak{X}}$ of Bochner integrable functions $f$ on $(\Omega, \mathscr{A}, \mu)$, taking values in a Banach space $\mathfrak{X}$ with norm $|\cdot|$. If $T$ is of the form $Tf = f \circ \tau$ for an endomorphism $\tau$ of the measure space the pointwise ergodic theorem for averages $A_n f$ shall be deduced from Birkhoff's ergodic theorem. However, the vector valued generalization of the Dunford-Schwartz ergodic theorem, due to Chacon [1962], is more subtle.

We begin with a brief description of the Bochner integral. (See also Yosida [1974], Neveu [1975]). Recall that $h\colon \Omega \to \mathfrak{X}$ has been called *finitely valued* if $h$ assumes only finitely many (distinct) values $x_1, \ldots, x_m$ and $\{h = x_i\}$ has finite measure for all $x_i \neq 0$. We can then define $\int h\, d\mu = \sum x_i \mu(h = x_i)$.

A function $f\colon \Omega \to \mathfrak{X}$ is called *strongly measurable*, if there exists a sequence $(h_k)$, $(k = 1, 2, \ldots)$, of finitely valued functions with $|h_k - f| \to 0$ $\mu$-a.e., and *Bochner integrable* if, in addition, $\int |h_k - f|\, d\mu$ tends to 0. We can then define the *Bochner integral* $\int f\, d\mu$ as the limit of $\int h_k\, d\mu$.

Passing to $h'_k(\omega)$ defined as $h_k(\omega)$ on $\{|h_k| \leqq 2|f|\}$ and as 0 on the complement of this set we see that we may always assume $|h_k| \leqq 2|f|$, because $h'_k$ is again finitely valued, $|h'_k - f|$ tends to 0, and $\int |h_k - f|$ tends to 0 iff $\int |h'_k - f|$ tends to 0. It is then easy to see that a strongly measurable $f$ is Bochner integrable iff $\int |f|\, d\mu$ is finite. As usual $\int_B f\, d\mu := \int 1_B f\, d\mu$. It is not hard to check

$$\Big|\int_B f\, d\mu\Big| \leqq \int_B |f|\, d\mu .$$

$L_p^{\mathfrak{X}} = L_p(\Omega, \mathscr{A}, \mu, \mathfrak{X})$ denotes the class of all strongly measurable $f$, for which the norm $\|f\|_p := (\int |f|^p\, d\mu)^{1/p}$ is finite $(1 \leqq p < \infty)$, and $L_\infty^{\mathfrak{X}}$ is the class of all strongly measurable $f$, for which $\|f\|_\infty := \inf\{\alpha\colon \mu(\{|f| > \alpha\}) = 0\}$ is finite.

As in the real and complex valued case these spaces are Banach spaces. (Again, we do not distinguish functions and equivalence classes of functions).

If $\mu$ is finite and $\mathscr{A}_0 \subset \mathscr{A}$ a sub-$\sigma$-algebra of $\mathscr{A}$, we may define the conditional expectation of a finitely valued function $h = \sum x_i 1_{\{h = x_i\}}$ by $E(h \mid \mathscr{A}_0) := \sum x_i E(1_{\{h = x_i\}} \mid \mathscr{A}_0)$. For general Bochner integrable $f$ we may define $E(f \mid \mathscr{A}_0)$ as the limit in the $\|\cdot\|_1$-norm of $E(h_k \mid \mathscr{A}_0)$, assuming $|h_k| \leqq 2|f|$ again. Passing to a subsequence of $(h_k)$ we can also assume that the limit exists a.e..

**Theorem 2.1.** *Let $\tau$ be an endomorphism of a $\sigma$-finite measure space $(\Omega, \mathscr{A}, \mu)$, $\mathfrak{X}$ a Banach space, and $f \in L_1^{\mathfrak{X}}$. Let $T$ be given by $Tg = g \circ \tau$. Then $A_n f$ converges a.e. in $|\cdot|$-norm to a function $\bar{f} \in L_1^{\mathfrak{X}}$ with $\bar{f} = \bar{f} \circ \tau$.*

*Proof.* It is enough to prove the theorem in the case (a): $\mu(\Omega) < \infty$, and in the case (b), where no non zero finite invariant measure $\nu \ll \mu$ exists. In case (a) let $\mathscr{I}$ denote the $\sigma$-algebra of $\tau$-invariant sets, and $T'$ the operator $g \to T'g = g \circ \tau$ in the *real* $L_1$. Set $A'_n = n^{-1} \sum_{\nu=0}^{n-1} (T')^\nu$. For any $x \in \mathfrak{X}$ and $B \in \mathscr{A}$ one has

$$|A_n(x1_B) - E(x1_B|\mathscr{I})| = |x(A'_n 1_B - E(1_B|\mathscr{I}))| = |x||A'_n 1_B - E(1_B|\mathscr{I})|.$$

As $h_k$ is finitely valued we see that by the Birkhoff theorem the second term on the right side of the inequality

$$|A_n f - E(f|\mathscr{I})| \leqq |A_n(f-h_k)| + |A_n h_k - E(h_k|\mathscr{I})| + |E(h_k|\mathscr{I}) - E(f|\mathscr{I})|$$

tends to 0 for any fixed $k$ when $n$ tends to $\infty$. Consequently, for all $k$,

$$\limsup |A_n f - E(f|\mathscr{I})| \leqq \limsup |A_n(f-h_k)| + |E(h_k|\mathscr{I}) - E(f|\mathscr{I})|.$$

By the choice of $(h_k)$ the last terms tends to 0 for $k \to \infty$. Finally, the triangle inequality for the norm, and the identity $|g \circ \tau^\nu| = |g| \circ \tau^\nu$ yield $|A_n(f-h_k)| \leqq A'_n|f-h_k|$. By the Birkhoff theorem $A'_n|f-h_k|$ tends to a limit function with integral $\int |f-h_k|\,d\mu$. As this can again be made arbitrarily small we have proved $|A_n f - E(f|\mathscr{I})| \to 0$ a.e.. $\bar{f} = E(f|\mathscr{I})$ is invariant under $\tau$. In the case (b) one can replace all conditional expectations in the proof of (a) by 0. Then the identical argument shows $A_n f \to 0$ a.e.. □

Apparently, the same argument can be applied in more general situations. E.g. the multiparameter ergodic theorem 6.2.8 yields an extension to Banach valued functions.

Our next aim is Chacon's vector valued ergodic theorem. We first describe a filling scheme for $\mathfrak{X}$-valued functions. Let $T$ be a linear operator acting in a linear space $L$ of strongly measurable $\mathfrak{X}$-valued functions such that, for all $f \in L$, all strongly measurable $f'$ with $|f'| \leqq |f|$ belong to $L$. For any measurable $a\colon \Omega \to \mathbb{R}^+$ set $f^{a-} = \operatorname{Min}(a, |f|) \cdot f \cdot |f|^{-1}$ on $\{f \neq 0\}$, $f^{a-} = 0$ on $\{f = 0\}$, and $f^{a+} = f - f^{a-}$. The filling scheme for $f \in L$ and a measurable $g\colon \Omega \to \mathbb{R}^+$ is defined inductively by

$$f_0 = f^{g+}, \qquad d_0 = f^{g-}, \qquad g_0 = g - |d_0|,$$
$$f_{n+1} = (Tf_n)^{g_n+}, \qquad d_{n+1} = (Tf_n)^{g_n-}, \qquad g_{n+1} = g_n - |d_{n+1}|.$$

Heuristically, $g_n$ still measures the size of the hole remaining after step $n$, the vector $d_n$ is the "matter" lost in the hole exactly at time $n$. It consumes $|d_n|$ of the hole $g_{n-1}$ left after step $n-1$. $f_n$ is the portion of the matter, which does not drop into the hole at time $n$ and will be mapped. It is clear that $(g_n)$ is a non negative decreasing sequence and that $g_n = 0$ holds on $\{|f_n| > 0\}$. As in the real valued filling scheme the assertion

$$(2.1) \qquad T^j f = f_j + \sum_{k=0}^{j} T^{j-k} d_k$$

is trivial for $j = 0$, and, if it is proved for $j$, then

$$T^{j+1} f = Tf_j + \sum_{k=0}^{j} T^{(j+1)-k} d_k = (Tf_j)^{g_j+} + (Tf_j)^{g_j-} + \sum_{k=0}^{j} T^{(j+1)-k} d_k$$

together with the definition of $f_{j+1}$ and $d_{j+1}$ completes the inductive proof of (2.1). Summing from $j=0$ to $j=n-1$ we arrive for $n \geqq 1$ at

$$(2.2) \qquad S_n f = \sum_{j=0}^{n-1} f_j + \sum_{j=0}^{n-1} \sum_{k=0}^{j} T^{j-k} d_k = \sum_{j=0}^{n-1} f_j + \sum_{j=0}^{n-1} T^j \sum_{k=0}^{n-j-1} d_k .$$

On $\{g_n = 0\}$ we have $d_m = 0$ for all $m > n$ and $g = \sum_{i=0}^{n} |d_i| = \sum_{i=0}^{\infty} |d_i|$. Hence, if $E$ is any subset of $\{\sum_{n=0}^{\infty} |f_n| > 0\}$, then $g = \sum_{i=0}^{\infty} |d_i|$ holds on $E$.

**Lemma 2.2.** *Let $T$ be a contraction in $L_1^{\mathfrak{X}}$ and $E \subset \{\sum_{n=0}^{\infty} |f_n| > 0\}$, then*

$$\int_E (g - |d_0|)\, d\mu \leqq \int |f_0|\, d\mu .$$

*Proof.* The left side is bounded by

$$\int_E \sum_{i=1}^{\infty} |d_i|\, d\mu \leqq \int \sum_{i=1}^{\infty} |d_i|\, d\mu .$$

Using $Tf_n = d_{n+1} + f_{n+1}$, $(n \geqq 0)$, and the contraction property, we find

$$\|f_0\|_1 \geqq \|d_1\|_1 + \|f_1\|_1 \geqq \|d_1\|_1 + \|d_2\|_1 + \|f_2\|_1 \geqq \dots \geqq \sum_{i=1}^{\infty} \|d_i\|_1 . \quad \square$$

If $T$ is defined on a space $L$ larger than $L_1^{\mathfrak{X}}$ and the restriction to $L_1^{\mathfrak{X}}$ contracts the $\|\cdot\|_1$-norm, the lemma still remains true: it is trivial if $f_0$ does not belong to $L_1^{\mathfrak{X}}$. A contraction $T$ in $L_1^{\mathfrak{X}}$ is called an $L_1^{\mathfrak{X}} - L_\infty^{\mathfrak{X}}$-contraction if also the norm $\|T\|_{1,\infty} := \sup\{\|Th\|_\infty : h \in L_1^{\mathfrak{X}} \cap L_\infty^{\mathfrak{X}} : \|h\|_\infty \leqq 1\}$ is bounded by 1. It is now simple to prove Chacon's maximal ergodic lemma.

**Theorem 2.3.** *Let $T$ be an $L_1^{\mathfrak{X}} - L_\infty^{\mathfrak{X}}$-contraction, $\alpha > 0$, $f \in L_1^{\mathfrak{X}}$, and $E_\alpha = \{\sup_{n \geqq 1} |A_n f| > \alpha\}$. Then*

$$(2.3) \qquad \int_{E_\alpha} (\alpha - |f^{\alpha -}|)\, d\mu \leqq \int |f^{\alpha +}|\, d\mu .$$

*Proof.* It remains to show $E_\alpha \subset \{\sum_{n=0}^{\infty} |f_n| > 0\}$ in the filling scheme with $g \equiv \alpha$. Now

$$\Big| \sum_{k=0}^{n-j-1} d_k \Big| \leqq \sum_{k=0}^{\infty} |d_k| \leqq g = \alpha$$

and $\|T\|_{1,\infty} \leqq 1$ yield $|n^{-1} \sum_{j=0}^{n-1} T^j \sum_{k=0}^{n-j-1} d_k| \leqq \alpha$ for all $n$.

It therefore follows from (2.2) that $\sum_{j=0}^{n-1} |f_j| > 0$ holds on $\{|A_n f| > \alpha\}$. □

Now note that an $L_1^{\mathfrak{X}} - L_\infty^{\mathfrak{X}}$-contraction $T$ may be extended in a natural way to $L_p^{\mathfrak{X}}$, $(1 \leqq p < \infty)$: If $f$ is in $L_p^{\mathfrak{X}}$ then, for any $\delta > 0$, $f^{\delta+}$ is in $L_1^{\mathfrak{X}}$. We define

$$Tf = \lim_{\delta \to 0+0} Tf^{\delta+}.$$

The almost sure existence of the limit follows from $\|f^{\delta_1+} - f^{\delta_2+}\|_\infty \leqq |\delta_1 - \delta_2|$ and $\|T\|_{1,\infty} \leqq 1$. The functions $f^{\delta+}$ belong to $L_1^{\mathfrak{X}}$. The Riesz-Thorin convexity theorem (see Thorin [1948]) implies that $T$ is a contraction in each $L_p^{\mathfrak{X}}$.

For the proof of a pointwise ergodic theorem $\mathfrak{X}$ must have some additional structure. After all, there exists a Banach space $\mathfrak{X}$ and a contraction $T_0$ in $\mathfrak{X}$, for which the *mean* ergodic theorem does not hold. Now for a 1-point probability space $L_1^{\mathfrak{X}}$ consists of all functions $x \cdot \mathbb{1}$, $(x \in \mathfrak{X})$. The operator $T(x \cdot \mathbb{1}) = (T_0 x)\mathbb{1}$ is an $L_1^{\mathfrak{X}} - L_\infty^{\mathfrak{X}}$-contraction for which $A_n(x\mathbb{1})$ need not converge strongly a.e..

**Theorem 2.4** (Chacon). *Let $\mathfrak{X}$ be a reflexive Banach space, and $T$ an $L_1^{\mathfrak{X}} - L_\infty^{\mathfrak{X}}$-contraction, then the averages $A_n f$ converge for all $f \in L_p^{\mathfrak{X}}$, $(1 \leqq p < \infty)$, a.e. in the $|\cdot|$-norm.*

*Proof.* We need a result of R.S. Phillips [1943], which says that for reflexive $\mathfrak{X}$ and $p > 1$ the space $L_p(\Omega, \mathscr{A}, \mu, \mathfrak{X})$ is a reflexive Banach space. (Its dual is $L_q(\Omega, \mathscr{A}, \mu, \mathfrak{X}^*)$, where $p^{-1} + q^{-1} = 1$). This together with theorem 2.1.3 shows that the set of functions $h_1 + (h_2 - Th_2)$ with $h_1, h_2 \in L_2^{\mathfrak{X}}$ and $h_1 = Th_1$ is dense in $L_2^{\mathfrak{X}}$. It is then easy to see that, for any $\eta > 0$, any $f \in L_p^{\mathfrak{X}}$ may be written as a sum $f = h_1 + h_2' - Th_2' + f' + f''$, where $h_1 = Th_1$, $h_2'$ belongs to $L_2^{\mathfrak{X}} \cap L_\infty^{\mathfrak{X}}$, and where $f', f''$ satisfy $\|f'\|_1 < \eta^2$, $\|f''\|_\infty < \eta$. Set

$$\Delta(h) = \lim_{n \to \infty} \sup_{m \geqq n} |A_n h - A_m h|.$$

Clearly $\Delta(h_1) = 0$, and $\Delta(h_2' - Th_2') = 0$, because of $h_2' \in L_\infty^{\mathfrak{X}}$ and cancellation. Also $\Delta(f'') < 2\eta$ is obvious. The set $\{\Delta(f') > 2\eta\}$ is contained in $E_\eta' = \{\sup |A_n f'| > \eta\}$. Using theorem 2.3 we can see that $\eta\mu(E_\eta')$ is bounded by $\|f'\|_1 < \eta^2$. This implies $\mu(\Delta(f') > 2\eta) < \eta$. Combining the estimates with $\Delta(f) \leqq \Delta(h_1) + \Delta(h_2' - Th_2') + \Delta(f') + \Delta(f'')$ we arrive at $\mu(\{\Delta(f) > 4\eta\}) < \eta$. As $\eta > 0$ was arbitrary we have proved $\Delta(f) = 0$, which is equivalent to the assertion of the theorem. □

## Notes

E. Mourier [1953] proved the strong law of large numbers for independent identically distributed Bochner integrable random variables taking values in a general Banach space. A. Beck [1963] showed that Kolmogorov's strong law for independent random variables holds in the $\mathfrak{X}$-valued case iff the Banach space $\mathfrak{X}$ satisfies a geometric condition, called *B*-

*convexity*. This led to a lot of further work connecting probability theory and the geometry of Banach spaces, see Padgett and Taylor [1973], Woyczyński [1978]. Beck and Schwartz [1957] proved a vector valued random ergodic theorem for reflexive $\mathfrak{X}$, now contained in theorem 2.4. Theorem 2.1 and several generalizations appear in the work of Tempel'man [1972], [1967]. The simple reduction to Birkhoff's theorem has been used in the paper of Landers and Rogge [1978] to prove a generalization where $f$ is allowed to assume values in a Fréchet space.

Call a sequence $(F_n)_{n\geqq 1}$ of Bochner integrable functions on a probability space $(\Omega, \mathscr{A}, \mu)$ with values in a Banach lattice $\mathfrak{X}$ a *subadditive vector valued process* (for the endomorphism $\tau$) provided for all natural $k, n$ we have $F_{n+k} \leqq F_n + F_k \circ \tau^n$. $\mathfrak{X}$ is called *countably order complete* if $\sup_{n\geqq 1} x_n$ exists for any sequence $(x_n)$ bounded above by some $y \in \mathfrak{X}$. $\mathfrak{X}$ is said to have *order continuous norm* if $\mathfrak{X}$ is countably order complete and every decreasing sequence $z_n \in \mathfrak{X}$ with $z_n \geqq 0$ is norm convergent.

**Theorem 2.5** (Ghoussoub-Steele [1980]). *For countably order complete Banach lattices $\mathfrak{X}$ the following are equivalent:*

*(i) For every $\mathfrak{X}$-valued subadditive process $F_n \geqq 0$ we have norm convergence of $F_n(\omega)/n$ for a.e. $\omega$;*

*(ii) $\mathfrak{X}$ has order continuous norm.*

For an alternative proof and for related results see Davis-Ghoussoub-Lindenstrauss [1981], and Derriennic-Krengel [1981].

The technique of the vector valued filling scheme has been extended to a more general situation, permitting a unified proof of Chacon's maximal theorem 2.3 and the (decreasing) martingale theorem by A. and C. Ionescu Tulcea [1963]. One of their results is:

**Theorem 2.6.** *Let $\mathfrak{X}$ be a Banach space. Let $I = T_0, T_1, \ldots, T_r$ be $L_1^{\mathfrak{X}} - L_\infty^{\mathfrak{X}}$-contractions with $T_{j+1} T_j = T_{j+1}$, $(j = 0, \ldots, r-1)$, $\alpha > 0$, and $f \in L_p^{\mathfrak{X}}$, $(1 \leqq p < \infty)$. If $F$ is a measurable set containing $\{|f| > \alpha\}$ and contained in*

$$\{\sup\{|(T_j^0 + \ldots + T_j^{n-1}) f/n| : 0 \leqq j < r, n \geqq 1\} > \alpha\}$$

*then*

$$\alpha\mu(F) \leqq \int_F |f| \, d\mu < \infty.$$

Hasegawa, Sato and Tsurumi [1978] derive, among other results, a continuous parameter extension of theorem 2.4 and replace the assumption $\|T\|_{1,\infty} \leqq 1$ by $\sup\{\|T^n\|_{1,\infty}\} < \infty$. Related papers are those of Chersi and Invernizzi [1976] and Kopp [1975].

Flytzanis [1976] has given the identification of the limit in the vector valued random ergodic theorem of Beck and Schwartz.

Pop-Stojanovic [1972] has proved a vector valued ergodic theorem for *weakly integrable* strictly stationary sequences.

Jajte [1968] showed: Let $U$ be a unitary operator in a Hilbert space $\mathscr{H}$, $X$ an $\mathscr{H}$-valued random variable with square integrable norm. Put $X_j(\omega) = U^j(X(\omega))$. If $Y_0, Y_1, \ldots$ is a sequence of independent $\mathscr{H}$-valued random variables, with $Y_j$ distributed like $X_j$, then $n^{-1}(Y_0 + \ldots + Y_{n-1})(\omega)$ converges in norm for a.e. $\omega$. This was generalized and simplified by Al-Hussaini [1974].

Yoshimoto [1979] has vector valued estimates of the $L \log^m L$ type; see §1.6.

## § 4.3 Power bounded operators and harmonic functions

We return to the study of positive linear operators $T$ in the real $L_1$. It is a simple exercise to show that such operators are bounded but we do not assume now that $T$ is a contraction.

By the usual monotone extension $Tf$ and $T^*f$ are well defined (possibly infinite valued) for all non negative measurable $f$; see § 1.6.

If $e$ is superharmonic, $E = \{e > 0\}$, and $f \in L_1^+(E^c)$, then $\int e(Tf) = \int (T^*e)f \leqq \int ef = 0$ shows $Tf \in L_1^+(E^c)$. In other words: $E^c$ is $T$-absorbing. For any $f \geqq 0$ we have $\int (Tf)\, d(e\mu) = \int (Tf) e\, d\mu = \int f(T^*e)\, d\mu \leqq \int fe\, d\mu = \int f d(e\mu)$. Hence $T$ is a positive contraction in $L_1(E, e\mu) = L_1(e\mu)$, and we can apply the results previously derived for positive contractions to $T$, although $T$ need not be a contraction in $L_1(\mu)$. E.g., we can infer that $S_n f / S_n g$ converges $(e\mu)$-a.e. on $\{S_\infty g > 0\}$, and hence $\mu$-a.e. on $E \cap \{S_\infty g > 0\}$ for any $f \in L_1(e\mu)$, $g \in L_1^+(e\mu)$. Here we have not assumed $e \in L_\infty$. But if $e$ is in $L_\infty$, any $\mu$-integrable $f$ is $(e\mu)$-integrable, and the Chacon-Ornstein theorem holds on $E$ in the usual formulation. This may have motivated the search for harmonic functions $e \geqq 0$ for which $E$ is as large as possible.

By an exhaustion argument we can find a decomposition of $\Omega$ into two disjoint measurable sets $Y$ and $Z$, such that the support $\{e' > 0\}$ of every harmonic $e' \in L_\infty^+$ is contained in $Y$ and there exists a harmonic $e \in L_\infty^+$ which is strictly positive on $Y$. Obviously the decomposition is determined uniquely mod nullsets. It is called the *Sucheston decomposition* of $\Omega$. For reasons which will be plain from theorem 3.1 below $Y$ is called the *remaining part* of $\Omega$, and $Z$ the *disappearing part*.

The full statement of the following theorem was first proved in the work of Derriennic and Lin [1973]:

**Theorem 3.1.** *Let $T$ be a Cesàro bounded positive linear operator in the real $L_1$. Then*

*(i) $f \in L_1(Z)$ implies* $\lim_{N \to \infty} N^{-1} \sum_{n=0}^{N-1} \|T^n f\|_1 = 0$;

*(ii)* $\liminf \|T^n f\|_1 > 0$ *holds for all non-zero $f \in L_1^+(Y)$;*

*(iii) the harmonic $e$ with $Y = \{e > 0\}$ can be chosen in such a way that $|h|$ is bounded by a multiple of $e$ for each harmonic $h \in L_\infty$;*

*(iv) for all $h \in L_\infty$, $A_n^* h$ converges to 0 a.e. in $Z$;*

*(v) $T^n f / n$ converges in $L_1$-norm to 0 for all $f \in L_1$.*

*If $T$ is even power bounded the assertions* (i) *and* (iv) *may be replaced by the stronger assertions* (i'): $\|T^n f\|_1 \to 0$ *for all $f$ all $f \in L_1(Z)$, and* (iv'): $\lim_n T^{*n} h = 0$ a.e. *in $Z$ for all $h \in L_\infty$.*

*Proof.* If $M$ is a bound for all norms $\|A_n\|$ the non negative functions $g_N := A_N^* \mathbb{1}$ are bounded by $M$, and $g = \limsup g_N$ is in $L_\infty$. Writing $g = \lim_k (\sup_{N \geqq k} g_N)$ we see $T^*g \geqq \limsup T^*g_N = \limsup N^{-1} \sum_{k=1}^{N} T^{*n}\mathbb{1} = g$. Let $e$ be the limit of the increasing sequence $T^{*k}g$. As $g$ is bounded by $M$ and $e$ is also the limit of the sequence $A_k^*g$, we obtain $\|e\|_\infty \leqq M^2$. It is clear that $e$ is harmonic. Set $Y = \{e > 0\}$. When we show (iii), $Y$ must be the remaining part of $\Omega$ as defined above. We may assume $|h| \leqq 1$. Now $T^*h = h$ implies $T^*h^+ \geqq h^+$ and $T^*h^- \geqq h^-$ and hence also $T^*|h| \geqq |h|$. (iii) therefore follows from $|h| \leqq \lim T^{*k}|h| = \lim A_N^*|h| \leqq g \leqq e$.

(ii): For any $f \in L_1^+(Y)$ with $f \neq 0$, we have $0 < \int fe\,d\mu = \int fT^{*n}e\,d\mu = \langle T^n f, e\rangle \leqq \|T^n f\|_1 \cdot \|e\|_\infty$.

(i): For any $f \in L_1(Z)$,

$$\limsup_{N\to\infty} N^{-1} \sum_{n=0}^{N-1} \|T^n f\|_1 \leqq \limsup \langle A_N|f|, \mathbb{1}\rangle$$
$$= \limsup \langle |f|, g_N\rangle \leqq \langle |f|, \limsup g_N\rangle \leqq \langle |f|, g\rangle \leqq \langle |f|, e\rangle = 0.$$

(iv): We may assume $|h| \leqq 1$. The definition of $e$ yields

$$0 \leqq \int_Z \limsup A_n^*|h|\,d\mu \leqq \int_Z \limsup g_n\,d\mu = \int_Z g\,d\mu \leqq \int_Z e\,d\mu = 0.$$

(v): We may assume $f \geqq 0$. Set $a_n = \|T^n f\|_1$. Then, for $k \leqq n$, we have

$$a_n/(k+1) \leqq (k+1)^{-1} \sum_{i=n-k}^{n} a_i = \|(k+1)^{-1} \sum_{i=0}^{k} T^i(T^{n-k}f)\|_1$$
$$\leqq M\|T^{n-k}f\|_1 = M a_{n-k}.$$

Summing over $k$ we obtain

$$(a_n/n) \sum_{k=0}^{n-1} (k+1)^{-1} \leqq Mn^{-1} \sum_{k=0}^{n-1} a_{n-k} \leqq M^2 a_1 < \infty.$$

As $\sum_{k=0}^{n-1} (k+1)^{-1}$ tends to infinity the assertion $a_n/n \to 0$ follows.

If $T$ is even power bounded $e$ is defined by

$$e = \lim_k T^{*k}(\limsup_n T^{*n}\mathbb{1})$$

and the same proof shows that (i′) and (vi′) hold. □

When $T$ is even a positive *contraction* in $L_1$ the restriction of $T$ to $L_1(C)$ is Markovian and (i) implies that $Z$ is contained in the dissipative part $D$. If $\tau$ is an endomorphism of $(\Omega, \mathscr{A}, \mu)$, and $T$ is given by $Tf = f \circ \tau$, we clearly have $\Omega = Y$ and see that $D = Y$ is possible.

Sucheston [1967b] has given an example of a power bounded $T$ with $\Omega = Z$ such that the ratio's $S_n h/S_n g$ need not converge a.e. on $\{S_\infty g > 0\}$ for $h, g \in L_1^+$. This may be seen also in the following simple example, which Fong [1970] gave in order to show that even the Hopf decomposition may fail on $Z$: Take $\Omega$

$= \{0, 1, 2, \dots\}$ and let $\mu$ be the counting measure. Define $T$ by $(Tf)(0) = \sum_{i=1}^{\infty} f(i)$, and $(Tf)(j) = f(j+1)$ for $j \geqq 1$. For all $n \geqq 1$, $(T^n f)(0) = \sum_{i=n}^{\infty} f(i)$ and $(T^n f)(j) = f(j+n)$ for $j \geqq 1$. It follows that $\|T^n f\|_1$ is bounded by $2 \sum_{i=n}^{\infty} |f(i)| \leqq 2\|f\|_1$. Hence $\|T^n\|_1 \leqq 2$, $(n \geqq 1)$, and $\lim \|T^n f\|_1 = 0$ yields $\Omega = Z$. It is simple to check that $(S_\infty g)(0) = \infty$ holds for $g$ defined by $g(i) = (i+1)^{-2}$, and $(S_\infty f)(0) < \infty$ for any $f$ with finite support. If $i_1 < i_2 < i_3 < \dots$ is increasing sufficiently fast and $h$ is defined by $h(i) = 0$ for $i$ in $[i_\nu, i_{\nu+1}[$ with $\nu$ even, and $h(i) = g(i)$ for $i$ in $[i_\nu, i_{\nu+1}[$ with $\nu$ odd, the ratios $S_n h / S_n g$ diverge in the point 0.

A. Ionescu Tulcea and M. Moretz [1969] proved that, for power bounded $T$ and $g \in L_1^+$, there is no subset $A$ of $Z \cap \{S_\infty g = \infty\}$ with $\mu(A) > 0$, such that $S_n f / S_n g$ converges a.e. on $A$ for all $f \in L_1$.

If $T$ is Cesàro bounded and $f \in L_1(\mu)$, $A_n f$ converges stochastically in $Y$ (Exercise). Fong [1979] proved that $n^{-1} F_n$ converges stochastically in $Y$ for superadditive $(F_n)$ with $\sup \|n^{-1} F_n\|_1 < \infty$, and gave an example of a power bounded $T$ and an $f \in L_1(Y)$ for which $A_n f$ does not converge stochastically in $Z$.

We shall now use the Sucheston decomposition to prove:

**Theorem 3.2.** *Let $T$ be a Cesàro bounded positive linear operator in $L_1$, and $p \in L_1$ strictly positive. The following conditions are equivalent:*

*(a) for every $h \in L_\infty$, $A_n^* h$ converges a.e. in $\Omega$;*

*(b) for every $f \in L_1$, $A_n f$ converges in $L_1$;*

*(c) for every $A \in \mathscr{A}$, the sequence $\langle p, A_n^* 1_A \rangle$ converges.*

*Proof.* Passing to the new reference measure $\mu' = p\mu$ we may assume $\mu(\Omega) < \infty$ and $p \equiv \mathbb{1}$. (a) $\Rightarrow$ (c) is trivial. (c) $\Rightarrow$ (b): (c) means that $A_n \mathbb{1}$ converges weakly in $L_1$. By (v) of the previous theorem $\|n^{-1} T^n \mathbb{1}\|_1 \to 0$. Now the mean ergodic theorem 2.1.1 shows that $A_n \mathbb{1}$ converges strongly. Therefore the sequence $(A_n \mathbb{1})$ is uniformly integrable. By approximation we find that for all $f \in L_1$ the sequence $(A_n f)$ is uniformly integrable and hence conditionally weakly compact. Theorem 2.1.1 implies (b).

(b) $\Rightarrow$ (a): Construct $g$, $e$ and $Y$ as in the proof of theorem 3.1. Let $u$ be the strong limit of $(A_n \mathbb{1})$. Clearly $0 \leqq u = Tu$. Because of

$$\int (T^* h) u \, d\mu = \int h (Tu) \, d\mu = \int h u \, d\mu$$

$T^*$ is a positive contraction in $L_1(u \cdot \mu)$. As $L_\infty$ is contained in $L_1(u\mu)$, the Chacon-Ornstein theorem implies the convergence $(u\mu)$-a.e. of $A_n^* h = e \cdot (S_n^* h / ne) = e(S_n^* h / S_n^* e)$ on $\{e > 0\}$, or, equivalently, the convergence $\mu$-a.e. on $Y \cap \{u > 0\}$. In analogy to the definition of $g$ and $e$ we now define

$$g^h = \limsup A_n^* h, \qquad g_h = \liminf A_n^* h,$$
$$e^h = \lim A_n^* g^h, \qquad e_h = \lim A_n^* g_h.$$

Then $e^h$ and $e_h$ are harmonic and $T^* g^h \geqq g^h$, $T^* g_h \leqq g_h$ yields $e^h \geqq g^h \geqq g_h \geqq e_h$. As $e^h - e_h$ is harmonic and therefore bounded by a multiple of $e$ we obtain $g^h = g_h$ on $Z = \{e = 0\}$. Together with the result above we have $g^h = g_h$ on $\{u > 0\}$. Observe that $\langle u, T^{*k} 1_{\{u=0\}}\rangle = \langle T^k u, 1_{\{u=0\}}\rangle = \langle u, 1_{\{u=0\}}\rangle = 0$ yields that $A_n^* v$ is 0 on $\{u > 0\}$ for all $v \in L_\infty(\{u = 0\})$. We can infer that $e^h = e_h$ holds on $\{u > 0\}$. Consequently we have $\langle \mathbb{1}, e^h - e_h\rangle = \langle \mathbb{1}, A_n^*(e^h - e_h)\rangle = \langle A_n \mathbb{1}, e^h - e_h\rangle$ $= \langle u, e^h - e_h\rangle = 0$. We have proved $e^h = e_h$ $\mu$-a.e., i.e., (a) holds. □

In the case of contractions theorem 3.2 follows from results of Helmberg [1972] and of Lin and Sine [1977]. For point maps, see Dowker [1947] and Wright [1960]. The general case is due to Sato [1978a]. It seems useful to state Helmberg's criterion separately:

**Theorem 3.3.** *If $T$ is a positive contraction in $L_1$ the condition* (a) *in the previous theorem is equivalent to*

*(d) $C \setminus \tilde{C}$ is empty and $T^{*k} 1_D$ tends to* 0.

*Proof.* It remains to prove the much simpler equivalence of (b) and (d). Assume (b). From the remarks on $L_1$-convergence in § 3.3 we know that (b) implies $C \setminus \tilde{C}$ $= \emptyset$. If the limit of the decreasing sequence $T^{*k} 1_D$ is not 0 there exists a non zero $f \in L_1^+(D)$ for which $\langle A_n f, 1_D\rangle = \langle f, A_n^* 1_D\rangle$ does not tend to 0. The $T$-invariant limit $\bar{f}$ of $A_n f$ has the property $\langle \bar{f}, 1_D\rangle > 0$. This contradicts $\tilde{C} \subset C$. On the other hand(d) implies (b) because of uniform integrability and a.e.-convergence of the sequence $A_n f$ on $\tilde{C}$ together with $\langle A_n f, 1_D\rangle = \langle f, A_n^* 1_D\rangle \to 0$, $(f \in L_1^+)$. □

## Notes

For power bounded $T$, Sucheston [1967a, b] proved (i′) and (ii), and A. Ionescu Tulcea and M. Moretz [1969] proved (iv′) in theorem 3.1. R. Sato [1973] derived the decomposition $Y + Z$ in the Cesàro-bounded case independently of Derriennic-Lin. Sucheston actually used the condition ≪$\sup \langle T^n f, h\rangle < \infty \; \forall f \in L_1^+$≫ for fixed $h \in L_\infty^+$ and constructed partitions $\Omega = Y^h + Z^h$. For $h = \mathbb{1}$ his condition reduces to power boundedness and the partition agrees with $\Omega = Y + Z$. Fong [1970] showed that his approach works also in $L_p$, $(1 \leqq p < \infty)$, and studied the existence of $T$-invariant functions. For this problem see also Derriennic-Lin [1973], Ornstein [1970], Fong and Lin [1976]. Sato [1977] studies the existence of invariant functions for ergodic semigroups in the sense of Eberlein and gives many references. Continuous parameter results appear in Fong and Sucheston [1971], Sato [1977a], and Fong [1979].

Using theorem 3.2, Sato [1978a] proved:

**Theorem 3.4.** *Assume $\mu(\Omega) < \infty$. Let $T$ be a bounded linear operator in $L_1$ such that the linear modulus* **T** *is Cesàro bounded both in $L_1$ and in $L_\infty$. Then $A_n f$ converges a.e. for all $f \in L_\infty$.*

Derriennic and Lin already proved this for $T \geqq 0$, $T\mathbb{1} = \mathbb{1}$ (and gave the identification of the limit) and they showed by example that $A_n f$ need *not* converge a.e. for all $f \in L_1$; see also Kubokawa [1972a], Dunford [1980].

Generalizing a result of Ornstein [1969], Fong and Sucheston [1973] have given an extension of the Jamison-Orey theorem to power bounded operators: If $T$ is a power bounded positive operator in $L_1$ with $\Omega = Y = \{e > 0\}$, and $f \in L_1^+$ implies $\{S_\infty f = \infty\} = \{S_\infty f > 0\}$, then $\lim \|T^n g\|_1 = 0$ holds for all $g \in L_1$ with $\lim \|e(T^n g)\|_1 = 0$. Derriennic and Lin have given a simpler proof.

# Chapter 5: Operators in $C(K)$ and in $L_p$, $(1 < p < \infty)$

The main results in our treatment of Markov operators in $C(K)$ are the theorem of Jamison deducing uniform convergence of $S^n f$ from pointwise convergence for irreducible $S$, and the strong law of Breiman for Markov processes with arbitrary initial distributions.

Concerning $L_p$, we present Akcoglu's pointwise ergodic theorem for positive contractions and the dominated estimates of Stein for self-adjoint operators.

## § 5.1 Markov operators in $C(K)$

**1. Mean ergodicity and unique ergodicity.** Let $K$ be a compact Hausdorff space. A linear operator $S$ in the Banach space $\mathfrak{X} = C(K)$ of real or complex continuous functions on $K$ with $\|f\| = \sup\{|f(\omega)|: \omega \in K\}$ will be called *Markov operator* if it is positive and satisfies $S\mathbb{1} = \mathbb{1}$. E.g., a continuous map $\tau: K \to K$ induces a Markov operator by $Sf(\omega) = f(\tau\omega)$.

We give only a brief introduction to the convergence theorems for Markov operators. For the basic notions in point set topology and functional analysis we refer to Dunford-Schwartz [1958] and Yosida [1974].

By the Riesz representation theorem each continuous linear functional $\nu$ on $\mathfrak{X}$ is of the form $\langle f, \nu\rangle = \int f d\nu$, where $\nu$ is a signed measure on the Baire $\sigma$-algebra $\mathscr{F}$ in $K$. Thus $\mathfrak{X}^*$ is the space $\mathscr{M}$ of signed measures on $(K, \mathscr{F})$. $\mathscr{M}_1^+$ denotes the set of probability measures on $\mathscr{F}$.

A Markov operator $S$ defines a stochastic kernel by $P(\omega, \cdot) = S^*\varepsilon_\omega$, where $\varepsilon_\omega$ is the point mass 1 in $\omega$. We have

$$(1.1) \qquad Sf(\omega) = \langle Sf, \varepsilon_\omega\rangle = \langle f, P(\omega, \cdot)\rangle = \int f(\eta)\, P(\omega, d\eta).$$

The continuity of $Sf$ implies that the map $\omega \to P(\omega, \cdot)$ of $K$ into $\mathscr{M}_1^+$ is continuous in the $w^*$-topology of $\mathfrak{X}^*$. Conversely, if an arbitrary stochastic kernel $\{P(\omega, A): \omega \in K, A \in \mathscr{F}\}$ is given, and we define $Sf$ by (1.1), then the $w^*$-continuity of the map $\omega \to P(\omega, \cdot)$ is sufficient for $Sf \in C(K)$. [We remark that, by (1.1), $S$ has the interpretation of the operator $T^*$ in section 3.1].

We use the notation $A_n f = A_n(S)f = n^{-1}(f + Sf + \ldots + S^{n-1}f)$ here.

**Proposition 1.1.** *Let $S$ be a Markov operator in $\mathfrak{X} = C(K)$ and $f \in \mathfrak{X}$. Then the following assertions are equivalent:*

*(i) $A_n f$ converges strongly;*

*(ii) $A_n f$ converges pointwise and the limit $\bar{f}$ is continuous;*

*(iii) The functions $A_n f$ are equi-continuous.*

*Proof.* The dominated convergence theorem implies that a bounded sequence $f_n$ in $\mathfrak{X}$ converges pointwise to a continuous limit function $\bar{f}$ iff it converges weakly to $\bar{f}$. Thus the equivalence of (i), (ii) and (iii) follows from the mean ergodic theorem, and the Arzela-Ascoli theorem. □

Any Markov operator $S$ admits at least one invariant probability measure: To see this take any Banach limit $L$ and any $\omega \in K$ and put $\mu(f) = L((\langle S^n f, \varepsilon_\omega \rangle)_{n=0}^\infty)$. The properties of Banach limits imply $\mu(\mathbb{1}) = 1$, $\mu \geqq 0$ and $\mu(Sf) = \mu(f)$. Thus the linear functional $\mu$ belongs to $\mathcal{M}_1^+$, and it satisfies $S^* \mu = \mu$.

$S$ is called *uniquely ergodic* if $S$ admits only one invariant probability measure, and *mean ergodic* if $A_n f$ converges strongly for all $f$.

**Proposition 1.2.** *Any uniquely ergodic $S$ is mean ergodic.*

*Proof.* Let $\nu \neq 0$ be a fixed point of $S^*$. $\nu^+$ satisfies $S^* \nu^+ \geqq S^* \nu = \nu$ and $S^* \nu^+ \geqq 0$. Hence $S^* \nu^+ \geqq \nu^+$. Now $\langle \mathbb{1}, \nu^+ \rangle = \langle S\mathbb{1}, \nu^+ \rangle = \langle \mathbb{1}, S^* \nu^+ \rangle$ yields $S^* \nu^+ = \nu^+$ and therefore also $S^* \nu^- = \nu^-$. It follows that $\nu^+$ and $\nu^-$ are non negative multiples of the unique invariant probability measure $\mu$. $\nu \neq 0$ then implies $\langle \nu, \mathbb{1} \rangle \neq 0$. Thus the fixed points of $S$ separate the fixed points of $S^*$. Now apply theorem 2.1.4. □

**Proposition 1.3.** *A mean ergodic $S$ is uniquely ergodic iff the limit $\bar{f}$ of $A_n f$ is constant for each $f$.*

*Proof.* If $\bar{f}$ is constant for all $f$ and $\nu_1, \nu_2$ are $S^*$-invariant probabilities, then

$$\langle f, \nu_i \rangle = \langle f, A_n^* \nu_i \rangle = \langle A_n f, \nu_i \rangle = \langle \bar{f}, \nu_i \rangle .$$

Thus $\langle \bar{f}, \nu_1 \rangle = \langle \bar{f}, \nu_2 \rangle$ implies $\langle f, \nu_1 \rangle = \langle f, \nu_2 \rangle$. As $f$ was arbitrary $\nu_1 = \nu_2$. Conversely, if an $\bar{f}$ and two points $\omega_1, \omega_2$ with $\bar{f}(\omega_1) \neq \bar{f}(\omega_2)$ exist, the sequences $A_n^* \varepsilon_{\omega_i}$ converge weak * to $S^*$-invariant probabilities $\nu_i$ with

$$\langle \bar{f}, \nu_i \rangle = \lim \langle \bar{f}, A_n^* \varepsilon_{\omega_i} \rangle = \langle \bar{f}, \varepsilon_{\omega_i} \rangle = \bar{f}(\omega_i) . \quad \square$$

A continuous $\tau : K \to K$ is called uniquely ergodic if the operator $Sf = f \circ \tau$ is uniquely ergodic. By theorem 1.2.7 the irrational rotations of the torus are examples. Many other examples can be found in the books on topological dynamics.

For general Markov operators $S$ in $C(K)$ even pointwise convergence of $A_n f$ may fail. E.g. let $K$ be the unilateral product space $\{0,1\} \times \{0,1\} \times \{0,1\} \times \ldots$ and $\tau$ the shift in $K$. If $g$ is the indicator function of the set of $\omega$'s with first coordinate 0 and $\eta$ an element of $K$ in which the relative frequency of zero's does not converge, then $A_n g(\eta)$ does not converge.

A point $\omega$ is called *quasi-regular* for $S$ if $A_n f(\omega)$ converges for all $f \in \mathfrak{X}$. If all points are quasi-regular and the limit function $\bar{f}$ of $A_n f$ is continuous, then $S$ is mean ergodic by proposition 1.1. However, there exist Markov operators for which all points are quasiregular but $\bar{f}$ is discontinuous for some $f$. [Take $K = [0,1]$ and $\tau\omega = \omega^2$].

*Induced contractions in* $L_p$. Let $\mu$ be an $S^*$-invariant probability measure on $(K, \mathscr{F})$ and $L_p = L_p(K, \mathscr{F}, \mu)$. Then $\langle |f|, \mu \rangle$ is the $L_1$-norm $\|f\|_1$ of $f \in \mathfrak{X}$. It follows from $\langle |f|, \mu\rangle = \langle |f|, S^*\mu\rangle = \langle S|f|, \mu\rangle \geqq \langle |Sf|, \mu\rangle$ that $\|Sf\|_1 \leqq \|f\|_1$. $S\mathbb{1} = \mathbb{1}$ implies $\|Sf\|_\infty \leqq \|f\|_\infty$. Thus $S$ is a positive $L_1 - L_\infty$-contraction on $\mathfrak{X} \subset L_1$. As $\mathfrak{X}$ is dense in $L_1$ by the Daniell-Stone theorem (see e.g. Bauer [1981]), the operator $S$ may be extended by continuity to $L_1$ and the extension is an $L_1 - L_\infty$-contraction $S_1$ (depending on $\mu$). Let $S_p$ be the restriction of $S_1$ to $L_p \subset L_1$. By lemma 1.7.4 $S_p$ is a contraction in $L_p$.

**Remark 1.4.** The ergodic theorem of Hopf (1.7.3) now implies that $A_n(S_1)f$ converges $\mu$-a.e. for all $f \in L_1$. In particular, $A_n f$ converges $\mu$-a.e. for all $f \in \mathfrak{X}$. If $K$ is compact metric, $\mathfrak{X}$ is separable. As the convergence of $A_n f(\omega)$ for all $f$ in a dense subset of $\mathfrak{X}$ implies the convergence of $A_n f(\omega)$ for all $f \in \mathfrak{X}$ the set of quasi-regular points has full $\mu$-measure for each invariant $\mu \in \mathscr{M}_1^+$.

**2. Irreducible Markov operators.** A Markov operator $S$ in $\mathfrak{X} = C(K)$ is called *irreducible* if for each non negative somewhere positive $f \in \mathfrak{X}$ and for each $\omega \in K$ there is an $n$ with $S^n f(\omega) > 0$.

It may help to see an equivalent probabilistic condition: Let $P^n(\cdot\,,\cdot)$ denote the kernel belonging to $S^n$ and describing the $n$-step transition probabilities. $S$ is irreducible iff for all non empty open $U$ and all $\omega \in K$, there exists an $n$ with $P^n(\omega, U) > 0$. [If the probabilistic condition is satisfied and $f$, $\omega$ are given, take $U = \{f > 0\}$. Then $S^n f(\omega) = \int f(\eta)\, P^n(\omega, d\eta)$ is positive. If $S$ is irreducible and $\omega$, $U$ are given find (by the normality of $K$) a continuous $f \geqq 0$ vanishing in $U^c$ and $= 1$ in some point of $U$. Next find $n$ with $S^n f(\omega) > 0$. Then $P^n(\omega, U)$ is positive.]

If $S$ is irreducible, only the constant functions are $S$-invariant. Otherwise there would exist a non negative non constant $f = Sf$ which is 0 in some point $\omega$. But then $S^n f(\omega) > 0$ would contradict $f = S^n f$. The transformation $\tau\colon \omega \to \omega^2$ in $[0,1]$ is an example showing that the converse does not hold. It is not irreducible, but its only $S$-invariant functions are the constants.

By proposition 1.3 a mean ergodic $S$ is uniquely ergodic iff $Sf = f$ implies $f$ constant. Thus a mean ergodic irreducible operator is uniquely ergodic.

[If $\tau: K \to K$ is a homeomorphism, then $S$ with $Sf = f \circ \tau$ is irreducible iff $\tau$ is minimal in the sense of topological dynamics, i.e., iff there exists no non empty proper closed subset $A \subset K$ with $\tau A = A$. There exist minimal transformations which are not uniquely ergodic and therefore not mean ergodic, and there exist uniquely ergodic transformations which are not minimal; see e.g. Oxtoby [1952], Keane [1968], Grillenberger [1976].]

**Theorem 1.5** (Jamison [1970]). *If $K$ is a compact Hausdorff space, $S$ an irreducible Markov operator in $\mathfrak{X} = C(K)$, and $f \in \mathfrak{X}$ satisfies*

(1.1) $\lim_n S^n f(\omega) = 0$ *for all* $\omega \in K$

*then* $\|S^n f\| \to 0$.

*Proof.* We first assume $K$ compact metric with a metric $d$. For a bounded sequence $(f_n)$ in $\mathfrak{X}$ put

$$u_m(\omega) = \limsup_n (\sup\{|f_n(\eta)|: d(\omega, \eta) \leqq 1/m\}),$$

and let $u(\omega)$ be the limit of the decreasing sequence $u_m(\omega)$. For $\varepsilon > 0$ and large $m$ we have $u(\omega) + \varepsilon \geqq u_m(\omega)$. If $(\omega_k)$ is a sequence in $K$ tending to $\omega$ and $k$ is so large that $d(\omega, \omega_k) < 1/2m$ holds, then $u_m(\omega) \geqq u_{2m}(\omega_k) \geqq u(\omega_k)$. Hence $\limsup_k u(\omega_k) \leqq u(\omega)$ follows and $u$ is upper semicontinuous. Thus $u$ assumes its supremum $\alpha$. We now prove

(1.2) $\|(|f_n| \vee \alpha) - \alpha\| \to 0, \quad (n \to \infty).$

If this fails there exist sequences $n_k \to \infty$ and $\omega_k$, and a number $\delta > 0$ such that $f_{n_k}(\omega_k) \geqq \alpha + \delta$. We may assume that $\omega_k$ converges to some $\omega$ by taking subsequences. Fix any $m$. For $k$ large $d(\omega_k, \omega) < 1/m$, and hence $\sup\{|f_{n_k}(\eta)|: d(\omega, \eta) \leqq 1/m\} \geqq \alpha + \delta$. Thus $u_m(\omega) \geqq \alpha + \delta$. As $m$ was arbitrary, $u(\omega) \geqq \alpha + \delta$, a contradiction.

We shall need

**Lemma 1.6.** *If $f_n(\omega) \to 0$ for each $\omega$, then $\{\omega: u(\omega) > 0\}$ is a set of the first category.*

*Proof.* For each $j = 1, 2, \dots$ the sets

$$F_j = \{\omega \in K: u(\omega) \geqq 1/j\}$$

are closed since $u$ is upper semicontinuous. Fix $j$ and suppose that $F_j$ has non empty interior $G$. Then there is a closed set $F \subset G$ with non empty interior. The sets

$$H_i = \{\omega \in F: |f_k(\omega)| \leqq 1/2j \text{ for all } k \geqq i\}, \quad (i = 1, 2, \dots)$$

are closed and their union is $F$. By the Baire category theorem applied to the complete metric space $F$ at least one of the sets $H_i$ must have a non empty interior $G'$. As each $|f_k|$ is bounded by $1/2j$ for $k \geqq i$ on $H_i$, we have $u(\omega) \leqq 1/2j$ on $G'$. This contradicts to $G' \subset F_j$.

It follows that each $F_j$ has empty interior. As the set $\{\omega: u(\omega) > 0\}$ is the union of the sets $F_j$, it is of the first category. □

We now put $f_n = S^n f$. If $\alpha$ is 0 the assertion of the theorem follows by (1.2). Suppose $\alpha > 0$. As $u$ is upper semicontinuous the set $E = \{\omega: u(\omega) = \alpha\}$ is non empty and closed.

**Lemma 1.7.** *$E$ is absorbing, i.e., $P(\omega, E^c) = 0$ holds for $\omega \in E$.*

*Proof.* Assume there exists an $\omega \in E$ with $P(\omega, E^c) > 0$. Then there exists a closed set $F' \subset E^c$ with $P(\omega, F') > 0$. Hence there exist two disjoint open sets $G, G'$ with $E \subset G$ and $F' \subset G'$.

It follows from $u(\omega) = \alpha$ that there are sequences $\omega_k \to \omega$ and $n_k \to \infty$ with $|S^{n_k+1} f(\omega_k)| \to \alpha$. The $w^*$-convergence of $P(\omega_k, \cdot)$ to $P(\omega, \cdot)$ yields $\lim_k P(\omega_k, G') \geqq P(\omega, G') > 0$. Let $F$ denote the closure of $G'$, and $\gamma = P(\omega, G')/2$. Then $P(\omega_k, F) \geqq \gamma > 0$ holds for all large enough $k$.

Put $\beta = \sup\{u(\eta): \eta \in F\}$. Since $F \subset E^c$, and the supremum is assumed by the upper semicontinuity of $u$, we have $\beta < \alpha$. Let $2\varepsilon = \alpha - \beta$. We have

$$S^{n_k+1} f(\omega_k) = \int_F S^{n_k} f(\eta)\, P(\omega_k, d\eta) + \int_{F^c} S^{n_k} f(\eta)\, P(\omega_k, d\eta). \tag{1.3}$$

Observe that $P(\omega_k, F^c) \leqq 1 - \gamma$ holds for large $k$ and that (1.2) implies $|S^n f(\eta)| \vee \alpha \to \alpha$.

Now assume that $|S^{n_k} f(\eta)| \vee (\alpha - \varepsilon)$ tends to $\alpha - \varepsilon$ uniformly in $F$. Then (1.3) leads to the contradiction

$$\alpha = \lim |S^{n_k+1} f(\omega_k)| \leqq \gamma(\alpha - \varepsilon) + (1-\gamma)\alpha = \alpha - \gamma\varepsilon < \alpha.$$

Thus we can find a subsequence $(r_j)$ of the sequence $(n_k)$ and a sequence $\eta_j \in F$ with $|S^{r_j}(\eta_j)| > \alpha - \varepsilon$. We may assume that the sequence $\eta_j$ converges. The limit $\eta$ necessarily belongs to $F$. But then $u(\eta) \geqq \limsup |S^{r_j}(\eta_j)| \geqq \alpha - \varepsilon > \alpha - 2\varepsilon = \beta$ leads to a contradiction to the definition of $\beta$. □

Now the proof of theorem 1.5 for compact metric $K$ follows easily: As $P(\omega, E) = 1$ holds for all $\omega \in E$, an inductive argument shows $P^n(\omega, E) = 1$ for all $\omega \in E$ and all $n$. By the irreducibility of $S$ the non empty, closed set $E$ must therefore coincide with $K$. Hence $u(\omega) \equiv \alpha > 0$, and this contradicts to lemma 1.6.

*The general case.* It remains to eliminate the additional assumption that $K$ is a metric space.

Put $A_1 = \{\mathbb{1}, f, Tf, T^2 f, \ldots\}$, and let $A_2$ be the set of all finite products

$g_1 \cdot g_2 \dots g_k$ with $g_i \in A_1$. If $A_3$ denotes the set of all rational linear combinations of elements of $A_2$ and $A$ the norm closure of $A_3$, then $A$ is a separable closed subalgebra of $\mathfrak{X}$ with $TA \subset A$.

Let $\tilde{K}$ be the space of equivalence classes of elements of $K$ with $\omega \sim \omega'$ iff $g(\omega) = g(\omega')$ holds for all $g \in A$. $\tilde{K}$ is a compact Hausdorff space with the quotient topology and to each $f \in A$ there corresponds an $\tilde{f} \in C(\tilde{K})$ in an obvious way. Let $\tilde{A}$ be the set of all $\tilde{f}$ with $f \in A$. Then $\tilde{A}$ is a closed subalgebra of $C(\tilde{K})$ containing $\tilde{A}$ and separating points. So $\tilde{A} = C(\tilde{K})$ by Stone-Weierstrass. As $\tilde{A}$ is separable, $\tilde{K}$ is metric. Since $SA \subset A$, the restriction of $S$ to $A$ lifts to a Markov operator $\tilde{S}$ on $C(\tilde{K})$ for which $\tilde{S}^n \tilde{f}$ is the image of $S^n f$ under the isomorphism. It remains to check that the assumptions for $S$ yield the analogous assumptions for $\tilde{S}$. So $\|\tilde{S}^n \tilde{f}\| \to 0$ by the metric case, and thus $\|S^n f\| \to 0$. □

As an application we show a strengthening of the Jacobs-Deleeuw-Glicksberg decomposition of $\mathfrak{X}$ for the present special case.

**Theorem 1.8.** *Let $K$ be a compact Hausdorff space and $S$ a weakly almost periodic irreducible Markov operator in $\mathfrak{X}$. If $f$ is a flight vector, then $\|S^n f\| \to 0$.*

*Proof.* There exists a sequence $n_i \to \infty$ with $S^{n_i} f \to 0$ weakly and hence pointwise. As $S$ is weakly almost periodic, $S$ is mean ergodic, and there exists a unique $S^*$-invariant probability $\mu$. Now $\int |S^{n_i} f| \, d\mu \to 0$, and the fact that $S$ contracts the $L_1(\mu)$-norm implies $\int |S^n f| \, d\mu \to 0$.

Assume $S^n f$ does not converge to 0 weakly. Then there exists $0 \neq g \in \mathfrak{X}$ and a sequence $m_i \to \infty$ with $S^{m_i} f \to g$ (weakly). Then $\int |S^{m_i} f| \, d\mu \to \int |g| \, d\mu$. Thus a contradiction results if we show $\int |g| \, d\mu > 0$. By the irreducibility of $S$, $\Omega$ is the union of the sets $\{S^n |g| > 0\}$. Hence $\langle S^n |g|, \mu \rangle > 0$ holds for at least one $n$. But the $S^*$-invariance of $\mu$ gives $\langle S^n |g|, \mu \rangle = \langle |g|, \mu \rangle$. Thus $S^n f$ converges to 0 weakly, and the assertion follows from theorem 1.5. □

**Corollary 1.9.** *Any weakly almost periodic irreducible Markov operator $S$ in $\mathfrak{X} = C(K)$ is strongly almost periodic.*

*Proof.* $\mathfrak{X} = \mathfrak{X}_{rev} \oplus \mathfrak{X}_{fl}$; see section 2.4. □

**3. Breiman's strong law.** Now we turn to the proof of Breiman's strong law of large numbers for processes of the form $f(X_0)$, $f(X_1), \dots$ where $f$ is continuous on the state space $K$ and $X_0, X_1, \dots$ a Markov process with *arbitrary* initial distribution. We shall use standard probabilistic concepts freely here.

Let $(K_n, \mathscr{F}_n) = (K, \mathscr{F})$, $(n = 0, 1, 2, \dots)$, $\Omega = K_0 \times K_1 \times \dots$, and for $\omega = (\omega_0, \omega_1, \dots) \in \Omega$ let $X_n(\omega) = \omega_n$ be its $n$-th coordinate. If $\mu$ is any probability measure on $(K_0, \mathscr{F}_0)$ there is a unique probability $P$ on the product $\sigma$-algebra $\mathscr{A} = \mathscr{F}_0 \otimes \mathscr{F}_1 \otimes \dots$ for which the process $X_0, X_1, \dots$ is Markov with transition probabilities $P(X_{n+1} \in B_{n+1} | X_n = \omega_n) = P(\omega_n, B_{n+1})$ and initial distribution $P(X_0 \in B_0) = \mu(B_0)$. The following result is essentially due to Breiman [1960]:

**Theorem 1.10.** *Let $K$ be a compact Hausdorff space, and $S$ a mean ergodic Markov operator in $\mathfrak{X} = C(K)$. For $f \in \mathfrak{X}$ let $\bar{f} = \lim A_n f$. Then $n^{-1} \sum_{k=0}^{n-1} f(X_k(\omega))$ converges P-a.e., and the distribution of the limit is the same as the distribution $\mu_f$ of $\bar{f}$ under $\bar{\mu}$, where $\bar{\mu}$ is the w*-limit of the sequence $A_n^* \mu$. (Recall $P(\omega_n, \cdot) := S^* \varepsilon_{\omega n}$).*

The proof will be split into some lemmas:

**Lemma 1.11.** *If $h = Sh \in \mathfrak{X}$, then $h(X_n)$ converges P-a.e. to a limit $Y$ distributed as $\mu_h$.*

*Proof.* As $P(\omega_{n-1}, B_n) = P(X_n \in B_n | X_{n-1} = \omega_{n-1})$ holds for all $n$, $\omega_{n-1}$, and $B_n$, we have $E(g(X_n) | X_{n-1}) = (Sg)(X_{n-1})$ for all $g \in \mathfrak{X}$. Using the Markov property we find

$$E(h(X_n) | X_0, \ldots, X_{n-1}) = E(h(X_n) | X_{n-1}) = (Sh)(X_{n-1}) = h(X_{n-1}) \text{ a.e..}$$

Thus $h(X_n)$ is a bounded martingale and converges a.e. to a limit $Y$. (See e.g., Loève [1960: 393]). For any bounded $g \in C(\mathfrak{X})$, $g(h(X_n)) \to g(Y)$ a.e.. Hence $n^{-1} \sum_{k=0}^{n-1} E(g(h(X_k))) \to E(g(Y))$. But

$$\begin{aligned} n^{-1} \sum_{k=0}^{n-1} E(g(h(X_k))) &= n^{-1} \sum_{k=0}^{n-1} \int E\{g \circ h(X_k) | X_0 = \omega_0\} \mu(d\omega_0) \\ &= n^{-1} \sum_{k=0}^{n-1} \int S^k(g \circ h) d\mu = \langle g \circ h, A_n^* \mu \rangle \to \langle g \circ h, \bar{\mu} \rangle . \end{aligned}$$

This implies $E(g(Y)) = \langle g \circ h, \bar{\mu} \rangle$ for all bounded $g \in C(\mathfrak{X})$, which is equivalent to the assertion that $Y$ is distributed under $P$ as $h = \bar{h}$ is distributed under $\bar{\mu}$. □

**Lemma 1.12.** *Let $\omega_0, \omega_1, \ldots$ be a sequence in $K$. Assume*

$$(1.4) \quad n^{-1} \sum_{k=0}^{n-1} \{Sg(\omega_k) - g(\omega_{k+1})\} \to 0 \quad \textit{for } g \in \{f, Sf, S^2 f, \ldots\}.$$

*Then*

$$(1.5) \quad \bar{f}(\omega_n) - n^{-1} \sum_{k=0}^{n-1} f(\omega_k) \to 0, \quad (n \to \infty).$$

*Proof.* Using $\sum_{k=0}^{n-1} \{S^{j+1} f(\omega_k) - f(\omega_{k+1})\} = \sum_{k=0}^{n-1} \{S^{j+1} f(\omega_k) - S^j f(\omega_{k+1})\}$ $+ \sum_{k=0}^{n-1} \{S^j f(\omega_k) - f(\omega_{k+1})\} + S^j f(\omega_n) - S^j f(\omega_0)$ an inductive argument yields

$n^{-1}\sum_{k=0}^{n-1}\{S^j f(\omega_k)-f(\omega_{k+1})\}\to 0$ for $j\geqq 1$. For $j=0$ this is trivial. Hence

$$n^{-1}\sum_{k=0}^{n-1}(A_m f(\omega_k)-f(\omega_{k+1}))\to 0$$

holds for each $m\geqq 1$. Since $A_m f$ converges uniformly to $\bar{f}$ we obtain (1.5). □

We can now derive the theorem from the following strong law for martingale differences: If $\mathscr{A}_0\subset\mathscr{A}_1\subset\ldots$ is an increasing sequence of $\sigma$-algebras in $\mathscr{A}$ and $Z_0, Z_1, \ldots$ an $L_2$-bounded sequence of random variables for which $Z_n$ is $\mathscr{A}_n$-measurable, then

$$n^{-1}\sum_{k=0}^{n-1}\{Z_{k+1}-E(Z_{k+1}\,|\,\mathscr{A}_k)\}\to 0 \quad \text{a.e.};$$

see e.g. Loève [1960: 387]. Let $\mathscr{A}_k$ be the $\sigma$-algebra generated by $X_0, \ldots, X_k$ and $Z_k=g(X_k)$. Then $E(Z_{k+1}\,|\,\mathscr{A}_k)=Sg(X_k)$ shows that (1.4) is satisfied for $P$-almost all sequences $\omega=(\omega_0,\omega_1,\omega_2,\ldots)$. Hence

$$(1.6)\qquad \bar{f}(X_n)-n^{-1}\sum_{k=0}^{n-1}f(X_k)\to 0, \quad (n\to\infty)$$

a.e. It remains to apply lemma 1.11 with $h=\bar{f}$. □

**Remark 1.13.** (1.6) shows that almost surely the sequences $\bar{f}(X_n)$ and $n^{-1}\sum_{k=0}^{n-1}f(X_k)$ have the same limit.

We close this section with a result of Choquet-Foias [1975] in which $S$ need not be Markovian or a contraction.

**Theorem 1.14.** *Let $K$ be a compact Hausdorff space, and let $S$ be a positive linear operator on $\mathfrak{X}=C(K)$. If $S^n\mathbb{1}$ converges to 0 pointwise, then $\|S^n\mathbb{1}\|\to 0$.*

*Proof.* By the compactness of $K$ there exists an $m$ and $\delta>0$ with $\inf\{S^k\mathbb{1}(\omega)\colon 1\leqq k\leqq m\}\leqq 1-2\delta$ for each $\omega\in K$. If $\theta>0$ is sufficiently small, the operator $R=S+\theta I$ satisfies $v:=\inf\{R^k\mathbb{1}\colon 1\leqq k\leqq m\}<1-\delta$. Now $v\leqq\inf\{R^k\mathbb{1}\colon 0\leqq k\leqq m-1\}$ implies $Rv\leqq v$, and hence $Sv\leqq(1-\theta)v$. By induction $S^n v\leqq(1-\theta)^n v$. On the other hand $v\geqq\theta^m\mathbb{1}$. Together we obtain $S^n\mathbb{1}\leqq\theta^{-m}S^n v\leqq\theta^{-m}(1-\theta)^n\mathbb{1}$ and this tends uniformly to 0 for $n\to\infty$. □

As abstract $M$-spaces with unit ( like $L_\infty$) have an isomorphic representation as spaces of continuous functions on a compact space, the results of this section may also be applied to them. Of course pointwise convergence to a continuous limit must be translated into weak convergence then.

## Notes

Krylov and Bogolioubov [1937b] have initiated the study of quasi-regular points and of the ergodicity of the measures $\mu_\omega = w^*\text{-lim}\, A_n^* \varepsilon_\omega$ for continuous transformations $\tau$ in a compact metric space $K$. In particular this leads to an ergodic decomposition of invariant probabilities. Much of the early work on this can be found in the paper of Oxtoby [1952]; see also Denker-Grillenberger-Sigmund [1976] and Furstenberg [1981].

Furstenberg [1961] has studied a set of homeomorphisms of the torus including affine transformations and he showed when all points are quasi-regular. This has led to a considerable development; see Dani [1981] and Dani-Muralidharan [1983].

The Krylov-Bogolioubov decomposition has been extended to Markov operators by Beboutov [1942] and sharper results have been obtained by Jamison [1965], Lloyd [1963], and Rosenblatt [1964] for mean ergodic $S$. Sine [1968] obtains a portion of the theory under weaker assumptions. Jamison's study of the sample paths has been generalized and sharpened in the work of Jamison-Sine [1974], and Sine [1976a].

Let us mention a few relatively elementary facts on restricting $S$ to subsets. If $B$ is closed and absorbing there exists an invariant probability on $B$. Conversely the (closed) support of an invariant probability is absorbing. The set of extreme points of the set of invariant probabilities $\mu$ consists the of the ergodic $\mu$'s.

Iwanik [1981] showed that for $w^*$ mean ergodic $S$ every minimal closed absorbing set carries a unique invariant probability. In general this may fail, see Raimi [1964].

For a discussion of mean ergodicity and quasi-compactness we refer to the notes of § 2.2.

Sine [1975a] studied the question of mean ergodicity on the center $Z$, i.e., on the closure of the union of the supports of invariant probabilities. Atalla [1974] showed: If $R$ is another Markov operator on $\mathfrak{X}$ and $A_n(S)f \to 0$ holds iff $Rf = 0$, then $S$ and $R$ induce operators $S_Z$, $R_Z$ on $C(Z)$ with $A_n(S_Z)g \to S_Z g$ for all $g \in C(Z)$. For generalizations of these results to semigroups see Sato [1979].

The book of Schaefer [1974] treats a fair amount of the spectral theory for Markov operators $S$, and results extending the Perron-Frobenius theory. Eigenfunctions and eigenvalues have also been studied by Rosenblatt [1964], and Jamison-Sine [1969]. The set of unimodular eigenfunctions forms a group under multiplication and so do the eigenvalues. The paper of Jamison-Sine is also of interest for its self-contained short development of the Jacobs-Deleeuw-Glicksberg decomposition of $C(K)$.

Rosenblatt showed that a weak flight vector for $S$ having a weakly compact orbit converges weakly. This and Jamison's theorem has been extended to some reducible operators by Sine [1974]. (Jamison had pointed out that the condition of irreducibility cannot be dropped).

Ando [1968] has studied problems on invariant measures of positive contractions in $C(K)$ when $K$ is *quasi-stonean* (i.e., $C(K)$ is a conditionally $\sigma$-complete lattice). In particular his results apply to $L_\infty$. They have been generalized by Sato [1980a]. Foguel, Horowitz [1969], and Lin [1970] have considered positive contractions in the space of bounded continuous functions on a locally compact space, and they have obtained some results similar to those in the theory of contractions in $L_1$. Typical problems are the existence of finite and $\sigma$-finite invariant measures, convergence of ratios etc. We refer to the survey of Foguel [1971]. Sine [1973] has pointed out that the restriction of $S$ to the conservative part in the sense of Foguel need not be conservative.

It does not seem difficult to extend the proof of Jamisons theorem to the multiparameter situation.

Breiman proved his strong law only for uniquely ergodic $S$ and Jamison [1965] pointed out that the proof extends to mean ergodic $S$.

Further related references: Atalla [1981], [1983], Atalla-Sine [1976], Chersi [1982], Davies [1982], Pakula-Sine [1977], Regnier [1970], Takahashi [1971a].

## § 5.2 Contractions in $L_p$, $(1 < p < \infty)$

**1. Akcoglu's theorem.** The main result in this section is the celebrated ergodic theorem of Akcoglu [1975], which asserts the convergence a.e. of $A_n f = A_n(T)f$ for positive contractions $T$ in $L_p = L_p(\Omega, \mathscr{A}, \mu)$. We assume $1 < p < \infty$, and set $q = p/(p-1)$. The Banach principle implies that it is enough to prove convergence a.e. for all $f$ in a dense subset of $L_p$ and a dominated estimate. The first of these steps is easy even under weaker assumptions:

**Lemma 2.1.** *Let $T$ be a power bounded linear operator in $L_p$. Then $A_n f$ converges a.e. for all $f$ in a dense subset of $L_p$.*

*Proof.* $L_p$ is the closure of the direct sum of the set of fixed points and the space $(I-T)L_p$, see § 2.1. If $f$ is a fixed point of $T$, we even have $A_n f = f$. If $f$ is of the form $g - Tg$ with $g \in L_p$, we have $A_n f = n^{-1}(g - T^n g)$. Let $M$ be a bound for the norms $\|T^n\|$. Then

$$\int \sum_{n=1}^{\infty} |A_n f|^p \, d\mu = \sum_{n=1}^{\infty} \|A_n f\|_p^p \leqq \sum_{n=1}^{\infty} n^{-p}(\|g\|_p + \|T^n g\|_p)^p$$
$$\leqq \sum_{n=1}^{\infty} n^{-p}((M+1)\|g\|_p)^p < \infty$$

yields $\sum_{n=1}^{\infty} |A_n f|^p < \infty$. Hence $A_n f \to 0$ a.e. □

The dominated estimate will first be proved for positive isometries. This case is due to A. Ionescu Tulcea [1964], but we follow the simplified proof of de la Torre [1976] and Charn-Huen Kan [1978]:

**Theorem 2.2.** *If $T$ is a positive isometry of $L_p$, and $f^* := \sup_{n \geqq 1} A_n|f|$, then $\|f^*\|_p \leqq q\|f\|_p$.*

*Proof.* First note that $T$ maps functions with disjoint supports to functions with disjoint supports. This follows from the fact that $\|g+h\|_p^p = \|g\|_p^p + \|h\|_p^p$ holds for $g, h \in L_p^+$ iff $g$ and $h$ have disjoint support.

We can assume $f \geqq 0$. Recall the notation $M_n f = \operatorname{Max}(A_1 f, \ldots, A_n f)$ and $M_n(T')f = \operatorname{Max}(A_1(T')f, \ldots, A_n(T')f)$. It suffices to prove $\|M_N f\|_p \leqq q\|f\|_p$ for all $N$. There exist disjoint subsets $E_1, \ldots, E_N$ of $\Omega$ with $M_N f = \sum_{i=1}^{N} 1_{E_i} A_i f$. For fixed $k \geqq 0$, the functions $T^k(1_{E_i} A_i f)$ have disjoint supports $D_{1k}, \ldots, D_{Nk}$. Hence

$$T^k M_N f = \sum_{i=1}^{N} 1_{D_{ik}} T^k(1_{E_i} A_i f) \leqq \sum_{i=1}^{N} 1_{D_{ik}} T^k A_i f \leqq M_N(T^k f), \text{ and}$$

$$\|M_N f\|_p = \|T^k M_N f\|_p \leqq \|M_N(T^k f)\|_p, \quad (k = 0, 1, 2, \dots).$$

Taking the $p$-th power and averaging over $k = 0, \dots, L-1$, we obtain

$$\text{(2.1)} \qquad \|M_N f\|_p^p \leqq L^{-1} \int \sum_{k=0}^{L-1} (M_N T^k f)^p \, d\mu.$$

We now apply the dominated ergodic theorem in the special case of the translation $k \to k+1$ on $\mathbb{Z}^+$ with counting measure. If $F$ is a non negative function on $\mathbb{Z}^+$ with $F(k) = 0$ for $k \geqq N + L - 1$, put $M_N^F(j) = \max_{1 \leqq i \leqq N} i^{-1}(F(j) + F(j+1) + \dots + F(j+i-1))$. Theorem 1.6.3. yields

$$\text{(2.2)} \qquad \sum_{k=0}^{L-1} M_N^F(k)^p \leqq q^p \sum_{j=0}^{N+L-2} F(j)^p.$$

If, for some fixed $\omega \in \Omega$, $F$ is the function with $F(j) = T^j f(\omega)$, $(0 \leqq j \leqq N + L - 2)$, and $F(j) = 0$, $(j \geqq N + L - 1)$, then the left hand side of (2.2) equals $\sum_{k=0}^{L-1} (M_N T^k f)^p(\omega)$, and the right hand side of (2.2) equals $q^p \sum_{j=0}^{N+L-2} (T^j f(\omega))^p$. Applying (2.1) and integrating we obtain

$$\|M_N f\|_p^p \leqq q^p L^{-1} \int \sum_{j=0}^{N+L-2} (T^j f)^p \, d\mu = q^p L^{-1}(N + L - 1) \|f\|_p^p.$$

Now let $L$ tend to $\infty$. □

The proof of the dominated estimate for positive contractions in $L_p$ will be attained by a reduction to theorem 2.2. We first consider the special case where $(\Omega, \mathscr{A}, \mu)$ is a probability space, $\mathscr{A}$ consists of all finite unions of disjoint sets $I_1, I_2, \dots, I_a$ with $\mu(I_k) =: m_k > 0$, $(k = 1, \dots, a)$. Moreover, we assume $\|T\| = 1$, and that the matrix $(T_{ij})$ with $T 1_{I_i} = \sum_{j=1}^{a} T_{ij} 1_{I_j}$ has strictly positive entries $T_{ij}$.

Using an isomorphism we may assume $\Omega = [0, 1[$, $\mu$ = restriction of the Lebesgue measure $\lambda$ to $\mathscr{A}$, and $I_1 = [0, m_1[$, $I_2 = [m_1, m_1 + m_2[, \dots, I_a = [1 - m_a, 1[$. If $\mathscr{B}$ denotes the $\sigma$-algebra of Borel sets, the conditional expectation $E_{\mathscr{A}}$ is a positive contraction mapping $L_p(\Omega, \mathscr{B}, \lambda)$ into $L_p(\Omega, \mathscr{A}, \mu)$. The plan is to construct a positive isometry $Q$ of $L_p(\Omega, \mathscr{B}, \lambda)$ such that $TE_{\mathscr{A}} f = E_{\mathscr{A}} Q f$ holds for all $f \in L_p(\mathscr{B})$. If this is done, induction yields

$$T^k g = E_{\mathscr{A}} Q^k g \quad \text{for all} \quad g \in L_p(\mathscr{A}) \quad \text{and all} \quad k \geqq 0,$$

and hence $|M_N g| \leqq M_N |g| \leqq E_{\mathscr{A}} M_N(Q)|g|$. Theorem 2.2 then implies

$$\text{(2.3)} \qquad \|M_N g\|_p \leqq \|M_N(Q)|g|\|_p \leqq q \|g\|_p,$$

the desired dominated estimate for $T$.

The construction of $Q$ depends on parameters $m_{ij}$, $h_{ij}$ determined later. We split each $I_i$ into half open disjoint subintervals $I_{ij}$ of length $\lambda(I_{ij}) = m_{ij} > 0$, $(j = 1, \ldots, a)$. Let $\tau$ denote the null preserving map of $\Omega$ into $\Omega$ which maps $I_{ij}$ affinely onto $I_j$.

*Formally*. If $a_j$ and $a_{ij}$ are the left end points of $I_j$ and $I_{ij}$, and $\omega \in I_{ij}$, then $\tau\omega = a_j + m_j(\omega - a_{ij})/m_{ij}$. Consider a positive function $h: \Omega \to \mathbb{R}$, assuming a constant value $h_{ij} > 0$ on each $I_{ij}$. We shall show that it is possible to choose $m_{ij}$, $h_{ij}$ in such a way, that the operator $Qf(\omega) = h(\omega)f(\tau\omega)$ has the desired properties.

It follows from

$$\|Qf\|_p^p = \sum_{i,j} \int_{I_{ij}} (f^p \circ \tau) h^p \, d\lambda = \sum_j \Big( \int_{I_j} f^p \, d\lambda \sum_i h_{ij}^p m_{ij}/m_j \Big)$$

that $Q$ is an isometry iff

$$(2.4) \qquad \sum_{i=1}^{a} h_{ij}^p m_{ij}/m_j = 1, \quad (j = 1, \ldots, a)$$

holds. Next let us check how we can obtain $TE_{\mathscr{A}} f = E_{\mathscr{A}} Qf$. Let $f_i = m_i^{-1} \int f 1_{I_i} \, d\lambda$ be the value of $E_{\mathscr{A}} f$ on $I_i$. $\sum_{i=1}^{a} T_{ik} f_i$ is the value of $TE_{\mathscr{A}} f$ on $I_k$, and this must be equal to $m_k^{-1} \sum_{i=1}^{a} \int 1_{I_{ki}} Qf \, d\lambda = m_k^{-1} \sum_i h_{ki} m_{ki} f_i$ for arbitrary $f_1, \ldots, f_a$. Thus

$$(2.5) \qquad T_{ik} = h_{ki} m_{ki}/m_k, \quad (i, k = 1, \ldots, a)$$

is necessary and sufficient. It remains to find $h_{ij}, m_{ij} > 0$ with (2.4), (2.5), and $\sum_j m_{ij} = m_i$. Note that any $f \in L_p(\mathscr{A}) = L_p$ can be identified with $(f_1, \ldots, f_a)$, where $f_i$ is the value of $f$ on $I_i$.

**Lemma 2.3.** *For any $f \in L_p^+$ there exists a unique $\hat{f} \in L_q^+$ with*

$$\langle f, \hat{f} \rangle = \|f\|_p^p = \|\hat{f}\|_q^q = \|f\|_p \|\hat{f}\|_q,$$

*and this $\hat{f}$* is given by $\hat{f}_i = f_i^{p-1}$.

*Proof*. This follows from Hölders inequality, together with the characterization of equality in Hölders inequality, see Hewitt-Stromberg [1969: 190–191]. □

$\|T\| = 1$ and a compactness argument implies the existence of an $f$ with $\|f\|_p = 1$ *and* $\|Tf\|_p = 1$. Put $u = |f|$. Then $\|u\|_p = 1$, and $1 \geqq \|Tu\|_p \geqq \|Tf\|_p = 1$, shows $\|u\|_p = \|Tu\|_p = 1$.

**Lemma 2.4.** *Put $v = Tu$. Then $\hat{u} = T^* \hat{v}$, and both $u$ and $v$ have strictly positive coordinates.*

*Proof.* $\|v\|_p = \|Tu\|_p = 1$ implies $\|\hat{v}\|_q = 1$, and $1 = \langle v, \hat{v}\rangle = \langle Tu, \hat{v}\rangle = \langle u, T^*\hat{v}\rangle$. Thus $\|T^*\hat{v}\|_q \leqq 1$ yields $\hat{u} = T^*\hat{v}$. As all $T_{ij}$ are strictly positive, $v = Tu$ has strictly positive coordinates $v_i$. It is simple to check that the $i$-th coordinate of $T^*g$ is given by $\sum_j (m_j/m_i)\, T_{ij} g_j$. As $\hat{v}$ has strictly positive coordinates, also $\hat{u} = T^*\hat{v}$ has strictly positive coordinates. Therefore the coordinates $u_i$ of $u$ are strictly positive. □

The identities $v = Tu$, $\hat{u} = T^*\hat{v}$ can be written in the form

$$v_i = \sum_j u_j T_{ji}, \quad m_i u_i^{p-1} = \sum_j m_j T_{ij} v_j^{p-1}. \tag{2.6}$$

Now put $h_{ij} = v_i/u_j$, and $m_{ij} = u_j m_i T_{ji}/v_i$. The first identity in (2.6) yields $\sum_j \mathrm{m}_{ij} = m_i$. (2.5) follows from the definition of $h_{ij}$ and $m_{ij}$, and the second identity in (2.6) leads to (2.4). We have completed the proof of (2.3) in the case of a finite dimensional space $L_2$ under the restrictions $\|T\| = 1$, $T_{ij} > 0$. This will be enough to prove:

**Theorem 2.5** (Akcoglu's dominated ergodic theorem). *If $T$ is a positive contraction in a space $L_p$ with $1 < p < \infty$, then $\|\sup_{n \geqq 1} A_n |f|\|_p \leqq q\|f\|_p$.*

*Proof.* It is simple to see that we can assume the underlying measure space $(\Omega, \mathscr{A}, \mu)$ $\sigma$-finite. Then there exists a strictly positive $\psi$ with $\int \psi d\mu = 1$. Let $\mu'$ be the probability measure with $d\mu'/d\mu = \psi$. The transformation

$$L_p(\mu) \ni f \to Vf := f/\psi^{1/p} \in L_p(\mu')$$

is a bijective order and norm preserving isomorphism of $L_p(\mu)$ and $L_p(\mu')$. $T' = VTV^{-1}$ is a positive contraction in $L_p(\mu')$, and the dominated ergodic theorem holds for $T$ if it holds for $T'$, since $(T')^k = VT^kV^{-1}$. Therefore we can assume $\mu(\Omega) = 1$ henceforth.

Assume there exists an $N$ and $f \in L_p^+$ with $\|M_N f\|_p > q\|f\|_p$. For any $\varepsilon > 0$ there exists a finite sub-$\sigma$-algebra $\mathscr{F} \subset \mathscr{A}$ (depending on $f$), with $\|T^i f - E_{\mathscr{F}} T^i f\|_p < \varepsilon$, $(i = 1, \ldots, N)$. Then

$$\|T^i f - (E_{\mathscr{F}} T)^i E_{\mathscr{F}} f\|_p < \varepsilon(i+1)$$

holds for $i = 0$, and the inductive step $i \to i+1$, $(i < N)$, follows from

$$\begin{aligned}\|T^{i+1} f - (E_{\mathscr{F}} T)^{i+1} E_{\mathscr{F}} f\|_p &\leqq \|T^{i+1} f - E_{\mathscr{F}} T^{i+1} f\|_p \\ &+ \|E_{\mathscr{F}} T^{i+1} f - (E_{\mathscr{F}} T)^{i+1} E_{\mathscr{F}} f\|_p \leqq \varepsilon + \|T^i f - (E_{\mathscr{F}} T)^i E_{\mathscr{F}} f\|_p,\end{aligned}$$

since $E_{\mathscr{F}} T$ is a contraction in $L_2(\mathscr{F})$. Choosing $\varepsilon > 0$ small enough the function $M_N f$ will be approximated arbitrarily well in $L_p$-norm by $M_N(E_{\mathscr{F}} T) E_{\mathscr{F}} f$, and we will have $\|M_N(E_{\mathscr{F}} T) E_{\mathscr{F}} f\|_p > q\|f\|_p \geqq q\|E_{\mathscr{F}} f\|_p$. This means that the dominated estimate fails for the positive contraction $E_{\mathscr{F}} T$ in $L_p(\Omega, \mathscr{F}, \mu)$. It is therefore sufficient to prove the theorem in the case when the $\sigma$-algebra $\mathscr{A}$ is finite.

Now represent $T$ by a matrix $(T_{ij})$ as after the proof of theorem 2.2. Again assume $\|M_N f\|_p > q\|f\|_p$ for some $f \geqq 0$. For $0 < \alpha < 1$ close enough to 1 we have $\|M_N(\alpha T)f\|_p > q\|f\|_p$. Clearly $\|\alpha T\| < 1$. For suitable $\beta > 0$ the operator $T'$ with $T'_{ij} = \alpha T_{ij} + \beta$ satisfies $\|T\| = 1$. $M_N(T')f \geqq M_N(\alpha T)f$ implies $\|M_N(T')f\|_p > q\|f\|_p$. But $T'$ is an operator which has all the properties used in the special case treated first. We have arrived at a contradiction. □

Combining lemma 2.1, theorem 2.5 and the Banach principle we have proved:

**Theorem 2.6** (Akcoglu's ergodic theorem). *If $T$ is a positive contraction in a space $L_p$, $(1 < p < \infty)$ then $A_n f$ converges a.e. for all $f \in L_p$.*

**2. Self-adjoint operators in $L_2$.** Our next aim is the dominated estimate of Stein [1961a] for the unaveraged sequence $T^n f$:

**Theorem 2.7.** *If $T$ is a self-adjoint positive contraction in $L_2$, and $f^{**} := \sup_{n \geqq 1} T^n|f|$, then $\|f^{**}\|_2 \leqq 6\|f\|_2$.*

*Proof.* We can assume $f \geqq 0$. $P = T^2$ is non negative definite and has a spectral resolution $P = \int_0^1 \lambda\, dE(\lambda)$. We begin with the identity

$$P^n f - A_{n+1}(P)f = (n+1)^{-1} \sum_{k=1}^{n} k(P^k - P^{k-1})f, \quad (n \geqq 0),$$

which can be easily checked. By the inequality $(a_1 + \ldots + a_n)^2 \leqq n(a_1^2 + \ldots + a_n^2)$ the square of the right hand side is bounded above by $h_f := \sum_{k=1}^{\infty} k(P^k f - P^{k-1} f)^2$. Now $|P^n f - A_{n+1}(P)f|^2 \leq h_f$ implies

$$\sup_{n \geqq 0} T^{2n} f \leqq \sup_{n \geqq 1} A_n(P)f + h_f^{1/2}.$$

By Akcoglu's estimate $\|\sup_{n \geqq 1} A_n(P)f\|_2 \leqq 2\|f\|_2$. $\sum_{k=1}^{\infty} k\lambda^{k-1}$ is the derivative $(1-\lambda)^{-2}$ of $(1-\lambda)^{-1}$. Hence $\sum_{k=1}^{\infty} k(\lambda^k - \lambda^{k-1})^2 = (1-\lambda)^2 \sum_{k=1}^{\infty} k\lambda^{2(k-1)}$ is bounded by 1 for $0 \leqq \lambda \leqq 1$. By the spectral resolution of $P$ we obtain

$$\|h_f^{1/2}\|_2^2 = \int h_f\, d\mu = \sum_{k=1}^{\infty} k\|P^k f - P^{k-1} f\|_2^2$$
$$= \int_0^1 \Big(\sum_k k(\lambda^k - \lambda^{k-1})^2\Big) d\|E_\lambda f\|_2^2 \leqq \int_0^1 d\|E_\lambda f\|_2^2 = \|f\|_2^2.$$

Thus $\|\sup T^{2n} f\|_2 \leqq 2\|f\|_2 + \|f\|_2$. Similarly $\|\sup T^{2n+1} f\|_2 \leqq 3\|Tf\|_2 \leqq 3\|f\|_2$. □

If $T$ is, in addition, non negative definite, and $P_1 f$ the projection of $f$ on the fixed space of $T$, then the spectral theorem shows that $f - P_1 f$ is the norm-limit of $f_n := E(1-1/n)f$. As the norm of the restriction of $T$ to $E(1-1/n)L_2$ is $\leqq 1 - 1/n$, we find $\sum_k \|T^k f_n\|_2 \leqq \sum_k (1-1/n)^k \|f_n\|_2 < \infty$. Hence $T^k f_n$ tends to 0 a.e. for $k \to \infty$. It follows that the set of elements $g \in L_2$, for which $T^k g$ converges a.e. is dense in $L_2$. The dominated estimate and the Banach principle then yield

**Theorem 2.8.** *If $T$ is a self-adjoint, non negative definite, positive contraction in $L_2$, then $T^k f$ converges a.e. to $P_1 f$ for all $f \in L_2$.*

**3. A counter-example.** The positivity of $T$, which is not needed for von Neumann's mean ergodic theorem, is essential for a pointwise ergodic theorem in $L_2$. This was recognized by Burkholder [1962a] who saw that an old example of Menchoff [1923] in harmonic analysis could be used to construct counter-examples in ergodic theory.

Menchoff showed the following: If $(\Omega, \mathscr{A}, \mu)$ is $[0, 1]$ with Lebesgue measure there exists an orthonormal basis $\{\varphi_n, n \geqq 1\}$ of $L_2$ and an $f_0 \in L_2$ such that the sequence $P_k f_0$ of projections of $f_0$ on the subspaces spanned by $\{\varphi_1, \ldots, \varphi_k\}$, diverges a.e..

We now use this (deep) fact to show that there exists a unitary operator $T$ in $L_2$, for which $A_n f$ diverges a.e.. An arbitrary $f \in L_2$ has a representation $\sum_{m=1}^{\infty} \alpha_m \varphi_m$, where the $\alpha_m$'s are complex numbers with $\sum_{m=1}^{\infty} |\alpha_m|^2 < \infty$.

If $\{\lambda_m\}$ is an arbitrary collection of real numbers, the map

$$f \to \sum_{m=1}^{\infty} e^{i\lambda_m} \alpha_m \varphi_m$$

is a unitary operator in $L_2$, and the powers $T^j$ are given by

$$T^j f = \sum_{m=1}^{\infty} e^{i\lambda_m j} \alpha_m \varphi_m .$$

We now have to choose $\{\lambda_m\}$ in a suitable way. Observe that $\sigma(\lambda, n) := n^{-1} \sum_{l=0}^{n-1} e^{i\lambda l}$ tends to 0 for fixed $\lambda > 0$ when $n \to \infty$, and that $\sigma(\lambda, n)$ tends to 1 for fixed $n$ when $\lambda \to 0$. Take $n_1 = 1 = \lambda_1$, and let $\varepsilon_1 > \varepsilon_2 > \varepsilon_3 > \ldots$ be a sequence tending to 0. If $\lambda_1, \ldots, \lambda_{k-1}$ have been determined, find $n_k$ such that $n \geqq n_k$ implies $|\sigma(\lambda_m, n)| < \varepsilon_k$ for $m = 1, \ldots, k-1$. Then find $\lambda_k > 0$ with $|\sigma(\lambda_k, n) - 1| < \varepsilon_k$ for $n \leqq n_k$. Using $\|A_{n_k} \varphi_m\|_2 = \|\sigma(\lambda_m, n_k) \varphi_m\|_2 < \varepsilon_k$ for $m \leqq k-1$, and $\|(A_{n_k} - I)\varphi_m\|_2 = \|(\sigma(\lambda_m, n_k) - 1)\varphi_m\|_2 < \varepsilon_k$ for $m \geqq k$, and the orthogonality of the orbits of all $\varphi_m$, we obtain

$$\|A_{n_k} f - (I - P_{k-1}) f\|_2^2 \leqq \sum_{m=1}^{k-1} |\alpha_m|^2 \|A_{n_k} \varphi_m\|_2^2$$

$$+ \sum_{m=k}^{\infty} |\alpha_m|^2 \|(A_{n_k} - I)\varphi_m\|_2^2 \leqq \varepsilon_k^2 \|f\|_2^2 .$$

If $\varepsilon_k$ tends to 0 fast enough, this yields $\sum_{k=1}^{\infty} |A_{n_k} f_0 - (I - P_{k-1}) f_0| < \infty$ a.e..

But then $A_{n_k} f_0$ has, up to nullsets, the same set of divergence as $(I - P_{k-1}) f_0$, and hence as $P_k f_0$.

## Notes

1. The proof of Akcoglu's theorem was the breakthrough after previous contributions by Chacon, McGrath, and Olsen, see e.g. Olsen [1973]. Our presentation was influenced by Derriennic [1979]. The key idea of Akcoglu was the construction of a dilation $Q$ of $T$. The oldest and best known dilation theorem, due to *Sz*. Nagy (see e.g. Sz. Nagy and Foiaş [1970]) asserts that if $T$ is a contraction of a Hilbert space $\mathfrak{H}'$, then there exists a larger Hilbert space $\mathfrak{H} = \mathfrak{H}' \oplus \mathfrak{H}''$ and a unitary operator $U$ in $\mathfrak{H}$, so that $T^n x = PU^n x$ for each $n \geqq 0$ and $x \in \mathfrak{H}'$, where $P$ is the projection from $\mathfrak{H}$ to $\mathfrak{H}'$. The main novelty of $Q$ was the preservation of the order structure. Akcoglu's $Q$ actually was invertible. He also constructed order preserving dilations in $L_1$; [1975*a*]. As the existence of positive dilations seems of independent interest we mention the following dilation theorem of Akcoglu-Sucheston [1977].

**Theorem 2.9.** *Let $T: L \to L$ be a positive contraction on an $L_p$-space $L$, $(1 \leqq p < \infty)$. Then there exists another $L_p$-space $\hat{L}$ and a positive invertible isometry $\hat{T}: \hat{L} \to \hat{L}$ so that $DT^n = P\hat{T}^n D$ for all $n = 0, 1, 2, \ldots$, where $D: L \to \hat{L}$ is a positive isometric imbedding of $L$ into $\hat{L}$ and $P: \hat{L} \to \hat{L}$ is a positive projection.*

Kern, Nagel, and Palm [1977] have a lattice theoretic approach to dilation theorems and investigate which properties of $T$ carry over to $\hat{T}$; see also Nagel-Palm [1982]. Peller [1978], [1981] obtains a dilation for operators in $L_p$ which have a norm contracting linear modulus, and gives applications to generalized von Neumann inequalities.

Burkholder's original example was only a contraction. Akcoglu-Sucheston [1975a] remarked that it could be made unitary, see also Akcoglu [1979]. Akcoglu-Miller [1976] have shown that for a unitary $T$ in $\mathfrak{H}$ the operator $\sum_{n=1}^{\infty} P_n A_n$ is bounded for every choice of orthogonal projections $P_n$ iff $z = 1$ is an isolated point of the spectrum. As Burkholders example assumes $p = 2$, it is unknown if $A_n f$ converges a.e. for non positive contractions in $L_p$ with $1 < p < \infty, p \neq 2$. Dominated estimates in $L_p$ under supplementary conditions appear in de la Torre [1977], Sato [1981b], Duncan [1977]. Gaposhkin [1981a] gives necessary and sufficient conditions for the pointwise ergodic theorem for normal contractions in $L_2$.

De la Torre [1978] has an extension of Akcoglu's ergodic theorem to functions $f$ taking values in the Banach spaces $\ell_r$, $(1 < r < \infty)$. Akcoglu showed that the assertion of the Chacon-Ornstein theorem does not extend to positive invertible isometries in $L_p$, $(1 < p < \infty)$. More elementary examples were given by Fong and Lin [1976].

2. Theorem 2.7 and relatives are discussed in Stein [1970]. Stein [1961a] also proved $\|f^{**}\|_p < C_p \|f\|_p$, with $C_p < \infty$ independent of $f$, $1 < p < \infty$, for self-adjoint $L_1 - L_\infty$-contractions $T$ which need not be positive; for the case $p = 2$ see Burkholder-Chow [1961]. It turned out that it is not important to have powers of a single $T$ in this result. Rota [1962] considered products $T_{10} = I$, $T_{1n} := T_n T_{n-1} \ldots T_1$, $(n \geqq 1)$, and proved a dominated estimate and convergence a.e. for $T_{1n}^* T_{1n} f$, $(f \in L_p, 1 < p < \infty)$, when the $T_i$ are bistochastic. This was generalized by Doob [1963]. Finally Starr [1966] proved the following mutual generalization of the results of Stein, Rota and Doob:

**Theorem 2.10.** *Let* $T_1, T_2, \ldots$ *be* $L_1 - L_\infty$*-contractions. Then* $\|\sup_{n \geqq 0} |T_{1n}^* T_{1n} f|\|_p \leqq q\|f\|_p$ *holds for* $1 < p < \infty$*. For* $f \in L_1$ *with* $\int |f| \log^+ |f| d\mu < \infty$*, and* $h \in L_1$ *with* $\mathbb{1} \geqq h \geqq \operatorname{Min}(|f|, \mathbb{1})$ *we have*

$$\int (\sup_{n \geqq 0} |T_{1n}^* T_{1n} f| h) d\mu \leqq \frac{e}{e-1} \int (h + |f| \log^+ |f|) d\mu < \infty.$$

*If the* $T_i$ *are also positive, then* $T_{1n}^* T_{1n} f$ *converges a.e. for* $f$ *belonging to some* $L_p$*,* $(1 \leqq p < \infty)$*, with* $\int |f| \log^+ |f| d\mu < \infty$*.*

Starr [1965] showed that there is no such result for the reversed products $T_1 T_2 \ldots T_n$ instead of $T_{1n}$. Burkholder [1962] showed that the theorem is false if $f$ belongs to $L_1$ only, see also Ornstein [1968]. For related material; see Rao [1979].

3. Lin [1974d] studied unaveraged convergence in $L_p$ for Markov operators: If $T$ is a conservative and ergodic positive contraction in $L_1$ with $\sigma$-finite invariant measure $\nu$ and $T^{*n} 1_A \to 0$ a.e. holds for some $A$ with $0 < \nu(A)$, then $\|T^{*n} f\|_p \to 0$ for $f \in L_p$, $(1 < p < \infty)$. $\|T^{*n} f\|_p \to 0$ for all $f \in L_p$ holds if the past remote $\sigma$-algebra in the corresponding bilateral product space has no sets of nonzero finite measure.

4. The following theorem follows from results of Sz. Nagy and Foias [1960], see also Foguel [1963], [1969]:

**Theorem 2.11.** *If* $T$ *is a positive contraction in* $L_2$ *the set*

$$K := \{f \in L_2 : \|T^n f\|_2 = \|T^{*n} f\|_2 = \|f\|_2 \quad \text{for all } n \geqq 1\}$$

*is a closed linear subspace invariant under* $T$ *and* $T^*$*, and on* $K$ *one has* $TT^* = T^*T = I$*. For* $f \perp K$*,* $w\text{-}\lim T^n f = w\text{-}\lim T^{*n} f = 0$*.*

5. Mme Bénozène [1981] proved the convergence a.e. of averages $n^{-1} \sum_{j=0}^{n-1} |T^j f|^\alpha \operatorname{sign}(T^j f)$ for $T$ a positive contraction in $L_p$, $(1 < p < \infty)$, $f \in L_p$, and $1 < \alpha < p$. Such sums are of interest, when one looks at averages of a type introduced by Beauzamy and Enflo, namely the elements $s_n^{(p)}(f)$ minimizing $n^{-1} \sum_{j=0}^{n-1} \|g - T^j f\|^p$. (In the case of Hilbert spaces or $p = 2$, one is led back to usual Cesàro averages). Bénozène studies the convergence of $s_n^{(p)}(f)$ also for nonlinear $T$; see also Guerre [1978].

6. The proof of Akcoglu's theorem for power bounded positive $T$ is an open problem. Brunel and Emilion [1984] showed that the power bounded case would imply even the Cesàro bounded case. Feder [1981a] has constructed a (non positive) power bounded $T$ with power bounded inverse in $L_p$, $(1 < p \leqq 2)$ for which $A_n f$ does not converge a.e.. Assani [1985] has examples for $1 < p < \infty$. Kan [1978], [1983] has positive results if $T$ is a "Lamperti operator" (maps functions with disjoint supports to functions with disjoint supports). Sato [1974], and Fong-Lin [1976] consider the generalization of the Sucheston decomposition of $\Omega$ into the remaining and disappearing parts. Other partial results have been obtained by Assani [1984] and Emilion [1984a]. Assani and Mesiar [1984] studied the a.e.-convergence of $n^{-\alpha} T^n f$ for power bounded $T$.

7. Hachem [1982] proved the convergence a.e. and in $L_p$-norm of $n^{-1} F_n$, when $(F_n) \subset L_p$ is a superadditive process for a positive contraction $T$ in $L_p$, $(1 < p < \infty)$, and $\liminf \|n^{-1} \sum_{i=1}^{n} (F_i - TF_{i-1})\|_p < \infty$.

# Chapter 6: Pointwise ergodic theorems for multiparameter and amenable semigroups

Some multiparameter pointwise ergodic theorems can be deduced by a simple inductive argument from the 1-parameter case. This is done in the first section. Section 2 treats the principal pointwise ergodic theorems for multiparameter additive and subadditive processes, and section 3 the maximal ergodic theorem and the pointwise ergodic theorem for multiparameter semigroups of contractions in $L_1$ and integrable functions. The chapter ends with a discussion of ergodic theorems for amenable semigroups.

## § 6.1 Unrestricted convergence for averages over *d*-dimensional intervals

Let $d \geqq 1$ be a fixed integer and let $\mathbb{V} = \{0, 1, 2, \dots\}^d$ be the additive semigroup of $d$-dimensional vectors with non negative integer coordinates. For $u = (u_i)$, $v = (v_i) \in \mathbb{V}$ we write $u \leqq v$ if $u_i \leqq v_i$, $(i = 1, \dots, d)$, $u < v$ if $u_i < v_i$, $(i = 1, \dots, d)$, and $[u, v[ = \{w \in \mathbb{V} : u \leqq w < v\}$. If $A$ is a finite set card $(A)$ is the number of elements of $A$. For $n = (n_1, \dots, n_d) \in \mathbb{V}$ let

$$\pi(n) = \prod_{\nu=1}^{d} n_\nu = \operatorname{card}([0, n[)$$

where $0 = (0, 0, \dots, 0)$ is the neutral element in $\mathbb{V}$.

For $n \in \mathbb{V}$ and operators $T_1, \dots, T_d$ we shall use the notation

$$T_n = T_1^{n_1} T_2^{n_2} \dots T_d^{n_d}$$
$$S_n = \textstyle\sum_{u \in [0, n[} T_u$$
$$A_n = \pi(n)^{-1} S_n .$$

$n \to \infty$ shall mean that $n_\nu$ tends to infinity for $\nu = 1, \dots, d$.

If $A_n f$ converges a.e. as $n \to \infty$ we say that we have unrestricted a.e.-convergence. Unfortunately unrestricted a.e.-convergence does not always hold for $f \in L_1$ even if the $T_i$ are given by commuting measure preserving transformations $\tau_1, \dots, \tau_d$. In this section we study unrestricted convergence. In the subsequent sections we will obtain convergence for all $f \in \mathrm{L}_1$ under some restrictions, e.g. by taking averages only over squares.

**Theorem 1.1** (Zygmund-Fava). *Let $T_1, \ldots, T_d$ be $L_1 - L_\infty$-contractions in $L_1$ of a finite measure space. Then* $\lim_{n\to\infty} A_n h$ *exists a.e. for $h \in L \log^{d-1} L$, and it equals $A_\infty(T_1) A_\infty(T_2) \ldots A_\infty(T_d) h$, where $A_\infty(T_i) f = \lim_{n_i \to \infty} A_{n_i}(T_i) f$.*

*Proof.* We first assume that the $T_i$ are positive. The case $d = 1$ is theorem 1.7.3. For the induction put $f_k = A_k(T_{d+1})h$. For $h \in L \log^d L$, theorem 1.6.4 implies $\sup |f_k| \in L \log^{d-1} L$. Apply theorem 1.7.5 with $L = L \log^{d-1} L$, $\Lambda = \mathbb{V}$, $T_\lambda = A_n$, $\lambda = n = (n_1, \ldots, n_d)$, $k = n_{d+1}$. The assertion follows from $A_n f_k = A_{n_1}(T_1) \ldots A_{n_{d+1}}(T_{d+1}) h$.

In the general case let $\mathbf{T}_i$ be the linear modulus of $T_i$. It is again an $L_1 - L_\infty$-contraction, and, hence, an $L_2$-contraction. It follows that $T_i$ is an $L_2$-contraction. Now note that theorem 1.7.3 remains true for non positive $T$ if the maximal estimate is done with $|T|$. Also theorem 1.6.4 holds for non positive operators because of $|A_n(T)f| \leqq |A_n(|T|)|f||$. Similarly, the argument in theorem 1.7.5 goes through with $T_\lambda = A_n$ by using $A_{n_1}(\mathbf{T}_1) \ldots A_{n_d}(\mathbf{T}_d)$ instead of $T_\lambda$ in the proof. It follows that theorem 1.1 remains true with the same proof. □

**Theorem 1.2.** *Let $T_1, \ldots, T_d$ be positive contractions in $L_p$, $(1 < p < \infty)$. (The measure may be σ-finite now). $A_n h$ converges a.e. for $n \to \infty$ and $h \in L_p$, and* $\|\sup\{A_n|h| : n \in \mathbb{V}\}\|_p \leqq (p/p-1)^d \|h\|_p$.

*Proof.* Put $h_{d+1} = |h|$ and $h_i = \sup\{A_{n_i}(T_i) h_{i+1} : n_i \in \mathbb{N}\}$, $1 \leqq i \leqq d$. Akcoglu's dominated estimate implies $\|h_i\|_p \leqq (p/p-1)\|h_{i+1}\|_p$. Therefore, the second assertion follows from $\sup\{A_n|h| : n \in \mathbb{V}\} = h_1$. The convergence of $A_n h$ is Akcoglu's ergodic theorem for $d = 1$, and it follows for general $d$ by the inductive argument in the previous proof. □

Theorem 1.1 was proved by Zygmund [1951] for the case of measure preserving transformations (and continuous time) and by Fava [1972] for operators. See also Dunford [1951] for the case $d = 2$. The simple argument with theorem 1.7.5 is due to Sucheston [1983]. The multiparameter form of Akcoglu's theorem was proved by McGrath [1980a]; see also Olsen [1983].

The integrability condition $h \in L \log^{d-1} L$ is in a sense weakest possible for unrestricted a.e.-convergence: To see this consider a family $\{X_u, u \in \mathbb{V}\}$ of independent identically distributed random variables. In the product space representation of multiparameter stationary processes, similar to that in § 1.4, we can assume that $X_u$ is the $u$-th coordinate variable in $\Omega = \mathbb{R}^{\mathbb{V}}$. If $\tau_u$ is the shift defined by $X_v(\tau_u \omega) = X_{u+v}(\omega)$, we can write $X_u = X_0 \circ \tau_u$. We obtain $A_n X_0 = \pi(n)^{-1} \sum_{u \in [0, n[} X_u$. By theorem 1.1, $X_0 \in L \log^{d-1} L$ implies the convergence a.e. of $A_n X_0$. Smythe [1973] proved that, conversely, the unrestricted convergence a.e. of $\pi(n)^{-1} \sum_{u \in [0, n[} X_u$ implies $X_0 \in L \log^{d-1} L$ when the $X_u$ are i.i.d..

**Infinite $\mu$.** We now turn to the infinite measure space result analogous to theorem

1.1. (It will not be needed in the rest of the book). For infinite $\mu$, $L \log^{d-1} L$ is clearly not the proper class of functions for the proof of a.e. convergence of $A_n f$, because it contains all $f$ with $|f| \leqq 1$. Fava [1972] introduced, for $k = 0, 1, 2, \ldots$, the class $\mathfrak{R}_k$ of all measurable functions $f\colon \Omega \to \mathbb{R}$ for which

$$(1.1) \qquad \int_{\{|f|>t\}} \frac{|f|}{t} \left( \log \frac{|f|}{t} \right)^k d\mu$$

is finite for all $t > 0$. Clearly, for finite $\mu$, this class coincides with $L \log^k L$. For infinite $\mu$ and $k \geqq 1$, $\mathfrak{R}_k$ is a proper subclass of $L \log^k L$. Obviously $L_1$ is contained in $\mathfrak{R}_0$ and, for $1 < p < \infty$, $L_p$ is contained in each $\mathfrak{R}_k$. It is fairly easy to check that each $\mathfrak{R}_k$ is a vector space and that among the classes $\mathfrak{R}_k$ we have the inclusions

$$\mathfrak{R}_0 \supset \mathfrak{R}_1 \supset \mathfrak{R}_2 \supset \ldots .$$

We shall now present Fava's proof of unrestricted a.e. convergence for $f \in \mathfrak{R}_{d-1}$.

Let $L_1 + L_\infty$ be the space of all measurable functions $f\colon \Omega \to \mathbb{R}$, which can be written as a sum $f = f_1 + f_2$ with $f_1 \in L_1$, $f_2 \in L_\infty$. We know from § 1.6. that a positive $L_1$-contraction extends uniquely to the space of non negative measurable functions and in particular to $(L_1 + L_\infty)^+$. If $T$ is a positive $L_1 - L_\infty$-contraction this extension maps $(L_1 + L_\infty)^+$ into itself in a linear way and therefore uniquely determines an extension to $L_1 + L_\infty$ also denoted by $T$, and linear in $L_1 + L_\infty$.

**Definition 1.3.** An operator $M$ mapping $L_1 + L_\infty$ into the space of real valued measurable functions is called an *abstract maximal operator* if it has the following properties:

(i) $f \geqq 0 \Rightarrow Mf \geqq 0$ (Positivity);

(ii) $|M(f+g)| \leqq |Mf| + |Mg|$; $|M(\alpha f)| = |\alpha|\,|Mf|$ $(f, g \in L_1 + L_\infty, \alpha \in \mathbb{R})$ (Sublinearity);

(iii) $0 \leqq f \leqq g \Rightarrow Mf \leqq Mg$;

(iv) $\|Mf\|_\infty \leqq \|f\|_\infty$, $(f \in L_1 + L_\infty)$;

(v) $M$ is of weak type $(1,1)$, i.e., there exists $c > 0$ with $\lambda > 0$, $f \in L_1 \Rightarrow \mu(|Mf| > \lambda) \leqq c\lambda^{-1} \|f\|_1$.

By lemma 1.6.1 the operators $M'_N = M'_N(T)$ with $M'_N f = M_N(T)\,|f|$ are abstract maximal operators, $(N \in \mathbb{N} \cup \{\infty\})$, when $T$ is a positive $L_1 - L_\infty$-contraction.

We shall only deal with finitely many abstract maximal operators and may use the same constant $c > 0$ for them. For $f \in L_1 + L_\infty$ and $t > 0$ we shall use the convenient notation $f^t = f 1_{\{f>t\}}$, $f_t = f 1_{\{f \leqq t\}} = f - f^t$.

**Lemma 1.4.** *If $M$ is an abstract maximal operator, $f \in (L_1 + L_\infty)^+$ and $t > 0$, then*

$$\mu(Mf > 2t) \leqq \frac{c}{t} \int_{\{f>t\}} f \, d\mu .$$

*Proof.* We may assume $f^t \in L_1 \cdot f = f^t + f_t \leqq f^t + t$ implies

$$\mu(Mf > 2t) \leqq \mu(Mf^t > t) \leqq ct^{-1} \|f^t\|_1 . \quad \square$$

**Lemma 1.5.** *For measurable $f \geqq 0$ and $E \in \mathscr{A}$*

$$\int_E f d\mu = \int_0^\infty \mu(E \cap \{f > t\}) dt .$$

*Proof.* Put $g = f 1_E$ and $G = \{(\omega, t) : t < g(\omega)\}$. Then

$$\int_\Omega g d\mu = \int_\Omega \int_0^\infty 1_G(\omega, t) dt \, \mu(d\omega) = \int_0^\infty \int_\Omega 1_G(\omega, t) \, \mu(d\omega) dt . \quad \square$$

The main step is the following:

**Theorem 1.6.** *Let $M_1, M_2, \ldots, M_k$ be abstract maximal operators.*

*(i) For any $f \in \mathfrak{R}_k$, $(k \geqq 1)$, the functions $M_j \cdot M_{j-1} \ldots M_1 f$, $(1 \leqq j \leqq k)$ belong to $L_1 + L_\infty$. In particular $M_k M_{k-1} \ldots M_1 f$ is well defined for $f \in \mathfrak{R}_{k-1}$;*

*(ii) There exists a constant $C_k > 0$ such that*

$$\mu(M_k \ldots M_1 f > 4t) \leqq C_k \int_{\{f > t\}} \frac{f}{t} \left( \log \frac{f}{t} \right)^{k-1} d\mu \tag{1.2}$$

*holds for all non negative $f \in \mathfrak{R}_{k-1}$ and all $t > 0$.*

*Proof.* The proof is by induction on $k$. Because of $L_1 + L_\infty \supset \mathfrak{R}_0 \supset \mathfrak{R}_1 \supset \ldots$, $M_1 f$ is always well defined.

Note that a measurable function belongs to $L_1 + L_\infty$ if and only if it is integrable over every set of finite measure.

For $k = 1$, (1.2) follows from Lemma 1.4. To prove (i) for $k = 1$ take some $f \in \mathfrak{R}_1^+$ and some set $E$ of finite measure. Then

$$\begin{aligned}
\int_E M_1 f d\mu &= \int_0^\infty \mu(E \cap \{M_1 f > t\}) dt \leqq 4\mu(E) + \int_4^\infty \mu(M_1 f > t) dt \\
&\leqq 4\mu(E) + \int_1^\infty \mu(M_1 f > 4s) 4 ds \leqq 4\mu(E) + 4c \int_1^\infty \int_{\{f > s\}} s^{-1} f d\mu ds \\
&\leqq 4\mu(E) + 4c \int_\Omega f 1_{\{f \geqq 1\}} \int_1^f s^{-1} ds d\mu \\
&= 4\mu(E) + 4c \int f (\log^+ f) d\mu < \infty .
\end{aligned}$$

As $E$ was arbitrary with $\mu(E) < \infty$ we see that $M_1 f$ belongs to $L_1 + L_\infty$.

Now assume that the theorem has been proved for $k$ operators and consider $k + 1$ abstract maximal operators $M_1, \ldots, M_{k+1}$. We first prove (ii) with $k$ replaced by

$k+1$. Take $f \in \mathfrak{R}_k^+$. By (i) of the induction hypothesis $M_{k+1} M_k \dots M_1 f$ is well defined. Using

$$M_{k+1} \dots M_1 f \leqq M_{k+1} \dots M_1 f^{2t} + 2t$$

and Lemma 1.4 we obtain

$$\begin{aligned} \mu(M_{k+1} \dots M_1 f > 4t) &\leqq \mu(M_{k+1} \dots M_1 f^{2t} > 2t) \\ &\leqq \frac{c}{t} \int_{\{M_k \dots M_1 f^{2t} > t\}} M_k \dots M_1 f^{2t} \, d\mu \\ &= c \int_{\{M_k \dots M_1 g > 1\}} M_k \dots M_1 g \, d\mu \quad \text{with } g = t^{-1} f^{2t}. \end{aligned}$$

Put $\varrho(s) = \mu(M_k \dots M_1 g > s)$. By Lemma 1.5 the last integral equals

$$(1.3) \qquad \int_0^\infty \mu(\{M_k \dots M_1 g > 1\} \cap \{M_k \dots M_1 g > s\}) \, ds = \varrho(1) + \int_1^\infty \varrho(s) \, ds.$$

As $\mathfrak{R}_k$ is contained in $\mathfrak{R}_{k-1}$, $g$ belongs to $\mathfrak{R}_{k-1}$ and the induction hypothesis gives us

$$(1.4) \qquad \varrho(s) \leqq 4 C_k \int_{\{4g > s\}} \frac{g}{s} \left( \log \frac{4g}{s} \right)^{k-1} d\mu.$$

This can now be used to estimate the right hand side of (1.3). We obtain

$$\begin{aligned} (1.5) \qquad \int_1^\infty \varrho(s) \, ds &\leqq \int_1^\infty [4 C_k s^{-1} \int_{\{4g > s\}} g \left( \log \frac{4g}{s} \right)^{k-1} d\mu] \, ds \\ &= 4 C_k \int_\Omega [g \int_1^{4g} \left( \log \frac{4g}{s} \right)^{k-1} s^{-1} \, ds] \, d\mu \\ &= 4 C_k k^{-1} \int_\Omega g (\log^+ 4g)^k \, d\mu. \end{aligned}$$

Applying the elementary inequalities $\log^+ ab \leqq \log^+ a + \log^+ b$ and $(a+b)^k \leqq 2^k (a^k + b^k)$ for $a, b \geqq 0$ we see that there are constants $A_k$, $B_k > 0$ such that the last term is

$$\begin{aligned} &\leqq A_k \int g \, d\mu + B_k \int g (\log^+ g)^k \, d\mu \\ &= A_k \int_{\{f > 2t\}} \frac{f}{t} \, d\mu + B_k \int_{\{f > 2t\}} \frac{f}{t} \left( \log^+ \frac{f}{t} \right)^k d\mu \\ &\leqq (A_k + B_k) \int_{\{f > t\}} \frac{f}{t} \left( \log^+ \frac{f}{t} \right)^k d\mu. \end{aligned}$$

Similarly (1.4) yields

$$\varrho(1) \leqq 4C_k \int_{\{4g>1\}} g(\log 4g)^{k-1}\,d\mu = 4C_k \int_{\Omega} g(\log^+ 4g)^{k-1}\,d\mu$$
$$\leqq A'_k \int g\,d\mu + B'_k \int g(\log^+ g)^{k-1}\,d\mu$$
$$\leqq (A'_k + B'_k) \int_{\{f>t\}} \frac{f}{t}\left(\log\frac{f}{t}\right)^k d\mu.$$

Combining our estimates we get (1.2) with $k$ replaced by $k+1$ und with $C_{k+1} = c(A_k + B_k + A'_k + B'_k)$.

To prove (i) with $k+1$ operators consider $f \in \mathfrak{R}^+_{k+1}$ and a set $E$ of finite measure. By Lemma 1.5 and by what we have proved above

$$\int_E M_{k+1} \ldots M_1 f\,d\mu = \int_0^\infty \mu(E \cap \{M_{k+1} \ldots M_1 f > t\})\,dt$$
$$\leqq 4\mu(E) + \int_4^\infty \mu\{M_{k+1} \ldots M_1 f > t\}\,dt$$
$$\leqq 4\mu(E) + C_{k+1} \int_4^\infty \Big[\int_{\{4f>t\}} \frac{4f}{t}\left(\log\frac{4f}{t}\right)^k d\mu\Big]\,dt.$$

The finiteness of this expression follows from $f \in \mathfrak{R}_{k+1} \subset L \log^{k+1} L$ by the same computation as in (1.5). □

For $n = (n_1, n_2, \ldots, n_d)$ put $M^d_n f = \sup\{A_u f : u \leqq n\}$, and let $e = (1, 1, \ldots, 1)$. Then theorem 1.6 and the estimate $M^d_n |f| \leqq M_{n_1}(T_1)\, M_{n_2}(T_2) \ldots M_{n_d}(T_d)\,|f|$ yield

**Corollary 1.7.** *Let $T_1, \ldots, T_d$ be positive $L_1 - L_\infty$-contractions. There exists a constant $C_d > 0$ such that*

$$\mu(M^d_{\infty e}|f| > 4t) \leq C_d \int_{\{|f|>t\}} \frac{|f|}{t}\left(\log\frac{|f|}{t}\right)^{d-1} d\mu$$

*holds for all $t > 0$ and all $f \in \mathfrak{R}_{d-1}$.*

**Theorem 1.8** (Zygmund-Fava). *Let $T_1, \ldots, T_d$ be positive $L_1 - L_\infty$-contractions, then $A_n f$ converges a.e. as $n \to \infty$ for all $f \in \mathfrak{R}_{d-1}$.*

*Proof.* We may assume $f \geqq 0$. Put

$$\Delta(g) = \limsup_{n\to\infty} A_n g - \liminf_{n\to\infty} A_n g.$$

By theorem 1.2 $\Delta(g)$ is 0 for $g \in L_p$, $(1 < p < \infty)$. Let $f_k$, $(k \geqq 1)$ be an increasing sequence in $L^+_p$ with $f_k \to f$ a.e.. Then

$$\Delta(f) \leqq \Delta(f - f_k) + \Delta(f_k) = \Delta(f - f_k) \leqq M^d_{\infty e}(f - f_k).$$

For any fixed $t > 0$,

$$\int_{\{f-f_k>t\}} \frac{f-f_k}{t}\left(\log\frac{f-f_k}{t}\right)^{d-1} d\mu$$

tends to 0 as $k\to\infty$, since the integrands tend to 0 a.e. and are bounded by the integrable function $1_{\{f>t\}}t^{-1}f(\log(f/t))^{d-1}$. Therefore Corollary 1.7 implies $\Delta(f)=0$. □

### Notes

For the continuous parameter form of the Zygmund-Fava theorem the reader is refered to Fava's paper and to Sato [1981a]. Sato also has further related estimates for maximal operators.

Yoshimoto [1979] derived dominated estimates for vector valued functions in $\mathfrak{R}_k$. Yoshimoto [1982] also studied similar estimates for $L_p-L_\infty$-contractions.

Converse maximal estimates were given by Dang-Ngọc [1980] for groups and by Sato [1983a] for semigroups.

## § 6.2 Multiparameter additive and subadditive processes

**1. Processes indexed by intervals.** We now consider $d \geqq 1$ commuting endomorphisms $\tau_1, \tau_2, \ldots, \tau_d$ of a measure space $(\Omega, \mathscr{A}, \mu)$. We continue to use the notation given in the beginning of the previous section, but now $T_i$ will always be the operator $f\to T_i f = f\circ\tau_i$ in $L_1$. We shall prove the a.e.-convergence of $A_n f = \pi(n)^{-1} S_n f$ for all $f\in L_1$, when $n$ tends to infinity along an increasing sequence in $\mathbb{V}=\{0,1,2,\ldots\}^d$. More generally, we shall obtain an analogous result when the additive process $S_n f$ is replaced by a subadditive process.

The commuting endomorphisms $\tau_i$ define a semigroup $\tau=\{\tau_u\colon u\in\mathbb{V}\}=\{\tau_u\}$ by $\tau_u=\tau_1^{u_1}\tau_2^{u_2}\ldots\tau_d^{u_d}$, where $u=(u_1,\ldots,u_d)$. Let $\mathscr{J}$ be the set of non empty intervals $[u,v[$, $(u,v\in\mathbb{V})$. For $I=[u,v[$, $I+w$ is the interval $[u+w, v+w[$. Put $e=(1,1,\ldots,1)$.

**Definition 2.1.** A *superadditive process* (with respect to $\tau$) is a set function $F\colon \mathscr{J}\ni I\to F_I\in L_1$ with the following properties:

(2.1) $F_I\circ\tau_u = F_{I+u}$ whenever $I\in\mathscr{J}$ and $u\in\mathbb{V}$;

(2.2) *if* $I_1,\ldots,I_n$ *are disjoint sets in* $\mathscr{J}$ *and if* $I=\bigcup_{i=1}^n I_i$ *is also in* $\mathscr{J}$, *then*

$$F_I \geqq \sum_{i=1}^n F_{I_i},$$

(2.3) $\gamma(F) := \sup\left\{\frac{1}{\operatorname{card}(I)}\int F_I\,d\mu\colon I\in\mathscr{J}\right\} < \infty.$

We may write $F$ in the form $F = \{F_I\} = \{F_I: I \in \mathscr{J}\}$. If we want to emphasize that $\tau$ is a discrete semigroup we call $F$ a discrete superadditive process. $\gamma(F)$ is called the *spatial constant* of $F$.

If $-F$ is superadditive then $F$ is called *subadditive*. In the definition of the spatial constant one must then write inf instead of sup. If both $F$ and $-F$ are superadditive then $F$ is called *additive*. As the semigroup $\tau$ was assumed discrete, the additive processes are just those of the form $F_I = \sum_{u \in I} T_u f$ with an $f \in L_1$.

Let us call $F$ *2-superadditive* if (2.1) and (2.3) hold and (2.2) holds for $n = 2$. For $d = 1$ superadditivity and 2-superadditivity coincide.

As in the case $d = 1$, the definition of a superadditive process can be given in an equivalent way without using endomorphisms. (2.1) is then replaced by the condition that for any finite collection $(I_1, \ldots, I_k)$ in $\mathscr{J}$ and for any $u \in \mathbb{V}$ the joint distribution of $(F_{I_1}, \ldots, F_{I_k})$ is the same as that of $(F_{I_1+u}, \ldots, F_{I_k+u})$. We shall use the notation $\bar{F}_I = \operatorname{card}(I)^{-1} F_I$.

Let us look at some examples of two-dimensional subadditive processes:

(a) Assume that random straight lines are given in the plane in a stationary way. E.g. their perpendicular distances from the origin are the points of a homogeneous Poisson process on $\mathbb{R}^+$ and their directions are uniformly and independently distributed in $[0, 2\pi[$. For $u, v \in \mathbb{V}$ let $R(u, v) = \{t \in \mathbb{R}^2: u \leqq t < v\}$. (As $[u, v[$ contains only points from $\mathbb{V}$ it differs from $R(u, v)$.) For $I = [u, v[$ let $F_I$ be the number of lines intersecting $R(u, v)$.

(b) (*Cluster process* of Grimmett [1976]). Let $X_u$, $(u \in \mathbb{V})$, be independent random variables taking the value 1 with probability $p$, $(0 < p < 1)$. We connect two elements $u, v \in \mathbb{V}$ with Euclidian distance 1 if $X_u = X_v = 1$. Thus, within each $I \in \mathscr{J}$ we have a random graph. Let $F_I$ be the number of connected components of this graph. $F_I$ is called the number of clusters in $I$.

The papers of Hammersley-Welsh [1965], Hammersley [1974], Smythe [1976], and Nguyen [1979] contain other interesting examples.

A real valued set function $g: \mathscr{J} \to \mathbb{R}$ is called 2-superadditive if $g(A) + g(B) \leqq g(A \cup B)$ holds for any disjoint $A, B \in \mathscr{J}$ with $A \cup B \in \mathscr{J}$, and $g(A) = g(A + u)$ holds for any $A \in \mathscr{J}$ and $u \in \mathbb{V}$. The proof of the following Lemma is an easy exercise similar to that of Lemma 1.5.1.

**Lemma 2.2.** *For any 2-superadditive* $g$, $\pi(n)^{-1} g([0, n[)$ *converges to* $\sup\{\operatorname{card}(I)^{-1} g(I): I \in \mathscr{J}\} \in \mathbb{R} \cup \{\infty\}$ *for* $n \to \infty$.

Clearly the set function $g$ defined by $g(I) = \int F_I \, d\mu$ is 2-superadditive if $F$ is 2-superadditive, and (2.3) then guarantees that the limit is finite. We see that the spatial constant both of 2-superadditive and of 2-subadditive processes can be written as

$$(2.4) \qquad \gamma(F) = \lim_{n \to \infty} \pi(n)^{-1} \int F_{[0, n[} \, d\mu = \lim_{n \to \infty} \int \bar{F}_{[0, n[} \, d\mu .$$

Let $\mathscr{I}$ denote the $\sigma$-algebra of all sets $A \in \mathscr{A}$, which are invariant under all $\tau_i$, $(i = 1, \ldots, d)$. As usual the *ergodicity* of $\tau$ means that $\mathscr{I}$ is trivial.

As its proof is similar to the 1-parameter case we only state the mean ergodic theorem for $F$ due to Smythe [1976]:

**Theorem 2.3.** *If $F$ is a 2-subadditive process in a finite measure space, then $\bar{F}_{[0,n[}$ converges in $L_1$-norm as $n \to \infty$ to an $\mathscr{I}$-measurable $\bar{f}$. For any $A \in \mathscr{I}$,*

$$\int_A \bar{f} d\mu = \lim_{n \to \infty} \int_A \bar{F}_{[0,n[} d\mu .$$

*In particular, for ergodic $\tau$, $\bar{f} = \mu(\Omega)^{-1} \gamma(F)$.*

We now turn to the question of a.e.-convergence. The first multiparameter pointwise ergodic theorems go back to Wiener [1939], who considered averages of functions $f \circ \tau_u$ with $u$ ranging over spheres in $\mathbb{R}^d$. His ideas will play an important role in what follows.

We shall prove that $\bar{F}_{I_r}$ converges a.e. for any subadditive process if $I_r$ tends to $\mathbb{V}$ in a regular way. This was shown by Akcoglu-Krengel [1981] and we follow their arguments.

**Definition 2.4.** A family $\{I_r\}_{r \in \mathbb{N}}$ (or a finite such family) is called *regular* (*with the constant* $C < \infty$) if there exists an increasing sequence $I'_1 \subset I'_2 \subset$ in $\mathscr{J}$ with $I_r \subset I'_r$ for all $r$ such that card $(I'_r) \leqq C$ card $(I_r)$ holds for all $r$.

If $\{I'_r\}$ can be chosen in such a way that the union of the sets $I'_r$ is $\mathbb{V}$, we shall write $\lim_{r \to \infty} I_r = \mathbb{V}$.

Obviously any increasing sequence of sets in $\mathscr{J}$ is regular with the constant 1. The notion of a regular sequence in $\mathscr{J}$ also allows to treat another interesting example: We say that a sequence $n(k)$ *remains in a sector* of $\mathbb{V}$ if there is a finite $C_0$ such that the ratios $n_i(k)/n_j(k)$ are bounded by $C_0$ for $1 \leqq i, j \leqq d$ and all $k$. If $n(1), n(2), \ldots$ is a sequence in $\mathbb{V}$ remaining in a sector and tending to infinity we may find a permutation $m(1), m(2), \ldots$ of it for which Max $\{m_i(k): 1 \leqq i \leqq d\}$ is increasing. The sequence $[0, m(r)[$ in $\mathscr{J}$ is regular with constant $C_0^d$. Therefore the a.e.-convergence of $\bar{F}_{I_r}$ for regular sequences tending to $\mathbb{V}$ will imply that $\bar{F}_{[0,n(k)[}$ converges a.e. for sequences $n(k) \to \infty$ remaining in a sector of $\mathbb{V}$. The definition of a regular sequence $I_1, I_2, \ldots$ also allows that the lower corner of the intervals tends to infinity.

In the next lemma we write $-C_1 + C_2$ for $\{v: \exists w \in C_1 \text{ with } w + v \in C_2\}$.

**Lemma 2.5.** *Let $I'_1 \subset I'_2 \subset \ldots I'_N$ be finitely many nested sets in $\mathscr{J}$. Let $A$ be a finite subset of $\mathbb{V}$ and let $k: u \to k(u)$ be a map of $A$ into $\{1, \ldots, N\}$. Then there exists a finite subset $A' \subset A$ for which the sets $D_u := I'_{k(u)} + u$, $(u \in A')$, are disjoint, and for which* $A$ is contained in the union of the sets $\tilde{D}_u := -\mathrm{I}'_{k(u)} + (\mathrm{I}'_{k(u)} + \mathrm{u})$, $(u \in A')$. *In particular,* card $(A) \leqq 2^d \sum_{u \in A'}$ card $(I'_{k(u)})$.

*Proof*. Let $\mathcal{M}_N$ be a maximal collection of disjoint sets of the form $I'_N + u$ with $k(u) = N$. When $\mathcal{M}_N, \ldots, \mathcal{M}_i$ have been constructed for some $i > 1$, let $\mathcal{M}_{i-1}$ be a maximal collection of sets of the form $D_u = I'_{i-1} + u$ with $k(u) = i - 1$, for which all sets in $\mathcal{M}_{i-1} \cup \ldots \cup \mathcal{M}_N$ are disjoint. Let $A'$ be the set of all $u \in A$ with $D_u \in \mathcal{M}_1 \cup \ldots \cup \mathcal{M}_N$. Take any $v \in A$. By the maximality of $\mathcal{M}_{k(v)}$ there exists some $j \geqq k(v)$ and $D_u \in \mathcal{M}_j$ with $D_v \cap D_u \neq \emptyset$. Then $k(u) \geqq k(v)$ and $I'_{k(u)} \supset I'_{k(v)}$. Now $(I'_{k(u)} + v) \cap (I'_{k(u)} + u) \neq \emptyset$ yields $v \in \tilde{D}_u$. The last assertion follows from the inequality $\operatorname{card}(-I_k + (I_k + u)) \leqq 2^d \operatorname{card}(I_k)$. □

**Theorem 2.6.** *Let $\{I_1, I_2, \ldots, I_N\}$ be a finite regular family in $\mathcal{J}$ with constant $C$. Let $B_1$, $B_2$ be other sets in $\mathcal{J}$ such that $B_2$ contains all sets $B_1 + I_k = \{v + w \in \mathbb{V}: v \in B_1, w \in I_k\}$, $(k = 1, \ldots, N)$. Let $F = \{F_I\}$ be a non negative superadditive process and define*

$$E = \{\omega \in \Omega: \max_{1 \leqq k \leqq N} \bar{F}_{I_k}(\omega) \geqq \alpha\}$$

*with $\alpha > 0$. Then*

$$\mu(E) - \mu(\Omega \setminus D) \leqq \frac{2^d C}{\alpha \operatorname{card}(B_1)} \int_D F_{B_2} \, d\mu \tag{2.5}$$

*holds for any $D \in \mathcal{A}$.*

*Proof*. For each $\omega \in \Omega$ put $A(\omega) = \{u \in B_1: \tau_u \omega \in E\}$. For any $D \in \mathcal{A}$

$$\int_D \operatorname{card}(A(\omega))\,\mu(d\omega) = \int_D \sum_{u \in B_1} 1_{\{\omega:\, \tau_u \omega \in E\}}\, d\mu = \sum_{u \in B_1} \mu(D \cap \tau_u^{-1} E) \tag{2.6}$$

$$\geqq \sum_{u \in B_1} [\mu(\tau_u^{-1} E) - \mu(\Omega \setminus D)] = \operatorname{card}(B_1)\,[\mu(E) - \mu(\Omega \setminus D)].$$

For each $u \in A(\omega)$ there is an integer $k(u)$, $1 \leqq k(u) \leqq N$, with

$$\mathrm{F}_{I_{k(u)}}(\tau_u \omega) = F_{u + I_{k(u)}}(\omega) \geqq \alpha \operatorname{card}(I_{k(u)}). \tag{2.7}$$

As the family $\{I_1, I_2, \ldots, I_N\}$ is regular there exists a nested family $I'_1 \subset I'_2 \subset \ldots \subset I'_N$ in $\mathcal{J}$ with $I_k \subset I'_k$ and $\operatorname{card}(I'_k) \leqq C \operatorname{card}(I_k)$. We may apply Lemma 2.5 to the set $A = A(\omega)$ and find finitely many elements $u^1, u^2, \ldots, u^l$ in $A(\omega)$ such that the sets $u^i + I'_{k(u^i)}$ are disjoint, and such that $2^d \sum_{i=1}^{l} \operatorname{card}(I'_{k(u^i)}) \geqq \operatorname{card}(A(\omega))$. Then the sets $u^i + I_{k(u^i)}$, $(i = 1, \ldots, l)$ are also disjoint and we have

$$2^d C \sum_{i=1}^{l} \operatorname{card}(I_{k(u^i)}) \geqq \operatorname{card}(A(\omega)).$$

As all these sets are contained in $B_2$ the nonnegativity and superadditivity of the process together with (2.7) imply

$$F_{B_2}(\omega) \geqq \sum_{i=1}^{l} F_{u^i + I_{k(u^i)}}(\omega) \geqq$$

$$\geqq \alpha \sum_{i=1}^{l} \operatorname{card}(I_{k(u^i)}) \geqq \frac{\alpha}{2^d C} \operatorname{card}(A(\omega)).$$

Integrating this inequality over $D$ and using (2.6) we obtain (2.5). □

Here we need only

**Corollary 2.7.** *Let $\{I_1, I_2, \dots\}$ be a finite or infinite regular family in $\mathscr{J}$ with constant $C$, and let $F = \{F_I\}$ be a non negative superadditive process. For any $\alpha > 0$ the measure of the set*

$$E = \{\omega \in \Omega: \sup_k \bar{F}_{I_k}(\omega) \geqq \alpha\}$$

satisfies

$$\mu(E) \leqq \frac{2^d C \gamma(F)}{\alpha}.$$

*Proof.* We may assume that the sequence $I_1, I_2, \dots$ is finite. If $B_2$ is the smallest element of $\mathscr{J}$ containing all sets $B_1 + I_k$, $(k = 1, \dots, N)$ and if $B_1$ in $\mathscr{J}$ is sufficiently large, then the ratio $\operatorname{card}(B_2)/\operatorname{card}(B_1)$ is arbitrarily close to 1. Taking $D = \Omega$, the estimate

$$\mu(E) \leqq \frac{2^d C}{\alpha \operatorname{card}(B_1)} \int_\Omega F_{B_2} d\mu \leqq \frac{2^d C}{\alpha \operatorname{card}(B_1)} \operatorname{card}(B_2)\gamma(F)$$

therefore yields the desired maximal inequality. □

To prove the ergodic theorem we must first study the additive case.

$$S_I f = \sum_{u \in I} f \circ \tau_u, \quad (I \in \mathscr{J})$$

is an additive process for integrable $f$.

**Theorem 2.8** (Tempel'man [1972]). *If $\tau_1, \tau_2, \dots, \tau_d$ are commuting endomorphisms of a measure space $(\Omega, \mathscr{A}, \mu)$ and if $I_1, I_2, \dots$ is a regular family in $\mathscr{J}$ tending to $\mathbb{V}$, then the sequence $A(r,f) = \operatorname{card}(I_r)^{-1} S_{I_r} f$ converges a.e. for all $f \in L_p$, $(1 \leqq p < \infty)$.*

*Proof.* Put

$$\Delta(f) = \limsup_{r \to \infty} A(r,f) - \liminf_{r \to \infty} A(r,f).$$

Lemma 1.1.3 implies that any $f \in L_p$ may be written as a sum $f = \sum_{j=1}^{d+3} f_j$ with functions $f_j$ having the following properties: For $1 \leqq j \leqq d$ there exists a $g_j \in L_\infty$ with $f_j = g_j - g_j \circ \tau_j$; $f_{d+1}$ belongs to $L_2$ and is $\tau_j$-invariant for $j = 1, \ldots, d$; $\|f_{d+2}\|_\infty < \varepsilon$, and $\|f_{d+3}\|_1 < \varepsilon^2$.

If $e^j$ is the $j$-th unit vector in $\mathbb{V}$ the sequence

$$\eta_{r,j} = \text{card}\,(I_r)^{-1}\,[\text{card}\,(I_r \Delta(I_r + e^j))]$$

tends to 0 for all $j$ since $I_r$ tends to $\mathbb{V}$. Therefore $\Delta(f_j) = 0$, $(j = 1, \ldots, d)$ is a consequence of the estimate $|A(r, f_j)| \leqq 2\|g_j\|_\infty \eta_{r,j}$. $\Delta(f_{d+1}) = 0$ is obvious. It is also clear that $\Delta(f_{d+2})$ is bounded by $2\|f_{d+2}\|_\infty < 2\varepsilon$.

Because of $\int S_I |f_{d+3}|\,d\mu = \text{card}\,(I)\|f_{d+3}\|_1$, the spatial constant of the additive process $\{S_I|f_{d+3}|\}$ is $\gamma(\{S_I|f_{d+3}|\}) = \|f_{d+3}\|_1$. By $\Delta(f_{d+3}) \leqq 2\sup_r A(r, |f_{d+3}|)$ the set $\{\Delta(f_{d+3}) > \varepsilon\}$ is contained in $\{\sup_r A(r, |f_{d+3}|) > \varepsilon/2\}$. Corollary 2.7 therefore gives us

$$\mu(\Delta(f_{d+3}) > \varepsilon) \leqq \frac{2^d C 2}{\varepsilon} \varepsilon^2 .$$

As $\varepsilon > 0$ was arbitrary $\Delta(f) \leqq \sum_{j=1}^{d+3} \Delta(f_j)$ yields $\Delta(f) = 0$. □

**Remark.** The limit $\hat{f}$ of the sequence $A(r, f)$ does not depend on the sequence $\{I_r\}$. We may identify it as follows, leaving the details as an exercise: We may assume that $\mu$ is $\sigma$-finite. By an exhaustion argument we can find a maximal set $\tilde{C}$ such that there exists a finite measure $\nu \ll \mu$ invariant under all $\tau_j$ with strictly positive density $h = d\nu/d\mu$ on $\tilde{C}$. The argument given above shows that $\hat{f}$ is arbitrarily close to a function which is invariant under all $\tau_j$ and so $\hat{f}$ must be invariant, too. If $f$ is in $L_1^+$, an application of Fatou's lemma shows $\|\hat{f}\|_1 \leqq \|f\|_1$. Thus, $\{\hat{f} > 0\}$ must be contained in $\tilde{C}$. As any $f \in L_p$ may be written as $f = f_\varepsilon + f'_\varepsilon$ with $|f'_\varepsilon| < \varepsilon$ and $f_\varepsilon \in L_1$ se see that $\hat{f}$ vanishes in $\tilde{C}^c$ for all $f \in L_p$, $(1 \leqq p < \infty)$.

As the sets $\{h \geqq 1\}$, $\{1 > h \geqq 1/2\}$, $\{1/2 > h \geqq 1/3\}, \ldots$ belong to $\mathscr{I}$ and partition $\tilde{C}$, we may determine $\hat{f}$ on these sets separately. Restricted to such a set $\mu$ is finite and $f$ integrable and by the usual $L_1$-convergence argument $\hat{f}$ is the unique $\mathscr{I}$-measurable function with $\int_A \hat{f}\,d\mu = \int_A f\,d\mu$ for all $A \in \mathscr{I}$.

We are now ready to complete the proof of the following multiparameter subadditive ergodic theorem of Akcoglu-Krengel:

**Theorem 2.9.** *If $F = \{F_I : I \in \mathscr{J}\}$ is a subadditive process and $I_1, I_2, \ldots$ a regular family in $\mathscr{J}$ tending to $\mathbb{V}$, then the sequence $\bar{F}_{I_r}$ converges a.e. as $r \to \infty$.*

*Proof.* The theorem is stated in the subadditive case because that seems to be the most frequent case of application. But for the proof we assume (passing to $-F$) that $F$ is superadditive.

The process $\{S_I F_{[0,e[}\}$ is additive and the process $\{F_I - S_I F_{[0,e[}\}$ is non negative and superadditive. As Tempelman's theorem takes care of the additive case, we may henceforth assume that $F$ is non negative and superadditive.

Let $I_r$ be the interval $[v(r), w(r)[$ and $I'_r = [v'(r), w'(r)[$. Replacing $C$ by a bigger constant (also denoted by $C$) we may enlarge the sets $I'_r$ and assume $v'(r) = 0$, and that for any $m \in \mathbb{N}$ there exists an $L(m)$ such that for $r \geqq L(m)$ all coordinates of $w'(r)$ are divisible by $m$.

Let $\mathbb{V}^{(m)} = m\mathbb{V} = \{(mi_1, \ldots, mi_d): (i_1, \ldots, i_d) \in \mathbb{V}\}$ and let $\mathscr{I}^{(m)}$ be the set of non empty intervals $[u, v[$ with $u, v \in \mathbb{V}^{(m)}$.

For $r \geqq L(m)$ let $I_{r,m}$ be the largest element of $\mathscr{I}^{(m)}$ contained in $I_r$ and let $J_{r,m}$ be the smallest element of $\mathscr{I}^{(m)}$ containing $I_r$. (If $L(m)$ is large enough, $I_r$ contains at least one non empty $\mathscr{I}^{(m)}$-interval.) As the length of the sides of $I_r$ tends to infinity the ratio card $(I_{r,m})/\text{card}\,(J_{r,m})$ tends to 1. The sequence $\ldots I_{r,m}, J_{r,m}, I_{r+1,m}, J_{r+1,m}, \ldots, (r \geqq L(m))$ is regular, as may be seen by comparing these intervals with $\ldots I'_r, I'_r, I'_{r+1}, I'_{r+1}, \ldots$. We call it $\zeta_m$.

Let $\underline{f}$ and $\bar{f}$ denote the lim inf and lim sup of the sequence $\bar{F}_{I_r}$, and let $\underline{f}_m$ and $\bar{f}_m$ be the lim inf and lim sup of $\bar{F}_I$ as $I$ ranges through $\zeta_m$. Comparing $F_{I_r}$ with $F_{J_{r,m}}$ and using card $(I_r)/\text{card}\,(J_{r,m}) \to 1$ we obtain $\bar{f} \leqq \bar{f}_m$. Similarly, a comparison with $F_{I_{r,m}}$ implies $\underline{f} \geqq \underline{f}_m$. The finiteness of $\bar{f}_m$ and hence that of $\bar{f}$, $\underline{f}$ and $\underline{f}_m$ follows from Corollary 2.7 when we apply it to $\zeta_m$ and let $\alpha$ tend to infinity.

As the set $\{\omega: \bar{f}(\omega) - \underline{f}(\omega) > \alpha\}$ is contained in the sets

$$E_{m,\alpha} = \{\omega: \bar{f}_m(\omega) - \underline{f}_m(\omega) > \alpha\}$$

it is now enough to show that $\mu(E_{m,\alpha})$ is small for fixed $\alpha > 0$ when $m$ is large.

Let $\varepsilon > 0$ be a given number. Lemma 2.2 shows that there is an integer $m$ with $\int \bar{F}_{[0,me[}\, d\mu > \gamma(F) - \varepsilon$.

We may define an additive process on $\mathscr{I}^{(m)}$ by

$$H_I^m = \sum_{u \in I \cap \mathbb{V}^{(m)}} F_{[0,me[} \circ \tau_u, \quad (I \in \mathscr{I}^{(m)}).$$

Subtracting this process from the restriction of $F$ to $\mathscr{I}^{(m)}$ we get a non negative superadditive process $F^m = \{F_I^m\}$ on $\mathscr{I}^{(m)}$. As the spatial constant of $\{H_I^m\}$ equals $\int \bar{F}_{[0,me[}\, d\mu$ the choice of $m$ implies $\gamma(F^m) < \varepsilon$.

As the sequence $\zeta_m$ is regular in $\mathscr{I}^{(m)}$ and $\{H_I^m\}$ additive, the sequence $\bar{H}_I^m$ converges a.e. as $I$ ranges through $\zeta_m$. But then $\bar{f}_m - \underline{f}_m$ must be bounded by $\sup\{\bar{F}_I^m: I \in \zeta_m\}$. Now corollary 2.7 yields

$$\mu(E_{m,\alpha}) \leqq \frac{2^d C \gamma(F^m)}{\alpha} \leqq \frac{2^d C \varepsilon}{\alpha}.$$

As $\varepsilon > 0$ was arbitrary the proof is complete. □

**2. Extensions to convex sets and to continuous parameters.** We now sketch some generalizations of results given above. The intervals will be replaced by convex sets.

The case of norm convergence is quite simple for additive processes. Let $\mathscr{B}^b$ denote the family of bounded Borel sets $A$ with positive Lebesgue measure $\lambda(A)$. Let $\mathscr{T} = \{T_u: u \in \mathbb{R}^{+d}\}$ be a bounded semigroup of linear operators in a Banach space $\mathfrak{X}$. An *additive process for* $\mathscr{T}$ *and* $\mathscr{B}^b$ is a map $F: \mathscr{B}^b \ni A \to F_A \in \mathfrak{X}$ with the properties:

(2.8) for disjoint $A_1, A_2 \in \mathscr{B}^b$, $F_{A_1 \cup A_2} = F_{A_1} + F_{A_2}$;

(2.9) $T_u F_A = F_{A+u}$, $(u \in \mathbb{R}^{+d}, A \in \mathscr{B}^b)$;

(2.10) there exists a constant $\gamma_0 < \infty$ with $\|F_A\| \leqq \gamma_0 \lambda(A)$; $(A \in \mathscr{B}^b)$.

In $\mathbb{V}$, intervals consisted only of lattice points, but, in $\mathbb{R}^{+d}$, $[u, v[$ is the set $\{w \in \mathbb{R}^{+d}: u \leqq w < v\}$. Let us say that a sequence $(A_n)_{n \in \mathbb{N}}$ of Borel sets in $\mathbb{R}^{+d}$ with $\lambda(A_n) \to \infty$ *consists asymptotically of unions of large intervals* if for any $m \in \mathbb{N}$ and any $\varepsilon > 0$ there is an $N(m, \varepsilon)$ such that for $n \geqq N(m, \varepsilon)$ there exists a set $A_n^{m,\varepsilon}$ which is a disjoint union of translates of the interval $[0, me[$ and for which $\lambda(A_n \Delta A_n^{m,\varepsilon}) < \varepsilon \lambda(A_n)$ holds.

For any $A \subset \mathbb{R}^{+d}$ we denote by $\bar{A}^m$ the union of all intervals $[u, u + me[$ with $u \in \mathbb{V}^{(m)}$ which intersect $A$, and by $\underline{A}^m$ the union of all intervals $[u, u + me[$ with $u \in \mathbb{V}^{(m)}$ which are contained in $A$.

**Theorem 2.10.** *In the situation just described assume that* $m^{-d} F_{[0, me[}$ *converges strongly to some* $\bar{f} \in \mathfrak{X}$ *as* $m \to \infty$. *Then* $\lambda(A_n)^{-1} F_{A_n}$ *converges strongly to* $\bar{f}$ *for any sequence* $(A_n)$ *in* $\mathscr{B}^b$ *with* $\lambda(A_n) \to \infty$ *which consists asymptotically of unions of large intervals.*

(The condition (2.10) may be replaced by the weaker assumption $\sup\{\|F_A\|: A \in \mathscr{B}^b,\ A \subset [0, e[\} < \infty$ if $\lambda(A_n)$ tends to $\infty$ for $n \to \infty$ and $\lambda(\bar{A}_n^m \setminus \underline{A}_n^m)/\lambda(A_n)$ tends to 0 for each $m \in \mathbb{N}$.)

The proof of this theorem, which generalizes a result of Fritz [1970], is straightforward and is left to the reader. It is clear that a similar theorem holds when $\mathbb{R}^{+d}$ is replaced by $\mathbb{V}$, $\mathbb{R}^d$ or $\mathbb{Z}^d$.

We now turn to the more subtle question of a.e.-convergence of sequences $\lambda(A_n)^{-1} F_{A_n}$. Tempel'man [1972] has given a very thorough treatment to this problem applying, e.g., to discrete additive processes and to integrals $F_A = \int_A f \circ \tau_u \lambda(du)$ even when $f$ takes values in a Banach space. He has given axiomatic properties for the sequence $(A_n)$; see § 6.4. We consider only convex sets $A_n$ and their restrictions to $\mathbb{V}$, but we treat subadditive processes.

For any Borel set $A$ in $\mathbb{R}^d$, $\varrho(A)$ denotes the supremum of all $r \geqq 0$ for which there exists a Euclidian sphere $S(x, r) = \{u \in \mathbb{R}^d: \|u - x\| \leqq r\}$ contained in $A$. $A + A'$ is the set $\{x + y: x \in A, y \in A'\}$.

We need the following simple lemma of Day [1942]:

**Lemma 2.11.** *If a convex set $E$ in $\mathbb{R}^d$ contains $S(0,r)$, then $S(0,\alpha)+E$ is contained in* $\left(\frac{r+\alpha}{r}\right)E$.

*Proof.* For any $y\in S(0,\alpha)$ there exists a $y'\in S(0,r)$ with $y=(\alpha/r)\,y'$. For any $y''\in E$

$$y+y''=\frac{r+\alpha}{r}\left[\frac{\alpha}{r+\alpha}\,y'+\frac{r}{r+\alpha}\,y''\right]=\frac{r+\alpha}{r}\,y_0$$

with $y_0\in E$. □

If $\partial_h E$ is the set of points in $\mathbb{R}^d$ having distance $\leqq h$ from the boundary of $E$, then $\partial_h E\cap E^c$ is contained in $(E+S(0,h))\setminus E$. For convex $E$, $E\setminus\partial_h E$ is convex and it contains a sphere of radius $\varrho(E)-2h$. Combining these observations with Day's lemma it is very simple to prove the following auxiliary result of Nguyen-Zessin [1976] and Fritz [1970]: If $A_n$ is a sequence of convex sets in $\mathbb{R}^d$ with $\varrho(A_n)\to\infty$, then $\lambda(\partial_h A_n)/\lambda(A_n)$ tends to 0 for all $h>0$, and $\lambda(\bar{A}_n^m\setminus \underline{A}_n^m)/\lambda(A_n)$ tends to 0 for all $m\in\mathbb{N}$.

A sequence of convex sets $A_n$ will be called *regular* if there exists an increasing sequence of intervals $I_n'$ in $\mathbb{R}^{+d}$ and a constant $C<\infty$ such that $I_n'$ contains $A_n$ and $\lambda(I_n')\leqq C\lambda(A_n)$ holds for all $n$.

**Theorem 2.12.** *If $\tau_1,\ldots,\tau_d$ are commuting endomorphisms of $(\Omega,\mathscr{A},\mu)$ and if $A_1,A_2,\ldots$ is a regular family of convex sets in $\mathbb{R}^{+d}$ with $\varrho(A_n)\to\infty$, then*

$$\operatorname{card}(A_n\cap\mathbb{V})^{-1}\sum_{u\in A_n\cap\mathbb{V}} f\circ\tau_u$$

*converges a.e. for all $f\in L_p$, $(1\leqq p<\infty)$. Here $A_n\cap\mathbb{V}$ may be replaced by any other subset of $\mathbb{V}$ differing from $A_n\cap\mathbb{V}$ only on $\partial_h A_n$ for some $h>0$.*

This is just another special case of the results of Tempel'man and the same proof applies. Note that $\lambda(\partial_h A_n)/\lambda(A_n)\to 0$ implies $\operatorname{card}(A_n\cap\mathbb{V})^{-1}[\operatorname{card}((A_n\cap\mathbb{V})\,\Delta\,((A_n\cap\mathbb{V})+e^j))]\to 0$ and the corresponding result for sets differing from $A_n\cap\mathbb{V}$ only on $\partial_h A_n$. Examples of such sets "close" to $A_n\cap\mathbb{V}$ are $\bar{A}_n^m\cap\mathbb{V}$ and $\underline{A}_n^m\cap\mathbb{V}$.

Let $X_{n,m}$ be the set of vectors $u\in\mathbb{V}^{(m)}$ with $[u,u+me[\,\subset A_n$ and let $Y_{n,m}$ be the set of vectors $u\in\mathbb{V}^{(m)}$ with $[u,u+me[\,\cap A_n\neq\emptyset$. Then we have

$$\bar{A}_n^m=\bigcup_{u\in Y_{n,m}}[u,u+me[\quad\text{and}\quad \underline{A}_n^m=\bigcup_{u\in X_{n,m}}[u,u+me[.$$

Applying the theorem to the sub-semigroup $\{\tau_u\colon u\in\mathbb{V}^{(m)}\}$ we find that

$$(2.11)\qquad \lambda(A_n)^{-1}\sum_{u\in Y_{n,m}\setminus X_{n,m}}\tilde{f}\circ\tau_u$$

converges to 0 a.e. for all integrable $\tilde{f}$.

It is now simple to prove the following ergodic theorem for additive processes with a continuous parameter due to Nguyen and Zessin [1979]:

**Theorem 2.13.** *Let $\tau = \{\tau_u: u \in \mathbb{R}^{+d}\}$ be a measurable measure preserving semi-flow in $(\Omega, \mathscr{A}, \mu)$ and let $F: \mathscr{B}^b \ni A \to F_A \in L_1$ satisfy* (2.8) *and* (2.9) *with $T_u f = f \circ \tau_u$. Assume that*

(2.12) *there exists an $\bar{f} \in L_1$ such that $|F_A| \leqq \bar{f}$ holds for all convex sets $A \in \mathscr{B}^b$ with $A \subset [0, e[$.*

*Then $\lambda(A_n)^{-1} F_{A_n}$ converges a.e. for any regular sequence $(A_n)$ of convex sets with $\varrho(A_n) \to \infty$.*

*Proof.* By the assumptions $|F_{A_n} - \sum_{u \in X_{n,1}} F_{[0,e[} \circ \tau_u|$ is bounded by $\sum_{u \in Y_{n,1} \setminus X_{n,1}} \bar{f} \circ \tau_u$. □

The assertion of theorem 2.13 remains true if the sequence $(A_n)$ is replaced by a family $(A_r)$ indexed by any countable linearly ordered set, but again one must add further conditions if $r$ shall range through $\mathbb{R}^+$.

Let $\mathscr{H}_{\mathbb{V}}$ denote the family of non empty finite subsets of $\mathbb{V}$. A subadditive process for $\{\tau_u: u \in \mathbb{V}\}$ defined on $\mathscr{H}_{\mathbb{V}}$ is a set function $F: \mathscr{H}_{\mathbb{V}} \ni A \to F_A \in L_1$ with the properties

(2.13) $F_A \circ \tau_u = F_{A+u}$ whenever $A \in \mathscr{H}_{\mathbb{V}}$ and $u \in \mathbb{V}$;

(2.14) if $A_1, A_2$ are disjoint sets in $\mathscr{H}_{\mathbb{V}}$ then $F_{A_1 \cup A_2} \leqq F_{A_1} + F_{A_2}$;

(2.15) $\gamma(F) = \inf\{\text{card}(A)^{-1} \int F_A \, d\mu: A \in \mathscr{J}\} > -\infty$.

Note that the spatial constant is again defined with intervals!

**Theorem 2.14.** *Let $\tau_1, \ldots, \tau_d$ be commuting endomorphisms of $(\Omega, \mathscr{A}, \mu)$ and let $F = \{F_A: A \in \mathscr{H}_{\mathbb{V}}\}$ be a subadditive process for $\tau = \{\tau_u: u \in \mathbb{V}\}$. If $(A_n)$ is a regular sequence of convex sets in $\mathbb{R}^{+d}$ with $\varrho(A_n) \to \infty$, then* card $(A_n \cap \mathbb{V})^{-1} F_{A_n \cap \mathbb{V}}$ *converges a.e..*

The proof is virtually identical with that of theorem 2.9. The role of $I_{r,m}$ is now played by $\underline{A}_n^m$ and the role of $J_{r,m}$ is played by $\bar{A}_n^m$.

In the continuous parameter version of this theorem, $F$ must be defined on $\mathscr{B}^b$ and (2.12) must be satisfied. Then one gets the convergence a.e. of $\lambda(A_n)^{-1} F_{A_n}$ as above.

## Notes

The idea of regularity of one sequence of sets with respect to another seems to appear first in the paper of Pitt [1942], and was most successfully extended by Tempel'man. The proof of lemma 2.5 uses an idea of Calderon [1953].

Also Fava and Nanclares [1978] and Becker [1981] have ergodic theorems for convex sets.

The ergodic theorem for additive processes in $\mathbb{R}^d$ not given by integral averages is of interest for the theory of point processes; see Nguyen-Zessin [1976], Rolski [1981a].

The question of a.e.-convergence for multiparameter subadditive processes was first treated by Smythe [1976]. Smythe considered 2-subadditive processes which are also *strongly subadditive*: For $d = 2$ this means that $s = (s_1, s_2) < t = (t_1, t_2) < u$ implies

$$F_{[s,u[} - F_{[(s_1,t_2),u[} - F_{[(t_1,s_2),u[} + F_{[t,u[} \leqq F_{[s,t[}.$$

Using this condition, Smythe proved an analogue of Kingman's decomposition theorem. He obtained a.e.-convergence of $\pi(n)^{-1} F_{[0,n[}$ for sequences remaining in a sector and some first results on unrestricted convergence.

Nguyen [1979] studied groups $\tau = \{\tau_t : t \in \mathbb{R}^d\}$ and subadditive processes $\{F_A\}$ with $A$ ranging through the class of bounded Borel sets. $\{F_A\}$ is strongly subadditive in his sense if $F_{A_1 \cup A_2} + F_{A_1 \cap A_2} \leqq F_{A_1} + F_{A_2}$ holds for all $A_1, A_2$ with $A_1 \cap A_2 \neq \emptyset$. He obtained a Kingman decomposition of processes with his property by a very different argument. He also proved a.e.-convergence for sequences $(A_n)$ of convex sets with $\varrho(A_n) \to \infty$, assuming his strong subadditivity.

Krengel and Pyke [1985] relaxed the convexity condition in theorem 2.14 and showed that the convergence to the limit $\bar{f}$ of the sequence $\bar{F}_{[0,ne[}$ is uniform in the following sense

$$\sup\{|n^{-d} F_{nB} - \lambda(B)\bar{f}| : B \in \mathscr{B}\} \to 0 \quad \text{a.e.}$$

when $\mathscr{B}$ is a family of sets in $[0, e[$ with "small boundaries".

The cluster process of Grimmett is not strongly subadditive in either sense. Akcoglu and Krengel [1981] have given an example of a subadditive process $\{F_I : I \in \mathscr{I}\}$ for $d = 2$ which does not admit a Kingman decompsition.

Grillenberger and Krengel [1982] showed: If $g$ is a real valued translation invariant superadditive set function on the system of bounded Borel sets in $\mathbb{R}^d$ and $\inf\{g(B): B \subset [0, e[\} > -\infty$, then $g(K_n)/\lambda(K_n)$ tends to the same $\gamma$ for all sequences $K_n$ of convex sets with $\varrho(K_n) \to \infty$. Thus, the limit property defining the spatial constant remains true for convex sets. For strongly superadditive $g$ this was shown by Robinson-Ruelle [1967] and Nguyen [1979].

Applications of the multiparameter subadditive ergodic theorem to Mathematical Physics have been given by Kirsch and Martinelli [1982] and by van Enter and van Hemmen [1983].

## § 6.3 Multiparameter semigroups of $L_1$-contractions

We return to the study of ergodic theorems for multiparameter semigroups of *operators*. We use the notation from § 6.1. Let $\{T_u : u \in \mathbb{V}\}$ be a semigroup of $L_1 - L_\infty$-contractions. We shall show that $A_n f$ converges a.e. for all integrable $f$ if $n$ tends to infinity in a sector of $\mathbb{V}$. Dunford-Schwartz [1956] proved the continuous parameter variant of this theorem by a method which they called "circuitous". We follow the much simpler direct approach to the discrete case given by Brunel [1973] and obtain the continuous parameter result as a corollary.

We also show by example that there is no such multiparameter generalization of the Chacon-Ornstein theorem.

Finally we sketch a multiparameter generalization of the stochastic ergodic theorem asserting *unrestricted* stochastic convergence of $A_n f$ for $f \in L_1$ when the $T_u$ are positive $L_1$-contractions which need not contract the $L_\infty$-norm.

**Lemma 3.1** *For $\xi(x) = 1 - \sqrt{1-x}$ the coefficients $\alpha_p^{(n)}$ in the expansion $[\xi(x)]^n = \sum_{p=0}^{\infty} \alpha_p^{(n)} x^p$ are given by*

$$(3.1) \qquad \alpha_p^{(n)} = \begin{cases} 0 & \text{for } p < n \\ \dfrac{n}{2p}\, 2^{n+1-2p} \dbinom{2p-n-1}{p-1} & \text{for } p \geqq n. \end{cases}$$

*Proof.* The case $n = 1$ follows directly from $(1-x)^{1/2} = \sum_{p=0}^{\infty} \binom{1/2}{p} (-x)^p$. It is then obvious that $\alpha_p^{(n)} = 0$, $(p < n)$, holds for all $n$. Assume that the assertion of the lemma has been proved for all $n \leqq N$. The trick in the induction step (due to N. Neumann) is to use the recursion

$$(3.2) \qquad [\xi(x)]^{N+1} = 2[\xi(x)]^N - x[\xi(x)]^{N-1}, \quad (N \geqq 1)$$

which follows from $(\xi(x))^2 = 2\xi(x) - x$. For $N = 1$ and $p \geqq 2$, (3.2) yields $\alpha_p^{(2)} = 2\alpha_p^{(1)}$ which easily implies (3.1) for $n = 2$. For $N > 1$ and $p \geqq N+1$ we obtain

$$\begin{aligned}
\alpha_p^{(N+1)} &= 2\alpha_p^{(N)} - \alpha_{p-1}^{(N-1)} \\
&= 2\frac{N}{2p} 2^{N+1-2p} \binom{2p-N-1}{p-1} - \frac{N-1}{2(p-1)} 2^{N+2-2p} \binom{2p-N-2}{p-2} \\
&= \frac{N+1}{2p} 2^{N+2-2p} \binom{2p-N-2}{p-1} \left[ \frac{N}{N+1} \cdot \frac{2p-N-1}{p-N} \right. \\
&\quad \left. - \frac{p}{p-1} \frac{N-1}{N+1} \cdot \frac{p-1}{p-N} \right].
\end{aligned}$$

Because of $[\ldots] = 1$ the induction is complete. □

**Lemma 3.2.** *For any $K \geqq 1$ there exists a constant $c(K) > 0$ such that $\alpha_p^{(N)} \geqq c(K) n p^{-3/2}$ holds for all $n, p$ with $n^2 \leqq pK$.*

*Proof.* Let us use $a_n \sim b_n$ as a shorthand for

$$\limsup a_n/b_n < \infty \quad \text{and} \quad \liminf a_n/b_n > 0\,.$$

By Stirlings formula $k! \sim e^{-k} k^{k+1/2}$. Because of (3.1)

$$\alpha_p^{(n)} \sim \frac{n}{p} 2^{n-2p} \frac{(2p-n-1)!}{(p-1)!\,(p-n)!} \sim \frac{n}{p^{3/2}} c(n, p)$$

with $c(n,p) = \left(1 - \frac{n}{2p}\right)^{2p-n-1} \left(1 - \frac{n}{p}\right)^{-p+n}$. Now

$$\log c(n,p) \sim (2p-n-1)\left(-\frac{n}{2p} - \frac{n^2}{8p^2} - \ldots\right) - (p-n)\left(-\frac{n}{p} - \frac{n^2}{2p^2} - \ldots\right)$$
$$= \left(-n - \frac{n^2}{4p} + \frac{n^2}{2p} + \frac{n^3}{8p^2} + \ldots\right) + \left(n + \frac{n^2}{2p} - \frac{n^2}{p} + \frac{n^3}{2p^2} + \ldots\right).$$

If $n$ and $p$ tend to infinity subject to $n^2 \leqq pK$ the last expression remains bounded. Hence $c(n,p)$ is bounded below by a positive constant. □

**Lemma 3.3.** *Let $\varphi(n)$ be the integer part of $\sqrt{n} + 1$. There exists a constant $c' > 0$ such that*

$$\frac{1}{\varphi(n)} \sum_{0 \leqq j < \varphi(n)} \alpha^{(j)}_{v+j} \alpha^{(j)}_{w+j} \geqq \frac{c'}{n^2}$$

*holds for $0 \leqq v, w < n$.*

*Proof.* Because of $\alpha^{(0)}_{00} = 1$ the term with $j = 0$ is large enough in the case $v = w = 0$. If $v = 0$ and $w > 0$, the term with $j = 1$ is $\geqq \varphi(n)^{-1} c(1)(w+1)^{-3/2}$, and this is $\geqq c' n^{-2}$ for small enough $c'$. By symmetry it is now sufficient to consider the case $0 < v \leqq w < n$. Using lemma 3.2 again and $\sum_{1}^{r-1} k^2 \sim r^3$ we then find

$$\varphi(n)^{-1} \sum_{j < \varphi(n)} \alpha^{(j)}_{v+j} \alpha^{(j)}_{w+j} \geqq \sum_{j=1}^{\varphi(v)-1} \text{const} \frac{\varphi(n)^{-1} j^2}{((v+j)(w+j))^{3/2}}$$
$$\geqq \varphi(n)^{-1} \text{const}' v^{-3} \varphi(v)^3 \geqq \text{const}'' n^{-2}. \quad \square$$

For $d > 1$ and $n \in \mathbb{N}$ let $n_d$ be the integer $\varphi^m(n)$ with $2^{m-1} < d \leqq 2^m$. The basic step in the argument of Brunel is:

**Theorem 3.4.** *For any $d > 1$ there exists a constant $\chi_d > 0$ and a family $\{a(u): u \in \mathbb{V}\}$ of strictly positive numbers summing to 1 such that the following holds: If $T_1, T_2, \ldots, T_d$ are commuting contractions of $L_1$ and $|T_i|$ denotes the linear modulus of $T_i$, then the operator $U = \sum_{u \in \mathbb{V}} a(u) |T_1|^{u_1} \ldots |T_d|^{u_d}$ satisfies for all $n \geqq 1$ and $f \in L_1^+$*

$$(3.3) \qquad n^{-d} \sum_{0 \leqq i_1 < n} \cdots \sum_{0 \leqq i_d < n} |T_1^{i_1} \ldots T_d^{i_d}| f \leqq \frac{\chi_d}{n_d} \sum_{j=0}^{n_d - 1} U^j f.$$

*Proof.* We may assume $d = 2^m$, because we can put $T_j = I$, $(d+1 \leqq j \leqq 2^m)$ if $d < 2^m$. We start with the case $T_i = |T_i| \geqq 0$.

Put $\bar{\xi}(x) = x^{-1}\xi(x) = \sum_{v=0}^{\infty} \alpha_{v+1}^{(1)} x^v$. For a contraction $T$, $\bar{\xi}(T)$ is defined by $\bar{\xi}(T) = \sum_{v=0}^{\infty} \alpha_{v+1}^{(1)} T^v$. Clearly $(\bar{\xi}(T))^j = \sum_{v=0}^{\infty} \alpha_{j+v}^{(j)} T^v$. For $m=1$ $(d=2)$ we may take

$$U = \bar{\xi}(T_1) \cdot \bar{\xi}(T_2).$$

Then

$$\varphi(n)^{-1} \sum_{j<\varphi(n)} U^j \geqq \varphi(n)^{-1} \sum_{0 \leqq i_1, i_2 < n} \sum_{j<\varphi(n)} \alpha_{j+i_1}^{(j)} \alpha_{j+i_2}^{(j)} T_1^{i_1} T_2^{i_2}$$
$$\geqq \frac{c'}{n^2} \sum_{0 \leqq i_1, i_2 < n} T_1^{i_1} T_2^{i_2},$$

i.e., (3.3) holds with $\chi_2 = c'$. For $m=2$, $(d=4)$, four contractions $T_1, \ldots, T_4$ are given and we put

$$U_1 = \bar{\xi}(T_1) \cdot \bar{\xi}(T_2), \quad U_2 = \bar{\xi}(T_3) \cdot \bar{\xi}(T_4)$$

and

$$U = \bar{\xi}(U_1) \cdot \bar{\xi}(U_2).$$

Then

$$\frac{1}{\varphi(\varphi(n))} \sum_{j<n_4} U^j \geqq \frac{c'}{\varphi(n)^2} \sum_{\substack{k_1 < n_2 \\ k_2 < n_2}} U_1^{k_1} U_2^{k_2}$$
$$\geqq \frac{(c')^3}{n^4} \sum_{j_1<n} \cdots \sum_{j_4<n} T_1^{j_1} \ldots T_4^{j_4}$$

and we may take $\chi_3 = \chi_4 = (c')^3$. In general, if we know the construction of $U$ for $d$ operators and $T_1, \ldots, T_{2d}$ are given, let $U_1$ be the operator constructed for $T_1, \ldots, T_d$, let $U_2$ be the operator constructed for $T_{d+1}, \ldots, T_{2d}$ and take $U = \bar{\xi}(U_1) \cdot \bar{\xi}(U_2)$.

$U$ is of the form $U = \sum_{u \in \mathbb{V}} a(u) T_1^{u_1} \ldots T_d^{u_d}$ with coefficients $a(u) > 0$ summing to 1 which do not depend on $T_1, \ldots, T_d$. Let $a^{(j)}(i_1, \ldots, i_d)$ be the coefficient of $T_1^{i_1} \ldots T_d^{i_d}$ in $U^j$. The inequality proved is due to the fact that

$$\chi_d n_d^{-1} \sum_{j<n_d} a^{(j)}(i_1, \ldots, i_d) \geqq n^{-d}$$

holds for all $(i_1, \ldots, i_d)$. The numbers $a^{(j)}(i_1, \ldots, i_d)$ are sums of coefficients of different products $T_{\alpha(1)} \cdot T_{\alpha(2)} \cdot \ldots \cdot T_{\alpha(s)}$ coming from the evaluation of $U^j = (\sum a(u) T_1^{u_1} \ldots T_d^{u_d})^j$, where $\alpha(v) = 1$ holds for $i_1$ values of $v$, $\alpha(v) = 2$ for $i_2$ values of $v$, etc. In other words: $a^{(j)}(i_1, \ldots, i_d) = \sum_\alpha a^{(j,\alpha)}(i_1, \ldots, i_d)$ where $a^{(j,\alpha)}(i_1, \ldots, i_d)$ is the *positive* coefficient of $T_{\alpha(1)} \cdots T_{\alpha(s)}$. The inequality (3.3) remains true for non positive operators and $f \geqq 0$ even though the operators $|T_1|, \ldots, |T_d|$ may not commute, because of

$$\chi_d n_d^{-1} \sum_{0 \leqq j < n_d} U^j f \geqq \chi_d n_d^{-1} \sum_{j<n_d} \sum_{i_1, \ldots, i_d < n} \sum_{\alpha} a^{(j,\alpha)}(i_1, \ldots, i_d) |T_{\alpha(1)}| \ldots |T_{\alpha(s)}| f$$

$$\geqq \chi_d n_d^{-1} \sum_{j<n_d} \sum_{i_1, \ldots, i_d < n} \sum_{\alpha} a^{(j,\alpha)}(i_1, \ldots, i_d) |T_{\alpha(1)} \cdot \ldots \cdot T_{\alpha(s)}| f$$

$$= \chi_d n_d^{-1} \sum_{j<n_d} \sum_{i_1, \ldots, i_d < n} a^{(j)}(i_1, \ldots, i_d) |T_1^{i_1} \cdot \ldots \cdot T_d^{i_d}| f$$

$$\geqq n^{-d} \sum_{i_1, \ldots, i_d < n} |T_1^{i_1} \cdot \ldots \cdot T_d^{i_d}| f. \quad \square$$

It is now easy to derive the following multiparameter ergodic theorem for $L_1 - L_\infty$-contractions:

**Theorem 3.5.** *If $T_1, \ldots, T_d$ are commuting $L_1 - L_\infty$-contractions, the averages $A_{u(k)} f$ converge a.e. for all $f \in L_1$ and all sequences $u(1), \ldots, u(k), \ldots$ tending to $\infty$ in a sector of $\mathbb{V}$*, in particular for $u(k) = ke = (k, k, \ldots, k)$.

*Proof.* We have to show that

$$\Delta_k(f) = \sup \{ |A_{u(l)} f - A_{u(k)} f| : l \geqq k \}$$

tends to 0 for all $f \in L_1^+$. By theorem 1.2 $\Delta_k(f')$ tends to 0 for all $f' \in L_2$. Given $\varepsilon > 0$ find $f' \in L_2^+$ with $f' \leqq f$ and $\|f - f'\|_1 < \varepsilon^2$. Clearly $\Delta^*(f) := \limsup_{k \to \infty} \Delta_k(f) \leqq \sup_k \Delta_k(f - f') \leqq 2 \sup_k |A_{u(k)}(f - f')|$. In the case $u(k) = k \cdot e$ we may apply (3.3) directly and obtain

$$\mu(\Delta^*(f) > \varepsilon) \leqq \mu(\sup_m \chi_d 2m^{-1} \sum_{i=0}^{m-1} U^i |f - f'| > \varepsilon).$$

As $U$ is a positive $L_1 - L_\infty$-contraction Lemma 1.6.1 now implies

$$\mu(\Delta^*(f) > \varepsilon) \leqq 2\chi_d \varepsilon^{-1} \|f - f'\|_1 < 2\chi_d \varepsilon.$$

As $\varepsilon > 0$ was arbitrary this completes the proof for the case $u(k) = ke$. In the general case put $n(k) = \max \{u_1(k), \ldots, u_d(k)\}$.

As the sequence $u(k)$ is in a sector of $\mathbb{V}$ there exists a constant $0 < c_0 < \infty$ with $n(k)^{-d} \leqq c_0 \pi(u(k))$ for all $k$.

By the previous theorem

$$|A_{u(k)}(f - f')| \leqq \pi(u(k))^{-1} \sum_{i_1 < u_1(k)} \cdots \sum_{i_d < u_d(k)} |T_1^{i_1} \ldots T_d^{i_d}(f - f')|$$

$$\leqq c_0 n(k)^{-d} \sum_{i_1 < n(k)} \cdots \sum_{i_d < n(k)} |T_1^{i_1} \ldots T_d^{i_d}(f - f')|$$

$$\leqq c_0 \chi_d (n(k))_d^{-1} \sum_{j < (n(k))_d} U^j (f - f').$$

The same argument as above now gives $\mu(\Delta^*(f) > \varepsilon) \leqq 2\chi_d c_0 \varepsilon$. $\square$

A similar proof shows that $\operatorname{card}(B_k \cap \mathbb{V})^{-1} \sum_{u \in B_k \cap \mathbb{V}} T_u f$ converges a.e. for all $f \in L_1$, when $B_k$ is a sequence of convex sets with $\varrho(B_k) \to \infty$ which is regular with respect to some increasing sequence of squares $I'_k = [0, n(k)e[$. We leave this as an exercise. One first has to extend theorem 1.2 to averages over convex sets. In the extension of theorem 1.2 regularity is permitted with respect to any sequence $I'_k = [0, u(k)[$ with $u(k) \to \infty$ in $\mathbb{V}$, which need not be increasing. The results of the previous section suggest that also for general $L_1 - L_\infty$-contrations $T_1, \ldots, T_d$ the averages $A_{u(k)} f$ converge a.e. for all $f \in L_1$ and all *increasing* sequences $u(k)$ in $\mathbb{V}$. This, however, cannot be proved with the method of majorization by $U$ and is unknown.

What can be said about the *identification of the limit* in theorem 3.5? We assume that the $T_i$ are positive: Let $\tilde{C}_U$ be the strongly conservative part of the positive contraction $U$. As $n^{-1} \sum_{i=0}^{n-1} U^i f$ converges to 0 for all $f \in L_1$ on the complement $\tilde{D}_U$ of $\tilde{C}_U$ the averages $A_{u(k)} f$ must also tend to 0 on $\tilde{D}_U$. There exists an element $p \in L_1^+$ with $Up = p$ and $\{p > 0\} = \tilde{C}_U$. Lemma 2.1.14 of Brunel-Falkowitz implies that $p$ is invariant under all $T_i$.

The next lemma shows that there is no influence of the dissipative part $D_U$:

**Lemma 3.6.** *If $C$ and $D$ are the conservative and dissipative parts of a positive $L_1 - L_\infty$-contraction $T$, then $T 1_D \leqq 1_D$.*

*Proof.* We know $T 1_C = 1_C$. Therefore $T\mathbb{1} \leqq \mathbb{1}$ implies $T 1_D \wedge 1_D = 0$. □

As there is also no influence of $C_U \setminus \tilde{C}_U$ in $\tilde{C}_U$ the limit behaviour of the averages depends only on the values of $f$ in $\tilde{C}_U$ and we may and do assume $\Omega = \tilde{C}_U$. Now the usual arguments show that the averages $A_u f$ are uniformly integrable and one has also $L_1$-convergence. The limit $\bar{f}$ must be invariant under all $T_u$, $(u \in \mathbb{V})$, and hence $\mathscr{I}_U$-measurable. $\mathscr{I}_U$ is the $\sigma$-algebra of sets invariant under all $T_i$. $\bar{f}$ is uniquely determined by $\int_F f\,d\mu = \int_F \bar{f}\,d\mu$ for all $F \in \mathscr{I}_U$.

We only sketch the extension to continuous parameter semigroups: If $\{T_u: u \in \mathbb{R}^{+d}\}$ is a semigroup of contractions in $L_1$ which is continuous in the interior of $\mathbb{R}^{+d}$ (i.e., $\| T_u f - T_v f \|_1 \to 0$ holds for $u \to v > 0$ for all $f \in L_1$) we may define suitable representatives of $T_u f$, $(u > 0)$, as in the 1-parameter case (§ 1.6): Put

$$f_n(t, \omega) = T_{(2^{-n} i_1, \ldots, 2^{-n} i_d)} f(\omega)$$
$$\text{for } t = (t_1, \ldots, t_d) \quad \text{with } t_\nu \in [2^{-n} i_\nu, 2^{-n}(i_\nu + 1)[.$$

The limit $f(t, \omega)$ of an appropriate subsequence defines representatives $f(t, .)$ of $T_t f$. Put

$$S_{0,v} f = \int_{[0,v[} T_u f\,du, \quad A_{0,v} f = \pi(v)^{-1} S_{0,v} f.$$

**Theorem 3.7** (Dunford-Schwartz). *If $\{T_u: u \in \mathbb{R}^{+d}\}$ is a semigroup of $L_1 - L_\infty$-contractions, continuous in the interior of $\mathbb{R}^{+d}$, the averages $A_{0,ae}f$ converge a.e. for all $f \in L_1$ as $a \to \infty$.*

*Proof.* An application of theorem 3.5 to the function $S_{0,e}f$ shows that there is a.e.-convergence when a takes only integer values. It remains to show

$$(3.4) \qquad \lim_{a\to\infty} (A_{0,ae}f - A_{0,[a]e}f) = 0 \quad \text{a.e.}.$$

For $f \in L_1 \cap L_\infty$ this is simple. The general case follows by approximation with bounded functions once we show

$$(3.5) \qquad \int \limsup_{a\to\infty} |A_{0,ae}f| \, d\mu \leqq \chi_d \|f\|_1 .$$

Recall that $e^j$ was $j$-th unit vector in $\mathbb{V}$. Put

$$\tilde{f}_n(t,\omega) = |T_{2^{-n}e^1}|^{i_1} \cdot |T_{2^{-n}e^2}|^{i_2} \dots |T_{2^{-n}e^d}|^{i_d} |f|(\omega)$$

for $t = (t_1, \dots, t_d)$ with $t_\nu \in [2^{-n} i_\nu, \dots, 2^{-n}(i_\nu + 1)[$. This sequence is increasing and $L_1$-bounded. If $\tilde{f}$ is the limit and

$$\tilde{F} = \int_{[0,e[} \tilde{f}(t,.)\,dt$$

then

$$|S_{0,ae}f| \leqq \sum_{0 \leqq i_1 \leqq [a]+1} \cdots \sum_{0 \leqq i_d \leqq [a]+1} |T_{(i_1,\dots,i_d)}| \tilde{F},$$

and theorem 3.4 yields

$$\limsup_{a\to\infty} |a^{-d} S_{0,ae}f| \leqq \chi_d \limsup_{n\to\infty} n^{-1} \sum_{i=0}^{n-1} U^i \tilde{F}.$$

(3.5) now is a consequence of $\|\tilde{F}\|_1 \leqq \|f\|_1$. □

*The following example, obtained jointly with A. Brunel, shows that the Chacon-Ornstein theorem does not extend to multiparameter semigroups.*

Let $d = 2$ and let $T_i$ for $i = 1, 2$ be given by a translation in $\Omega = \mathbb{Z}$ (with counting measure), i.e., $T_i f(\omega) = f(\omega + 1)$. $S_{ne}f$ has the form

$$S_{ne}f(\omega) = f(\omega) + 2f(\omega+1) + \dots + nf(\omega+n-1) + (n-1)f(\omega+n) + \dots + f(\omega+2n-2).$$

We sketch the construction of non negative integrable functions $f, g$ with $g(0) > 0$, for which $S_{ne}f/S_{ne}g$ diverges in $\omega = 0$: Let $h_k$ be the function which equals $1/\sqrt{k}$ on $\{k-1\}$ and 0 everywhere else. For $n \geqq k$, $S_{ne}h_k(0) = \sqrt{k}$, and for $n < k/2$, $S_{ne}h_k(0) = 0$. Let $(m_j)_{j\geqq 1}$ and $(n_j)_{j \geqq 1}$ be two sequences of integers with $m_1 = 1, n_j > 2m_j, m_{j+1} > 2n_{j+1}, (j \geqq 1)$. Then $f = \sum_{j=1}^{\infty} h_{n_j}$ and $g = \sum_{j=1}^{\infty} h_{m_j}$ are integrable, and

$$S_{m_j} g(0) = S_{n_j} g(0) = \sum_{k=1}^{j} \sqrt{m_k},$$

$$S_{m_j} f(0) = \sum_{k=1}^{j-1} \sqrt{n_k},$$

$$S_{n_j} f(0) = \sum_{k=1}^{j} \sqrt{n_k}.$$

The sequences may therefore be chosen in such a way that the lim inf of $S_{ne} f / S_{ne} g$ at 0 is 0 and the lim sup is $\infty$.

With some more effort one can construct functions $f, g$ for which the same ratios diverge everywhere. Using stacking constructions (see Friedman [1970]) it is possible to find a *conservative* $T_1 = T_2$ and functions $f, g$ for which the ratios diverge everywhere.

Using the same operators as above it is also easy to construct integrable and strictly positive functions $f, g$ with

$$\sum_{u \in \mathbb{V}} T_1^{u_1} T_2^{u_2} f = \infty \quad \text{and} \quad \sum_{u \in \mathbb{V}} T_1^{u_1} T_2^{u_2} g < \infty$$

on $\Omega = \mathbb{Z}$. Thus, there is no Hopf decomposition.

**Proposition 3.8.** *Let $T_1, \ldots, T_d$ be commuting positive contractions in $L_1$. If $C_i$ denotes the conservative part of $\Omega$ for $T_i$, then $C^d = \bigcap_{i=1}^{d} C_i$ is $T_j$-absorbing for $j = 1, \ldots, d$.*

*Proof.* We may assume $d = 2$. If $C^2$ is not $T_1$-absorbing there exists $f \in L_1^+$ with $\{f > 0\} \subset C^2$ and a set $A \subset C_2^c \cap C_1$ of positive measure with $T_1 f > 0$ on $A$. As $\sum_{i=0}^{N} T_2^i f$ diverges on $\{f > 0\}$ we see that

$$\|((\sum_{i=0}^{N} T_2^i f) \wedge Kf) - Kf\|_1 \to 0, \quad (N \to \infty)$$

for all $K > 0$. Hence, on $A$

$$T_1(\sum_{i=0}^{N} T_2^i f) = \sum_{i=0}^{N} T_2^i (T_1 f) \to \infty, \quad (N \to \infty).$$

On the other hand $T_1 f \in L_1$ and $A \subset C_2^c$ imply $\sum_{i=0}^{\infty} T_2^i (T_1 f) < \infty$ on $A$, a contradiction. □

For the multiparameter generalization of the stochastic ergodic theorem we need a multiparameter generalization of Neveu's decomposition of $\Omega$ into the strongly conservative part $\tilde{C}$ and its complement:

If $T_1, T_2, \ldots, T_d$ are commuting positive contractions in $L_1$ a function $h \in L_\infty^+$ is called *weakly wandering* for $\{T_u : u \in \mathbb{V}\}$ if there exists a sequence $u(1) < u(2) < \ldots$

in $\mathbb{V}$ with $\|\sum_{k=1}^{\infty} T^*_{u(k)} h\|_\infty < \infty$. $h$ is called *m-weakly wandering* if there exists such a sequence having the stronger property

$$\sup_{v \in \mathbb{V}} \|\textstyle\sum_{u(k) \geqq v} T^*_{u(k)-v} h\|_\infty < \infty .$$

**Theorem 3.9.** *For the semigroup $\{T_u: u \in \mathbb{V}\}$ there exists a decomposition of $\Omega$ into disjoint sets $\tilde{C}$, $\tilde{D}$ which is determined uniquely* mod $\mu$ *by the properties:*
*(i) There exists $p_0 \in L_1^+$ with $\{p_0 > 0\} = \tilde{C}$ and $T_i p_0 = p_0$ for $i = 1, \ldots, d$;*
*(ii) There exists an m-weakly wandering $h_0 \in L_\infty^+$ with $\{h_0 > 0\} = \tilde{D}$.*

We do not give the proof though it contains some technical difficulties, because in essence it is very similar to the 1-parameter special case treated in § 3.4. The role of the conservative part of a single operator is now played by $C^d$.

**Theorem 3.10.** *Let $T_1, T_2, \ldots, T_d$ be commuting $L_1$-contractions. Assume that also the linear moduli $|T_1|, |T_2|, \ldots, |T_d|$ commute. Then, for $f \in L_1$, the averages $A_u f$ converge stochastically as $u \to \infty$.*

*Sketch of proof.* We apply the previous theorem to the semigroup generated by $|T_1|, \ldots, |T_d|$. Let $E$ and $F$ denote the sets with $E = \{p_0 > 0\} = F^c$. $E$ is absorbing for the contractions $|T_i|$ and hence also for $T_i$. For $f \in L_1$ with support in $E$ the existence of $p_0 \in L_1$ with $|T_i p_0| \leqq p_0$ implies even convergence in the $L_1$-norm. To prove stochastic convergence to 0 in $F$ we may assume $T_i \geqq 0$ because of $|T_u f| \leqq |T_1|^{u_1} \ldots |T_d|^{u_d} |f|$. As in the 1-parameter case one can then show that $\|A_u^* h\|_\infty \to 0$ for weakly wandering $h$ and this again yields the desired stochastic convergence on $F$. The only new difficulty is to get control of the influence of $F$ in $E$. Put $U = T_1 \cdot T_2 \ldots T_d$. For $f \in L_1$ the functions $d_k = d_k(f)$ and $r_k = r_k(f)$ are defined inductively by $d_0 = 1_E f$, $r_0 = 1_F f$, $d_{k+1} = 1_E U r_k$, $r_{k+1} = 1_F U r_k$. The maps $f \to K^n f := \sum_{i=0}^{n} d_i$, $(n = \infty, 0, 1, 2, \ldots)$ are contractions mapping $L_1(\Omega)$ into $L_1(E)$. It suffices to prove $\|1_E A_u (f - K^\infty f)\|_1 \to 0$ for $u \to \infty$.

We write $f \rightsquigarrow_1 f'$ if there exists a measurable $g$ with $0 \leqq g \leqq 1$ and an integer $i$ with $1 \leqq i \leqq d$ and $f' = gf + T_i((1-g)f)$. $f \rightsquigarrow f'$ shall mean that for some $n$ there are functions $f_1, f_2, \ldots, f_n$ with $f \rightsquigarrow_1 f_1 \rightsquigarrow_1 f_2 \rightsquigarrow \ldots \rightsquigarrow_1 f_n = f'$. By a cancellation argument one easily sees that $f \rightsquigarrow_1 f'$ implies $\|A_u(f - f')\|_1 \to 0$. By induction this remains true for $f \rightsquigarrow f'$. One can check that $f \rightsquigarrow (K^N f + r_N)$ holds for all $N$. Because of $\|K^N f - K^\infty f\|_1 \to 0$ it is therefore enough to show that $\|1_E A_u r_N\|_1$ tends to 0 uniformly in $u$ for $N \to \infty$.

The sequence $\|r_N\|_1$ decreases to some $\eta \geqq 0$. We show that $\|r_N\|_1 - \eta < \varepsilon$ implies $\|1_E T_v r_N\|_1 \leqq \varepsilon$ for all $v \in \mathbb{V}$. Pick some $L \geqq \operatorname{Max}\{v_1, v_2, \ldots, v_d\}$. As $E$ is $T_i$-absorbing for all $i$ we have

$$r_{N+L} = 1_F U^L r_N = 1_F T_1^{L-v_1} T_2^{L-v_2} \ldots T_d^{L-v_d} (1_F T_1^{v_1} \ldots T_d^{v_d} r_N).$$

Hence $\eta \leqq \|r_{N+L}\|_1 \leqq \|1_F T_v r_N\|_1 = \|T_v r_N\|_1 - \|1_E T_v r_N\|_1 \leqq \|r_N\|_1 - \|1_E T_v r_N\|_1$ and $\|1_E T_v r_N\|_1 \leqq \|r_N\|_1 - \eta$. □

## Notes

Dunford and Schwartz have used a construction of continuous time 1-parameter semigroups from 2-parameter semigroups which has inspired the construction of $U$ in theorem 3.4. They do not give an inequality between operators as in (3.3). But in principle, their construction could also be used to prove theorem 3.5 and it remains useful for local ergodic theorems.

The assertion of theorem 3.5 holds also for positive $L_1$-contractions $T_1, \ldots, T_d$ with $\|T_i\|_p \leqq 1$ for some $p > 1$. This was shown for $d = 1$ by Akcoglu-Chacon [1965] and the general case was reduced to the case $d = 1$ by McGrath [1980a]; see also Sato [1983].

Combining theorem 3.4 with theorem 1.6.3 we obtain the following theorem which goes back to Wiener in the special case of measure preserving transformations.

**Theorem 3.11.** *Let $\{T_u: u \in \mathbb{V}\}$ be a semigroup of $L_1 - L_\infty$-contractions in $L_1$ of a finite measure space. For non negative $f \in L \log L$, $f^* = \sup\{n^{-d} \sum_{u<ne} |T_u f|\}$ is integrable.*

Extending Ornstein's result of § 1.6., K. Petersen [personal communication, 1977] and Nghiêm Dang Ngoc [1980] have proved a converse for this theorem for ergodic $d$-parameter groups of measure preserving transformations: For non negative $f$ the integrability of $f^*$ implies $f \in L \log L$.

Derriennic and Krengel [1981] proved the multiparameter $L_1$-mean ergodic theorem for superadditive processes:

**Theorem 3.12.** *Let $\{T_u, u \in \mathbb{V}\}$ be a d-parameter semigroup of positive $L_1$-contractions with $\Omega = \tilde{C}$ and let $F = \{F_I: I \in \mathscr{J}\} \subset L_1$ satisfy $T_u F_I = F_{I+u}$, $(I \in \mathscr{J}, u \in \mathbb{V})$; (2.2) (with $n = 2$); and (2.3), then $\bar{F}_{[0,n[} = \pi(n)^{-1} F_{[0,n[}$ converges in $L_1$-norm to an invariant limit $\bar{f}$ for $n \to \infty$.*

Akcoglu and Sucheston [1983] proved a multiparameter superadditive stochastic ergodic theorem: If $\{T_u: u \in \mathbb{V}\}$ is a $d$-parameter semigroup of Markovian operators and $F$ is as above, then $\bar{F}_{[0,n[}$ converges stochastically. The limit is 0 in $\tilde{D}$. It is easy to deduce this from theorems 3.10 and 3.12 (to which they had access), but they have a direct argument which replaces the characterization of $\tilde{C}$ in theorem 3.9 by a different characterization for general families of contractions which admits a shorter proof.

An application of theorem 3.12 can replace the argument for the convergence on $\tilde{C}$ in theorem 3.10. The original argument was prefered here, because it gives access to the identification of the limit.

Emilion and Hachem [1983] have proved a.e. convergence of $\bar{F}_{[0,ke[}$ for Markovian $T_u$ when $\{F_I\}$ is "strongly superadditive". They also discuss the submarkovian case and processes in $L_p$, $(1 < p < \infty)$.

**Finite invariant measures.** Blum and Friedman [1967] and Hanson and Wright [1970] started the search for conditions for the existence of finite invariant measures for Abelian semigroups of nonsingular transformations, and Granirer [1971] gave an extension of the first four conditions in Corollary 3.4.7 to the case of amenable semigroups of transformations. Sachdeva [1971] and Takahashi [1971] proved for general left amenable semigroups of positive contractions in $L_1$ that there exists a strictly positive invariant $f \in L_1$ if and only if there exists no weakly wandering $h \neq 0$. Takahashi also constructed a strictly positive weakly wandering $h$ for left amenable semigroups admitting no invariant $f \in L_1^+$ with $f \neq 0$. In the special case of semigroups $\{T_u: u \in \mathbb{V}\}$ their results follow from theorem 3.9. They do not give the decomposition $\Omega = \tilde{C} \cup \tilde{D}$ needed for the proof of theorem 3.10. The omitted proof of theorem 3.9 along the lines of the 1-dimensional case uses tools of their papers and has been carried out in detail by my students, Mrs. Kleinknecht and Mr. Pelanchon.

Hiai and Sato [1977] have studied $\tilde{C}$ and mean ergodic theorems for semigroups of positive contractions in $L_1$.

Hajian and Ito [1969] have extended the result of Hajian-Kakutani on weakly wandering sets to *general* groups $\{\tau_g: g \in G\}$ of nonsingular transformations: There exists an invariant probability measure $\nu$ equivalent to the given measure $\mu$ if and only if there exists no weakly wandering $A$ with $\mu(A) > 0$. Here, $A$ is called weakly wandering if there exists a sequence $g_1, g_2, \ldots$ in $G$ for which the sets $\tau_{g_i}^{-1} A$ are disjoint. A related reference is Dang Ngoc Ngiem [1973]. J.M. Rosenblatt [1974] derives sufficient conditions for the existence of an invariant $\nu \sim \mu$ for finitely generated groups.

M. Wolff [1971] has considered the existence of vector valued invariant measures for amenable semigroups of operators.

J.M. Rosenblatt [1981] proved for some compact Abelian groups $X$ that the group $G$ of topological automorphisms of $X$ has Haar integral as unique $G$-invariant mean. (This has been used by Sullivan [1981] and Margulis [1980] to show that for $n > 3$ Lebesgue measure is the only finitely additive normalized measure on the $n$-sphere invariant under rotations.)

Further references are given in the notes of § 3.4 and in § 4.3.

## § 6.4 Amenable semigroups

We now consider a class of semigroups of operators more general than the $d$-parameter semigroups, and containing many non Abelian semigroups. This class constitutes a natural general setting for *mean* ergodic theorems. We need further restrictions for a pointwise ergodic theorem for measure preserving transformations.

Throughout this section, $G$ will be an abstract semigroup and $\mathcal{G}$ a $\sigma$-algebra of subsets of $G$ such that for $u \in G$ and $B \in \mathcal{G}$ the sets $Bu$, $Bu^{-1} := \{v \in G: vu \in B\}$ are $\mathcal{G}$-measurable. We assume the existence of a $\sigma$-finite measure $\nu$ on $(G, \mathcal{G})$ which is right invariant ($\nu(Bu) = \nu(B)$). E.g. $G$ can be a locally compact group and $\nu$ the right Haar measure.

**1. The mean ergodic theorem.** We first derive a mean ergodic theorem. Let $\mathfrak{X}$ be a (real or complex) Banach space, and let $\mathcal{S} = \{T_u: u \in G\}$ be a bounded semigroup of continuous linear operators in $\mathfrak{X}$. (We tacitly assume $T_{uv} = T_u T_v$). Put $M := \sup\{\|T_u\|: u \in G\} < \infty$. $\mathcal{S}$ is called *weakly measurable* if the maps $u \to \langle T_u x, h\rangle$ from $G$ into the scalar field of $\mathfrak{X}$ are measurable for all $x \in \mathfrak{X}$, $h \in \mathfrak{X}^*$. For $B$ with $\nu(B) < \infty$ let $S(B)x$ denote the element of $\mathfrak{X}^{**}$ with

$$(4.1) \qquad \langle S(B)x, h\rangle = \int_B \langle T_u x, h\rangle\, \nu(du).$$

If $x$ is an element of $\mathfrak{X}$ for which the weak (= strong) closed convex hull $C_x := \overline{\mathrm{co}}\, \mathscr{S} x$ is weakly compact, then $S(B)x$ belongs to $\mathfrak{X} \subset \mathfrak{X}^{**}$: To check this we may assume $v(B) = 1$. $C_x$, considered as a subset of $\mathfrak{X}^{**}$, is compact in the $\sigma(\mathfrak{X}^{**}, \mathfrak{X}^*)$-topology. If $S(B)x$ does not belong to $C_x$, there exists an $h \in \mathfrak{X}^*$ with

$$\mathrm{Re}\langle S(B)x, h\rangle > \beta_x := \sup\{\mathrm{Re}\langle y, h\rangle : y \in C_x\},$$

(see e.g. theorem V.2.10 in Dunford-Schwartz [1958]). This contradicts (4.1).

For any measurable $E \subset G$ we have $(B \cap Eu^{-1})u = Bu \cap E$ and hence $v(B \cap Eu^{-1}) = v(Bu \cap E)$. Because of the identity $1_{Eu^{-1}}(v) = 1_E(vu)$ this means that

$$(4.2) \qquad \int_B f(vu)\, v(dv) = \int_{Bu} f(v)\, v(dv)$$

holds for indicator functions $f = 1_E$. Thus (4.2) holds for all bounded measurable $f$ on $G$. Applying (4.2) with $f(v) = \langle T_v x, h\rangle$ we find

$$(4.3) \qquad S(B)\, T_u x = S(Bu)x.$$

We shall consider sequences $I = (I_n)$ of measurable subsets of $G$ having some or all of the following properties:

(P1) $0 < v(I_n) < \infty$, $(n = 1, 2, \ldots)$;
(P2) For all $u \in G$, $\lim_n v(I_n \Delta I_n u)/v(I_n) = 0$;
(P3) $I_n \subset I_{n+1}$, $(n = 1, 2, \ldots)$;
(P4) There exists $K_1 < \infty$ with $\lim_n v^*(I_N I_n)/v(I_n) \leqq K_1$ for all $N$;
(P5) There exists $K_2 < \infty$ with $v^*(I_n^{-1} I_n) \leqq K_2\, v(I_n)$, $(n = 1, 2, \ldots)$.

(Here $v^*$ denotes the outer measure, and $B_1^{-1} B_2$ the set $\{v \in G : \exists u \in B_1 \text{ with } uv \in B_2\}$.)

If (P1) holds, put $A_n^I x = v(I_n)^{-1} S(I_n)x$. If $C_x$ is weakly compact for all $x$, the $A_n^I$ are linear operators in $\mathfrak{X}$ with norm $\leqq M$, and $A_n^I x$ belongs to $C_x$. For any $x$ and $u$ there exists an $h \in \mathfrak{X}^*$ with $\|h\| = 1$ and

$$\begin{aligned}\|A_n^I T_u x - A_n^I x\| &= \langle A_n^I T_u x - A_n^I x, h\rangle \\ &= v(I_n)^{-1}\Big(\int_{I_n u} \langle T_v x, h\rangle\, v(dv) - \int_{I_n} \langle T_v x, h\rangle\, v(dv)\Big) \\ &= v(I_n)^{-1}\, v(I_n u \Delta I_n)\, 2M\|x\|.\end{aligned}$$

Applying Eberlein's theorem (Theorem 2.1.5) and the splitting theorem 2.1.9, we obtain the following mean ergodic theorem:

**Theorem 4.1.** *Let $v$ be right invariant and let $\mathscr{S}$ be a weakly measurable bounded semigroup in $\mathfrak{X}$. Let $(I_n)$ satisfy* (P1) *and (P2). If $C_x = \overline{\mathrm{co}}\, \mathscr{S} x$ is weakly compact for all $x$, the $A_n^I$ form a right-$\mathscr{S}$-ergodic net, and the strong limit* $\lim A_n^I x = \bar{x}$ *exists for each $x$. $\bar{x}$ belongs to the fixed space $F(\mathscr{S}) = \{y \in \mathfrak{X}: T_u y = y$* for all $u \in G\}$, *and $x - \bar{x}$ belongs to the closure of the linear subspace $N = \mathrm{lin}\{(T_u - I)z : u \in G, z \in \mathfrak{X}\}$.*

**Remarks.** 1) The compactness condition is satisfied if $\mathfrak{X}$ is reflexive. It is also satisfied if $\mathfrak{X}$ is the space $L_1$ of a finite measure space and the $T_u$ are $L_1 - L_\infty$-contractions.

2) $\|A_n^I T_u x - A_n^I x\| \to 0$ implies $A_n^I y \to \bar{x}$ for any $y = T_u x$, and hence for all $y \in C_x$. In particular $\bar{x}$ is the unique fixed point in $C_x$. If follows that if $(J_n)$ is another sequence of sets with the same properties, then $\lim A_n^I x = \lim A_n^J x$.

Let $C_b(G)$ denote the space of bounded continuous functions on $G$.

The existence of sequences $(I_n)$ with the properties (P1) and (P2) has been studied in the theory of topological groups. If $G$ is a semitopological semigroup and $f$ belongs to the space $C_b(G)$ we put $f_u(v) = f(vu)$ and ${}_uf(v) = f(uv)$. A positive functional $\varrho \in C_b(G)^*$ is called *right* (left) *invariant mean* if it satisfies $\langle \mathbb{1}, \varrho\rangle = 1$ and $\langle f_u, \varrho\rangle = \langle f, \varrho\rangle$ ($\langle {}_uf, \varrho\rangle = \langle f, \varrho\rangle$) for all $u \in G$ and all $f$. $G$ is called *right amenable* if a right invariant mean exists, and *amenable* if a (right *and* left) invariant mean exists. A concise introduction to the theory of amenable groups has been given by Greenleaf [1969]. Abelian, and, more generally, solvable groups, and compact groups are amenable. If $G$ is a $\sigma$-compact locally compact right amenable group (with right Haar measure) there exists an increasing sequence $(I_n)$ of compact sets with union $G$ which satisfies (P1) and (P2), and conversely the existence of such a sequence implies the right amenability of $G$ (see Emerson and Greenleaf [1967]). If one allows more general avering procedures, more can be said:

A right amenable $\mathcal{S}$ (with the weak operator topology) always admits a right $\mathcal{S}$-ergodic net, c.f. Sato [1978], Day [1969]. However, we shall need averages over *sets* $I_n$ in the discussion of pointwise convergence.

**2. The maximal ergodic theorem and pointwise convergence.** A semigroup $\{\tau_u\} = \{\tau_u : u \in G\}$ of endomorphisms $\tau_u$ of a $\sigma$-finite measure space $(\Omega, \mathcal{A}, \mu)$ is called measurable if the map $(u, \omega) \to \tau_u\omega$ is $(\mathcal{G} \otimes \mathcal{A}, \mathcal{A})$-measurable. Then the semigroup $\mathcal{S} = \{\tau_u\}$ defined in $L_p$, $(1 \leqq p \leqq \infty)$, by $T_u f = f \circ \tau_u$ is measurable.

Let $\mathcal{F}$ be the family of sets $I \in \mathcal{G}$ with $0 < \nu(I) < \infty$. A process $F = \{F_I : I \in \mathcal{F}\}$ is called *superadditive* for $\{\tau_u\}$ if each $F_I$ belongs to $L_1$, $I \cap J = \emptyset$ implies $F_I + F_J \leqq F_{I \cup J}$, $F_{Iu} = F_I \circ \tau_u$, $(u \in G, I \in \mathcal{F})$, and $\gamma(F) = \sup\{\nu(I)^{-1} \int F_I \, d\mu : I \in \mathcal{F}\} < \infty$. $F$ is called subadditive if $-F$ is superadditive, and additive if $F$ is sub- and superadditive. For any $f \in L_1$ the process $F^f = \{F_I^f : I \in \mathcal{F}\}$ with

$$F_I^f(\omega) = \int_I f(\tau_v \omega)\, \nu(dv)$$

is additive with $\gamma(F^f) = \int f \, d\mu$.

**Theorem 4.2** (Maximal ergodic theorem). *Let $G$ be a semigroup with right invariant measure $\nu$, and $\{\tau_u : u \in G\}$ a measurable semigroup of endomorphisms of the $\sigma$-finite measure space $(\Omega, \mathcal{A}, \mu)$. Let $(I_n)$ be a sequence of subsets of $G$ with* (P1)–(P5), *and let $F$ be a non negative superadditive process. Put $\bar{F}_k = \nu(I_k)^{-1} F_{I_k}$ and*

$E := \{\omega \in \Omega \colon \sup(\bar{F}_k(\omega) \colon 1 \leqq k \leqq N) > \alpha\}$, *where* $N$ *is an arbitrary integer and* $\alpha > 0$. *Then* $\mu(E) \leqq K_1 K_2 \gamma(F)/\alpha$.

The proof is similar to that of theorem 2.6. We begin with

**Lemma 4.3.** *Let* $A$ *be a measurable subset of* $G$ *for which* $I_N A$ *is contained in a set* $B$ *of finite measure. Let* $k \colon u \to k(u)$ *be a map of* $A$ *into* $\{1, \ldots, N\}$. *Then there exists a finite subset* $A' \subset A$ *for which the sets* $D_u := I_{k(u)} u$ *are disjoint such that* $A$ *is contained in the union of the sets* $\tilde{D}_u := I_{k(u)}^{-1} I_{k(u)} u$, $(u \in A')$.

The proof of this lemma is identical to that of lemma 2.5. The collections $\mathcal{M}_j$ are finite because of $\nu(D_u) \geqq \nu(I_j)$ and $D_u \subset B$.

*Proof of theorem 4.2.* First fix some $\omega$ and $n$ and put $A = A(\omega) := \{u \in I_n \colon \tau_u \omega \in E\}$. For any $u \in A$ there exists $k(u) \in \{1, \ldots, N\}$ with $\bar{F}_{k(u)}(\tau_u \omega) > \alpha$. Let $B$ be a set of minimal measure containing $I_N I_n$. If $M(\omega)$ is the union of the sets $D_u$, $(u \in A')$, obtained from lemma 4.3, property (P5) yields $\nu(A(\omega)) \leqq \nu^*(\bigcup_{u \in A'} \tilde{D}_u) \leqq K_2 \nu(M(\omega))$. As the sets $D_u$ are contained in $B$, the properties of $F$ imply

$$F_B(\omega) \geqq \sum_{u \in A'} F_{D_u}(\omega) = \sum \nu(D_u) \bar{F}_{k(u)}(\tau_u \omega) \geqq \alpha \nu(M(\omega)),$$

and hence $\nu(A(\omega)) \leqq \alpha^{-1} K_2 F_B(\omega)$. If we integrate the left side and use Fubini we obtain

$$\int \nu(A(\omega)) \mu(d\omega) = \int_{I_n} \mu\{\omega \colon \tau_u \omega \in E\} \nu(du) = \nu(I_n) \mu(E).$$

On the other hand we have $\int F_B(\omega) \mu(d\omega) \leqq \gamma(F) \nu(B) = \gamma(F) \nu^*(I_N I_n)$. Hence $\mu(E) \leqq \alpha^{-1} K_2 \gamma(F) \nu^*(I_N I_n)/\nu(I_n)$. The assertion now follows from (P4). □

Theorem 4.2 is due to Tempel'man for additive $F$. The proof of a pointwise ergodic theorem for superadditive processes with general $G$ is an open problem, but the additive case is now easy:

**Theorem 4.4** (Tempel'man). *Let* $G$, $\{\tau_u\}$, $(I_n)$ *satisfy the assumptions of theorem 4.2, and* $f \in L_p(\mu)$ *for some* $p$ *with* $1 \leqq p < \infty$. *Then* $A_n^I f$ *converges a.e..*

*Proof.* Using theorem 4.1 and 4.2 we can extend the proof of theorem 2.9 to the present situation. □

It is clear that theorem 4.2 yields also a dominated ergodic theorem for additive $F$. We only sketch the arguments. Put $M(f) = f \vee \sup_n A_n^I f$ and $c = K_1 K_2 + 1$. Then $\mu(M(f) > \alpha) \leqq c\alpha^{-1} \int f^+$ holds for $f \geqq 0$, and hence for all $f$ with $f^+ \in L_1$. In particular, $\mu(M(f) > 2\alpha) = \mu(M(f - \alpha) > \alpha) \leqq c\alpha^{-1} \int (f - \alpha)^+$. Replacing $\alpha$

by $\alpha/2$ we obtain for $f \geqq 0$:

$$(4.4) \qquad \mu(M(f) > \alpha) \leqq 2c\alpha^{-1} \int_{\{2f > \alpha\}} f d\mu \leqq 2c\alpha^{-1} \int_{\{2M(f) > \alpha\}} f d\mu .$$

If we write $X = f$ and $Y = M(f)$, the computation in the proof of lemma 1.6.2, carried out with the new estimate (4.4), leads to

$$(4.5) \qquad \int \psi(Y) d\mu \leqq 2c \int \{X(\omega) \int_0^{2Y(\omega)} t^{-1} \psi(dt)\} \mu(d\omega).$$

Now the argument in the beginning of the proof of theorem 1.6.2 implies

$$(4.6) \qquad \| Y \|_p \leqq 2^{p-2} \frac{cp}{p-1} \| f \|_p, \quad (1 \leqq p < \infty).$$

For the $L_1$-norm we obtain the estimate

$$\| Y \|_1 \leqq \mu(\Omega) + \int_1^\infty \mu(Y > \alpha) d\alpha \leqq \mu(\Omega) + 2c \int_1^\infty \alpha^{-1} \int_{\{2f > \alpha\}} f(\omega) \, d\mu \, d\alpha$$
$$\leqq \mu(\Omega) + 2c \int f \int_1^{2f \vee 1} \alpha^{-1} \, d\alpha \, d\mu = \mu(\Omega) + 2c \int f \log^+ (2f) \, d\mu .$$

## Notes

Day [1942] proved mean ergodic theorems for Abelian semigroups. Calderon [1953], Tempel'man [1962–1974], Greenleaf [1973], Chatard [1970], Bewley [1971], and others have studied mean ergodic theorems for amenable and more general semigroups, some of them unaware of the work of Eberlein. Implications between measurability and continuity properties of semigroups are discussed in the memoir of Moore [1968].

Nagel [1973] used theorem 2.1.11 to deduce the existence of an $\mathscr{S}$-absorbing projection in $\overline{\text{co}}\,\mathscr{S}$ for amenable bounded $\mathscr{S}$ on a reflexive Banach space from the existence of fixed points (due to Day [1961]). This does not involve the construction of $\mathscr{S}$-ergodic nets – which may be an advantage in some respects. Along similar lines, Nagel-Palm-Derndinger [1984] proved:

**Theorem 4.5.** *If $\mathscr{S}$ is bounded right amenable, then the following assertions are equivalent:*
*(a) $\overline{\text{co}}\,\mathscr{S}$ contains an $\mathscr{S}$-absorbing projection,*
*(b) $\overline{\text{co}}\,\mathscr{S} x$ contains a fixed point for each $x$,*
*(c) The fixed points of $\mathscr{S}$ separate the fixed points of $\mathscr{S}^*$.*

Related results appear in Takahashi [1972], and Kijima-Takahashi [1973].

Amenable semigroups play an important role in the Jacobs-Deleeuw-Glicksberg theory. It was shown by Deleeuw-Glicksberg that the sets $\mathfrak{X}_{rec}$ and $\mathfrak{X}_{fl}$ are $\mathscr{S}$-invariant subspaces iff $\mathscr{S}$ is amenable. An application to Banach lattices appears in Schaefer [1974].

Itoh, Körezlioğlu and Takahashi [1973] have characterized extremal invariant means for semigroups $\mathscr{S}$ on the space of bounded measurable functions by ergodicity.

The problem of constructing avering sequences $(I_n)$ and averaging measures (weighted averages) has been discussed by Oseledeč [1965] and Tempel'man [1967].

Blum and Eisenberg [1974] call a sequence $(\varrho_n)$ of probability measures on a locally compact Abelian group $G$ *generalized summing* if, for every group $\{T_u, u \in G\}$ of unitary operators $T_u$ on a Hilbert space $\mathfrak{H}$ and every $x \in \mathfrak{H}$ the averages $\int T_u x \varrho_n(du)$ converge strongly to an invariant element of $\mathfrak{H}$. They show that this property is equivalent to each of the following conditions

(i) For every character $\chi$ on $G$ with $\chi \not\equiv 1$ the Fourier transforms $\hat{\varrho}_n(\chi)$ converge to 0;

(ii) $\varrho_n$ considered as restrictions of measures on the Bohr compactification $\bar{G}$ of $G$ converge weakly to Haar measure on $\bar{G}$.

If $(I_n)$ satisfies (P1) and $\varrho_n^I(B) = \nu(B \cap I_n)/\nu(I_n)$ then (P2) clearly implies that $(\varrho_n^I)$ is generalized summing.

But one can construct sequences $k_1 < k_2 < \ldots$ so that $I_n = \{k_1, \ldots, k_n\}$ does not satisfy (P2), but $(\varrho_n^I)$ is generalized summing, see Blum-Eisenberg-Hahn [1973], Blum-Cogburn [1975]. For related results, see Blum-Reich [1976], [1982].

The class of groups, for which sequences $(I_n)$ with the properties (P1)–(P5) exist, is fairly small. (If $G$ is countable, essentially the only such groups are those with polynomial growth, i.e., finitely generated groups for which the size of the sets $W_n$ of elements representable by a product of at most $n$ generators is bounded by a fixed power of $n$. Then one can take $I_n = W_n$.)

On the other hand, (P1) is essential for the definition of the averages, and examples of Tempel'man [1972] show that (P2), (P3) cannot even be omitted for $G = \mathbb{Z}$, and (P5) not for $G = \mathbb{Z}^2$. (For $G = \mathbb{Z}^d$, (P4) is a consequence of (P2)). In these examples $f$ is bounded and $\mu$ a probability measure. Greenleaf [1973] has shown that without (P5) a.e.-convergence of $A_n^I f$ may fail even for $G = \mathbb{Z}$, using an unbounded integrable $f$. The following theorem provides many examples having both properties, $G = \mathbb{Z}$ and $|f| \leqq 1$. In addition, it sharpens an example of Krengel [1971], showing that the individual ergodic theorem for subsequences fails for all aperiodic automorphisms; see § 8.2.

**Theorem 4.6.** *There exists a sequence* $(I_n)$ *of subsets of* $\mathbb{Z}$ *with* (P1)–(P4) *such that for all n the elements of* $I_{n+1} \setminus I_n$ *are all larger than the elements of* $I_n$, *and such that for all aperiodic automorphisms* $\tau$ *of a probability space there exists an* $f = 1_B$, *for which the sequence* $A_n^I f$ *(with* $\tau_u = \tau^u$*) fails to converge a.e..*

*Sketch of proof.* $I_1 := \{0\}$. Step $n+1$: $I_i$, $(i = 1, \ldots, N_n)$ are known. Write $N = N_n$. Choose $k$ larger than the absolute value of all elements of these $I_i$, $l = 2^{n+2}$, and an extremely big $m$. Rohlin's lemma yields a $D$ $(= D(n))$ for which $D_i = \tau^i D$, $(i = 0, \ldots, 2k(l+1)m - 1)$, are disjoint and cover most of $\Omega$. $C = C(n)$ shall contain the first $2k$ sets $D_i$, then it omits $2kl$ of them, then it contains the next $2k$ (second block), omits $2kl$ again, etc. Choose the first $k$ elements of $I_{N+1} \setminus I_N$ consecutively so that $\tau^j$ carries the first half of the $\nu$-th block into block $\nu + 1$. Then choose the next $k$ new elements $j$ so that the first half of block $\nu$ goes into block $\nu + 2$ (except for some blocks at the end). If $I_{N+1}$ is obtained in this way adding sufficiently often $k$ elements, then $A_{N+1}^I 1_C(\omega)$ is close to 1 for $\omega$ in the first halves of most blocks. The next set $I_{N+2}$ is constructed similarly to make the averages large in the second halves of most blocks. Then choose $I_{N+3}$ so as to produce big averages on the first stretch of length $k$ of the gaps between blocks. After the construction of $I_{N_{n+1}}$ with $N_{n+1} = N + 2(l+1)$ we know that on most of $\Omega$ at least one of the averages constructed in this step is close to 1.

After completing all steps put $B = \cup C(n)$. Then $\limsup A_k^I f = 1$ a.e.. But $\int f d\mu < 1$ and the mean ergodic theorem shows that we cannot have a.e.-convergence. □

The condition on the $I_{n+1} \setminus I_n$ could be replaced by the condition $\mathbb{Z} = \cup I_n$, and one could obtain divergence a.e. by minor modifications of the construction.

The proof above is similar to that of Krengel [1971]. While it was written, A. Bellow [1984] showed that one could take $I_n \setminus I_{n-1} = \{2^n, 2^n + 1, \ldots, 2^n + n\}$.

Although (P5) cannot be omitted in Tempelman's theorem it is not a necessary condition for a.e.-convergence. In fact, for some groups, Greenleaf and Emerson [1974] have succeeded in constructing sequences $(I_n)$ for which a.e.-convergence of $A_n^I f$ holds for $f \in L_1$ (and (P5) fails) by completely different techniques. They appeal to the non commutative ergodic theorems of section 6.1 and show that for certain groups there exist sequences $(I_n)$ for which $A_n^I$ has the form of the averages discussed there.

Chatard [1970], [1972] and Bewley [1971] obtained results similar to Tempelman's after the announcement of his main results but before his proofs were available. Further related results were obtained by Emerson [1974], [1974a], and Tempel'man [1974].

The proof of theorem 4.2 uses arguments on $G$ for fixed $\omega$ and then a passage to $\Omega$ via Fubini. Calderon [1968] has shown that this idea (appearing already in Wiener's work) admits a generalization to general principles, showing that maximal estimates in $G$ translate into more general estimates for measure preserving transformations.

The paper of Ornstein-Weiss [1983] contains a short proof of theorem 4.4 for countable groups $G$.

Lin [1977] has considered operators of the form $U = \int_G T_g \nu(dg)$, where $\nu$ is a regular probability measure on the Borel set of a locally compact Abelian semigroup $G$ and $\{T_g: g \in G\}$ a strongly continuous semigroup of contractions in a Banach space $\mathfrak{X}$. He related the convergence of $U^n x$ to assumptions on the support of $\nu$.

Luczak [1974] has given a spectral representation of left stationary processes over a separable locally compact group $G$ of type I, and he has proved a mean ergodic theorem when $G$ is the group of linear transformations of the line.

Finally, we mention some interesting *ergodic theorems for the group of rational rotations.* Let $f$ be measurable on $\{z \in \mathbb{C}: |z| = 1\}$, $z_n = e^{2\pi i/n}$ and $f_n(z) = n^{-1} \sum_{k=0}^{n-1} f(z z_n^k)$. Jessen [1934] proved the convergence a.e. of $f_{n_i}$ for $f \in L_1(\lambda)$ and sequences $\mathbf{n} = (n_i)$ with $n_{i+1}/n_i \in \mathbb{N}$, called *chains*. Rudin [1964] showed that $f_n$ need not converge a.e. for $f \in L_\infty$; more specifically, he showed that for any subsequence $(p_i)$ of the primes there exists an $f \in L_\infty$ such that $f_{p_i}$ diverges a.e.. Jessen's result has been improved by Baker [1976] and by Dubins and Pitman [1979]. For $K, L \subset \mathbb{N}$ let $K \vee L = \{k \vee l: k \in K, l \in L\}$ where $k \vee l$ is the smallest common multiple of $k$ and $l$. The *dimension* $d$ of $K \subset \mathbb{N}$ is the smallest integer such that there are chains $K_1, \ldots, K_d$ with $K \subset K_1 \vee \ldots \vee K_d$. Dubins and Pitman showed that $f_{n_i}$ converges a.e. for all $f \in L \log^{d-1} L$ if $(n_i)$ has dimension $d$.

A related problem deals with the convergence of $B_n^{\mathbf{k}} f(\omega) = n^{-1} \sum_{i=1}^{n} f(k_i \omega)$, where $\mathbf{k} = (k_i)$ is a strictly increasing sequence of integers and $f$ is periodic on $\mathbb{R}$ with period 1. The averages $B_n^{\mathbf{k}} f$ converge a.e. for Riemann integrable $f$, but Marstrand [1970] showed that $B_n^{\mathbf{k}} f$ may diverge for bounded measurable $f$ even in the special case $k_i = i$.

Both Rudin's and Marstrand's examples may be considered as counterexamples to ergodic theorems for Abelian semigroups with countably many generators.

Kieffer [1975a] and Emerson [1975/76] considered the problem of convergence of $\lambda(A_n)^{-1} S(A_n)$ for strongly subadditive set functions $S$ on unimodular amenable groups with Haar measure $\lambda$.

# Chapter 7: Local ergodic theorems and differentiation

Let $\{T_t, t > 0\}$ be a semigroup of linear operators in $L_p$. Under suitable assumptions

$$S_{a,b} f = \int_a^b T_s f ds$$

is well defined for all $f \in L_p$. Put $A_{0,t} = t^{-1} S_{0,t}$. Local ergodic theorems assert that

(0.1) $\quad A_{0,t} f$ converges a.e. for all $f \in L_p$ as $t \to 0$

under various conditions, or they give an analogous conclusion for the multi-parameter case. They may be considered as results on the differentiability of indefinite integrals.

We also look at a more general setting. Recall that a family $F = \{F_t, t > 0\}$ of elements of $L_p$ is called an *additive process* for the semigroup if it satisfies: $F_{t+s} = F_t + T_t F_s$, $(t, s > 0)$. Clearly $F_t = S_{0,t} f$ is an example. If $F_t = (T_t - I) f$, then $\lim_{t \to 0+0} t^{-1} F_t$ – if it exists – is the usual derivative of $f$, when $\{T_t\}$ is the semigroup of translations in $\mathbb{R}$. In § 1 of this chapter we study the a.e.-convergence of $t^{-1} F_t$ as $t \to 0 + 0$ for additive processes and semigroups of positive operators. In § 2 we discuss extensions to non positive operators, multiparameter semigroups and functions $f$ taking values in a Banach space.

## § 7.1 Positive 1-parameter semigroups

**1. Two decompositions of $\Omega$.** Let $\mathcal{T} = \{T_t, t > 0\}$ be a locally bounded strongly continuous semigroup of linear operators in $L_p$ of a $\sigma$-finite measure space $(\Omega, \mathcal{A}, \mu)$, where $1 \leqq p < \infty$ is fixed. Recall from § 1.6 that the integrals $S_{a,b} f$ may be defined either using suitable representatives $f_\infty(t, \cdot)$ of $T_t f$ or as the limit of the Riemann sums. E.g., $S_{0,b} f$ is the limit of

$$R_b(n) f = (b/n) \sum_{i=0}^{n-1} T_{b/n}^i f.$$

We do not assume that the semigroup is strongly continuous at 0.

By a *process* in $L_p$ we mean a family $F = \{F_t, t > 0\}$ of elements of $L_p$. $F$ is called *positive* if $F_t \in L_p^+$, $(t > 0)$, and *strongly continuous* if $\| F_t - F_s \|_p \to 0$, $(t \to s > 0)$.

As the $F_t$ are equivalence classes mod $\mu$ the question of a.e.-convergence of $t^{-1}F_t$ makes sense only when $t$ ranges through a countable set or when we refer to specific representatives. A function $\tilde{F}$: $]0, \infty[ \times \Omega$ is a *representative* for $F$ if $\tilde{F}$ is measurable with respect to the obvious product-$\sigma$-algebra and if, for each $t > 0$, $\tilde{F}(t, \cdot)$ is a representative of $F_t$. It is easy to see that every strongly continuous process has a representative. Note that an additive process $F$ with $\|F_t\|_p \to 0$, $(t \to 0)$, is strongly continuous.

In this section we shall assume that the operators $T_t$ are positive. A positive additive process must then be non decreasing, i.e., $F_t \leqq F_s$ holds for $t < s$. It is easy to see that a strongly continuous non decreasing process $F$ has a non decreasing representative $\tilde{F}$: $\tilde{F}$ satisfies $\tilde{F}(t, \omega) \leqq \tilde{F}(s, \omega)$ whenever $0 < t \leqq s$ and $\omega \in \Omega$.

If $Q$ is a countable dense subset of $\mathbb{R}^+ \setminus \{0\}$ we say that $Q$-$\lim_{t\to 0} f_t$ exists a.e. if the limit exists a.e. when $t$ ranges through $Q$. If $\tilde{F}$ is a non decreasing representative of $F$ the assertions «$Q$-$\lim_{t\to 0} t^{-1} F_t$ exists a.e.» and «$\lim_{t\to 0} \tilde{F}(t, \omega)$ exists for a.e. $\omega$» are entirely equivalent. It will therefore be convenient to forget about the representatives and to speak only about $Q$-limits. (Another way of dealing with these technicalities is to use separability as in § 1.6 or essential limits; see the Notes. For local ergodic theorems they do not appear since $S_{0,a}f(\omega)$ is a continuous function of $a$.) Before we can state the main result we must describe two decompositions of $\Omega$:

Let $h' \in L_p^+$ with $h' > 0$ a.e., and let $h = S_{0,1}h'$. $C_0 := \{h > 0\}$ is called the *initially conservative part of* $\Omega$, and $D_0 = \{h = 0\}$ is called the *initially dissipative part*. It follows from the following lemma, that this decomposition does not depend on $h'$:

**Lemma 1.1.** *$T_t f$ vanishes on $D_0$ for all $f \in L_p$ and all $t > 0$. If $C_0$ is contained in $\{g > 0\}$ for some $g \in L_p^+$, then $S_{0,t}g$ is strictly positive on $C_0$ for all $t > 0$.*

*Proof.* We may assume $f \geqq 0$ and $t \leqq 1$. The continuity of the semigroup implies that $T_t h'$ vanishes on $D_0$. Now $T_t f = \lim_n T_t (f \wedge nh')$. If the second assertion is wrong there exists a subset $A$ of $C_0$ having positive measure, and an $\eta > 0$ with $T_\varepsilon g = 0$ on $A$ for $0 < \varepsilon < \eta$. For $t > \varepsilon$, $T_{t-\varepsilon}h'$ vanishes in $D_0$. Hence $T_t h' = T_\varepsilon(\lim_n (ng \wedge T_{t-\varepsilon}h')) = 0$ on $A$. As $\varepsilon$ was arbitrarily small we obtain $T_t h' = 0$ in $A$ for all $t > 0$, a contradiction to $A \subset C_0$. □

We say that *all mass in $A$ disappears instantly* if $\|S_{0,t}f\|_p$ is $= 0$ for all $f \in L_p^+(A)$ and all $t > 0$. Clearly there is a maximal set $D^*$ mod $\mu$ in which all mass disappears instantly. Put $C^* = \Omega \setminus D^*$. By the strong continuity of the semigroup, $f \in L_p(D^*)$ implies $T_t f = 0$ for all $t > 0$.

Obviously $D^*$ is empty if the enlarged semigroup $\{T_t, t \geqq 0\}$ is strongly continuous at $t = 0$ and $T_0$ is the identity. For other choices of $T_0$ the strong continuity of the semigroup at 0 need not imply $D^* = \emptyset$.

In general the sets $D_0 \cap D^*$, $D_0 \cap C^*$, $C_0 \cap D^*$, and $C_0 \cap C^*$ may all be non empty; see Lin [1982], Suzuki [1983]. It is simple to see that $D^*$ is the initially dissipative part of the adjoint semigroup.

**Lemma 1.2.** *$D^* \subset D_0$ holds when the $T_t$, $(t > 0)$, are contractions, and also when $\{T_t, t \geqq 0\}$ is strongly continuous at $t = 0$ with $\| T_0 \| \leqq 1$.*

*Proof.* Take $h = S_{0,1} h'$ as above. In the first case $\| h 1_{C^*} \|_p \geqq \| T_\varepsilon (h 1_{C^*}) \|_p = \| T_\varepsilon h \|_p \to \| h \|_p$, $(\varepsilon \to 0)$, implies $h = h 1_{C^*}$. In the second case $T_\varepsilon (h 1_{C^*}) = T_\varepsilon h$ tends in norm to $T_0 h = h$ and to $T_0 (h 1_{C^*})$. Hence $\| h 1_{C^*} \|_p \geqq \| T_0 (h 1_{C^*}) \|_p = \| h \|_p$. Thus $h$ must vanish on $D^*$. □

The decomposition of $\Omega$ into $C_0$ and $D_0$ has been introduced by Akcoglu-Chacon [1970], and that into $C^*$ and $D^*$ by Sato [1978b].

**2. Positive additive processes.** The following "differentiation theorem" (Akcoglu-Krengel [1979]) unifies and extends several previous local ergodic theorems:

**Theorem 1.3.** *Let $\{T_t, t > 0\}$ be a locally bounded strongly continuous semigroup of positive linear operators in $L_p$, $(1 \leqq p < \infty)$, and $F = \{F_t, t > 0\}$ a positive additive process in $L_p$ with $\| F_t \|_p \to 0$, $(t \to 0 + 0)$. Then $Q$-$\lim_{t \to 0+0} t^{-1} F_t$ exists a.e. in $C^* \cup D_0$ for each countable dense subset $Q$ of $\mathbb{R}^+$. In particular, the $Q$-limits exist a.e. in $\Omega$ under the additional assumption that the $T_t$ are contractions.*

The proof of this theorem will be obtained after several lemmas. We first define, for $t > 0$, a class $\mathscr{P}_t = \mathscr{P}_t(F)$ of $L_p$-functions as follows: $f \in \mathscr{P}_t(F)$ if and only if there exist two numbers $r > 0$ and $s > 0$ and an integer $n \geqq 0$ and functions $f_0, f_1, \ldots, f_n \in L_p^+$ such that

(i) $\quad r + ns \leqq t$;

(ii) $\quad T_s^m \dfrac{1}{r} F_r \geqq \sum\limits_{i=0}^{m} T_s^{m-i} f_i \quad$ for $m = 0, 1, \ldots, n$, and

(iii) $\quad f = \sum\limits_{i=0}^{n} f_i$.

In order to understand the intuitive meaning of this definition the reader may like to recall the filling scheme of Chacon-Ornstein. If a hole $f$ is filled with starting mass $g$ by repeated applications of $T_s$, we obtain $T_s^m g \geqq \sum_{i=0}^{m} T_s^{m-i} f_i$, where $f_i$ is the part of the hole filled with a portion of $T_s^i g$. If the additive process is of the form $F_t = S_{0,t} g$ we actually have a density $g$ at the beginning. But in general we have to replace it by $r^{-1} F_r$. Thus, the functions $f \in \mathscr{P}_t(F)$ are the holes we can fill up to time $t$ if we reserve a short initial interval of length $r$ in order to define an "approximate density" $r^{-1} F_r = \bar{F}_r$ with which we want to do the job.

**Lemma 1.4.** *If $f \in \mathscr{P}_t(F)$ then $S_{0,u} f \leqq F_{t+u}$ for each $u > 0$.*

*Proof.* Let $f = f_0 + \ldots + f_n$ with $\sum_{i=0}^{m} T_s^{m-i} f_i \leqq T_s^m \bar{F}_r$ for $m = 0, \ldots, n$ and $r + sn \leqq t$. Then

$$\begin{aligned}
S_{0,u} f &= \sum_{i=0}^{n} \int_0^u T_\alpha f_i \, d\alpha \leqq \sum_{i=0}^{n} \int_0^{u+(n-i)s} T_\alpha f_i \, d\alpha \\
&\leqq \sum_{m=0}^{n-1} \int_0^s T_\alpha \sum_{i=0}^{m} T_s^{m-i} f_i \, d\alpha + \int_0^u T_\alpha \sum_{i=0}^{n} T_s^{n-i} f_i \, d\alpha \\
&\leqq \sum_{m=0}^{n-1} \int_0^s T_\alpha (T_s^m \bar{F}_r) \, d\alpha + \int_0^u T_\alpha (T_s^n \bar{F}_r) \, d\alpha \\
&= \sum_{m=0}^{n-1} \int_{ms}^{(m+1)s} T_\alpha \bar{F}_r \, d\alpha + \int_{ns}^{u+ns} T_\alpha \bar{F}_r \, d\alpha \\
&= \int_0^{u+ns} T_\alpha \bar{F}_r \, d\alpha = \int_0^{u+ns} \frac{1}{r} (F_{\alpha+r} - F_\alpha) \, d\alpha \\
&= \frac{1}{r} \int_r^{r+ns+u} F_\alpha \, d\alpha - \frac{1}{r} \int_0^{u+ns} F_\alpha \, d\alpha \leqq \frac{1}{r} \int_{u+ns}^{u+ns+r} F_\alpha \, d\alpha \\
&\leqq F_{u+ns+r} \leqq F_{u+t}. \quad \square
\end{aligned}$$

For $p = 1$, $\sup \{ \int_E f \, d\mu : f \in \mathscr{P}_t(F) \}$ can be used successfully to measure how much we can fill into a set $E$ up to time $t$. But for general $p$ with $1 \leqq p < \infty$ we have to use

$$\Psi_E^t(F) := \sup \{ \alpha \geqq 0 : \forall \varepsilon > 0 \, \exists f \in \mathscr{P}_t(F) \quad \text{with } \| \alpha 1_E - f \|_p < \varepsilon \}.$$

Note that $0 \leqq g \leqq f \in \mathscr{P}_t(F)$ implies $g \in \mathscr{P}_t(F)$. Hence, if $\alpha_0 = \Psi_E^t(F)$, we can fill a hole of size $\alpha_0 1_E$ up to time $t$ arbitrarily well, but no hole $\beta 1_E$ with $\beta > \alpha_0$.

It is clear that $0 \leqq \Psi_E^{t'}(F) \leqq \Psi_E^t(F)$ if $t' \leqq t$. Hence

$$\Psi_E(F) := \lim_{t \to 0+0} \Psi_E^t(F) \geqq 0$$

exists.

**Lemma 1.5.** *Let $h' \in L_p^+$, $h' > 0$ a.e. and $h = S_{0,1} h'$. Assume that $h > \alpha > 0$ on $E \in \mathscr{A}$ and $\mu(E) > 0$. Then given $\varepsilon > 0$ there is an $E' \subset E$ so that $\mu(E \setminus E') < \varepsilon$ and so that the additive process $H = \{H_t : t > 0\}$ given be $H_t = S_{0,t} h$ has the property that $\Psi_{E''}(H) \geqq \alpha - \varepsilon$ for all $E'' \subset E'$.*

*Proof.* The definition of $h$ implies that $\bar{H}_t = t^{-1} H_t$ tends to $h$ in norm as $t \to 0$. Hence $\| 1_E \bar{H}_t - 1_E h \|_p \to 0$. Find $\varepsilon_i > 0$ with $\sum_{i=1}^{\infty} \varepsilon_i < \varepsilon$ and for each $i$ choose $t_i > 0$ so that $t_i > t_{i+1} \to 0$ and so that

$$\mu(E \cap \{ \bar{H}_{t_i} \leqq \alpha - \varepsilon \}) < \varepsilon_i .$$

This uses the finiteness of $\mu(E)$, which follows from $h \geqq \alpha 1_E$. Let

$$E' = \bigcap_{i=1}^{\infty} E \cap \{\bar{H}_{t_i} > \alpha - \varepsilon\}.$$

Then $\mu(E \setminus E') < \varepsilon$, and if $E'' \subset E'$ then $\bar{H}_{t_i} 1_{E''} > (\alpha - \varepsilon) 1_{E''}$ for all $t_i$. This means that $(\alpha - \varepsilon) 1_{E''} \in \mathscr{P}_{t_i}(H)$ for all $t_i$, because we can take $r = t_i, n = 0, f_0 = \bar{H}_{t_i}$ to fill the hole $(\alpha - \varepsilon) 1_{E''}$. Now $(\alpha - \varepsilon) \leqq \Psi_{E''}^{t_i}(H)$ and this tends to $\Psi_{E''}(H)$. □

**Lemma 1.6.** *Let $E \subset C^*$, $0 < \mu(E)$, then $\Psi_E(F) < \infty$.*

*Proof.* We may assume $\mu(E) < \infty$ because smaller holes are filled more easily. By the local boundedness of the semigroup we have $K := \sup\{\| T_t \| : 0 < t \leqq 1\} < \infty$. Let $\delta := \sup\{\| A_{0,t} 1_E \|_p : 1/2 \leqq t \leqq 1\}$. $E \subset C^*$ and $0 < \mu(E)$ imply $\delta > 0$. Choose an integer $N \geqq 1$ so that $N^{-1} K \| 1_E \|_p < \delta/2$, and let $t_1 = (N+1)^{-1}$ We shall even prove the finiteness of $\Psi_E^{t_1}(F)$.

If for some $\alpha > 0$ there exists an $f \in \mathscr{P}_{t_1}(F)$ with $\| \alpha 1_E - f \|_p < \varepsilon \leqq 1$, then there exists $r > 0$, $s > 0$, $n \geqq 0$ with $r + ns \leqq t_1$ and $f$ is a sum of functions $f_i$, $(i = 0, \ldots, n)$, with $\sum_{i=0}^{n} T_s^{n-i} f_i \leqq T_s^n \bar{F}_r$. There also exists an $e \in L_p^+$ with $\| e \|_p < \varepsilon$ and $\alpha 1_E \leqq f + e$. We need an upper estimate for $\alpha$. Now

$$\begin{aligned} \alpha 1_E \leqq f + e &= \sum_{i=0}^{n} (f_i - T_s^{n-i} f_i) + \sum_{i=0}^{n} T_s^{n-i} f_i + e \\ &\leqq \sum_{i=0}^{n} (f_i - T_s^{n-i} f_i) + T_{sn} \bar{F}_r + e . \end{aligned}$$

Hence

$$\begin{aligned} \alpha S_{0,Nt_1} 1_E &\leqq \sum_{i=0}^{n} (S_{0,Nt_1} f_i - T_s^{(n-i)} S_{0,Nt_1} f_i) + \frac{1}{r} S_{0,Nt_1} T_{sn} F_r + S_{0,Nt_1} e \\ &\leqq \sum_{i=0}^{n} S_{0,(n-i)s} f_i + \frac{1}{r} \int_0^{Nt_1} T_{\alpha+ns} F_r \, d\alpha + S_{0,Nt_1} e . \end{aligned}$$

Using $T_{\alpha+ns} F_r = F_{r+\alpha+ns} - F_{\alpha+ns}$ we can continue

$$\begin{aligned} &\leqq S_{0,ns} \sum_{i=0}^{n} f_i + \frac{1}{r} \int_{Nt_1+ns}^{Nt_1+sn+r} F_\alpha \, d\alpha + S_{0,Nt_1} e \\ &\leqq \alpha S_{0,ns} 1_E + S_{0,ns} e + F_{Nt_1+ns+r} + S_{0,Nt_1} e \\ &\leqq \alpha S_{0,ns} 1_E + 2 S_{0,Nt_1} e + F_1 . \end{aligned}$$

Dividing by $Nt_1$, we obtain

$$\frac{\alpha}{Nt_1} S_{0,Nt_1} 1_E \leqq \alpha \frac{ns}{Nt_1} \cdot \frac{1}{ns} S_{0,ns} 1_E + \frac{2}{Nt_1} S_{0,Nt_1} e + \frac{N+1}{N} F_1$$

and

$$\alpha \| A_{0,Nt_1} 1_E \|_p \leqq \frac{\alpha}{N} \| A_{0,ns} 1_E \|_p + 2 \| A_{0,Nt_1} e \|_p + 2 \| F_1 \|_p .$$

The definition of $\delta$ and the inequalities $1/2 \leqq Nt_1 \leqq 1$ now imply

$$\alpha\delta \leqq \alpha \frac{K}{N} \| 1_E \|_p + 2K\varepsilon + 2 \| F_1 \|_p \leqq \alpha \frac{\delta}{2} + 2K\varepsilon + 2 \| F_1 \|_p .$$

Hence $\alpha \leqq 2\delta^{-1}(2K\varepsilon + 2\| F_1 \|_p)$. We have obtained an upper bound for $\alpha$ which does not depend on $f \in \mathscr{P}_{t_1}(F)$. This means that $\Psi_E^{t_1}(F)$ is finite. □

In order to prove the a.e.-convergence of $t^{-1} F_t$, $(t \to 0+0)$, when $t$ ranges through a countable dense subset $Q \subset \mathbb{R}^+ \setminus \{0\}$, we may assume that $Q$ contains the positive rationals. For $t_0 > 0$ let $Q(t_0)$ denote the set $\{s \in Q : 0 < s < t_0\}$.

**Lemma 1.7.** *Let $F = \{F_t\}$ be a positive additive process as in theorem 1.3 and let $g \in L_p^+$. If* $\sup\{(F_q - S_{0,q}g) : q \in Q(t_0)\} > 0$ *on $E \in \mathscr{A}$*, then for any $\varepsilon > 0$ there exists an $f \in \mathscr{P}_{t_0}(F)$ so that $\| f - g1_E \|_p < \varepsilon$.

*Proof.* Let $\varepsilon > 0$ be given. By the strong continuity of the semigroup we can find finitely many rational $q_i \in Q(t_0)$, $(1 \leqq i \leqq n)$, and an $\alpha > 0$ so that the sets $E_i := E \cap \{(F_{q_i} - S_{0,q_i}g) > \alpha\}$ have the property that

$$\int_{E \setminus \bigcup_{i=1}^{n} E_i} g^p \, d\mu < \left(\frac{\varepsilon}{4}\right)^p .$$

Let $\delta > 0$ be so small that $\| g1_H \|_p < \varepsilon/4n$ for all $H \in \mathscr{A}$ with $\mu(H) < \delta$. Choose $M$ so large that $m \geqq M$ implies $\| R_{q_i}(m)g - S_{0,q_i}g \|_p < \alpha\delta^{1/p}$ for $i = 1, \ldots, n$. As the $q_i$ are rational numbers we can find $r > 0$ and integers $m_1, \ldots, m_n \geqq M$ so that $q_i = rm_i$ for all $i$. Let $B_i = E_i \cap \{(F_{q_i} - R_{q_i}(m_i)g) \leqq 0\}$. Then $B_i$ is contained in $\{| R_{q_i}(m_i)g - S_{o,q_i}g | > \alpha\}$. The choice of $M$ implies $\alpha^p \mu(B_i) < \alpha^p \delta$ and hence $\| g1_{B_i} \|_p < \varepsilon/4n$. Let $E' = \bigcup_{i=1}^{n} (E_i \setminus B_i)$, then $\| g1_{E \setminus E'} \|_p < \varepsilon$. It now suffices to show that $g1_{E'} \in \mathscr{P}_{t_0}(F)$. On $E_i \setminus B_i$ we have

$$F_{q_i} - R_{q_i}(m_i)g = r \sum_{j=0}^{m_i - 1} T_r^j \left(\frac{1}{r} F_r - g\right) > 0 .$$

Hence, letting $K = \max(m_1, m_2, \ldots, m_n)$ we have $Kr < t_0$ and

$$\sup_{0 < k \leqq K} \sum_{j=0}^{k-1} T_r^j \left(\frac{1}{r} F_r - g\right) > 0 \quad \text{on } E' .$$

We may therefore apply the construction of the Chacon-Ornstein filling scheme: Lemma 3.7.1 asserts that there exist functions $d_0, d_1, \ldots, d_{K-1} \in L_p^+$ such that

$$\sum_{j=0}^{k} T_r^{k-j} d_j \leqq T_r^k \frac{1}{r} F_r \quad \text{for } k = 0, \ldots, K-1$$

and $d := d_0 + d_1 + \ldots + d_{K-1} = g$ on $E'$. We have $d \in \mathscr{P}_{t_0}(F)$ and, because of $0 \leqq g 1_{E'} \leqq d$, also $g 1_{E'} \in \mathscr{P}_{t_0}(F)$. □

**Lemma 1.8.** *Let $\{F_t\}$ and $\{G_t\}$ be two positive additive processes with $\|F_t\|_p \to 0$ resp. $\|G_t\|_p \to 0$, $(t \to 0+0)$, and assume* $\sup\{(F_t - G_t): t \in Q(t_0)\} > 0$ *on $E \in \mathscr{A}$ with $\mu(E) < \infty$. Then, for each $\varepsilon > 0$, there exists a set $E' \subset E$ and a $\delta > 0$ such that $\mu(E \setminus E') < \varepsilon$, and such that, for all $r \leqq \delta$,*

$$\sup\{(F_t - S_{0,t} g): t \in Q(t_0)\} > 0 \text{ on } E' \quad \text{for all } g \in \mathscr{P}_r(G).$$

*Proof.* We find finitely many $t_i \in Q(t_0)$, $1 \leqq i \leqq n$, and an $\alpha > 0$ such that if $E_i = E \cap \{F_{t_i} - G_{t_i} > \alpha\}$ then $\mu(E \setminus \bigcup_{i=1}^{n} E_i) < \varepsilon/2$.

Since $G$ is strongly continuous and $G_{t_i+s}$ decreases to $G_{t_i}$ when $s$ decreases to 0, there exists a $\delta > 0$ such that the sets $B_i = \{G_{t_i+\delta} - G_{t_i} > \alpha\}$ have measure $\mu(B_i) < \varepsilon/2n$, $(i = 1, \ldots, n)$. If $0 < r \leqq \delta$ and $g \in \mathscr{P}_r(G)$, then $g \in \mathscr{P}_\delta(G)$. By Lemma 1.4

$$F_{t_i} - S_{0,t_i} g \geqq F_{t_i} - G_{t_i+\delta} \geqq (F_{t_i} - G_{t_i}) - (G_{t_i+\delta} - G_{t_i}) > 0 \quad \text{on } E_i \setminus B_i.$$

Therefore $E' = \bigcup_{i=1}^{n} (E_i \setminus B_i)$ has the desired property. □

**Lemma 1.9.** *Assume that* $\sup\{(F_t - G_t): t \in Q(t_0)\} > 0$ *for all $t_0 > 0$ on a set $E \in \mathscr{A}$ with $\mu(E) < \infty$. Then, for each $\varepsilon > 0$, there exists a set $E' \subset E$ with $\mu(E \setminus E') < \varepsilon$ so that $\Psi_{E''}(F) \geqq \Psi_{E''}(G)$ whenever $E'' \in \mathscr{A}$ and $E'' \subset E'$.*

*Proof.* We choose $\varepsilon_i > 0$ with $\sum_{i=1}^{\infty} \varepsilon_i < \varepsilon$ and also $t_i > 0$ with $t_i \downarrow 0$. By the previous lemma we find for each $i$ a set $E_i'$ and a number $\delta_i > 0$ such that $\mu(E \setminus E_i') < \varepsilon_i$ and $\sup\{(F_t - S_{0,t} g): t \in Q(t_i)\} > 0$ on $E_i'$ for all $g \in \mathscr{P}_{\delta_i}(G)$. Let $E' = \bigcap_{i=1}^{\infty} E_i'$. Then $\mu(E \setminus E') < \varepsilon$. Let $E'' \subset E'$ be a fixed set. Let $\alpha = \Psi_{E''}(G)$. Given any $\xi > 0$ and $i$ there exists a $g \in \mathscr{P}_{\delta_i}(G)$ with $\|g - \alpha 1_{E''}\|_p < \xi$. Hence, as we have $\sup\{(F_t - S_{0,t} g): t \in Q(t_i)\} > 0$ on $E''$, there exists an $f \in \mathscr{P}_{t_i}(F)$ with $\|f - 1_{E''} g\|_p < \xi$. Now $\|f - \alpha 1_{E''}\|_p < 2\xi$. As $\xi$ was arbitrary $\Psi_{E''}^{t_i}(F) \geqq \alpha$, and hence $\Psi_{E''}(F) = \lim_{i \to \infty} \Psi_{E''}^{t_i}(F) \geqq \alpha = \Psi_{E''}(G)$. □

*Proof of theorem 1.3.* For $0 < s < t$, $F_t = F_s + T_s F_{t-s}$. By $\|F_s\|_p \to 0$, $(s \to 0)$ and Lemma 1.1, we see that $F_t = 0$ on $D_0$. Thus, the convergence of $t^{-1} F_t$ on $D_0$ is trivial.

Let $h' \in L_p^+$, $h' > 0$ a.e., $h = S_{0,1}h'$, and $H_t = S_{0,t}h$ as in Lemma 1.5. It follows from

$$S_{t,1}h' \leqq T_s h \leqq S_{0,1+t}h', \quad (0 \leqq s \leqq t)$$

that $S_{t,1}h' \leqq t^{-1}H_t \leqq S_{0,1+t}h'$, and therefore that $Q$-$\lim t^{-1}H_t = h$ exists a.e.. Assume theorem 1.3 fails. Then $Q$-$\lim_{t\to 0+0} F_t H_t^{-1}$ fails to exist on some set $E \in \mathscr{A}$ with $0 < \mu(E) < \infty$ and $E \subset C^* \cap \{h > 0\}$. Passing to a smaller set with these properties we may assume that there exist numbers $\alpha > 0$ and $0 < \beta_1 < \beta_2$ so that $h > \alpha$ on $E$ and

$$0 \leqq Q\text{-}\liminf_{t\to 0+0} F_t H_t^{-1} < \beta_1 < \beta_2 < Q\text{-}\limsup_{t\to 0+0} F_t H_t^{-1}$$

on $E$. Using Lemma 1.5 with $\varepsilon = \alpha/2$ we find $E' \subset E$ with $0 < \mu(E')$ so that $\Psi_{E''}(H) > \alpha/2 > 0$ for all $E'' \subset E'$. Observe that

$$\sup_{t\in Q(t_0)} (F_t - \beta_2 S_{0,t}h) > 0 \quad \text{and} \quad \sup_{t\in Q(t_0)} (\beta_1 S_{0,t}h - F_t) > 0$$

on $E$ for all $t_0 > 0$. Using Lemma 1.9 we can find $E_1' \subset E'$ and $E_2' \subset E'$ so that $\Psi_{E''}(F) \geqq \Psi_{E''}(\beta_2 H) = \beta_2 \Psi_{E''}(H)$ and $\beta_1 \Psi_{E''}(H) \geqq \Psi_{E''}(F)$ for all $E'' \subset E^* := E_1' \cap E_2'$. Choosing the epsilons in the application of Lemma 1.9 small we may assume $\mu(E^*) > 0$. For $E'' = E^*$ we get $\beta_1 \Psi_{E^*}(H) \geqq \beta_2 \Psi_{E^*}(H)$ and this contradicts $\beta_1 < \beta_2$ because Lemma 1.5 and Lemma 1.6 imply $0 < \Psi_{E^*}(H) < \infty$. This completes the proof of the theorem. □

**3. Non positive additive processes.** If $\{T_t, t > 0\}$ is a semigroup of operators in $L_p$ as in theorem 1.3, the a.e.-convergence of $A_{0,t}f$ follows also for non positive $f \in L_p$ because of $A_{0,t}f = A_{0,t}f^+ - A_{0,t}f^-$.

We now discuss a similar but less obvious decomposition for a wider class of additive processes; see Akcoglu-Krengel [1978], [1979].

An additive process satisfying

$$(1.1) \qquad b(F) := \sup_{0<t} t^{-1} \|F_t\|_p < \infty.$$

will be called *linearly bounded.*

**Theorem 1.10.** *Let $\{T_t, t > 0\}$ be a locally bounded strongly continuous semigroup of positive linear operators in $L_p$, $(1 \leqq p < \infty)$. The family of all linearly bounded additive processes is a vector lattice. In particular, any linearly bounded additive process $F$ is of the form $F_t = F_t^{(1)} - F_t^{(2)}$, where $F^{(1)}$, $F^{(2)}$ are positive linearly bounded additive processes.*

*Proof.* For dyadic rational $t = k2^{-n}$ put

$$F_t^{(n,1)} = \sum_{j=0}^{k-1} T_{2^{-n}}^j (F_{2^{-n}})^+ \quad \text{and} \quad F_t^{(n,2)} = \sum_{j=0}^{k-1} T_{2^{-n}}^j (F_{2^{-n}})^-.$$

From

$$(F_{2^{-n}})^+ = (F_{2^{-(n+1)}} + T_{2^{-(n+1)}}F_{2^{-(n+1)}})^+$$
$$\leqq F^+_{2^{-(n+1)}} + T_{2^{-(n+1)}}F^+_{2^{-(n+1)}}$$

it follows that $F_t^{(n,1)} \leqq F_t^{(n+1,1)}$. Similarly, $F_t^{(n,2)} \leqq F_t^{(n+1,2)}$. For dyadic rational $t>0$ put $F_t^{(i)} = \lim_{m\to\infty} F_t^{(m,i)}$. The additivity and the linear boundedness can be verified in a straightforward way for dyadic rationals.

The new processes satisfy $\| F_t^{(i)} - F_s^{(i)} \|_p \leqq b(F)\cdot|t-s|$. Therefore we can define $F_t^{(i)}$ for general $t>0$ by $F_t^{(i)} = \lim_{s\to t} F_t^{(i)}$, where $s$ ranges through the dyadic rationals. The additivity and linear boundedness of the processes $F_t^{(i)}$ follow by continuity. Clearly, $F^{(1)} = F \vee 0$. □

If $\{T_t, t \geqq 0\}$ is strongly continuous at $t=0$ with $T_0 = I$, then the decomposition of $F_t = S_{0,t}f$ has the form $F_t^{(1)} = S_{0,t}f^+$. Theorem 1.10 is also related to the decomposition of functions of bounded variation: Let $\Omega$ be the real line with $\mu =$ Lebesgue measure. For a measurable $f$: $\Omega = \mathbb{R} \to \mathbb{R}$ define $F_t$ by $F_t(\omega) = f(\omega+t) - f(\omega)$. The translation operator $T_t g(\omega) = g(\omega + t)$ is a contraction in $L_1$. For contractions the supremum in (1.1) is a limit as $t \to 0+0$. We define the *essential total variation* $\|f\|_{\text{ess.t.v.}}$ of $f$ by

$$\|f\|_{\text{ess.t.v.}} = \lim_{t\to 0+0} \frac{1}{t} \int |f(\omega+t) - f(\omega)|\,\mu(d\omega) = b(F).$$

Recall that the total variation $\|f'\|_{\text{t.v.}}$ of a function $f'$: $\mathbb{R}\to\mathbb{R}$ is the supremum of all sums $\sum_{i=1}^{n} |f'(t_i) - f'(t_{i-1})|$, where the sup is taken over all $t_0 < t_1 < \ldots < t_n$ in $\mathbb{R}$. If $f'$ has finite total variation, $f'$ is the difference $u - v$ of two bounded increasing functions $u$, $v$ and $\|f\|_{\text{t.v.}} = \|u\|_{\text{t.v.}} + \|v\|_{\text{t.v.}}$. For increasing functions the essential total variation coincides with the total variation because

$$\begin{aligned}\frac{1}{t}\int_{-\infty}^{+\infty} |u(x+t) - u(x)|\,dx &= \lim_{\substack{a\to-\infty\\ b\to+\infty}} \frac{1}{t}\int_a^b (u(x+t) - u(x))\,dx \\ &= \lim_{\substack{a\to-\infty\\ b\to+\infty}} \left\{\frac{1}{t}\int_b^{b+t} u(x)\,dx - \frac{1}{t}\int_a^{a+t} u(x)\,dx\right\} \\ &= \sup u(x) - \inf u(x) = \|u\|_{\text{t.v.}}.\end{aligned}$$

It follows that $\|f\|_{\text{ess.t.v.}} \leqq \|f'\|_{\text{t.v.}}$ holds for all measurable $f$ and for all $f'$ with $f' = f$ a.e.. The terminology "essential total variation", in analogy to the "essential supremum" as the minimal supremum in the equivalence class, is justified by the following result of Akcoglu and Krengel [1978]: For any measurable $f$ with $\|f\|_{\text{ess.t.v.}} < \infty$ there exists an $f'$ with $f' = f$ a.e. and $\|f\|_{\text{ess.t.v.}} = \|f'\|_{\text{t.v.}}$. In this example the decomposition of $F_t$ has the form $F_t^{(1)}(\omega) = u(\omega+t) - u(\omega)$, $F_t^{(2)}(\omega) = v(\omega+t) - v(\omega)$.

Stadje [1984] has given a different proof and observed that one can take $f'(x) = \lim_{t\to 0+0} t^{-1} \int_x^{x+t} f(u)\,du$.

**4. Continuity at the origin.** By a theorem of Dunford, a semigroup $\{T_t, t>0\}$ of bounded linear operators in a Banach space $\mathfrak{X}$ is strongly continuous, when the map $t \to T_t f$ is strongly measurable for all $f \in \mathfrak{X}$; see § 1.6. Thus, the strong continuity of the semigroup at $t>0$ is a rather weak assumption. For the questions of convergence with $t \to 0+0$ considered here, it is of interest to know when there exists a bounded linear operator $T_0$ for which $\{T_t, t \geqq 0\}$ is a semigroup which is strongly continuous also at $t=0$. We shall then say that $\{T_t, t>0\}$ is strongly continuous at $t=0$.

The following theorem (R. Sato [1975a]) implies that for $\mathfrak{X} = L_p$, $1<p<\infty$, a locally bounded semigroup strongly continuous at $t>0$ is also strongly continuous at $t=0$:

**Theorem 1.11.** *Let $\{T_t, t>0\}$ be a semigroup of bounded linear operators on a Banach space $\mathfrak{X}$ which is strongly continuous at $t>0$. Assume that, for each $f \in \mathfrak{X}$, $\{T_t f\colon 0<t<1\}$ is weakly sequentially compact. Then $\{T_t, t>0\}$ is also strongly continuous at $t=0$. The same conclusion holds if, for each $f \in \mathfrak{X}$, $\{A_{0,t} f\colon 0<t<1\}$ is weakly sequentially compact.*

*Proof.* As each set $\{T_t f\colon 0<t<1\}$ is norm bounded, the uniform boundedness principle implies that the semigroup is locally bounded. It follows that

$$(1.2) \qquad \{f \in \mathfrak{X}\colon \lim_{t\to 0+0} \|T_t f - f\| = 0\} = cl(\bigcup_{t>0} T_t \mathfrak{X}).$$

For any $f$ there exists a closed separable subspace $\mathfrak{X}_f$ of $\mathfrak{X}$ containing $f$ such that $T_t \mathfrak{X}_f \subset \mathfrak{X}_f$, $(t>0)$. Therefore we may assume without loss of generality that $\mathfrak{X}$ is separable. Let $\{f_i, i \geqq 1\}$ be a dense subset of $\mathfrak{X}$. By the diagonal argument there exists a strictly decreasing sequence $t_n \to 0$ such that $T_{t_n} f_i$ converges weakly for all $i$. As the semigroup is locally bounded it follows that $T_{t_n} f$ converges weakly for all $f \in \mathfrak{X}$. Let $T_0 f$ denote the limit. Clearly $T_0$ is a linear operator and it is bounded because the semigroup is locally bounded. It is straightforward to verify $T_t T_0 = T_0 T_t = T_t$ for $t \geqq 0$. Hence, putting $f_1 = T_0 f$, we see that each $f \in \mathfrak{X}$ can be written as a sum $f = f_1 + f_2$ with $T_0 f_1 = f_1$ and $T_0 f_2 = 0$.

We claim that $T_t f_1$ converges strongly to $f_1$ for $t \to 0+0$. Otherwise, by (1.2) and Hahn-Banach, there exists an $h \in \mathfrak{X}^*$ with $0 \neq \langle f_1, h\rangle$ and with $\langle T_t f, h\rangle = 0$ for all $t>0$ and $f \in \mathfrak{X}$, a contradiction to $f_1 = w\text{-}\lim T_{t_n} f$. Thus, $T_0$ is the limit in the strong operator topology of the $T_t$. The proof of the second assertion is the same. □

**Corollary 1.12** (Akcoglu-Chacon [1970]). *If $\{T_t, t>0\}$ is a locally bounded semi-*

*group of positive linear operators in $L_1$, strongly continuous at $t > 0$, and if $D_0$ is empty, then $\{T_t, t > 0\}$ is strongly continuous at $t = 0$.*

*Proof.* For strictly positive $h' \in L_1$, $h = S_{0,1} h'$ is strictly positive. Any $f \in L_1^+$ is the limit of $f^{(n)} = f \wedge nh$. Now $0 \leqq T_t f^{(n)} \leqq nS_{t,t+1} h' \leqq n(h + T_1 h)$, $(0 < t < 1)$, and the local boundedness of semigroup imply that $\{T_t f: 0 < t < 1\}$ is uniformly integrable and hence weakly sequentially compact. □

The following example of Akcoglu-Chacon shows that for $D_0 \neq \emptyset$ continuity at 0 need not hold:
Take $\Omega = \mathbb{R} \cup \{\omega_0\}$ with $\omega_0 \notin \mathbb{R}$ and let $\mu$ be the Lebesgue-measure on $\mathbb{R}$ and $\mu\{\omega_0\} = 1$. For $f \in L_1$ and $t > 0$ put $(T_t f)(\omega_0) = 0$ and

$$(T_t f)(\omega) = f(\omega_0)(2\pi t)^{-1/2} \exp\{-\omega^2/2t\} + \int_{\mathbb{R}} (2\pi t)^{-1/2} \exp(-(\omega - \eta)^2/2t) f(\eta)\, \mu(d\eta),$$

for $\omega \neq \omega_0$. If $f = 1_{\{\omega_0\}}$, then $A_{0,\varepsilon} f$ converges to 0 a.e. on $\Omega$ for $\varepsilon \to 0$. Consequently, the strong continuity of $\{T_t, t \geqq 0\}$ at $t = 0$ would imply $T_0 f = 0$. But $\| T_t f \|_1 = 1$ for all $t > 0$. Heuristically, the contractions $T_t$ are given by Brownian motion plus a contribution of the mass $f(\omega_0)$, which is transformed as if it was located in $\{0\}$ at time 0.

If a semigroup $\{T_t, t \geqq 0\}$ is strongly continuous at $t = 0$, $T_0 f$ is the limit of $A_{0,\varepsilon} f$ in the local ergodic theorem. If the operators $T_t$ are positive contractions in $L_1$, and $D_0$ is empty, $T_0 f$ may be described as a conditional expectation by passing to an appropriate reference measure: Let $\nu$ be the measure having density $h$ with respect to $\mu$. By the isomorphism $L_1(\mu) \ni f \leftrightarrow f/h \in L_1(\nu)$ the semigroup induces a semigroup $\{T_t', t \geqq 0\}$ in $L_1(\nu)$: $T_t' g = h^{-1} T_t(h \cdot g)$. This new semigroup has the property $T_0' \mathbb{1} = \mathbb{1}$. It is known that a positive contraction $T_0'$ in $L_1(\nu)$ with $(T_0')^2 = T_0'$ and $T_0' \mathbb{1} = \mathbb{1}$ is a conditional expectation operator. (See Ando [1966], or Neveu [1965: 123]). Call $A \in \mathscr{A}$ *initially invariant* if $T_0' 1_A = 1_A$. The initially invariant sets form a $\sigma$-algebra $\mathscr{I}_0$, and we have $T_0' g = E_\nu(g \mid \mathscr{I}_0)$. Thus, we may say that using a suitable reference measure $\nu$ the limit in the local ergodic theorem is a conditional expectation just as in Birkhoff's theorem. This "identification of the limit" is due to Lin [1972]. It is not hard to show that $A$ is initially invariant if and only if there is some $f \in L_1^+(\mu)$ with $A = \{f > 0\}$ and $\int_A T_t f \, d\mu \to \|f\|_1$.

The case, where $T_0$ is the identity operator in $L_1$ may merit special attention. It is characterized in the following theorem of Lin [1972]:

**Theorem 1.13.** *Let $\{T_t, t > 0\}$ be a semigroup of positive contractions in $L_1$, strongly continuous at $t > 0$. Then the following statements are equivalent:*

*(i) For every $f \in L_1$ we have $\lim_{t \to 0+0} A_{0,t} f = f$ a.e.;*

*(ii) The $T_t$ converge to the identity in the strong operator topology* $(t \to 0+0)$;
*(iii) The $T_t$ converge to the identity in the weak operator topology* $(t \to 0+0)$;
*(iv)* $\mu(A) > 0$ *implies* $\lim_{t \to 0+0} \| T_t^* 1_A \|_\infty = 1$.

*Proof.* Clearly (i) implies $D_0 = \emptyset$, so that (ii) follows from corollary 1.12. Conversely (ii) implies (i) by theorem 1.3.

(i) ⇒ (iv): Take $f \in L_1^+$ with $\|f\|_1 = 1 = \int_A f d\mu$. Using Fatou's lemma we find

$$1 = \langle f, 1_A \rangle = \langle \lim_{\varepsilon \to 0} A_{0,\varepsilon} f, 1_A \rangle \leqq \liminf_{\varepsilon \to 0} \langle A_{0,\varepsilon} f, 1_A \rangle$$
$$\leqq \liminf_{\varepsilon \to 0} \varepsilon^{-1} \int_0^\varepsilon \| T_t^* 1_A \|_\infty dt \leqq 1 .$$

Now (iv) follows because $\| T_t^* 1_A \|_\infty$ is a decreasing function of $t$. The same $f$ may be used to prove (iii) ⇒ (iv):

$$1 = \langle f, 1_A \rangle = \lim_{\varepsilon \to 0} \langle T_\varepsilon f, 1_A \rangle = \lim_{\varepsilon \to 0} \langle f, T_\varepsilon^* 1_A \rangle$$
$$\leqq \|f\|_1 \lim_{\varepsilon \to 0} \| T_\varepsilon^* 1_A \|_\infty \leqq 1 .$$

As (ii) ⇒ (iii) is trivial, it remains to show (iv) ⇒ (ii):

Let $L$ be the closed subspace of $\mathfrak{X} = L_1$ described in (1.2). If (ii) fails then $L \neq L_1$. Then by Hahn-Banach there exists a $g \in L_\infty = \mathfrak{X}^*$ with $g \neq 0$ and $\langle f, g \rangle = 0$ for $f \in L$. In particular, $\langle T_t f, g \rangle = 0$, $(t > 0, f \in L_1)$. Hence $T_t^* g = 0$, $(t > 0)$. We may assume $\|g\|_\infty \leqq 1$. For $\delta > 0$ put $A = \{\omega : g(\omega) \geqq \delta\}$. $-\mathbb{1} \leqq g \leqq \mathbb{1}$ yields $\mathbb{1} + g \geqq (1+\delta) 1_A$ and

$$(1+\delta) T_t^* 1_A \leqq T_t^* (\mathbb{1} + g) = T_t^* \mathbb{1} \leqq \mathbb{1} .$$

This implies $\| T_t^* 1_A \|_\infty \leqq 1/(1+\delta)$ for all $t > 0$ and, by (iv), $\mu(A) = 0$. As $\delta > 0$ was arbitrary we obtain $\mu(g > 0) = 0$. Applying the same argument to $-g$ we also obtain $\mu(g < 0) = 0$. But $g = 0$ contradicts the choice of $g$. □

**5. The local ergodic theorem in $L_\infty$.** Let $\{T_t, t > 0\}$ be a strongly continuous locally bounded semigroup of positive linear operators in $L_1$. If $A_{0,t}^* g$ converges a.e. (as $t \to 0$) for all $g \in L_\infty$, then $\langle f, A_{0,t}^* g \rangle$ converges for all $f \in L_1$ by Lebesgue's bounded convergence theorem. Hence $\{A_{0,t} f : 0 < t \leqq 1\}$ is weakly sequentially compact in $L_1$ for all $f \in L_1$, and $\{T_t, t > 0\}$ is strongly continuous at $t = 0$ by theorem 1.11. We now prove that conversely the strong continuity at $t = 0$ implies the a.e.-convergence of $A_{0,t}^* g$, $(t \to 0)$, for all $g \in L_\infty$. This was shown in the case $\Omega = C_0$ by Sato [1978 b], and in general by Lin [1982]. (For the case of null preserving transformations; see Krengel [1969 a]).

**Theorem 1.14.** *If* $\{T_t, t \geqq 0\}$ *is a strongly continuous locally bounded semigroup of positive linear operators in* $L_1$ *and* $g \in L_\infty$, *then* $A_{0,t}^* g$ *converges a.e. as* $t \to 0+0$.

*Proof.* The proof is a reduction to the local ergodic theorem in $L_1$: For any $f \in L_1$ we have $S_{0,t} f = 0$ on $D_0$. Therefore $A^*_{0,t} g$ is $\equiv 0$ for $g \in L_\infty(D_0)$, and it is enough to prove the convergence of $A^*_{0,t} g$ for $g \in L_\infty(C_0)$.

We first prove the convergence on $C_0$. $L_1(C_0)$ is invariant under $\{T_t, t \geqq 0\}$. Let $R_t$ be the restriction of $T_t$ to $L_1(C_0)$. Then, for $f \in L_1(C_0)$, $g \in L_\infty(C_0)$, we have $\langle f, R_t^* g\rangle = \langle R_t f, g\rangle = \langle T_t f, g\rangle = \langle f, T_t^* g\rangle$, so that $R_t^* g = T_t^* g$ on $C_0$. We may therefore assume $\Omega = C_0$ for this part of the proof.

As $\{T_t, t \geqq 0\}$ is strongly continuous there exist $M < \infty$ and $\alpha > 0$ with $\| T_t \| \leqq M e^{t\alpha/2}$, (see e.g. Yosida [1974: 232]). If $h' \in L_1$ is strictly positive in $\Omega = C_0$ then $h = S_{0,1} h'$ and hence also

$$u = \int_0^\infty e^{-\alpha t} T_t h' \, dt$$

is strictly positive in $\Omega$. The choice of $\alpha$ implies $u \in L_1$. For $t \geqq 0$,

$$T_t u = \int_0^\infty e^{-\alpha t} T_{t+s} h' \, ds \leqq e^{\alpha t} u.$$

Hence, for $t \geqq 0$ and $g \in L_\infty$, we see that

$$\int (e^{-\alpha t} T_t^* g) u \, d\mu = \int e^{-\alpha t} g \, T_t u \, d\mu \leqq \int g u \, d\mu,$$

so that $e^{-\alpha t} T_t^*$ is a contraction in $L_1(\Omega, \mathscr{A}, u \cdot \mu)$. For $f \in L_\infty$, $g \in L_\infty$ and $t \to 0$,

$$\int f (e^{-\alpha t} T_t^* g) u \, d\mu = e^{-\alpha t} \int T_t (f \cdot u) g \, d\mu \to$$
$$\to \int T_0 (f \cdot u) g \, d\mu = \int f (T_0^* g) u \cdot d\mu.$$

By approximation this shows that $e^{-\alpha t} T_t^*$ is weakly continuous at 0 in $L_1(\Omega, \mathscr{A}, u\mu)$ and hence also strongly continuous. As $L_\infty$ is contained in $L_1(\Omega, \mathscr{A}, u \cdot \mu)$ theorem 1.3 implies the convergence a.e. of $\varepsilon^{-1} \int_0^\varepsilon e^{-\alpha t} T_t^* g \, dt$, $(\varepsilon \to 0)$. (Actually the simpler special case of continuous semigroups of positive $L_1$-contractions is enough). As $e^{-\alpha t}$ tends to 1 for $t \to 0$, we can conclude that also $\varepsilon^{-1} \int_0^\varepsilon T_t^* g \, dt$ converges a.e.. As the integrals $S_{0,t} f$ and $\int_0^\varepsilon T_t^* g \, dt$ can be obtained as limits of approximating Riemann sums in $L_1(\mu)$ and in $L_1(u \cdot \mu)$ it is simple to check $\int_0^\varepsilon T_t^* g \, dt = S^*_{0,\varepsilon} g$ and the proof of the convergence on $C_0$ is complete.

Now let $Q$ be the set of positive rationals, $g \in L_\infty^+(C)$, and let $\bar{g} = Q\text{-}\limsup_{\varepsilon \to 0} A^*_{0,\varepsilon} g$, $\tilde{g} = Q\text{-}\liminf_{\varepsilon \to 0} A^*_{0,\varepsilon} g$. $\| S^*_{0,\varepsilon} \|_\infty \leqq M e^{\varepsilon\alpha/2}$ implies $\|\tilde{g}\|_\infty \leqq \|\bar{g}\|_\infty \leqq M \|g\|_\infty < \infty$. Using $T_0^* (g_1 \vee g_2 \vee \ldots \vee g_n) \geqq (T_0^* g_1) \vee \ldots \vee (T_0^* g_n)$ and a representation of $\bar{g}$ as limit of suprema of averages $A^*_{0,\varepsilon} g$ it may be seen that

$$T_0^* \bar{g} \geqq Q\text{-}\limsup_{\varepsilon \to 0} \frac{1}{\varepsilon} \int_0^\varepsilon T_0^* T_t^* g \, dt = \bar{g}.$$

Similarly $T_0^* \tilde{g} \leqq \tilde{g}$. Hence $T_0^*(\bar{g} - \tilde{g}) \geqq \bar{g} - \tilde{g} (\geqq 0)$. But $\bar{g} - \tilde{g}$ has support in $D_0$. Therefore $T_0^* 1_D = 0$ yields $T_0^*(\bar{g} - \tilde{g}) = 0$, and $\bar{g} = \tilde{g}$. □

## Notes

The local ergodic theorem for a strongly continuous semigroup $\{T_t, t \geqq 0\}$ of positive contractions in $L_1$, obtained independently by Krengel [1969b] and Ornstein [1970], has a much simpler proof than theorem 1.3; see the proof of theorem 2.4. Akcoglu and Chacon [1970] showed that the continuity at $t = 0$ was not needed. Fong and Sucheston [1971] and Kubokawa [1972] replaced the contraction hypothesis by weaker assumptions. Kubokawa [1974] proved the local ergodic theorem for a locally bounded strongly continuous semigroup $\{T_t, t \geqq 0\}$ of positive operators in $L_p$, $(1 \leqq p < \infty)$, with $T_0 = I$. Theorem 1.3 unifies these results and extends them to additive processes. The proof uses ideas of Akcoglu-Chacon and of Kubokawa. Sato [1978b] proved local ergodic theorems in which the assumption of local boundedness is replaced by the assumption that for each $f \in L_p$ the vector valued function $t \to T_t f$ is Lebesgue integrable over every finite interval.

Emilion [1984b] has a (nontrivial) argument showing that theorem 1.3 (for $p = 1$) can be deduced from the special case due to Krengel and Ornstein. Moreover, he shows for $F_t \geqq 0$, that $\tilde{f} = Q\text{-lim}\, t^{-1} F_t$ is the largest function $f$ with $F_t \geqq S_{0,t} f$, $(t > 0)$.

Sato [1978b] has given an example of a strongly continuous semigroup $\{T_t; t \geqq 0\}$ of positive operators in $L_1$ with $T_t \mathbb{1} = \mathbb{1}$ and $\| T_t \|_1 = 1 + \delta$ for which $A_{0,\varepsilon} f$ fails to converge a.e.. Thus, convergence need not hold on $D^*$; see also Akcoglu and Krengel [1979a].

If $\mathfrak{X}$ is a reflexive Banach space and $\{F_t, t > 0\}$ a linearly bounded additive process for a strongly continuous locally bounded semigroup $\{T_t, t > 0\}$, then $F_t$ is of the form $F_t = S_{0,t} f$; Akcoglu and Krengel [1979].

For a technique of deducing local ergodic theorems in $L_p$, $(1 < p \leqq \infty)$ from theorems in $L_1$ see McGrath [1976], [1976a], Lin [1982], Sato [1978b], Emilion [1984d].

Feyel [1982] proved the following superadditive differentiation theorem:

**Theorem 1.15.** *Let $\{T_t, t > 0\}$ be a strongly continuous semigroup of positive contractions in $L_1$. Let $F = \{F_t, t > 0\}$ satisfy*

*(i) $\|F_t^-\|_1 \leqq Mt$ for some $M < \infty$, and*

*(ii) $F_{t+s} \geqq F_t + T_t F_s$, $(t, s > 0)$.*

*Then $Q\text{-lim}_{t \to 0+0}\, t^{-1} F_t$ exists a.e..*

Akcoglu-Krengel [1978] had proved this for Markovian $\{T_t\}$ and $\|F_t\|_1 \leqq Mt$, $(t \leqq 1)$, (holding for $F \geqq 0$); see also Emilion-Hachem [1982], and Akcoglu [1983].

Again, generalizations in $L_p$ follow from theorem 1.15; see Kataoka, Sato, Suzuki [1983], Emilion [1982].

A way of expressing the existence a.e. of $Q$-limits for arbitrary countable dense $Q \subset \mathbb{R}^+$ is to use *essential limits*; see Feyel [1982]:

For a family $(f_t)$ of equivalence classes of measurable functions and $I \subset \mathbb{R}^+$ let ess-sup $\{f_t, t \in I\}$ be the supremum in the set of equivalence classes. (It is the supremum over a suitable countable subset $I_0 \subset I$; see Neveu [1965], § 2.4.) Put $\text{ess-limsup}_{t \to 0+0} f_t = \text{ess-inf}_{s > 0} (\text{ess-sup}\, \{f_t: 0 < t < s\})$. ess-lim is said to exist if ess-limsup = ess-liminf.

## § 7.2 Local ergodic theorems for multiparameter and non positive semigroups, and for vector valued functions

**1. Multiparameter local ergodic theorems.** $\mathbb{P}_d$ denotes the set of vectors $u = (u_1, u_2, \ldots, u_d) \in \mathbb{R}^d$ with $u_i > 0$, $(i = 1, \ldots, d)$, and $\mathbb{P}_d^0$ denotes $\{0\} \cup \mathbb{P}_d$, where $0 = (0, 0, \ldots, 0)$. $u < v$, $(u \leqq v)$, means that $u_i < v_i$, $(u_i \leqq v_i)$, for all $i$. Let $[u, v] = \{t \in \mathbb{R}^d : u \leqq t \leqq v\}$, $e = (1, 1, \ldots, 1)$, and $\pi(u) = u_1 \cdot u_2 \ldots u_d$. A semigroup $\mathscr{T} = \{T_u : u \in \mathbb{P}_d\}$ of linear operators in $L_p$ is called *locally bounded* if $\sup\{\|T_u\| : u < v\}$ is finite for one and hence for all $v > 0$. If $\mathscr{T}$ is strongly continuous and locally bounded we may define the integrals

$$S_{u,v} f = \int_u^v T_t f\, dt = \int_{[u,v]} T_t f\, dt$$

as in the 1-parameter case (§ 1.6) by approximating Riemann sums or by a suitable choice of the representatives. Put $A_{u,v} = \pi(v-u)^{-1} S_{u,v}$, and $\mathfrak{M}_p = \{A_{0,v} g : 0 < v, g \in L_p\}$. As usual $u \to 0+0$ means that $u_i \to 0+0$ for all $i$.

**Lemma 2.1.** *Let $\mathscr{T}$ be a locally bounded strongly continuous semigroup of linear operators in $L_p = L_p(\Omega, \mathscr{A}, \mu)$, $(1 \leqq p < \infty)$. Then $A_{0,u} f$ converges a.e. to $f$ as $u \to 0+0$ for all $f \in \mathfrak{M}_p$.*

*Proof.* Let $\lambda$ denote the Lebesgue measure in $\mathbb{P}_d$. For a suitable choice of the representatives $|T_s g(\omega)|^p$ is $\mu \times \lambda$-integrable on $\Omega \times [0, u]$. Hence $|T_s g(\omega)|$ is $\mu \times \lambda$-integrable on $A \times [0, u]$ for $\mu(A) < \infty$. For $0 < w < v$ consider

$$d_{w,v}(\omega) = \int_{[0, v+w] \setminus [w, v]} |T_s g(\omega)|\, ds.$$

When $w$ decreases to 0 the sets $[0, v+w] \setminus [w, v]$ decrease to a $\lambda$-nullset and $d_{w,v}(\omega)$ decreases a.e. to 0. For $0 \leqq t \leqq u \leqq w$ we have

$$|T_t S_{0,v} g - S_{0,v} g| \leqq d_{w,v}.$$

Hence $|A_{0,u} S_{0,v} g - S_{0,v} g| \leqq d_{w,v}$. If $u$ tends to 0 we obtain

$$\limsup_{u \to 0+0} |A_{0,u} S_{0,v} g - S_{0,v} g| \leqq d_{w,v}.$$

Because of $d_{w,v} \to 0$, $(w \to 0)$ the assertion of the lemma is proved for $S_{0,v} g$, and hence for $f = A_{0,v} g$. □

To derive a local ergodic theorem we also need a maximal inequality. First notice that Lemma 1.6.1 together with Theorem 6.3.4 yields the following discrete parameter inequality:

**Lemma 2.2.** *There exists a constant $c(d) > 0$ with the following property: If $T_1, T_2, \ldots, T_d$ are commuting $L_1 - L_\infty$-contractions then*

$$\mu(\sup_{n\geqq 1}|A_{ne}f|>\alpha)\leqq\frac{c(d)}{\alpha}\int_{\{|2f|>\alpha\}}|f|\,d\mu$$

*holds for all* $\alpha>0$ *and* $f\in L_1$.

Applying this to the operators $T_{2^{-k}(0,0,\ldots,1,0,..)}$ and to the function $T_0 f$ one obtains an inequality of Dunford-Schwartz by discrete approximation:

**Lemma 2.3.** *If* $\{T_u\colon u\in\mathbb{P}_d^0\}$ *is a semigroup of* $L_1-L_\infty$*-contractions which is strongly continuous (also at* $u=0$*!), then*

$$\mu(\sup_{0<\varepsilon<\infty}|A_{0,\varepsilon e}f|>\alpha)\leqq\frac{c(d)}{\alpha}\int_{\{|2T_0f|>\alpha\}}|T_0f|\,d\mu$$

*holds for all* $\alpha>0$ *and* $f\in L_1$.

The detailed argument is the same as in the proof of Theorem 1.6.11. It is now easy to derive the first local ergodic theorem of Terrell [1970]; (see also Terrell [1972]):

**Theorem 2.4.** *Let* $\{T_u\colon u\in\mathbb{P}_d^0\}$ *be a strongly continuous semigroup of* $L_1-L_\infty$*-contractions, then* $\lim_{\varepsilon\to 0}A_{0,\varepsilon e}f=T_0f$ *a.e. for all* $f\in L_1$.

*Proof.* We may assume $T_0f=f$ because $A_{0,\varepsilon e}f=A_{0,\varepsilon e}T_0f$. Now $f_v=A_{0,v}f$ tends in norm to $T_0f$, since the semigroup is continuous at 0. By lemma 2.1

$$\begin{aligned}\limsup_{\varepsilon\to 0+0}|A_{0,\varepsilon e}f-T_0f|&\leqq\limsup_{\varepsilon\to 0+0}|A_{0,\varepsilon e}f-A_{0,\varepsilon e}f_v+f_v-T_0f|\\&\leqq|f_v-T_0f|+\sup_{0<\varepsilon<\infty}|A_{0,\varepsilon e}(T_0f-f_v)|.\end{aligned}$$

Both terms can be made small by a small choice of $v$ using $\|f_v-T_0f\|_1\to 0$ and lemma 2.3. □

Let us briefly discuss more general averages. For measurable $B$ with $0<\lambda(B)<\infty$ put

$$A(B,f)=\lambda(B)^{-1}\int_B T_uf\,du.$$

Essentially the same proof yields convergence a.e. of $A(B_\varepsilon,f)$ for $f\in L_1$, if $\{B_\varepsilon\}$ is a family of sets with $B_\varepsilon\subset[0,\varepsilon e]$ and there is some $\beta>0$ with $\lambda(B_\varepsilon)\geqq\beta\varepsilon^d$.

In the special case of translations $T_uf(x)=f(x+u)$ in $\mathbb{R}^d$ the question of convergence $\lambda$-a.e. of averages $A(B(q),f)$ for suitable families $\{B(q)\}$ has been investigated very thoroughly starting with a famous theorem of Lebesgue [1910]: He showed that $A(C_\varepsilon,f)$ converges $\lambda$-a.e. for $\lambda$-integrable $f$ if $C_\varepsilon$ is $[-\varepsilon e,+\varepsilon e]$ or $\{x\in\mathbb{R}^d\colon\sum_{i=1}^d x_i^2\leqq\varepsilon\}$. (It is easy to see that this classical theorem follows from the assertion above about convergence of $A(B_\varepsilon, f)$). Saks [1934], and Busemann and Feller [1934] have shown that $A_{0,u}f$ may diverge for $\lambda$-integrable $f$ for $u \to 0+0$.

The next theorem gives a sufficient condition for *unrestricted* convergence, due to Zygmund (in the translation case), and to McGrath [1976]; (see also theorem 2.13):

**Theorem 2.5.** *For some p with* $1 < p < \infty$ *let* $\{T_u: u \in \mathbb{P}_d^0\}$ *be a strongly continuous semigroup of positive contractions in* $L_p$. *Then* $A_{0,u}f$ *converges a.e. to* $T_0 f$ *as* $u \to 0+0$ *for* $f \in L_p$.

*Proof.* Lemma 2.1 yields convergence for $f \in \mathfrak{M}_p$, and by the strong continuity $\mathfrak{M}_p$ is dense in $T_0 L_p$. Now the $d$-dimensional extension of the maximal inequality of Akcoglu (Theorem 6.1.2) is used to prove an analogous inequality for the continuous parameter case by discrete approximation. Then the proof can be completed as above. □

**Theorem 2.6** (Terrell [1970]). *If* $\{T_u: u \in \mathbb{P}_d^0\}$ *is a strongly continuous semigroup of positive contractions in* $L_1$ *and* $f \in L_1$, *then* $A_{0,\varepsilon e}f$ *converges a.e. for* $\varepsilon \to 0+0$.

*Proof* (See McGrath [1976]). For some $\alpha > 0$ define a new semigroup by $T_t' f = e^{-\alpha(t_1+t_2+\ldots+t_d)} T_t f$, and define $A_{0,u}'$ and $S_{0,u}'$ with $T_t'$ as $A_{0,u}$ and $S_{0,u}$ were defined with $T_t$. Because of $e^{-\alpha(t_1+t_2+\ldots+t_d)} \to 1$, $(t \to 0)$, it is enough to prove the a.e.-convergence of $A_{0,\varepsilon e}' f$.

For some strictly positive $h' \in L_1$ put $C_0 = \{T_0 h' > 0\}$. For $f \geqq 0$, $T_0 f = \lim_{n\to\infty} T_0(nh' \wedge f)$. Thus $T_0 f$ vanishes for all $f \geqq 0$ in $C_0^c$ and this implies that $T_t f = T_0 T_t f$ vanishes in $C_0^c$ for all $t$ and all $f \in L_1$. Because of $A_{0,\varepsilon e} f = A_{0,\varepsilon e} T_0 f$ we may assume $\Omega = C_0$ for the rest of the proof. Now the function

$$h = \int_{\mathbb{P}_d^0} T_t' h' \, dt$$

is strictly positive in $\Omega$ and integrable. Moreover

$$(2.1) \qquad T_u' h = \int_{\{t \geqq u\}} T_t' h' \, dt \leqq h$$

for all $u$. The measure $\nu$ with density $h$ with respect to $\mu$ is equivalent with $\mu$. Put $T_t'' g = (T_t'(g \cdot h))/h$ for $g \in L_1(\nu)$. Using (2.1) it is easy to see that $\{T_t'': u \in \mathbb{P}_d^0\}$ is a strongly continuous semigroup of positive $L_1 - L_\infty$-contractions in $L_1(\nu)$. The proof is completed by an application of theorem 2.4 to $A_{0,\varepsilon e}'' g = A_{0,\varepsilon e}' f$ with $g = f/h$. □

**2. Non positive operators.** For the proof of a local ergodic theorem for a non positive semigroup $\{T_t\}$ of contractions in $L_1$ we begin with the construction of a minimal semigroup $\{\tilde{T}_t\}$ of positive contractions dominating $\{T_t\}$.

The linear modulus $|T|$ of a bounded linear operator $T$ in a real or complex space $L_1$, as discussed in § 4.1, is the minimal linear operator with $|Tg| \leqq |T| \, |g|$ for all $g \in L_1$. The operators $|T_t|$ in general do not form a semigroup: the properties of the linear modulus only imply $|T_{t+s}| \leqq |T_t| \, |T_s|$. Therefore $\tilde{T}_t$ has to be defined differently. Put $\mathbf{T}_t = |T_t|$.

**Theorem 2.7** (Kubokawa-Kipnis). *Let $\{T_t: t > 0\}$ be a semigroup of contractions in $L_1$. There exists a semigroup $\{\tilde{T}_t: t > 0\}$ of positive contractions with the following properties:*

*(a) $|T_t g| \leqq \tilde{T}_t |g|$ holds for all $g \in L_1$ and for all $t > 0$,*

*(b) if $\{U_t: t > 0\}$ is a semigroup of contractions such that $|T_t g| \leqq U_t |g|$ holds for all $t > 0$ and for all $g \in L_1$, then $\tilde{T}_t \leqq U_t$.*

*If $\{T_t: t > 0\}$ is strongly continuous, then $\{\tilde{T}_t: t > 0\}$ is strongly continuous. The same result holds for semigroups defined also at $t = 0$.*

*Proof.* For $t > 0$ let $\mathscr{D}_t$ denote the family of all finite subdivisions $0 = s_0 < s_1 < s_2 < \ldots < s_n = t$ of $[0, t]$. If $\mathbf{s} = (s_i)$ and $\mathbf{s}' = (s'_j)$ are two elements of $\mathscr{D}_t$ we write $\mathbf{s} < \mathbf{s}'$ if $\mathbf{s}'$ is a refinement of $\mathbf{s}$. With this partial order $\mathscr{D}_t$ is an increasingly filtered set. For $f \in L_1^+$ put

$$\Phi(\mathbf{s}, f) = \mathbf{T}_{s_1} \cdot \mathbf{T}_{s_2 - s_1} \ldots \mathbf{T}_{s_n - s_{n-1}} f.$$

It follows from $\mathbf{T}_{\alpha+\beta} \leqq \mathbf{T}_\alpha \mathbf{T}_\beta$ that $\mathbf{s} < \mathbf{s}'$ implies $\Phi(\mathbf{s}, f) \leqq \Phi(\mathbf{s}', f)$. As all operators $\mathbf{T}_u$ are contractions, $\|\Phi(\mathbf{s}, f)\|_1 \leqq \|f\|_1$ and $\tilde{T}_t$ can be defined by

$$\tilde{T}_t f = \sup\{\Phi(\mathbf{s}, f) : \mathbf{s} \in \mathscr{D}_t\} = \lim_{\mathbf{s} \in \mathscr{D}_t} \Phi(\mathbf{s}, f),$$

using the monotone convergence theorem for increasingly filtered families. It is simple to check $\tilde{T}_t(f_1 + f_2) = \tilde{T}_t f_1 + \tilde{T}_t f_2$, $\tilde{T}_t \alpha f = \alpha \tilde{T}_t f$, $(\alpha \geqq 0)$, and $\|\tilde{T}_t f\|_1 \leqq \|f\|_1$. Therefore $\tilde{T}_t$ can be defined for all $f \in L_1$ as a positive contraction in $L_1$.

If $0 = s_0 < \ldots < s_n = t$ and $0 = s'_0 < \ldots < s'_m = t'$ are subdivisions of $[0, t]$ and $[0, t']$, then $0 = s_0 < \ldots < s_n = s_n + s'_0 < s_n + s'_1 < \ldots < s_n + s'_m = t + t'$ is a subdivision of $[0, t + t']$. Conversely every subdivision of $[0, t + t']$ which is fine enough to contain $t$ is of this form. This yields $\tilde{T}_{t+t'} = \tilde{T}_t \tilde{T}_{t'}$ and $\{\tilde{T}_t: t > 0\}$ is a semigroup.

The proof of the assertions (a) and (b) is straightforward from the construction of $\tilde{T}_t$.

If the semigroup is also defined for $t = 0$, then $T_0 T_t = T_t$ implies $\mathbf{T}_0 \mathbf{T}_t \geqq \mathbf{T}_t$. On the other hand $\|\mathbf{T}_0 \mathbf{T}_t f\|_1 \leqq \|\mathbf{T}_t f\|_1$ for $f \geqq 0$. Hence $\mathbf{T}_0 \mathbf{T}_t = \mathbf{T}_t$, $(t \geqq 0)$. We may then put $\tilde{T}_0 = \mathbf{T}_0$ and use $\Phi'(\mathbf{s}, f) = \mathbf{T}_{s_1} \mathbf{T}_{s_2 - s_1} \ldots \mathbf{T}_{s_n - s_{n-1}} \mathbf{T}_0 f$ instead of $\Phi(\mathbf{s}, f)$ to define $\tilde{T}_t$ for $t > 0$. Then one can proceed as above.

It remains to prove the assertion on strong continuity. As the $\tilde{T}_u$ are contractions it is enough to prove $\|\tilde{T}_s f - \tilde{T}_t f\|_1 \to 0$ for $s \to t + 0$ and $f \geqq 0$. By the proof of theorem 4.1.1 there exist for any $\varepsilon > 0$ finitely many $g_1, g_2, \ldots, g_k$ with $|g_i| \leqq f$ and $\|\mathbf{T}_t f - \max_{1 \leqq i \leqq k} |T_t g_i|\|_1 < \varepsilon$. Hence

$$\|(\mathbf{T}_s f - \mathbf{T}_t f)^-\|_1 \leqq \|(\mathbf{T}_s f - \max_{1\leqq i\leqq k} |T_t g_i|)^-\|_1 + \varepsilon$$
$$\leqq \|(\max_{1\leqq i\leqq k} |T_s g_i| - \max_{1\leqq i\leqq k} |T_t g_i|)^-\|_1 + \varepsilon < 2\varepsilon$$

when $s$ is so close to $t$ that $\| T_s g_i - T_t g_i\|_1$ is smaller than $\varepsilon/k$. As $\varepsilon$ was arbitrary

$$\lim_{s\to t+0} \|(\mathbf{T}_s f - \mathbf{T}_t f)^-\|_1 = 0, \quad (f \in L_1^+).$$

Using $\mathbf{T}_{s-t}\mathbf{T}_t f \geqq \mathbf{T}_s f$ and $\mathbf{T}_{s-t}\mathbf{T}_t f = (\mathbf{T}_{s-t}\mathbf{T}_t f - \mathbf{T}_t f)^+ - (\mathbf{T}_{s-t}\mathbf{T}_t f - \mathbf{T}_t f)^- + \mathbf{T}_t f$ we obtain

$$\|(\mathbf{T}_s f - \mathbf{T}_t f)^+\|_1 \leqq \|(\mathbf{T}_{s-t}\mathbf{T}_t f - \mathbf{T}_t f)^+\|_1$$
$$\leqq \|(\mathbf{T}_{s-t}\mathbf{T}_t f - \mathbf{T}_t f)^-\|_1 + \|\mathbf{T}_{s-t}\mathbf{T}_t f\|_1 - \|\mathbf{T}_t f\|_1$$
$$\leqq \|(\mathbf{T}_s f - \mathbf{T}_t f)^-\|_1 \to 0.$$

Hence

$$\lim_{s\to t+0} \|\mathbf{T}_s f - \mathbf{T}_t f\|_1 = 0. \tag{2.2}$$

For each $\mathbf{s} = (s_i)$ in $\mathscr{D}_t$ with $t > 0$ and $f \in L_1^+$

$$(\tilde{T}_t f - \tilde{T}_s f)^+ \leqq (\tilde{T}_t f - \mathbf{T}_{(s-t)+s_1}\mathbf{T}_{s_2-s_1}\cdots\mathbf{T}_{s_n-s_{n-1}} f)^+.$$

For given $\varepsilon > 0$ there exists $\mathbf{s} \in \mathscr{D}_t$ with

$$\|\tilde{T}_t f - \mathbf{T}_{s_1}\mathbf{T}_{s_2-s_1}\cdots\mathbf{T}_{s_n-s_{n-1}} f\|_1 < \varepsilon.$$

Then

$$\|(\tilde{T}_t f - \tilde{T}_s f)^+\|_1 \leqq \|\tilde{T}_t f - \mathbf{T}_{(s-t)+s_1}\mathbf{T}_{s_2-s_1}\cdots\mathbf{T}_{s_n-s_{n-1}} f\|_1$$
$$\leqq \varepsilon + \|(\mathbf{T}_{s_1} - \mathbf{T}_{(s-t)+s_1})\mathbf{T}_{s_2-s_1}\cdots\mathbf{T}_{s_n-s_{n-1}} f\|_1$$

and $\lim_{s\to t+0} \|(\tilde{T}_t f - \tilde{T}_s f)^+\|_1 = 0$ follows from (2.2). Because of $\int \tilde{T}_t f d\mu \geqq \int \tilde{T}_s f d\mu$ this even implies $\|\tilde{T}_t f - \tilde{T}_s f\|_1 \to 0$ for $s \to t+0$. The proof for a semigroup $\{T_t, t \geqq 0\}$ defined also at $t = 0$ is analogous. □

A combination of this result with the local ergodic theorem for positive contractions now yields a local ergodic theorem for non positive contractions:

**Theorem 2.8.** *Let* $\{T_t, t \geqq 0\}$ *be a strongly continuous semigroup of contractions in* $L_1$, *then* $A_{0,\varepsilon} f$ *converges a.e.,* $(\varepsilon \to 0+0)$, *for all* $f \in L_1$.

*Proof.* We may assume $f \in T_0 L_1$. For any $\varepsilon > 0$ there exists a small $u > 0$ with $\|f - A_{0,u} f\|_1 < \varepsilon$. Convergence of $A_{0,\varepsilon}(A_{0,u} f)$ for $\varepsilon \to 0$ follows from Lemma 2.1. Put $g = f - A_{0,u} f$ and $\tilde{A}_{o,\varepsilon} g = \varepsilon^{-1} \int_0^\varepsilon \tilde{T}_t g\, dt$. Then

$$\limsup_{\varepsilon\to 0} |A_{0,\varepsilon}g| \leqq \limsup_{\varepsilon\to 0} |\tilde{A}_{0,\varepsilon}|g||.$$

But by the local theorem for positive contractions (Theorem 1.3 or Theorem 2.6) the second lim sup is a limit and the integral of the limit is $\leqq \|g\|_1 < \varepsilon$. As $\varepsilon$ was arbitrary the theorem is proved. □

**3. Local ergodic theorems for vector valued functions.** We now study local ergodic theorems for semigroups of operators acting in the space $L_1(\Omega, \mathscr{A}, \mu, \mathfrak{X})$ of Bochner integrable functions on $\Omega$ taking values in a Banach space $\mathfrak{X}$ with norm $|\cdot|$. As many arguments are similar to the real valued case the proofs will be only sketched or even omitted. We shall use some notations and results from § 4.3.

If $\{T_u: u\in[0,\infty)^d\}$ is a strongly continuous semigroup of contractions in $L_1(\Omega, \mathscr{A}, \mu, \mathfrak{X}) = L_1^{\mathfrak{X}}$ the strong continuity may be used as in § 1.6 to show that, for any $f\in L_1^{\mathfrak{X}}$ there exists a map $f_\infty: [0,\infty)^d \times \Omega \to \mathfrak{X}$ such that $f_\infty(u,\cdot)$ is a representative of $T_u f$ for each $u$, and such that, for a.e. $\omega$, $f_\infty(\cdot,\omega)$ is Bochner integrable on each interval $[0,u)$ with respect to Lebesgue measure. We may put

$$S_{0,u}f(\omega) = \int_0^u f_\infty(v,\omega)\,dv, \quad A_{0,u}f = \pi(u)^{-1} S_{0,u}f.$$

Lemma 2.1 extends to the present situation with the same proof.

**Lemma 2.9.** *Let $\{T_t, t\geqq 0\}$ be a strongly continuous 1-parameter semigroup of $L_1 - L_\infty$-contractions in $L_1^{\mathfrak{X}}$, $f\in L_1^{\mathfrak{X}}$, and $a>0$. Then*

$$\mu(\sup_{0<t<\infty} |A_{0,t}f| > 2a) \leqq \frac{1}{a}\int |(T_0 f)^{a+}|\,d\mu.$$

The proof may be obtained by discrete approximation from the maximal ergodic lemma of Chacon in § 4.3. As in the proof of theorem 2.4 this yields

**Theorem 2.10.** *Let $\{T_t, t\geqq 0\}$ be a strongly continuous 1-parameter semigroup of $L_1 - L_\infty$-contractions in $L_1^{\mathfrak{X}}$ and $f\in L_1^{\mathfrak{X}}$, then $|A_{0,t}f - T_0 f| \to 0$, $(t\to 0+0)$, holds $\mu$-a.e..*

The full multiparameter generalization of theorem 2.10 (and also of Chacon's Banach valued ergodic theorem 4.3) seems to require new ideas. However, in the case of measure preserving transformations the maximal inequalities appearing in the paper of Tempel'man [1972] are enough to prove:

**Theorem 2.11.** *Let $\{\tau_u: u\in[0,\infty)^d\}$ be a measurable semigroup of measure preserving transformations in $(\Omega, \mathscr{A}, \mu)$, $\tau_0 =$ identity, and $\mathfrak{X}$ a Banach space. The semigroup $\{T_u\}$ in $L_1^{\mathfrak{X}}$ defined by $T_u f = f\circ\tau_u$ is strongly continuous. For all $f\in L_1^{\mathfrak{X}}$*

$$\lim_{\varepsilon\to 0} |A_{0,\varepsilon e}f - f| = 0 \quad \mu\text{-a.e..}$$

The proof of the first assertion is similar to that of the special case treated in § 1.6. Therefore, in view of the Banach valued version of lemma 2.1, the proof can be completed with the help of the following inequality

$$\mu\big(\sup_{0<\varepsilon<\infty} |A_{0,\varepsilon e} g| > a\big) \leqq \frac{\text{const}}{a} \int |g| \, d\mu, \quad (a > 0),$$

which is a special case of the results of Tempel'man. The averages $A_{0,\varepsilon e}$ may be replaced by more general "regular" averages. Complete proofs and further generalizations have been given in the Diplomarbeit of U. Wacker [1982].

## Notes

**1. Multiparameter semigroups.** De Guzmán [1975], Bruckner [1971], and Hayes and Pauc [1970] are references concerning the differentiation of integrals.

Akcoglu and del Junco [1981] have proved a generalization of theorem 2.6 for multiparameter additive processes: Let $\mathscr{I}_d$ be the class of intervals $I = [u, v[$ in $\mathbb{P}_d$ having positive Lebesgue measure $\lambda(I)$. A *subadditive process* $F = \{F_I: I \in \mathscr{I}_d\}$ for a semigroup $\{T_u: u \in \mathbb{P}_d\}$ of positive linear operators is a set function $F: \mathscr{I}_d \ni I \to F_I \in L_1$ with the following properties:

(i) $T_u F_I = F_{I+u}$, $(u \in \mathbb{P}_d, I \in \mathscr{I}_d)$;

(ii) if $I_1, \ldots, I_n \in \mathscr{I}_d$, are disjoint and their union $I$ belongs to $\mathscr{I}_d$, then $\sum_{i=1}^{n} F_{I_i} \geqq F_I$;

(iii) $\sup\{\|F_I\|_1/\lambda(I): I \in \mathscr{I}_d\} < \infty$.

**Theorem 2.12** (Akcoglu-del Junco). *If $F$ is an additive process for a strongly continuous semigroup $\{T_u: u \in \mathbb{P}_d\}$ of positive contractions in $L_1$, then $\varepsilon^{-d} F_{[0,\varepsilon e[}$ converges a.e. along each countable sequence $\varepsilon \to 0$.*

Note that $\{T_u\}$ need not be continuous at 0.

The proof uses a reduction to Terrell's theorem and to theorem 1.3.

Lin [1982] has used theorem 2.12 to derive a full multiparameter generalization of theorem 1.3. For multiparameter local ergodic theorems in $L_\infty$ see Lin [1982], Emilion [1984d].

Akcoglu and Krengel [1981] have proved a local ergodic theorem for multiparameter *superadditive* processes: Let $\{\tau_u: u \in \mathbb{P}_d^0\}$ be a measurable semigroup of measure preserving transformations, $\tau_0 = $ identity, and $T_u f = f \circ \tau_u . \pi(u)^{-1} F_{[0,u[}$ converges a.e. for all decreasing (and other "regular") sequences $u \to 0+0$.

Emilion [1983] has a generalization of the theorem of Akcoglu – del Junco to "strongly subadditive" processes $\{F_I\}$.

Emilion [1984b] considered the problem of continuous extension at 0 for a strongly continuous locally bounded semigroup $\{T_u, u \in \mathbb{P}_d\}$ of positive operators in $L_1$. The restriction to $L_1(C_0 \cup D^*)$ admits a continuous extension at 0. Sato [1984] studied this problem (and local ergodic theorems) for null preserving semiflows $\{\tau_u\}$ and the induced contractions $T_u$ in $L_1$. $\lim_{u\to 0} T_u = T_0$ exists and $T_0 = I$ iff $\mu(E) > 0$ implies $\mu(\tau_u^{-1} E) > 0$ for some $u > 0$.

Conze and Dang-Ngoc [1978] have local ergodic theorems for semigroups indexed by a Lie group.

**2. Non positive semigroups.** Theorem 2.7 was obtained by Kubokawa [1975] assuming strong convergence of $T_t$ to the identity, $(t \to 0)$, and independently by Kipnis [1974] without this condition, Sato [1978] also gave a proof. Kipnis [1974] and Sato [1978c] obtained a representation for non positive contraction semigroups analogous to the results which Akcoglu-Brunel gave in the discrete parameter case, see § 4.1. They showed that the construction of $\tilde{T}_t$ may break down if the contraction hypothesis is dropped.

Theorem 2.8 is due to Kubokawa-Kipnis. Akcoglu and Falkowitz [1983] showed that the continuity at 0 is not needed. Emilion [1984c], [1984e] has a complete generalization of the Akcoglu-del Junco theorem to the non positive case.

Concerning unresticted convergence, McGrath [1980] used Fava's dominated estimates for non commuting $L_1 - L_\infty$-contractions (§ 6.1) to prove:

**Theorem 2.13.** *Let* $\{T_k(t): t \geqq 0\}$, $(k = 1, \dots, d)$, *be strongly continuous semigroups of* $L_1 - L_\infty$*-contractions in* $L_1$ *of a* $\sigma$*-finite measure space;* $T_k(0) = I$. *If*

$$\int_{|f|>t} |f/t| (\log |f/t|)^{d-1} \, d\mu < \infty \quad \text{for all } t > 0$$

*then*

$$\frac{1}{\pi(u)} \int_0^{u_d} \dots \int_0^{u_1} T_d(t_d) \dots T_1(t_1) f \, dt_1 \dots dt_d$$

*converges to* $f$ *a.e. for* $u \to 0 + 0$.

(*Remark.* The argument in the middle of p. 215 of McGrath's paper should be applied to $g_k(\beta_k) = \sup\{|A(T_k, \alpha) f - f| : 0 < \alpha < \beta_k\}$ rather than to $|A(T_k, \alpha_k) - f|$, where $\beta_k > 0$ is so small that $\|g_k(\beta_k)\|_p^p < \varepsilon^{p+1}$.)

Theorem 2.13 goes back to Jessen-Marcinkiewicz-Zygmund [1935] for the semigroup of translations in $\mathbb{R}^d$; see also Fava [1972], de Guzmàn [1975]. Sato [1981a] remarked that the continuity at 0 is not essential. For further generalizations see Yoshimoto [1982].

Akcoglu-Krengel [1979a] have constructed a strongly continuous semigroup $\{T_t, t \geqq 0\}$ of unitary operators in $L_2$, and an $f_0 \in L_2$ for which $\varepsilon^{-1} S_{0,\varepsilon} f$ diverges a.e..

Thus in $L_2$, in contrast to $L_1$, the positivity of the $T_t$ is essential for a local ergodic theorem. Feder [1981] has related examples in $L_p$, $(1 \leqq p < 2)$. Gaposhkin [1981] has given necessary and sufficient conditions for the local ergodic theorem for a unitary group $\{T_t, t \in \mathbb{R}\}$ in $L_2$ in terms of the spectral measure. Baxter and Chacon [1974] proved a local ergodic theorem for non positive contractions $\{T_t, t \geqq 0\}$ in $L_p$, $(1 < p < \infty)$, assuming the existence of a strictly positive $h$ on $[0, \infty) \times \Omega$ such that $|f| \leqq h(t, \cdot)$ implies $|T_s f| \leqq h(t + s, \cdot)$.

**3. Vector valued functions.** Theorem 2.10 is a slight variant of a theorem of Hasegawa-Sato-Tsurumi [1978]. They assume strong continuity only at $t > 0$, but take $\mathfrak{X}$ reflexive. They use only $\sup\{\|T_t\|_\infty : 0 < t < 1\} < \infty$ instead of $\|T_t\|_\infty \leqq 1$.

For further related results, see Yoshimoto [1979].

# Chapter 8: Subsequences and generalized means

We now replace the usual Cesàro averages by weighted averages. A related problem is the convergence of averages of $T^{k_i}f$, when $k_1 < k_2 < \ldots$ is a subsequence of the integers. The first section treats mean convergence and the second pointwise convergence.

## § 8.1 Strong convergence and mixing

**1. Weighted averages for eigenvectors and flight vectors.** Let $T$ be a power bounded linear operator in a Banach space $\mathfrak{X}$. We consider averages

$$A_N^\alpha x = \sum_{k=0}^{\infty} \alpha_{Nk} T^k x, \quad (N = 1, 2, \ldots)$$

where the $\alpha_{Nk}$ are real numbers (weights) satisfying

$$\text{(W1)} \qquad \alpha_{Nk} \geqq 0, \quad \sum_{k=0}^{\infty} \alpha_{Nk} = 1 \quad \text{for all } N.$$

We also consider the family $\mathfrak{K}$ of all strictly increasing sequences $0 \leqq k_1 < k_2 < \ldots$ of integers. If $\alpha = (\alpha_{Nk})$ is the matrix of "canonical" weights associated with $\mathbf{k} = (k_i)$ by

$$\text{(1.1)} \qquad \alpha_{Nk} = N^{-1} \text{ if } k \in \{k_1, k_2, \ldots, k_N\}, \quad \text{and } = 0 \text{ else,}$$

then $A_N^\alpha x$ is the $N$-th average $A_N^{\mathbf{k}} x := N^{-1} \sum_{i=1}^{N} T^{k_i} x$ along the subsequence. Many matrices of weights have the properties

$$\text{(W2)} \qquad \lim_N \alpha_{Ni} = 0 \quad \text{for all } i \quad \text{and}$$

$$\text{(W3)} \qquad \lim_{k \to \infty} \sum_{i=k}^{\infty} |\alpha_{N,i+1} - \alpha_{N,i}| = 0 \text{ uniformly in } N.$$

For these $\alpha$, $(A_N^\alpha)$ is an $\mathscr{S}$-ergodic net in the sense of § 2.1. (We could even replace (W1) by: $\sum_0^\infty |\alpha_{Nk}| \leqq K$ for all $N$ and $\lim \sum_0^\infty \alpha_{Nk} = 1$.) By Eberlein's theorem $A_N^\alpha x$ converges strongly if a subsequence converges weakly. Thus, this yields many examples – but adds nothing new.

Recall that under fairly general conditions there is a splitting $\mathfrak{X} = \mathfrak{X}_{uds} \oplus \mathfrak{X}_{fl}$, where $\mathfrak{X}_{uds}$ is the closure of the linear span of the eigenvectors with unimodular eigenvalue, and $\mathfrak{X}_{fl}$ the space of "flight vectors" having 0 in their weak orbit closure. It is very simple to give necessary and sufficient conditions for the strong convergence of $A_N^\alpha x$ for all $x \in \mathfrak{X}_{uds}$. If $x$ is an eigenvector of $T$ with unimodular eigenvalue $\lambda$, we have $A_N^\alpha x = (\sum_{k=0}^{\infty} \alpha_{Nk} \lambda^k) x$.

Hence, the condition

$$\text{(W4)} \quad \lim_{N\to\infty} \sum_{k=0}^{\infty} \alpha_{Nk} \lambda^k = 0 \quad \text{for all } \lambda \text{ with } |\lambda| = 1 \quad \text{and} \quad \lambda \neq 1$$

implies the strong convergence of $A_N^\alpha x$ to a $T$-invariant limit for all $x \in \mathfrak{X}_{uds}$. Conversely, if $A_N^\alpha x$ shall converge strongly to a $T$-invariant limit for all $T$ and all eigenvectors $x$ with unimodular eigenvalue, (W4) must hold.

[Hanson and Pledger [1969] have used the equivalent condition (W4′), that

$$(1.2) \quad \lim_{N\to\infty} \sum_{k=0}^{\infty} \alpha_{N,kl+j} = l^{-1} \quad \text{for } l \geqq 2 \quad \text{and} \quad j = 0, \ldots, l-1$$

and

$$(1.3) \quad \lim_{N\to\infty} \sum_{\{k:\, k\gamma \bmod 1 \in [a,b)\}} \alpha_{N,k} = b - a \quad \text{for } 0 \leqq a < b < 1$$

and irrational $\gamma$,

hold. (W4′) $\Rightarrow$ (W4) is easy, but the converse is a bit cumbersome.]

At present no condition seems to be known which is necessary *and* sufficient for the strong convergence of $A_N^\alpha x$ for all flight vectors. We discuss a sufficient condition for subsequences. Recall that $D_*(F)$ was the lower density, $D^*(F)$ the upper density, and $D(F)$ the density of a set $F \subset \mathbb{Z}^+$ as defined in § 2.3. For $\mathbf{k} = (k_i)$ put $D_*(\mathbf{k}) := D_*(\{k_1, k_2, \ldots\})$, etc.

**Theorem 1.1** (Jones-Lin). *Let $T$ be power bounded, $B = \{h \in \mathfrak{X}^*: \|h\| \leq 1\}$, and $x \in \mathfrak{X}$. The following conditions are equivalent:*

$$(1.4) \quad D\text{-}\lim_N \langle T^j x, h\rangle = 0 \quad \textit{for all } h \in \mathfrak{X}^*;$$

$$(1.5) \quad \lim_N A_N^{\mathbf{k}} x = 0 \quad \textit{for all } \mathbf{k} \textit{ with } D_*(\mathbf{k}) > 0;$$

$$(1.6) \quad \sup\{N^{-1} \sum_{j=1}^{N} |\langle T^{k_j} x, h\rangle| : h \in B\} \to 0 \textit{ as } N \to \infty$$

*for all* $\mathbf{k}$ *with* $D_*(\mathbf{k}) > 0$.

*Proof.* (1.6) together with $\|y\| = \sup\{\langle y, h\rangle : h \in B\}$ implies (1.5). To show that (1.4) implies (1.6) we use proposition 2.4.9, which says that (1.4) is equivalent to

$$(1.7) \quad \sup\{N^{-1} \sum_{i=0}^{N-1} |\langle T^i x, h\rangle| : h \in B\} \to 0 \quad (N \to \infty).$$

$D_*(\mathbf{k}) > 0$ is equivalent to $C := \sup k_i/i < \infty$. For any $h \in B$ we have

$$N^{-1} \sum_{j=1}^{N} |\langle T^{k_j} x, h\rangle| \leqq N^{-1} \sum_{i=1}^{k_N} |\langle Tx, h\rangle| \leqq C k_N^{-1} \sum_{i=1}^{k_N} |\langle T^i x, h\rangle|.$$

Therefore $k_N \to \infty$ and (1.7) yield (1.6).

(1.5) $\Rightarrow$ (1.4): If (1.4) fails there exists an $\varepsilon > 0$, an $h \in B$, and a sequence $\mathbf{l}$ with positive upper density and $|\langle T^{l_i} x, h\rangle| > 2\varepsilon$, $(i = 1, 2, \ldots)$. As the union of finitely many sequences of density 0 has density 0 we can assume $\operatorname{Re}(\langle T^{l_i} x, h\rangle) > \varepsilon$ by passing to a subsequence of positive upper density $2\alpha > 0$ and multiplying the original $h$ with a complex number of modulus 1. Take any $p \in \mathbb{N}$ and form the sequence $\mathbf{k}$ consisting of the elements of $\mathbf{l}$ and all multiples of $p$. As $\mathbf{k}$ has positive lower density, (1.5) implies $A_N^{\mathbf{k}} x \to 0$. On the other hand, there are infinitely many $n$ with $\alpha l_n < n$. If $M$ is a bound for the $\|T^i\|$, and $k_N$ is such an $l_n$, we have

$$\begin{aligned} \operatorname{Re}(\langle A_N^{\mathbf{k}} x, h\rangle) &= N^{-1} \sum_{i=1}^{N} \operatorname{Re}(\langle T^{k_i} x, h\rangle) \geqq N^{-1}[\alpha k_N \varepsilon - M\|x\| k_N/p] \\ &\geqq \alpha\varepsilon - M\|x\| C/p. \end{aligned}$$

For large $p$ we have a contradiction. □

For $x \in \mathfrak{X}_{fl}$, condition (1.4) holds by the results of Jones and Lin at the end of section 2.4, and (1.5) follows.

**2. Weak convergence of $T^i x$ implies strong convergence of $A_N^{\mathbf{k}} x$.** We now derive a convergence theorem for rather general weights imposing the condition of weak convergence of $T^i x$. We first show that this condition is necessary.

**Proposition 1.2.** *Let $T$ be power bounded in a Banach space $\mathfrak{X}$. If $A_N^{\mathbf{k}} x$ converges strongly for all $\mathbf{k} \in \mathfrak{K}$, the sequence $T^i x$ converges weakly to a fixed point.*

*Proof.* The limit is necessarily the same for all $\mathbf{k}$. Otherwise we could form a third sequence $\mathbf{l}$, consisting of pieces of sequences $\mathbf{k}$, $\mathbf{k}'$ with different limits, for which $A_N^{\mathbf{l}} x$ diverges. Looking at $k_i = i$ we see that the limit is $T$-invariant. Subtracting it from $x$ we may assume that it is 0. If $T^i x$ does not converge weakly to 0 there exists $h \in B$, $\varepsilon > 0$, and $k_1 < k_2 < \ldots$ with $\operatorname{Re}(\langle T^{k_i} x, h\rangle) > \varepsilon$ for all $i$. Then $\operatorname{Re}(\langle A_N^{\mathbf{k}} x, h\rangle) > \varepsilon$, $(N \geqq 1)$, contradicts the strong convergence of $A_N^{\mathbf{k}} x$ to 0. □

The following theorem was proved by Blum-Hanson [1960] for measure preserving transformations, and by Akcoglu-Sucheston [1972] and Jones-Kuftinec [1971] for general contractions in Hilbert space. (These papers deal only with subsequences).

**Theorem 1.3.** *Let $T$ be a contraction in a Hilbert space $\mathfrak{X}$. If $T^i x$ converges weakly, the averages $A_N^{\alpha} x$ converge strongly to the same limit for all weights $(\alpha_{Ni})$ with*

(W5) $c := \sup_N |\alpha_{Ni}| < \infty$; $\lim_N \sum_i \alpha_{Ni} = 1$; $\lim_N \sup_i |\alpha_{Ni}| = 0$.

*In particular,* $A_N^{\mathbf{k}} x$ *converges strongly for all* $\mathbf{k} \in \mathfrak{K}$.

*Proof.* Necessarily, the limit is fixed, and we may assume it is 0. Since $T$ is a contraction $\lim \|T^n x\|$ exists. Hence, given an $\varepsilon > 0$, one can choose an integer $K > 0$ such that $k \geqq K$ and $j \geqq 0$ imply $\|T^k x\| - \|T^{k+j} x\| < \varepsilon^2$, and also $|\langle T^k x, x\rangle| \leqq \varepsilon$. We apply Lemma 2.3.3 to $x' = T^k x$, $y' = x$, and $S = T^j$, and obtain that

$$|\langle T^k x, x\rangle - \langle T^{k+j} x, T^j x\rangle| \leqq \varepsilon \|x\|$$

and

$$|\langle T^{k+j} x, T^j x\rangle| \leqq \varepsilon(1 + \|x\|)$$

hold for all $k \geqq K$ and all $j$. This means that $|i - j| \geqq K$ implies $|\langle T^i x, T^j x\rangle| \leqq \varepsilon(1 + \|x\|)$. Given $\eta > 0$, we have $\max_i |\alpha_{Ni}| < \eta$ for large enough $N$. Clearly

$$\Big\|\sum_{i=1}^{\infty} \alpha_{Ni} T^i x\Big\|^2 \leqq \sum_{i=1}^{\infty} \sum_{j=1}^{\infty} |\alpha_{Ni}\alpha_{Nj} \langle T^i x, T^j x\rangle|.$$

For fixed $i$, there are at most $2K - 1$ indices $j$ with $|i - j| < K$. Thus, the sum of these terms can be estimated by $(2K - 1)|\alpha_{Ni}| \|x\|^2 \eta$. The sum of the remaining terms is bounded by $\sum\sum |\alpha_{Ni}\alpha_{Nj}| \varepsilon(1 + \|x\|) \leqq \varepsilon(1 + \|x\|)c^2$. As $\varepsilon$ and $\eta$ were arbitrarily small the proof is complete. □

**3. Mixing and complete mixing.** It follows from the spectral mixing theorem that an endomorphism $\tau$ of a probability space is weakly mixing iff $D$-$\lim \langle f \circ \tau^n, g\rangle = 0$ holds for all $f \in L_2$ with $\int f = 0$. It is even simpler to check that $\tau$ is mixing iff $f \circ \tau^n$ tends weakly to 0 for all $f \in L_2$ with zero integral. Thus, the results above are characterizations of weak mixing and mixing.

It has been a problem dating back to Hopf [1937] to find an appropriate notion of mixing for endomorphisms of infinite measure spaces. Let $L_1^0$ *be the subspace of* $L_1$ consisting of the functions with $\int f d\mu = 0$. We call a positive contraction $T$ in $L_1$ *mixing* if $T^n f$ tends weakly to 0 for all $f \in L_1^0$, *weakly mixing* if $D$-$\lim \langle T^n f, g\rangle = 0$ holds for all $f \in L_1^0$, $g \in L_\infty$, and *completely mixing* if $\|T^n f\|_1$ tends to 0 for all $f \in L_1^0$. An endomorphism $\tau$ (or, more generally, a null preserving $\tau$) is called mixing (weakly mixing, ...), if the $L_1$-contraction corresponding to $\tau$ (i.e., with $\langle Tf, g\rangle = \langle f, g \circ \tau\rangle$ for $f \in L_1$, $g \in L_\infty$) is mixing (weakly mixing, ...). It is simple to check that mixing and weak mixing agree with the classical notions for endomorphisms of finite measure spaces.

The following theorem of Krengel and Sucheston [1969a] was (and perhaps remains) quite surprising. [It provided a negative answer to the problem whether $K$-automorphisms of $\sigma$-finite measure spaces are mixing, and showed that, in fact, *invertible* mixing measure preserving transformations of a $\sigma$-finite infinite space do not exist. On the other hand, it was shown in the same paper that exact endomorphisms are mixing.]

**Theorem 1.4.** *Let $T$ be a positive contraction in $L_1$, for which there exists no non zero $p \in L_1^+$ with $Tp = p$. If $f$ is integrable and $T^n f$ converges weakly to some element of $L_1$, then $\| T^n f \|_1 \to 0$. In particular, if $T$ is mixing, then $T$ is completely mixing.*

*Proof.* Let $L$ be a Banach limit. Define a positive linear functional $\gamma$ on $L_\infty$ by $\gamma(h) = L((\int |T^n f| h\, d\mu))$. As $(T^n f)$ converges weakly, this sequence is conditionally weakly compact and hence uniformly integrable by the Vitali-Hahn-Saks theorem. It follows that if $A_n$ is a sequence of sets with $A_n \downarrow \emptyset$, then $\gamma(1_{A_n})$ tends to 0. Thus $\gamma$ is $\sigma$-additive and there exists $p \in L_1^+$ with $\gamma(h) = \int ph\, d\mu$. For $h \geqq 0$ we have

$$\begin{aligned}\int (Tp) h\, d\mu = \gamma(T^* h) = L((\int |T^n f| T^* h\, d\mu)) = L((\int T |T^n f| h\, d\mu))\\ \geqq L((\int |T^{n+1} f| h\, d\mu)) = \gamma(h) = \int ph\, d\mu .\end{aligned}$$

Hence $Tp \geqq p$. As $T$ is a contraction, $Tp = p$. The assumption implies $p = 0$. In particular, $0 = \gamma(\mathbb{1}) = \lim \| T^n f \|_1$. □

**Remarks.** Only the conditional weak compactness of $(T^n f)$ was used in the proof. If $T$ is an arbitrary contraction in $L_1$ and $|T|$ its modulus, the argument shows $|T| p = p$. Hence, if $|T|$ has no positive non zero fixed points, and $(T^n f)$ is conditionally weakly compact, again $\| T^n f \|_1 \to 0$ follows.

## Notes

Cohen [1940] considered mean ergodic theorems with weights now covered by Eberlein's theorem; see also Kurtz and Tucker [1968]. Hanson and Pledger [1969] showed that (W4′) is necessary and sufficient for the strong convergence of $A_N^\alpha x$ for all isometries in Hilbert space using a reduction to the unitary case in which this follows from the spectral theorem. They also showed for the isometries in $L_2$ induced by endomorphisms of a probability space that $A_N^\alpha x$ converges strongly for all flight vectors if the weights satisfy

$$\text{(W6)} \qquad \sum_{k \in F} \alpha_{Nk} \to 0 \quad \text{for all } F \subset \mathbb{Z}^+ \text{ with } D(F) = 0 .$$

It seems to have been overlooked that this contains the special case of measure preserving transformations of Jones' [1971] result, that the weak convergence of $T^{n_i} x$ along a subsequence of density 1 implies (1.5). This implication follows from the easy half of

**Proposition 1.5.** *If $(\alpha_{Ni})$ is the matrix of canonical weights for* **k**, *the condition* (W6) *is equivalent to $D_*(\mathbf{k}) > 0$.*
Jones [1972] proved

**Theorem 1.6.** *If $T$ is a contraction in a uniformly convex space $\mathfrak{X}$ the following are equivalent:*

(1.8) $\mathfrak{X} = \mathfrak{X}_{fl}$;

(1.9) *Given* $x \neq 0$ *there exists* $\beta_x > 0$ *with* $D_*(\{n: \|T^n x - x\| < \beta_x\}) = 0$;

(1.10) *Given* $x$, *there exists* $F$ *with* $D(F) = 1$ *and* $\liminf_{n \in F} \|T^n x - x\| \geqq \|x\|$.

For related material, see Jones [1973], Nagel [1974], Patil [1977], Blum-Mizel [1971].

Akcoglu, Huneke and Rost [1974] showed that theorem 1.3 does not extend to arbitrary Banach spaces. Their counterexample is a contraction in $C(K)$ for compact $K$. It was further analysed by Sine [1974].

Akcoglu and Sucheston [1972], [1975] proved the equivalence of

(A) $w\text{-}\lim T^n x = y$

and

(B) $A_N^\alpha x \to y$ *for all weights with* (W5)

for contractions in $L_1$, in $L_2$, and for positive contractions in $L_p$, $(1 < p < \infty)$.

Weights with (W5) were first used by Fong-Sucheston [1974] for these problems. Mrs. Bellow [1975], [1976] gave a proof for $L_p$, $(1 < p < \infty)$, of (A) $\Leftrightarrow$ (B) avoiding the dilation argument of Akcoglu-Sucheston. Hiai [1978], Iwanik [1979], and Blum-Eisenberg [1979] have generalizations for certain groups of operators. Mrs. Millet [1980] proved (A) $\Leftrightarrow$ (B) for positive power bounded $T$ in $L_1$, for certain non positive operators in $L_1$, and for positive contractions in Orlicz spaces having a uniformly convex dual; see also Sato [1975b], [1976]. These papers contain more references.

Mixing for Markov operators was studied by Lin [1971].

Lin [1972a] extended theorem 1.4 to power bounded $T$ and showed that $T$ is mixing iff $T \times T$ is mixing. (The terminology in these papers is different).

Aaronson, Lin and Weiss [1979] showed that the following conditions are equivalent to weak mixing for a positive contraction in $L_1$:

(1.11) $T$ is ergodic (i.e., $\|A_N f\|_1 \to 0$ for all $f \in L_1^0$), and $T^*$ has no unimodular eigenvalues $\neq 1$;

(1.12) For every ergodic positive contraction $S$ in $L_1$ with finite invariant measure $T \times S$ is ergodic.

Lin [1981] has characterized weak mixing by

(1.13) For every $h \in L_\infty$ there is a sequence $n_i$ of density 1 such that all $w^*$-cluster points of $\{T^{*n_i} h\}$ are constants.

Foguel [1973] studied the convergence of $R^n x$, where $R = \int_0^\infty \varphi(t)\, T_t\, dt$ is an "average" over a semigroup $\{T_t, t \geqq 0\}$.

Friedman [1983] calls an endomorphism $\tau$ of a probability space $(\Omega, \mathscr{A}, \mu)$ mixing along **k** if $\mu(A \cap \tau^{-k_i} B) \to \mu(A)\,\mu(B)$ holds for all $A, B \in \mathscr{A}$ and studies the Blum-Hanson theorem along such sequences.

## § 8.2 Pointwise convergence

**1. Weighted and Abelian averages.** Some ergodic theorems and maximal inequalities deal with Abelian averages like $\lambda \int_0^\infty e^{-\lambda s} T_s f ds$ or $(1-\lambda) \sum_{k=0}^\infty \lambda^k T^k f$. We shall see in the Notes that the investigation of such averages can make sense. Frequently, however, ergodic theorems about Abelian averages do not deserve a separate proof. As Cesàro convergence implies Abel convergence, the ordinary ergodic theorems imply Abelian-ergodic theorems as corollaries. We derive a variant of this classical result which should suffice for most ergodic theoretic applications, including ratio theorems.

Consider non negative weights $\alpha_{\lambda k}$ with $\sum_{k=0}^\infty \alpha_{\lambda,k} = 1$ for all $\lambda \in \Lambda$, $\Lambda$ an arbitrary directed set. For a sequence $(f_k)$ put $F_n = \sum_{k=0}^n f_k$, and $F_\lambda^\alpha = \sum_{k=0}^\infty \alpha_{\lambda k} f_k$, whenever this sum is well defined. Similarly, $(g_k)$ defines $G_n$ and $G_\lambda^\alpha$.

**Theorem 2.1.** *Assume the weights* $(\alpha_{\lambda k})$ *satisfy*

(W7) $\quad \alpha_{\lambda k} \geqq \alpha_{\lambda, k+1}$ *for all* $\lambda, k$ *and*

(W8) $\quad \lim_\lambda (\alpha_{\lambda,k} - \alpha_{\lambda,k+1}) = 0$ *for all* $k$.

*If* $(f_k)$ *is a sequence of vectors in a Banach space which satisfies* $\sum_{k=0}^\infty \|\alpha_{\lambda k} f_k\| < \infty$ *for all* $\lambda$, *and if* $(g_k)$ *is a sequence of non negative numbers, not all* 0, *then* $F_n/G_n \to x$ *implies* $F_\lambda^\alpha / G_\lambda^\alpha \to x$.

*Proof.* The assumptions permit the use of the Abel transformation

$$(2.1) \qquad F_\lambda^\alpha = \sum_{k=0}^\infty (\alpha_{\lambda,k} - \alpha_{\lambda,k+1}) F_k, \quad G_\lambda^\alpha = \sum_{k=0}^\infty (\alpha_{\lambda,k} - \alpha_{\lambda,k+1}) G_k.$$

First assume $G_n \to \infty$. Then (W8) and (2.1) imply $G_\lambda^\alpha \to \infty$. Put $\beta_k = F_k/G_k - x$. Then

$$(2.2) \qquad \|F_\lambda^\alpha - x G_\lambda^\alpha\| \leqq \sum_{k=0}^\infty (\alpha_{\lambda,k} - \alpha_{\lambda,k+1}) \|\beta_k\| G_k.$$

For $\varepsilon > 0$ there exists $K$ with $\|\beta_k\| < \varepsilon$ for $k \geqq K$. The sum of the first $K$ terms on the righthand side of (2.2) tends to 0 after division by $G_\lambda^\alpha$, and

$$\sum_{k=K}^\infty (\alpha_{\lambda,k} - \alpha_{\lambda,k+1}) \|\beta_k\| G_k / G_\lambda^\alpha \text{ is } \leqq \varepsilon.$$

Now assume that the limit $G$ of $(G_n)$ is finite. Then $F_n$ tends to a limit $F$. For $\varepsilon > 0$ there exists $L > 0$ with $\|F_k - F\| < \varepsilon$ for $k \geqq L$. Hence

$$\|F_\lambda^\alpha - F\| \leqq \sum_{i=0}^{L} (\alpha_{\lambda,k} - \alpha_{\lambda,k+1}) \sup_j \|F_j\| + \varepsilon,$$

and (W8) shows $F_\lambda^\alpha \to F$. Similarly, $G_\lambda^\alpha \to G$. □

**Examples.** 1) The weights $\alpha_{\lambda k} = (1-\lambda)\lambda^k$, $(\lambda \in \Lambda = ]0,1[)$ satisfy (W7) and (W8) for $\lambda \to 1$. If $T$ is a contraction in some $L_p$, then $\sum_k |T^k f \lambda^k| < \infty$ for all $\lambda$ holds a.e.. If $A_n(T)f$ converges a.e., $(1-\lambda) \sum_{k=0}^{\infty} \lambda^k T^k f$ converges a.e. for $\lambda \to 1$.

2) Let $w_0 \geqq w_1 \geqq \ldots$ be a decreasing sequence of positive numbers with divergent sum. Then the weights with $\alpha_{Nk} = w_k / \sum_0^N w_k$ for $k \leqq N$, $= 0$ for $k > N$, satisfy (W7) and (W8).

The next lemma will show that maximal inequalities and dominated ergodic theorems extend to many weighted averages:

**Lemma 2.2.** *If $f_0, \ldots, f_n$ are real numbers, $g_0 > 0$, $g_1, \ldots, g_n \geqq 0$ and $w_0 \geqq w_1 \geqq \ldots \geqq w_n > 0$, then*

$$\max_{0 \leqq k \leqq n} \frac{f_0 + \ldots + f_k}{g_0 + \ldots + g_k} \geqq \max_{0 \leqq k \leqq n} \frac{w_0 f_0 + \ldots + w_k f_k}{w_0 g_0 + \ldots + w_k g_k} =: \varrho_n \tag{2.3}$$

*Proof.* Let $k$ be the first index for which the quotient on the right hand side equals $\varrho_n$. Then

$$w_j f_j + \ldots + w_k f_k \geqq \varrho_n (w_j g_j + \ldots + w_k g_k) \tag{2.4}$$

for all $j$ with $0 \leqq j \leqq k$. Find $\xi_0, \ldots, \xi_k$ successively by solving the equations $w_j \sum_{i=0}^{j} \xi_i = 1$ for $j = 0, \ldots, k$. The monotonicity of the $w_j$ implies $\xi_j \geqq 0$. Multiply the inequalities in (2.4) by $\xi_j$ and sum over $j = 0, \ldots, k$. Then $(f_0 + \ldots + f_k) \geqq \varrho_n (g_0 + \ldots + g_k)$ follows. □

The lemma, applied with $g_k = 1$ for all $k$, shows that dominated ergodic theorems for Cesàro averages $A_n(T)f$ immediately yield dominated ergodic theorems for Abelian averages as corollaries. If $T$ is a positive contraction in $L_1$ and $f$ is integrable, the lemma implies

$$E' := \{\sup_{0<\lambda<1} (1-\lambda) \sum_{k=0}^{\infty} \lambda^k T^k f > 0\} \subset \{\sup_{n \geqq 1} A_n(T) f > 0\}. \tag{2.5}$$

As $\{f > 0\}$ is contained in $E'$, (2.5) and Hopf's maximal ergodic theorem yield the Abelian maximal ergodic theorem

$$\int_{E'} f \, d\mu \geqq 0.$$

**2. Good and bad subsequences.** Recall that the Blum-Hanson theorem asserts the $L_2$-norm-convergence of $A_n^{\mathbf{k}} f = A_n^{\mathbf{k}}(\tau) f := n^{-1} \sum_{i=1}^{n} f \circ \tau^{k_i}$ for all $\mathbf{k} \in \mathfrak{K}$ and all $f \in L_2$, when $\tau$ is a mixing endomorphism of a probability space $(\Omega, \mathscr{A}, \mu)$. By a simple approximation argument, $L_p$-norm-convergence holds for all $f \in L_p$, $(1 \leqq p < \infty)$. The problem of finding sufficient conditions for a.e.-convergence has generated great interest in spite of the present lack of applications. Friedman and Ornstein [1972] showed by example that the assumptions of the Blum-Hanson theorem are not sufficient. Then Friedman [1970] asked whether a.e.-convergence holds for $K$-automorphisms, and also whether a.e.-convergence holds for mixing automorphisms when $\mathbf{k}$ has positive density. Krengel [1971] answered the first of these questions negatively even for Bernoulli shifts by showing that there exists a *universally bad sequence* $\mathbf{k} \in \mathfrak{K}$, in the sense that for every aperiodic automorphism $\tau$ of a probability space there exists an $f \in L_1$ for which $A_n^{\mathbf{k}} f$ diverges a.e.; see Theorem 6.4.6. The second question of Friedman is still open, but Conze [1973] showed that there exists a $\mathbf{k}$ with $D_*(\mathbf{k}) > 0$, a weakly mixing $\tau$, and an $f \in L_1$ for which $A_n^{\mathbf{k}} f$ does not converge a.e.. He also showed that the combination of the conditions proposed in the two questions is sufficient. To prove this we begin with some general considerations.

Define $C(\mathbf{k})$ as the infimum of all $C > 0$ with the following property: For any automorphism $\tau$ of any probability space $(\Omega, \mathscr{A}, \mu)$, any $f \in L_1$, and any $\lambda > 0$

$$(2.6) \qquad \mu(\sup_{n \geqq 1} A_n^{\mathbf{k}} |f| > \lambda) \leqq \frac{C}{\lambda} \|f\|_1 .$$

If no finite $C$ with this property exists, put $C(\mathbf{k}) = \infty$.

Note that $D_*(\mathbf{k}) > 0$ is equivalent to $c := \sup (k_i + 1)/i < \infty$. Using

$$A_n^{\mathbf{k}} |f| = n^{-1} \sum_{i=1}^{n} |f| \circ \tau^{k_i} \leqq \frac{k_n + 1}{n} \cdot \frac{1}{k_n + 1} \sum_{i=0}^{k_n} |f| \circ \tau^i \leqq c A_{k_n+1} |f|$$

and the maximal ergodic inequality, we obtain $C(\mathbf{k}) \leqq c$. Thus $D_*(\mathbf{k}) > 0$ implies $C(\mathbf{k}) < \infty$.

An automorphism $\tau$ is said to have *Lebesgue spectrum* if the subspace $L_2^0$ of $L_2$ orthogonal to the constant function $\mathbb{1}$ is the closure of the linear span of a family $\{f_{in}: i \in I, n \in \mathbb{Z}\}$ of orthogonal vectors with $f_{i,n+1} = f_{i,n} \circ \tau$ for all $i \in I$, $n \in \mathbb{Z}$. ($I$ is an arbitrary index set). It is well known that $K$-automorphisms have Lebesgue spectrum.

**Theorem 2.3** (Conze). *Let $\tau$ be an automorphism with Lebesgue spectrum. Then $A_n^{\mathbf{k}} f$ converges a.e. for all $f \in L_1$ and all $\mathbf{k} \in \mathfrak{K}$ with $D_*(\mathbf{k}) > 0$.*

*Proof.* The theorem of Rajchman (see e.g. Chung [1968: 97]) asserts that the sequence of Cesàro averages of any bounded sequence of orthogonal vectors in $L_2^0$ convergences a.e. to 0. Thus, $A_m^{\mathbf{k}} f_{i,n}$ convergences a.e. to 0 for all $f_{i,n}$. Clearly,

$A_m^{\mathbf{k}}\mathbb{1} = \mathbb{1}$. Thus, $A_m^{\mathbf{k}} f$ converges a.e. for all $f$ in a subset of $L_1$ which is dense in $L_1$. The estimate (2.6) implies that the set of all $f \in L_1$ for which $A_n^{\mathbf{k}} f$ converges a.e. is closed (Theorem 1.7.2). □

Obviously, periodic sequences $\mathbf{k}$ are *universally good* in the sense that $A_n^{\mathbf{k}} f$ converges a.e. for all automorphisms $\tau$ of an arbitrary probability space and all integrable $f$. The first non trivial class of universally good sequences $\mathbf{k}$ has been described by Brunel-Keane [1969] as follows:

Let $\sigma$ be a homeomorphism of a compact metric space $\Sigma$ with metric $\varrho$ such that all powers $\sigma^n$ are equicontinuous, and assume that there exists $z \in \Sigma$ with dense orbit $\{\sigma^n z : n \in \mathbb{N}\}$. Then there exists a unique (and hence ergodic), $\sigma$-invariant probability measure $\nu$ on $\mathscr{B}$, the $\sigma$-algebra of Borel sets, see § 1.2. Each non empty open set has positive $\nu$-measure. A sequence $\mathbf{k} \in \mathfrak{K}$ is called *uniform* if there exists such $(\Sigma, \mathscr{B}, \nu, \sigma)$, and a set $Y \in \mathscr{B}$ with $\nu(\partial Y) = 0$ and $\nu(Y) > 0$, and a point $y \in \Sigma$ with $k_i = i$-th entry time of the orbit of $y$ into $Y$. (Formally: $k_i = \inf\{l \geqq 0 \colon \sum_{j=0}^{l} 1_Y(\sigma^j y) = i\})$. $\partial Y$ is the boundary of $Y$.

**Theorem 2.4** (Brunel-Keane). *Uniform sequences are universally good.*

*Proof.* Let $\varepsilon > 0$ be given. For small $\delta > 0$ and very small $\delta' = \delta'(\delta) > 0$ the open sets $Y' = (x \in Y \colon \varrho(x, \partial Y) > \delta\}$ and $W = \{x \in \Sigma \colon \varrho(x, y) < \delta'\}$ satisfy $\mu(Y \setminus Y') < \varepsilon$ and

$$(2.7) \qquad 1_{Y'}(\sigma^n x) \leqq 1_Y(\sigma^n y) \quad \text{for all } x \in W, n \geqq 0.$$

Consider the Cartesian product $(\Omega', \mathscr{A}', \mu') = (\Omega, \mathscr{A}, \mu) \times (\Sigma, \mathscr{B}, \nu)$ and the product automorphism $\tau' = \tau \times \sigma$ defined by $\tau'(\omega, x) = (\tau\omega, \sigma x)$. For given $f \in L_1(\mu)$ the functions $g, g'$ on $\Omega'$ defined by $g(\omega, x) := f(\omega) 1_Y(x)$ and $g'(\omega, x) := f(\omega) 1_{Y'}(x)$ are $\mu'$-integrable. Thus, $\bar{g}' = \lim A_n(\tau') g'$ exists $\mu'$-a.e.. Using the uniform integrability of the $A_n(\tau')g'$ and the ergodicity of $\sigma$ we find

$$\begin{aligned} \int_{\Omega \times W} \bar{g}' \, d\mu' &= \lim_{n \to \infty} n^{-1} \sum_{k=0}^{n-1} \iint_{\Omega \times W} f(\tau^k \omega) 1_{Y'}(\sigma^k x) \, \mu(d\omega) \, \nu(dx) \\ &= \int f d\mu \lim \int_W n^{-1} \sum_{k=0}^{n-1} 1_{Y'}(\sigma^k x) \, \nu(dx) = \nu(W) \nu(Y') \int f d\mu. \end{aligned}$$

Now (2.7) implies that $\underline{S}(\omega) := \liminf (A_n(\tau')g)(\omega, y)$ satisfies $\bar{g}'(\omega, x) \leqq \underline{S}(\omega)$ for $\omega \in \Omega$, $x \in W$. Hence

$$\begin{aligned} \int \underline{S} \, d\mu &= \nu(W)^{-1} \iint_{\Omega \times W} \underline{S}(\omega) \, d\mu' \geqq \nu(W)^{-1} \iint_{\Omega \times W} \bar{g}' \, d\mu' \\ &= \nu(Y') \int f d\mu \geqq (\nu(Y) - \varepsilon) \int f d\mu. \end{aligned}$$

A symmetric argument shows that $\bar{S}(\omega) := \limsup (A_n(\tau')g)(\omega, y)$ satisfies $\int \bar{S} \, d\mu \leqq (\nu(Y) + \varepsilon) \int f d\mu$. Together these estimates yield $\int (\bar{S} - \underline{S}) \, d\mu \leqq 2\varepsilon \|f\|_1$.

As $\varepsilon > 0$ was arbitrary, $\underline{S} = \bar{S}$ $\mu$-a.e.. Thus

$$(2.8) \qquad (A_n(\tau')g)(\omega, y) = n^{-1} \sum_{i=0}^{n-1} f(\tau^i \omega) 1_Y(\sigma^i y)$$
$$= n^{-1} \sum_{\{i:\, k_i \leqq n-1\}} f(\tau^{k_i} \omega)$$

converges $\mu$-a.e.. Setting $f \equiv 1$, we see that $n^{-1}$ card $\{i: k_i \leqq n-1\}$ converges to $\nu(Y)$, and then (2.8) yields the convergence $\mu$-a.e. of $A_n^{\mathbf{k}} f$. □

**Remark.** If $\tau$ is weakly mixing, $\tau'$ is ergodic and the limit is constant.

**3. Random ergodic theorems.** While the ergodic theorems considered so far deal with iterates of a fixed operator or with semigroups, random ergodic theorems deal with sequences of operators for which the operator applied at time $k$ depends on the state in which a stochastic process is at time $k$. Again, applications seem scarce, which allows us to restrict attention to the simplest case.

Let $\sigma$ be an automorphism of a probability space $(\Sigma, \mathscr{B}, \nu)$, and let $(\Omega, \mathscr{A}, \mu)$ be another probability space. Assume that for each $x \in \Sigma$ there is an endomorphism $\tau(x)$ of $\Omega$, such that the *skew product* $\tau'$ defined in the product measure space $(\Omega', \mathscr{A}', \mu') = (\Omega, \mathscr{A}, \mu) \times (\Sigma, \mathscr{B}, \nu)$ by $\tau'(\omega, x) := (\tau(x)\omega, \sigma x)$ is measurable. For any $x \in \Sigma$ put $\Theta_0(x) =$ identity in $\Omega$, and $\Theta_{n+1}(x) = \tau(\sigma^n x) \circ \Theta_n(x)$.

**Theorem 2.5** (Random ergodic theorem for endomorphisms). *For any $f \in L_1(\mu)$ there exists a set $N_f \in \mathscr{B}$ with $\nu(N_f) = 0$ such that $n^{-1} \sum_{i=0}^{n-1} f \circ \Theta_i(x)$ converges $\mu$-a.e. for $x \in N_f^c$.*

*Proof.* The assumptions imply that $\tau'$ is an endomorphism of $\Omega'$. Put $g(\omega, x) = f(\omega)$. Induction yields $(\tau')^i(\omega, x) = (\Theta_i(x)\omega, \sigma^i x)$. Hence $g((\tau')^i(\omega, x)) = (f \circ \Theta_i(x))(\omega)$. Birkhoff's theorem implies the convergence $\mu'$-a.e. of $A_n(\tau')g$. Let $N'$ be the $\mu'$-nullset on which there is divergence. Then $N_f := \{x \in \Sigma: \mu(\{\omega: (\omega, x) \in N'\} > 0)\}$ is a $\nu$-nullset. For $x \in N_f^c$, $n^{-1} \sum_{i=0}^{n-1} g((\tau')^i(\omega, x)) = n^{-1} \sum_{i=0}^{n-1} (f \circ \Theta_i(x))(\omega)$ converges $\mu$-a.e.. □

**Example.** Let $\sigma$ be the shift in $\Sigma = \{x = (x_0, x_1, \ldots): x_i \in \{1, 2\}\}$, where the coordinates are independent under $\nu$ with $\nu(\{x: x_i = 1\}) = 1/2$. Let $\tau$ be a fixed automorphism of $(\Omega, \mathscr{A}, \mu)$ and $\tau(x) = \tau^{x_0}$. Then $\Theta_i(x) = \tau^{k_i}$ with $k_i = x_0 + \ldots + x_{i-1}$. We obtain convergence $\mu$-a.e. of $A_n^{\mathbf{k}} f$ for almost all subsequences $\mathbf{k} = \mathbf{k}(x)$.

## Notes

The books of Hardy [1949] and Zeller-Beekmann [1970] are well known references for the theory of summability. It is interesting to note that one can also deduce Cesàro convergence from Abelian ergodic theorems using the following Tauberian theorem of Hardy and Littlewood (cf. Hardy [1949: 155]):

**Theorem 2.6.** *If $(a_n)$ is a bounded or non negative sequence, and $(1-\lambda)\sum_{k=0}^{\infty}\lambda^k a_k$ tends to $x$ for $\lambda \to 1-0$, then $n^{-1}\sum_{k=0}^{n-1} a_k$ tends to $x$.*

Rota [1963] remarked that the proof of the Abelian maximal inequality is somewhat simpler than the proof of Hopf's inequality. But the Tauberian theorem is not so easy, and thus a direct proof of Cesàro convergence seems preferable.

Several authors have produced counterexamples to show that certain conditions cannot be weakened in Abelian ergodic theorems. By theorem 2.6, these examples are superfluous when the corresponding Cesàro type examples exist.

Krengel [1967] used a variant of theorem 2.1 and lemma 2.2 to point out that weighted ratio ergodic theorems and maximal inequalities for decreasing weights do not require a separate proof. The argument can be modified for continuous parameter semigroups. If weights of the form $\alpha_{\lambda s} = \int_s^{\infty} \varrho_\lambda(u)\,du$ with integrable $\varrho_\lambda \geqq 0$ are given, one can perform a transformation

$$\int_0^{\infty} \alpha_{\lambda s} f(s)\,ds = \int_0^{\infty} \varrho_\lambda(s) \int_0^{s} f(u)\,du\,ds.$$

E.g. for $\alpha_{\lambda s} = \lambda e^{-\lambda s}$ one takes $\varrho_\lambda(u) = \lambda^2 e^{-\lambda u}$, and finds: If $e^{-\lambda s} f(s)$ is integrable for all $\lambda$ with $0 < \lambda < \infty$, and $g \geqq 0$ has positive integral, then $\lim_{t\to\infty} \left(\int_0^t f(s)\,ds \big/ \int_0^t g(s)\,ds\right) = x$ implies $\lim_{\lambda\to 0} \left(\int_0^{\infty} e^{-\lambda s} f(s)\,ds \big/ \int_0^{\infty} e^{-\lambda s} g(s)\,ds\right) = x$. This remark yields ergodic theorems for resolvents. Similarly, Abelian local ergodic theorems may be deduced from ordinary local ergodic theorems. Details have been given by Sato [1980], [1981c]. (These papers contain more references.)

The following Tauberian theorem is a special case of the results of Karamata [1931]; see Widder [1946: 192].

**Theorem 2.7.** *If $F(t)$ is non decreasing, and such that the integral $r_\lambda = \int_0^{\infty} e^{-\lambda t}\,dF(t)$ converges for $\lambda > 0$, then $\lambda r_\lambda \to x$, $(\lambda \to 0+0)$, implies $t^{-1}F(t) \to x$, $(t \to \infty)$, and $\lambda r_\lambda \to x$, $(\lambda \to \infty)$, implies $t^{-1}F(t) \to x$, $(t \to 0+0)$.*

Feyel has systematically employed this theorem to deduce local and global ergodic theorems for continuous parameter semigroups. He proves his results even for pseudo-resolvents; see § 2.1 (Notes). He calls a pseudo-resolvent for which the $\lambda V_\lambda$ are positive contractions of $L_1$ *proper* if $V_\lambda \infty = \infty$, (i.e., $0 \leqq f_n \uparrow \infty$ implies $V_\lambda f_n \uparrow \infty$). The following maximal ergodic theorem of Feyel [1977] does not follow from Hopf's maximal ergodic theorem even for resolvents:

**Theorem 2.8.** *Let $(\lambda V_\lambda)_{\lambda>0}$ be a pseudo-resolvent for which the $\lambda V_\lambda$ are positive contractions in $L_1$, and which is proper. For $f \in L_1$ and $D \subset \mathbb{R}^+$ put $E = \{\sup_{\lambda\in D} \lambda V_\lambda f > 0\}$, where* sup *is the "essential" supremum in $L_1$. Then $\int_E f\,d\mu \geqq 0$.*

The subsequent papers of Feyel [1977a], [1977b], [1979], [1982], [1982a] contain pseudo-resolvent forms of the Chacon-Ornstein theorem, the Dunford-Schwartz theorem, Brunel's lemma, and of local ergodic theorems for superadditive processes. The Abelian analogue of a superadditive process is a *superabelian process*, i.e., a family of measurable functions $(u_\lambda)_{\lambda>0}$ with $\|u_\lambda^-\|_1 \leqq M/\lambda$ for some $M<\infty$, and $u_\lambda \geqq u_\varrho + (\varrho-\lambda) V_\varrho u_\lambda$, $(0<\lambda \leqq \varrho<\infty)$. Using an extension of the notion of the modulus for pseudo-resolvents (c.f. Kipnis [1974], Feyel [1978]), also non positive pseudo-resolvents have been investigated by Feyel [1984].

If $(\lambda V_\lambda)$ is bounded and continuous at $\infty$, and $T_0 f = \lim_{\lambda\to\infty} \lambda V_\lambda f$, the Hille-Yosida theorem implies that there exists a strongly continuous semigroup on $T_0 L_1$, such that $(\lambda V_\lambda)$ is the resolvent of this semigroup when restricted to $T_0 L_1$. As the Dunford-Schwartz theorem, the Chacon-Ornstein theorem, etc. hold with the same proofs for the subspace, they yield in turn ergodic theorems for pseudo-resolvents. Feyel [1979] showed that proper pseudo-resolvents of positive contractions in $L_1$ are continuous at $\infty$.

Feyel [1982a] does not need the assumption that the pseudo-resolvents are proper for his results on superabelian processes. Also, the class of superabelian processes is strictly larger than the class of Laplace transforms of superadditive processes.

Tauberian theorems for sequences of ratios do not seem to be known. Thus, at present, it is not clear if the Abelian form of the Chacon-Ornstein theorem, which is simpler to prove, could be used to deduce the Chacon-Ornstein theorem.

Baxter [1964], [1965] has proved weighted ergodic theorems with "recurrent" weights $(w_k)_{k\geqq 0}$ given recursively by $w_0=1$ and $w_n = p_1 w_{n-1} + p_2 w_{n-2} + \ldots + p_n w_0$, where $(p_k)_{k\geqq 1}$ is a probability distribution on $\mathbb{N}$. Chacon [1964] has shown that the a.e.-convergence of ratios $\sum_{i=0}^{n} w_i T^i f / \sum_{i=0}^{n} w_i T^i g$ for positive contractions $T$ in $L_1$ and $f, g \in L_1^+$ can also be deduced from the Chacon-Ornstein theorem. Berk [1968] has proved continuous parameter ergodic theorems of this type.

Irmisch [1980] has studied $(C,\alpha)$-averages

$$A_n(f,\alpha) := \binom{n+\alpha}{n}^{-1} \sum_{i=0}^{n} \binom{n-i+\alpha-1}{n-i} T^i f.$$

He proved the a.e.-convergence of these averages, when $T$ is a positive contraction in $L_p$ with $1<p<\infty$, $0<\alpha\leqq 1$, and $\alpha p>1$. He also gave a counterexample with $\alpha p = 1$.

Akcoglu and del Junco [1975] showed that $A_n^\alpha(\tau) f$ need not converge a.e. for weights $(\alpha_{Nk})$ with (W4).

In the situations above, the convergence of Cesàro averages was known. A different point of view is that weighted averages should be used in cases where Cesàro averages fail to converge: Krengel [1967] showed that for any positive contraction $T$ in $L_1$ there exists $w_0 \geqq w_1 \geqq \ldots$ with $\sum_{i=0}^{\infty} w_i = \infty$, such that $\sum_{i=0}^{n} w_i T^i f / \sum_{i=0}^{n} w_i$ converges a.e. for all $f \in L_1$. Grillenberger and Krengel [1976] showed that for any such $T$ there exists a sequence $n_1 < n_2 < \ldots$ such that $A_{n_i}(T) f$ converges a.e. for all $f$. On the other hand, it is shown in the same paper that there does not exist any matrix summation method compatible with the Cesàro method which enforces a.e. convergence of the averages of $T^n f$ for *all* positive contractions $T$ in $L_1$.

Assani [1985a] constructed weighted averages of $L_1 - L_\infty$-contractions for which pointwise convergence holds, but which are not of strong type $(p,p)$.

**Good and bad sequences.** Conze [1973] proved that a subsequence $\mathbf{k}$ is bad universal iff $C(\mathbf{k}) = \infty$ holds, and used this result to give simple examples. Mrs. Bellow [1982], [1984] proved $C(\mathbf{k}) = \infty$ for sequences $\mathbf{k}$ for which there exists a $\lambda > 1$ with $k_{i+1}/k_i > \lambda$ for all $i$, and she has constructed other bad universal sequences. Bellow and Losert [1984] showed the existence of good universal sequences $\mathbf{k}$ with $D(\mathbf{k}) = 0$.

The theorem of Brunel-Keane has been extended to more general operators by Sato, Olsen, and others. The paper of Baxter-Olsen [1983] contains references, gives a more systematic approach, and proposes new classes of universally good sequences. Here the original special case has been presented for reasons of simplicity. Ryll-Nardzewski [1975] showed that, for uniform $\mathbf{k}$, the sequence $(\alpha_j)$ with $\alpha_j = 1$ if there exists an $i$ with $k_i = j$, and $\alpha_j = 0$ else, is a bounded Besicovich sequence. (A bounded sequence $(\alpha_j) \subset \mathbb{C}$ is called bounded *Besicovich sequence* if, for any $\varepsilon > 0$, there exists a trigonometric polynomial $w_\varepsilon(j) = \sum \gamma_s \exp(i\Theta_s j)$ with $\limsup n^{-1} \sum_{j=1}^{n} |\alpha_j - w_\varepsilon(j)| \leqq \varepsilon$.) Ergodic theorems for uniform sequences are special cases of ergodic theorems for averages $n^{-1} \sum_{i=0}^{n-1} \alpha_j T^j f$. B. Schmitt [1972] showed that there is uniform convergence in the Brunel-Keane theorem if $\tau$ is weakly mixing and strictly ergodic in a compact $\Omega$, $f \in C(\Omega)$.

Blum and Reich [1977] gave another proper extension of the notion of uniform sequences by introducing "$p$-sequences". J. Reich [1977] showed convergence a.e. of $A^{\mathbf{k}}_{nJ} f$ for $f$ in a dense subset of $L_2$, "saturated" sequences $\mathbf{k}$ and weakly mixing $\tau$. He used this to prove a random ergodic theorem in which the exceptional nullset of sequences $\mathbf{k}$ does not depend on $\tau$. Related results appear in Blum-Reich [1977a].

Brunel-Keane [1969] showed that a.e.-convergence can be enforced in the Blum-Hanson theorem by using suitable decreasing weights.

Bellow and Losert [1985] have written a long partly expository article on this circle of problems.

**Random ergodic theorems.** The first random ergodic theorems were proved by Pitt [1942], Ulam and von Neumann [1945], and Kakutani [1951]. The idea of applying a "non-random" ergodic theorem to a product space is a recipe which has furnished numerous random ergodic theorems. Jacobs [1962/63] has summarized and extended most of the random ergodic theorems appearing before 1962 by proving

**Theorem 2.9.** *Let $\sigma$ be an endomorphism of a probability space $(\Sigma, \mathscr{B}, \nu)$, and $\{T_x, x \in \Sigma\}$ a family of contractions in $L_1$ of a $\sigma$-finite space $(\Omega, \mathscr{A}, \mu)$. For any $f \in L_1$ let the map $\varphi_f: x \to T_x f$ of $\Sigma$ into $L_1$ be strongly measurable* (i.e., $\nu \circ \varphi_f^{-1}$ *has separable support*). *Let $p_0, p_1, \dots$* on $\Omega \times \Sigma$ *be $\mathscr{A} \otimes \mathscr{B}$-measurable and admissible* (i.e., $|g| \leqq p_k(\sigma x, \cdot)$ *implies* $|T_x g| \leqq p_{k+1}(x, \cdot)$ *for $k \geqq 0$ and $\nu$-almost all $x$). Then there exists, for any $f \in L_1$, a $\nu$-nullset $N$ such that for $x \in N^c$ the ratios*

$$\sum_{k=0}^{n} T_x T_{\sigma x} \dots T_{\sigma^{k-1} x} f \Big/ \sum_{k=0}^{n} p_k(x, \cdot)$$

*converge* $\mu$-a.e. *to a finite limit on* $\{\omega: \sum_{k=0}^{\infty} p_k(x, \omega) > 0\}$.

We refer to Yoshimoto [1976], [1977], Kin [1972], Tsurumi [1972], Wós [1982], Delasnerie [1977] and J. Reich [1981] for further random ergodic theorems (multiparameter, multidimensional, local, etc.) and for references.

Révész [1960/61] has random mean and pointwise ergodic theorems for averages of functions $f \circ \tau_k \circ \tau_{k-1} \dots \circ \tau_1$, where the $\tau_i$ are chosen independently, but not necessarily with identical distributions.

Saleski [1980] has a random mean ergodic theorem with the exceptional nullset $N$

independent of $\tau$ and $f$: Let $S_n$ be an integer valued random walk on $(\Sigma, \mathscr{B}, \nu)$: There exists $N$ with $\nu(N) = 0$ such that, for all weakly mixing $\tau$ and all $f \in L_p$, $(1 \leqq p < \infty)$, $x \in N^c$ implies the $L_p$-convergence of $n^{-1} \sum_{k=1}^{n} f \circ \tau^{S_i(x)}$.

**Ergodic theorems for sub-orbits and super-orbits.** In the random ergodic theorem, the transformation $\tau(\sigma^n x)$ applied at time $n$ is independent of $\omega$. Recently, Kieffer-Dunham [1983], and Kieffer-Rahe [1982] have proved ergodic theorems, in which the subsequence, along which averages are taken, may depend on $\omega$. The setting is the following: $\tau$ and $\xi$ are measurable transformations of a probability space $(\Omega, \mathscr{A}, \mu)$, and $\tau$ is invertible with measurable inverse. For each $\omega \in \Omega$, $\xi\omega$ belongs to the bilateral orbit $\mathcal{O}_\tau(\omega) = \{\tau^i \omega: i \in \mathbb{Z}\}$.

Thus, $\{\omega, \xi\omega, \xi^2\omega, \ldots\}$ is a subset of $\mathcal{O}_\tau(\omega)$. Write $\omega' > \omega$ if there is some $n \geqq 1$ with $\omega' = \tau^n \omega$.

**Theorem 2.10** (Kieffer-Rahe). *If $\tau$ is measure preserving and the sets $\{\omega' \in \Omega: \omega' < \omega$ and $\xi\omega' \geqq \omega\}$ and $\{\omega' \in \Omega: \omega < \omega'$ and $\xi\omega' \leqq \omega\}$ are finite for $\mu$-almost all $\omega$, then $\mu$ is asymptotically mean stationary for $\xi$. (In other words: $A_n(\xi)f$ converges $\mu$-a.e. for bounded measurable $f$.)*

There even exists $A$ with $\mu(A) > 0$ such that $\mu_A$ with $\mu_A(B) = \mu(A)^{-1}\mu(A \cap B)$ is invariant under $\xi$ and asymptotically dominates $\mu$ with respect to $\xi$.

There is a similar theorem when $\xi$ is measure preserving:

**Theorem 2.11** (Kieffer-Rahe). *If $\xi$ is measure preserving and aperiodic, and the map $L$ with $\xi\omega = \tau^{L(\omega)}\omega$ satisfies $\int |L| d\mu < \infty$, then there exists a $\tau$-invariant $\nu \gg \mu$ on $(\Omega, \mathscr{A})$. In particular, $A_n(\tau)f$ converges $\mu$-a.e. for bounded measurable $f$.*

Kieffer and Dunham consider a stationary sequence $(X_i)$ with values in a space $\mathfrak{A}$, another space $\mathfrak{B}$, and a function $F: \mathfrak{A} \times \mathfrak{B} \to \mathfrak{B}$. A $\mathfrak{B}$-valued process $(S_i)$ is generated recursively by $S_0 = b^*$ (fixed $\in \mathfrak{B}$), $S_{i+1} = F(X_i, S_i)$. They give a sufficient condition (trivially satisfied for finite $\mathfrak{B}$) for the sequence $\{(X_i, S_i)_{i=0,1,\ldots}\}$ to be asymptotically mean stationary. There are applications to information theory and queueing theory.

# Chapter 9: Special topics

This chapter deals with a number of results which require special methods and do not fit well into the previous chapters. The first section treats ergodic theorems in von Neumann algebras, the second ergodic theorems for entropy and information and the third ergodic theorems for nonlinear operators. In the last section we report on miscellaneous related topics.

## § 9.1 Ergodic theorems in von Neumann algebras

This section deals with questions of convergence of averages

$$A_n(T)x = n^{-1}(T^0 + T^1 + \ldots + T^{n-1})x$$

when $T$ is a *-automorphism of a von Neumann algebra, and with similar problems for more general operators and semigroups. Here we cannot develop all the tools from the theory of von Neumann algebras required for the proofs. But all notions will be defined, and the meaning of the statements should become clear also to non-specialists. References will be given for the auxiliary results, except for some basic facts which can be found in the early parts of the monographs of Takesaki [1979], Pedersen [1979], Sakai [1971] or Dixmier [1957].

The motivation for the results in this section arose in quantum statistical mechanics and quantum field theory. An introduction is given in the books of Ruelle [1969] and Bratteli-Robinson [1979]. Roughly speaking, states in a "classical" system may be described mathematically as linear functionals $\mu$ (Radon measures) on the space $C(K)$ of continuous functions on a compact hyper-Stonean space $K$. Homeomorphisms of $K$ induce *-automorphisms of the commutative von Neumann algebra $C(K)$. States of quantum systems may be described by positive normed functionals $\varrho$ on a von Neumann algebra $\mathfrak{A}$. Invariant states $\varrho$ under *-automorphisms $T$ may therefore be regarded as generalizations of invariant measures to von Neumann algebras which need not be commutative. Their study has therefore been called non commutative ergodic theory. The words "non commutative" refer to the multiplication in $\mathfrak{A}$; this must be distinguished from ergodic theorems for non commutative semigroups.

**1. Basic notions.** A complex Banach space $\mathfrak{A}$ with norm $\|\cdot\|$ is called a *Banach algebra* if it is an algebra satisfying $\|xy\| \leqq \|x\|\,\|y\|$. A mapping $x \to x^*$ of $\mathfrak{A}$ into itself is called *involution* if $(x^*)^* = x$, $(\lambda x)^* = \bar{\lambda}x^*$, $(\lambda \in \mathbb{C})$, $(x + y)^* = x^*$

$+ y^*$, and $(xy)^* = y^* x^*$. A Banach algebra with involution is called *$C^*$-algebra* if $\|x^* x\| = \|x\|^2$. $\mathfrak{A}$ is called *unital* if it contains a multiplicative unit 1. It seems unnecessary and unusual to distinguish $1 \in \mathfrak{A}$ from the real number 1 in the notation. Clearly, $1^* = 1$, $\|1\| = 1$.

A mapping $T$ of a $C^*$-algebra $\mathfrak{A}$ into a $C^*$-algebra $\mathfrak{C}$ is called a *-homomorphism if $T(\lambda x) = \lambda Tx$, $T(x+y) = Tx + Ty$, $T(xy) = (Tx)(Ty)$, and $Tx^* = (Tx)^*$ holds for all $x, y, \lambda$. Then $\|Tx\| \leqq \|x\|$, with equality for injective $T$; see Pedersen [1979: 16]. The notions*-isomorphism, *-automorphism, etc. are now obvious.

As any "abstract" $C^*$-algebra is *-isomorphic to a subalgebra $\mathfrak{A}$ of the space $\mathscr{L}(\mathfrak{H})$ of bounded linear operators on a complex Hilbert space $\mathfrak{H}$, we may assume throughout $\mathfrak{A} \subset \mathscr{L}(\mathfrak{H})$. The involution $x \to x^*$ in $\mathfrak{A}$ is the passage to the adjoint operator. A subset $M$ of a $C^*$-algebra $\mathfrak{A}$ is called *self-adjoint* if $x \in M$ implies $x^* \in M$. $x$ is called self-adjoint if $x^* = x$. $\mathfrak{A}_{sa}$ denotes the set of self-adjoint elements of $\mathfrak{A}$. Any $x \in \mathfrak{A}$ has a unique representation $x = y + iz$ with $y, z \in \mathfrak{A}_{sa}$.

$x \in \mathfrak{A}$ is called *positive* (in symbols $x \geqq 0$) if there exists $y \in \mathfrak{A}$ with $x = y^* y$. The set $\mathfrak{A}_+$ of positive elements of $\mathfrak{A}$ is a convex cone. $x_1 \leqq x_2$ iff $x_2 - x_1 \geqq 0$ defines a partial order in $\mathfrak{A}$. Any $x \in \mathfrak{A}_{sa}$ has a unique representation $x = y - z$ with $yz = 0$ and $y, z \geqq 0$. A linear functional $\varphi\colon \mathfrak{A} \to \mathbb{C}$ is called positive if $x \geqq 0$ implies $\varphi(x) \geqq 0$.

A $C^*$-subalgebra of the algebra $\mathscr{L}(\mathfrak{H})$ is called a *von Neumann algebra* if it contains the identity $1 = I$ and is closed in the weak operator topology (or – equivalently – in the strong operator topology).

A linear functional $\varphi$ on a von Neumann algebra $\mathfrak{A}$ is called *normal* if for each norm bounded monotone increasing net $\{x_\alpha\}$ in $\mathfrak{A}_{sa}$ with limit $x$ (in the strong operator topology) the net $\varphi(x_\alpha)$ converges to $\varphi(x)$. $\varphi$ is normal iff the restriction of $\varphi$ to the unit sphere $\mathfrak{A}_1$ of $\mathfrak{A}$ is continuous in the weak operator topology. The set $\mathfrak{A}_*$ of normal linear functionals is a Banach space and $\mathfrak{A}$ is the dual $(\mathfrak{A}_*)^*$. Therefore $\mathfrak{A}_*$ is called the *predual* of $\mathfrak{A}$; see Pedersen [1979: 55].

We remark that, by a theorem of Haagerup [1975], a positive linear functional on $\mathfrak{A}$ is normal iff $\varphi(\sum x_i) = \sum \varphi(x_i)$ holds for any set $\{x_i\} \subset \mathfrak{A}_+$ for which $\sum x_i$ is defined.

Positive linear functionals on $\mathfrak{A}$ having norm 1 are called *states*. A state $\varphi$ is called *invariant* under a *endomorphism $T$ if $\varphi(Tx) = \varphi(x)$ holds for all $x$.

To gain some first orientation let us consider the $C^*$-algebra $L_\infty(\mu)$ of a probability space $(\Omega, \mathscr{A}, \mu)$. $f^*$ is the function taking the complex conjugate values of $f$. $L_\infty$ is *-isomorphic to a von Neumann algebra $\mathfrak{A} \subset \mathscr{L}(L_2)$ by $f \to x_f$ with $x_f g = f \cdot g$. If we identify $L_\infty$ with $\mathfrak{A} = \{x_f\colon f \in L_\infty\}$, we see that $L_1$ is the predual $\mathfrak{A}_*$. $h \in L_1$ corresponds to the functional $\psi_h(f) = \int fh\,d\mu$. Any automorphism $\tau$ of $(\Omega, \mathscr{A}, \mu)$ defines a *-automorphism $T$ of $L_\infty$ by $Tf = f \circ \tau$. The functional $\psi_1$ corresponding to the constant function $\mathbb{1}$ is $T$-invariant. For suitable $\tau$ and $f$ the sequence $n^{-1}(f + f \circ \tau + \ldots + f \circ \tau^{n-1})$ does not converge in $L_\infty$-norm. Thus, the existence of normal invariant states does not imply *norm*-convergence of $A_n(T)x$

for $x \in \mathfrak{A}$. However, in this example we do have norm-convergence of Cesàro averages for the operator induced in the predual and therefore $w^*$-convergence in $L_\infty$. The extension of this to the non commutative case has been initiated by Kovács-Szücs [1966]. Birkhoff's theorem in this example asserts a.e.-convergence which, by Egoroff, is equivalent to the existence of sets $E$ with $\mu(E)$ arbitrarily close to 1 for which there is $L_\infty$-norm-convergence on $E$. In this form, Birkhoff's theorem has been extended to the non commutative case by the remarkable work of C. Lance [1976].

Before we proceed to these results let us remark that the existence of a $T$-invariant normal state is not enough for $w^*$-convergence unless it has an additional property: Assume above that $\mu$ is $\sigma$-finite, but infinite and that there exists a $\tau$-invariant set $F$ with $\mu(F) = 1$ such that the restrictions of $\tau$ to $F$ and to $F^c$ both are ergodic. Then the functional corresponding to $1_F$ is a $T$-invariant normal state, but we do not have $L_1$-convergence or $w^*$-convergence on $F^c$.

A positive normal state $\varphi$ is called *faithful* if $x \geqq 0$, $\varphi(x) = 0$ implies $x = 0$, i.e., $\varphi$ is strictly positive on $\mathfrak{A}_+ \setminus \{0\}$. Above, $\psi_h$ is faithful iff $h$ is strictly positive a.e..

This is what we need. A family $\Phi = \{\varphi_i \cdot i \in I\}$ of positive normal states will be called faithful if $x \geqq 0$, $\varphi_i(x) = 0$ for all $i$ implies $x = 0$.

In addition to the norm topology, which is also called *uniform topology*, there is a host of other topologies in $\mathfrak{A}$: The $\sigma(\mathfrak{A}, \mathfrak{A}_*)$-topology, called *$\sigma$-topology* for short, is the weakest topology in $\mathfrak{A}$ with respect to which all $\varphi \in \mathfrak{A}_*$ are continuous. (In the terminology of functional analysis this would be the $w^*$-topology). On norm-bounded sets the $\sigma$-topology coincides with the weak operator topology. Clearly, a linear functional on $\mathfrak{A}$ is $\sigma$-continuous iff it is normal.

Let $\mathfrak{A}_*^+$ be the set of all positive normal linear functionals on $\mathfrak{A}$. Each $\varphi \in \mathfrak{A}_*^+$ defines a seminorm $\alpha_\varphi$ on $\mathfrak{A}$ by $\alpha_\varphi(x) = \varphi(x^*x)^{1/2}$. The *s-topology* (or *strong topology*) in $\mathfrak{A}$ is the topology generated by these seminorms. In other words, it is the weakest topology with respect to which all $\alpha_\varphi$ are continuous. It is stronger than the $\sigma$-topology. On norm-bounded sets it coincides with the strong operator topology. (See Sakai [1971: 20, 33–35] and Dixmier [1957: 32–36].)

The *$s^*$-topology* in $\mathfrak{A}$ is the topology generated by the seminorms $\alpha_\varrho$ and $\alpha_\varrho^*$, ($\varrho \in \mathfrak{A}_*^+$), where $\alpha_\varrho^*(z) = \varrho(zz^*)^{1/2}$. Thus, $s^*$-lim $b_k = 0$ holds iff $s$-lim $b_k = 0$ and $s$-lim $b_k^* = 0$. On $\mathfrak{A}_{sa}$ the $s^*$-topology and the $s$-topology coincide.

Recall that the *$w^*$-operator topology* in $\mathscr{L}(\mathfrak{A})$ is the weakest topology which renders the maps $T \to \langle \varphi, Tx \rangle := \varphi(Tx)$ continuous for all $\varphi \in \mathfrak{A}_*$ and all $x \in \mathfrak{A}$. An element $p$ of $\mathfrak{A}_{sa}$ is called *projection* if $p^2 = p$. For $\varphi \in \mathfrak{A}_*^+$ the set $\mathfrak{L}_\varphi = \{x \in \mathfrak{A}: \varphi(x^*x) = 0\}$ is a $\sigma$-closed left ideal. Since $\mathfrak{A}$ is a von Neumann algebra there exists a projection $q_\varphi$ with $\mathfrak{L}_\varphi = \mathfrak{A}q_\varphi$. $q_\varphi$ is the greatest of all projections $q$ such that $\varphi(q) = 0$. The projection $s_\varphi = 1 - q_\varphi$ is called the *support* of $\varphi$. For all $x \in \mathfrak{A}$ one has $\varphi(x) = \varphi(s_\varphi x) = \varphi(xs_\varphi)$. $\varphi$ is faithful iff $s_\varphi = 1$. A family $\Phi = \{\varphi_i\} \subset \mathfrak{A}_*^+$ is faithful iff $\mathfrak{H}$ is the closure of lin$\{s_{\varphi_i}\mathfrak{H}: \varphi_i \in \Phi\}$. On norm-bounded sets, $\alpha_\varphi(x_n)$ tends to 0 iff $x_n s_\varphi$ tends to 0 in the strong operator topology. (See Sakai [1971: 31], and Dixmier [1957: 61–62].)

After these preliminaries we can turn to the proof of ergodic theorems.

**2. The mean ergodic theorem.** A bounded linear operator $T \in \mathscr{L}(\mathfrak{A})$ is the adjoint of a continuous linear operator $T_* \in \mathscr{L}(\mathfrak{A}_*)$ (called its *pre-adjoint*) iff $T$ is $\sigma$-continuous. We may define $T_*\psi$ for $\psi \in \mathfrak{A}_*$ by $T_*\psi(x) = \psi(Tx)$, and use the fact that elements of the bidual $\mathfrak{X}^{**}$ of a Banach space $\mathfrak{X}$ belong to $\mathfrak{X}$ iff they are $w^*$-continuous functionals on $\mathfrak{X}^*$. If $T$ is a *-endomorphism of $\mathfrak{A}$, then $T$ is $\sigma$-continuous because for normal $\psi$ also $\psi \circ T$ is normal. Thus, *-endomorphisms have a pre-adjoint.

A linear operator $T \in \mathscr{L}(\mathfrak{A})$ is called an *$\alpha_\varphi$-contraction* (for $\varphi \in \mathfrak{A}_*^+$) if $\alpha_\varphi(Tx) \leqq \alpha_\varphi(x)$ holds for all $x \in \mathfrak{A}$. For $\Phi \subset \mathfrak{A}_*^+$ we call $T$ an *$\alpha_\Phi$-contraction* when it is an $\alpha_\varphi$-contraction for all $\varphi \in \Phi$.

Two independent approaches to mean ergodic theorems in $\mathfrak{A}$ will be described below. The first works for semigroups of $\alpha_\varphi$-contractions without any algebraic conditions. The second will be simpler. It relies on Eberlein's theorem and works for example for $d$-parameter semigroups of *positive* operators $T$ with $T_*\varphi \leqq \varphi$ for a faithful $\varphi$. Only this case will be used in the subsequent treatment of the almost uniform ergodic theorem.

**Theorem 1.1** (Mean ergodic theorem for von Neumann algebras). *Let $\Phi$ be a faithful family of normal states for a von Neumann algebra $\mathfrak{A}$ and let $\mathscr{S}$ be a bounded semigroup of $\sigma$-continuous $\alpha_\Phi$-contractions, then the closure of* co $\mathscr{S}$ *in the $w^*$-operator topology contains an $\mathscr{S}$-absorbing projection $P$. $Px$ belongs to the s-closure of* co $\mathscr{S}x$ *for all $x \in \mathfrak{A}$.*

*Proof.* (1) First fix some $\varphi \in \Phi$. Clearly $\mathfrak{L}_\varphi = \{x: \alpha_\varphi(x) = 0\}$ is $\mathscr{S}$-invariant. Therefore, we can define a semigroup $\mathscr{S}_\varphi = \{T_\varphi: T \in \mathscr{S}\}$ of linear operators in $\mathfrak{K}_\varphi = \mathfrak{A}s_\varphi$ by

$$T_\varphi(xs_\varphi) := (T(xs_\varphi))s_\varphi = (Tx)s_\varphi.$$

We now apply the Gelfand-Naimark-Segal construction: $\mathfrak{K}_\varphi$ is a pre-Hilbert-space with scalar product $\langle x, y\rangle_\varphi = \varphi(y^*x)$ and norm $\alpha_\varphi$. Let $\mathfrak{H}_\varphi$ be its completion. As each $T_\varphi$ contracts the norm in $\mathfrak{K}_\varphi$ it may be uniquely continued to $\mathfrak{H}_\varphi$. Call this extension $\tilde{T}_\varphi$. $\tilde{\mathscr{S}}_\varphi := \{\tilde{T}_\varphi: T \in \mathscr{S}\}$ is a semigroup of contractions in the Hilbert space $\mathfrak{H}_\varphi$. By the Alaoglu-Birkhoff theorem (2.1.10) there exists an $\tilde{\mathscr{S}}_\varphi$-absorbing projection $P_\varphi$ in $\mathscr{L}(\mathfrak{H}_\varphi)$ and $P_\varphi x$ is contained in the $\alpha_\varphi$-closure of $C_x := \text{co}\, \tilde{\mathscr{S}}_\varphi x$ for any $x \in \mathfrak{H}_\varphi$.

Now recall that the elements of $\mathfrak{A}$ are operators in a Hilbert space $\mathfrak{H}$ and that $\mathscr{S}$ is bounded. Put $M = \sup\{\|T\|: T \in \mathscr{S}\}$. For $x \in \mathfrak{K}_\varphi$ there exists a sequence $z_n \in C_x \subset \{z \in \mathfrak{K}_\varphi: \|z\| \leqq M\|x\|\}$ with $\alpha_\varphi(z_n - P_\varphi x) \to 0$. For norm-bounded sequences $(x_n)$ in $\mathfrak{A}$ the convergence to 0 of $x_n s_\varphi$ in the strong operator topology is equivalent to $\alpha_\varphi(x_n) \to 0$, see Dixmier [1957: 62]. Using $z_n = z_n s_\varphi$ we find $\|z_n\xi - z_m\xi\| \to 0$, $(m, n \to \infty)$, for any $\xi \in \mathfrak{H}$. Put $z_\infty\xi = \lim z_n\xi$. It is then simple to check that $z_\infty$ belongs to $\mathfrak{R}_\varphi$. Hence, $\alpha_\varphi(z_n - z_\infty) \to 0$, and $z_\infty$ must agree with $P_\varphi x$. Thus, $P_\varphi x$ belongs to the strong operator closure of $C_x$. Convex sets in $\mathscr{L}(\mathfrak{H})$ have the same closure in the weak as in the strong operator topology, see Dunford-Schwartz [1958: 477].

We have proved that for any $x \in \mathfrak{R}_\varphi$ the $\mathscr{S}_\varphi$-fixed point $P_\varphi x$ belongs to $\sigma - \overline{\text{co}}\, \mathscr{S}_\varphi x$. It is the unique $\mathscr{S}_\varphi$-fixed point in this set because $\mathfrak{R}_\varphi$ is $\sigma$-closed and a second fixed point would be a second fixed point for $\mathscr{S}_\varphi$ in $\mathfrak{H}_\varphi$, contradicting the uniqueness in the Alaoglu-Birkhoff theorem.

(2) The second step in the proof is the passage to general $\Phi$. It is unnecessary if $\Phi$ consists of a single $\varphi$.

Let us first remark that for projections $p$, $q$ the inequality $p \leqq q$ means that $p$ is a projection on a smaller subspace of $\mathfrak{H}$ than $q\mathfrak{H}$. (This may be deduced from an alternative definition of positivity in $\mathscr{L}(\mathfrak{H})$: $z \geqq 0$ is equivalent to $\langle z\xi, \xi\rangle \geqq 0$ for all $\xi \in \mathfrak{H}$, see e.g., Dixmier [1977: 17]). For $x \in \mathfrak{A}$ and $\varphi \in \Phi$ set

$$Q_\varphi(x) := \{ y \in \mathfrak{A}\colon P_\varphi(x s_\varphi) = y s_\varphi \quad \text{and} \quad \|y\| \leqq M\|x\|\}.$$

If $\varphi$, $\psi$ have supports $s_\varphi \leqq s_\psi$ then these projections satisfy $s_\psi s_\varphi = s_\varphi$. We obtain $(T_\psi(x s_\psi)) s_\psi s_\varphi = (Tx) s_\psi s_\varphi = (Tx) s_\varphi = (T_\varphi(x s_\varphi)) s_\varphi$. It follows that if $y s_\psi$ belongs to the convex hull of $\mathscr{S}_\psi(x s_\psi)$, then $y s_\varphi$ must belong to the convex hull of $\mathscr{S}_\varphi(x s_\varphi)$. As the map $u \to u s_\varphi$ is $\sigma$-continuous (Dixmier [1957: 18]), we deduce that $y s_\psi \in \sigma - \overline{\text{co}}\, \mathscr{S}_\psi(x s_\psi)$ implies $y s_\varphi \in \sigma - \overline{\text{co}}\, \mathscr{S}_\varphi(x s_\varphi)$. If $y s_\psi$ is $\mathscr{S}_\psi$-fixed, $y s_\varphi$ is $\mathscr{S}_\varphi$-fixed. Using the uniqueness proved above we see $Q_\varphi(x) \supset Q_\psi(x)$.

For any $\varphi$, $\psi$ the inequality $s_\varphi \leqq s_\psi$ holds if and only if $\varphi(x) = 0$ holds for all $x \in \mathfrak{A}_+$ with $\psi(x) = 0$; (see Dixmier [1957: 62]). As we may assume $\Phi$ convex we see that for any $\varphi_1, \varphi_2 \in \Phi$ there exists $\psi$ $(= (\varphi_1 + \varphi_2)/2)$ with $Q_\psi(x) \subset Q_{\varphi_1}(x) \cap Q_{\varphi_2}(x)$. The family $\{Q_\psi(x)\colon \psi \in \Phi\}$ is therefore filtered downwards. Each $Q_\psi(x)$ is non empty by the previous step and $w^*$-compact. It follows that there exists a point $y_x$ in the intersection of all $Q_\varphi(x)$. As $(Ty_x - y_x)s_\varphi = T_\varphi(y_x s_\varphi) - y_x s_\varphi = 0$ holds for all $\varphi$, $y_x$ is an $\mathscr{S}$-fixed point. $y_x$ belongs to the $\alpha_\varphi$-closure of co $\mathscr{S}x$ for all $\varphi$. As $\mathfrak{H}$ is the closure of lin $\{s_\varphi \mathfrak{H}\colon \varphi \in \Phi\}$ and as we deal with norm-bounded sets, $y_x$ belongs to the closure of co $\mathscr{S}x$ in the strong and hence in the weak operator topology, and thus also in the $\sigma$-topology. We have proved that $\sigma$-$\overline{\text{co}}\, \mathscr{S}x$ contains an $\mathscr{S}$-fixed point for all $x$.

(3) Finally, let us show that $F(\mathscr{S}_*) = \{\psi \in \mathfrak{A}_*\colon T_*\psi = \psi \,\forall\, T_* \in \mathscr{S}_*\}$ separates $F(\mathscr{S}) = \{u \in \mathfrak{A}\colon Tu = u \,\forall\, T \in \mathscr{S}\}$. For any $0 \neq u \in F(\mathscr{S})$ there exists $\varphi \in \Phi$ with $u_\varphi := u s_\varphi \neq 0$ because $\Phi$ is faithful. Clearly $u_\varphi$ belongs to $\mathfrak{H}_\varphi$ and is fixed under $\tilde{\mathscr{S}}_\varphi$. By the Alaoglu-Birkhoff theorem the condition (iii) in the theorem 2.1.11 is satisfied for the semigroup $\tilde{\mathscr{S}}_\varphi$ and thus (iv) holds. Hence there exists a continuous linear form $\tilde{\psi}$ on $\mathfrak{H}_\varphi$ with $\tilde{\psi}(u_\varphi) \neq 0$ and $T_\varphi^*\tilde{\psi} = \tilde{\psi}$ for all $T_\varphi \in \tilde{\mathscr{S}}_\varphi$. Put $\psi(y)$

$= \tilde{\psi}(ys_\varphi)$. Then $\psi$ is $s$-continuous on $\mathfrak{A}$ and therefore $\sigma$-continuous (Sakai [1971: 21]).

It then follows from $\langle T_* \psi, y\rangle = \langle \psi, Ty\rangle = \langle \tilde{\psi}, (Ty)s_\varphi\rangle = \langle \tilde{\psi}, T_\varphi(ys_\varphi)\rangle = \langle T_\varphi^* \tilde{\psi}, ys_\varphi\rangle = \langle \tilde{\psi}, ys_\varphi\rangle = \ldots = \langle \psi, y\rangle$ that $\psi$ belongs to $F(\mathscr{S}_*)$. Clearly $\psi(u) \neq 0$. As $0 \neq u$ has been arbitrary in $F(\mathscr{S})$ the fixed space $F(\mathscr{S}_*)$ indeed separates $F(\mathscr{S})$.

Theorem 2.1.11 now implies the first assertion. As $\sigma\text{-}\overline{\mathrm{co}}\, \mathscr{S} x$ is norm-bounded it agrees with the closure of $\mathrm{co}\, \mathscr{S} x$ in the weak operator topology. As $\mathrm{co}\, \mathscr{S} x$ is convex it also agrees with the closure in the strong operator topology. □

Theorem 1.1 is due to Kovács-Szücs [1966] for *-automorphisms and to Kümmerer-Nagel [1979] for general $\alpha_\Phi$-contractions with the similar but simpler proof worked out here.

**Corollary 1.2.** *Under the conditions of theorem* 1.1 *the closure of* $\mathrm{co}\, \mathscr{S}_*$ *in the strong operator topology contains an* $\mathscr{S}_*$*-absorbing projection* $P_*$.

*Proof.* Apply theorem 2.1.11. □

We shall be interested in the particular case of $d$-parameter semigroups, and use the notation from § 6.1.

**Corollary 1.3.** *Assume that a commutative d-parameter semigroup* $\mathscr{S} = \{T_u \colon u \in \mathbb{V}\}$ *satisfies the assumptions in theorem* 1.1, *then*

*(a)* $(A_n)_* \psi$ *converges in norm to* $P_* \psi$ *for all* $\psi \in \mathfrak{A}_*$

*(b)* $Px = s\text{-}\lim A_n x$ *for all* $x \in \mathfrak{A}$ *as* $n \to \infty$ *in* $\mathbb{V}$.

*Proof.* (a) follows from theorem 2.1.9.

(b) We may assume $Px = 0$. For any $\varepsilon > 0$ and any $\varphi \in \Phi$ there exists a finite convex combination $\sum_u \beta_u T_u$ with $\alpha_\varphi(\sum_u \beta_u T_u x) < \varepsilon$. For $n \to \infty$, $\|A_n T_u - A_n\|$ tends to 0 for all $u \in \mathbb{V}$. Hence, for large $n$,

$$\begin{aligned} \alpha_\varphi(A_n x) &\leqq \alpha_\varphi(\sum_u \beta_u (A_n T_u - A_n)x) + \alpha_\varphi(A_n \sum \beta_u T_u x) \\ &\leqq \sum \beta_u \|A_n T_u - A_n\| \, \|x\| + \alpha_\varphi(\sum \beta_u T_u x) < 2\varepsilon . \end{aligned}$$

Thus, $A_n x s_\varphi$ tends to 0 in the strong operator topology of $\mathscr{L}(\mathfrak{H})$ for all $\varphi$. As $\Phi$ is faithful, $s\text{-}\lim A_n x = 0$. □

We now begin with the second approach.

**Lemma 1.4.** *Let* $\Phi$ *be a convex faithful family of normal states in* $\mathfrak{A}_*$, *then the union U of all order intervals* $[-k\varphi, +k\varphi]$ $(k \in \mathbb{N}, \varphi \in \Phi)$ *is norm-dense in the self-adjoint part* $(\mathfrak{A}_*)_{sa}$ *of* $\mathfrak{A}_*$.

*Proof.* $U$ is a linear subspace of $(\mathfrak{A}_*)_{sa}$ because $\Phi$ is convex. If $U$ is not norm dense there exists by Hahn-Banach an $x \in \mathfrak{A}_{sa}$ with $x \neq 0$ and $x(\psi) = 0$ for $\psi \in U$. For $\psi \geqq 0$ we have

$$x^+(\psi) = \sup \{ \sum_{i=1}^{n} x(\psi_i)^+ : 0 \leqq \psi_i \quad \text{and} \quad \sum \psi_i = \psi \}.$$

Thus $x^+(\psi)$ vanishes for $\psi \in U$. As $\Phi$ is faithful, $x^+ = 0$. The same argument applied to $-x$ gives $x^- = 0$. Contradiction. □

**Theorem 1.5.** *Let $\Phi$ be a faithful family of normal states of a von Neumann algebra $\mathfrak{A}$ and let $\mathscr{S} = \{T_u : u \in \mathbb{V}\}$ be a d-parameter semigroup of $w^*$-continuous positive contractions in $\mathfrak{A}$ with $T_{u*}\varphi \leqq \varphi$ for all $u \in \mathbb{V}$ and $\varphi \in \Phi$. Then*

*(i) $(A_n)_*\psi$ converges for $n \to \infty$ in norm to $P_*\psi$ for all $\psi \in \mathfrak{A}_*$, where $P_*$ is a positive contraction in $\mathfrak{A}_*$;*

*(ii) $\sigma$-$\lim_{n\to\infty} A_n x = Px$ for all $x \in \mathfrak{A}$, where $P$ is the adjoint of $P_*$;*

*(iii) any $x \in \mathfrak{A}$ is the $s^*$-limit of a norm-bounded net $(x_\lambda)$ such that each $x_\lambda - Px$ is a finite convex combination of elements of the form $(I - T_u)x$.*

*Proof.* We may assume $\Phi$ convex. A subset $D \subset \mathfrak{A}_*$ is relatively weakly compact iff it is bounded and $\lim \varphi(p_k) = 0$ uniformly for $\varphi \in D$ holds for every decreasing sequence $p_k$ of projections with $\inf p_k = 0$, see Takesaki [1979: 149]. Thus the intervals $[0, \psi]$ are relatively weakly compact for $0 \leqq \psi \in \mathfrak{A}_*$. If $\psi$ belongs to an interval $[0, k\varphi]$ with $\varphi \in \Phi$ then $A_n\psi$ belongs to this interval. By Eberlein's theorem $A_n\psi$ converges in norm for $\psi \in U$. (i) then follows from the lemma. (ii) is a consequence of (i).

(iii) By (ii), $Px$ belongs to the closed convex hull of $\mathscr{S}x$ in the weak operator topology of $\mathscr{L}(\mathfrak{H})$. But this coincides with the closed convex hull in the strong operator topology and in the $s^*$-topology; Takesaki [1979], p. 70. Hence there is a net $(z_\lambda)$ in co $\mathscr{S}x$ with $s^*$-$\lim z_\lambda = Px$. If $z_\lambda$ is the convex combination $\sum \alpha_{\lambda u} T_u x$, then $x - z_\lambda + Px = \sum \alpha_{\lambda u}(x - T_u x) + Px$. $x$ is the $s^*$-limit of the norm-bounded net $x_\lambda := x - z_\lambda + Px$. □

**3. Almost uniform convergence.** We now turn to the non commutative extension of the pointwise ergodic theorem. The following concept is a suitable generalization of the notion of convergence a.e.:

**Definition 1.6.** A sequence or net $(x_n)$ of elements of $\mathfrak{A}$ *converges* to $x_0$ *almost uniformly* if for every normal state $\psi$ and for every $\varepsilon > 0$ there exists a projection $p_{\varepsilon,\psi}$ with $\psi(p_{\varepsilon,\psi}) > 1 - \varepsilon$ and $\|(x_n - x_0)p_{\varepsilon,\psi}\| \to 0$.

**Remark.** If $\varphi$ is a faithful normal state and $p_k$ a sequence of projections with $\varphi(p_k) \to 1$, then $\alpha_\varphi(1 - p_k) = \varphi(1 - p_k)$ tends to 0. Then $s$-$\lim (1 - p_k)s_\varphi = 0$ (use Dixmier [1957: 62] again). By $s_\varphi = 1$ we obtain $s$-$\lim p_k = 1$ and hence $\sigma$-$\lim p_k$

$= 1$. We conclude that the existence of $p_{\varepsilon,\varphi}$ for all $\varepsilon > 0$ already implies the existence of $p_{\varepsilon,\psi}$ for all normal states $\psi$.

Our aim is

**Theorem 1.7** (Almost uniform ergodic theorem). *Let $\mathscr{S} = \{T_u: u \in \mathbb{V}\}$ be a d-parameter commutative semigroup of $w^*$-continuous positive operators in a von Neumann algebra $\mathfrak{A}$ with $T_u 1 \leqq 1$, $(u \in \mathbb{V})$. Assume that there exists a faithful normal state $\varphi$ with $(T_u)_* \varphi \leqq \varphi$, $(u \in \mathbb{V})$. Then, for each $x \in \mathfrak{A}$, the net $A_v x$ converges almost uniformly to $Px$, $(v \to \infty)$, where $P$ is the $\mathscr{S}$-absorbing projection of theorem* 1.5.

The breakthrough work for this theorem was done by Lance [1976], who proved it for $d = 1$ and *-automorphisms $T_u$. Kümmerer [1978] obtained the proof for $d = 1$ under the present assumptions and Conze and Dang-Ngoc [1978] gave a proof for $d \geqq 1$ with the added assumption that the $T_u$ are $\alpha_\varphi$-contractions with $T_u 1 = 1$ and $T^* \varphi = \varphi$. Here their arguments will be combined and the case $d > 1$ simplified.

For the proof of theorem 1.7 we will have to rely on some results on von Neumann algebras, which go beyond the usual textbook material. They will be stated as theorem 1.11 and 1.15 below. Moreover, we shall need a basic theorem from the theory of compact convex sets:

For compact convex $K$ in a locally convex Hausdorff space let $A(K)$ be the set of continuous real valued affine functions on $K$. The barycenter $\beta_\mu$ of a probability Radon measure $\mu$ on $K$ is the element of $K$ with $f(\beta_\mu) = \int f d\mu$ for all $f \in A(K)$. For $\beta \in K$ let $\Omega_\beta$ be the set of all probability Radon measures $\mu$ with $\beta_\mu = \beta$. For $f \in C(K)$ put $\bar{f} = \inf\{g \in A(K): g \geqq f\}$. The result we need is:

**Theorem 1.8.** *For any $f \in C(K)$ the function $\bar{f}$ is concave and upper semicontinuous on $K$ and satisfies $\bar{f}(\beta) = \sup\{\int f d\mu: \mu \in \Omega_\beta\}$.*

We refer to the book of Alfsen [1971], 1.3.6, for a proof.

If $\mathfrak{A}$ is a von Neumann algebra the state space

$$\mathfrak{S} := \{\psi \in \mathfrak{A}^*: \psi \geqq 0 \quad \text{and} \quad \|\psi\| = 1\}$$

with the $w^*$-topology is compact and convex and we shall apply theorem 1.8 with $K = \mathfrak{S}$. Each $x \in \mathfrak{A}_{sa}$ induces a continuous affine function $\xi_x: \psi \to \xi_x(\psi) := \psi(x)$ on $\mathfrak{S}$. Conversely, each $w^*$-continuous affine $\xi$ on $\mathfrak{S}$ determines a unique $x \in \mathfrak{A}$ with $\xi = \xi_x$, because it may be extended to a $w^*$-continuous linear function on $\mathfrak{A}^*$. To avoid overly heavy notation we identify $x$ with $\xi_x$.

The first step in the proof of theorem 1.7 is the proof of a maximal lemma. We shall assume $d = 1$ for a while. Fix $n \geqq 1$, $x \in \mathfrak{A}$, and $\gamma > 0$, and put $S_j = \sum_{u=0}^{j-1} T_u$,

$$g = \sup\{S_j x - j\gamma 1: 1 \leqq j \leqq n\}, \ h = g^+ \quad \text{and} \quad E = \{\psi \in \mathfrak{S}: g(\psi) \geqq 0\}.$$

**Theorem 1.9** (Maximal ergodic lemma). *Under the assumptions of theorem* 1.7 (*with* $d = 1$) *let* $\mu \in \Omega_\varphi$ *be a measure with* $\int h\,d\mu = \bar{h}(\varphi)$, *then* $\int_E (x - \gamma 1)\,d\mu \geqq 0$.

(The existence of such a $\mu$ follows from the compactness of $\Omega_\varphi$ in the topology of weak convergence of measures ($\mu_n \to \mu$ iff $\int f\,d\mu_n \to \int f\,d\mu \,\forall f \in C(K)$).

*Proof.* Put $T := T_1$ and define an operator $\tilde{T}$ on $C(\mathfrak{S})$ by

$$(\tilde{T}f)(\psi) = \|T^*\psi\| f(T^*\psi/\|T^*\psi\|)$$

for $\psi$ with $\|T^*\psi\| \neq 0$, and $= 0$ else. It is simple to check $\tilde{T}f \in C(\mathfrak{S})$ for $f \in C(\mathfrak{S})$. $\tilde{T}$ is a positive linear operator, and $\|T^*\psi\| = (T^*\psi)(1) = \langle T1, \psi\rangle \leqq \langle 1, \psi\rangle = 1 = \|\psi\|$ shows that $\tilde{T}$ is a continuous operator in $C(\mathfrak{S})$. The restriction of $\tilde{T}$ to $A(\mathfrak{S})$ coincides with $T$.

We now argue as in Garsia's proof of Hopf's maximal ergodic theorem: From $T \geqq 0$ and $T1 \leqq 1$ we obtain

$$\begin{aligned}\tilde{T}h &= \tilde{T}\big(\sup_{1 \leqq j \leqq n} (S_j x - j\gamma 1)\big)^+ \geqq \sup_{1 \leqq j \leqq n} (\tilde{T}(S_j x - j\gamma 1))^+ \\ &\geqq \sup_{1 \leqq j \leqq n} (S_{j+1} x - (j+1)\gamma 1 - (x - \gamma 1))^+ \\ &= \sup_{1 \leqq j \leqq n+1} (S_j x - j\gamma 1 - (x - \gamma 1))^+ \geqq g - (x - \gamma 1).\end{aligned}$$

Hence

$$\int_E (x - \gamma 1)\,d\mu \geqq \int_E (g - \tilde{T}h)\,d\mu = \int h\,d\mu - \int_E \tilde{T}h\,d\mu \geqq \int h\,d\mu - \int \tilde{T}h\,d\mu.$$

Put $\alpha = \langle \tilde{T}1, \mu\rangle = \langle 1, \tilde{T}^*\mu\rangle$. For $\alpha = 0$ the assertion of the theorem follows from $\int \tilde{T}h\,d\mu = 0$. For $\alpha \neq 0$ the measure $\nu = \alpha^{-1}\tilde{T}^*\mu$ has total mass 1. For any $z \in A(\mathfrak{S})$,

$$\int z\,d\nu = \alpha^{-1}\int Tz\,d\mu = \alpha^{-1}(Tz)(\varphi) = z(\alpha^{-1}T^*\varphi). \tag{1.1}$$

The element $\vartheta = \alpha^{-1}T^*\varphi$ belongs to $\mathfrak{S}$ because $\|\alpha^{-1}T^*\varphi\| = \alpha^{-1}\langle T^*\varphi, 1\rangle = \alpha^{-1}\langle \varphi, \tilde{T}1\rangle = \alpha^{-1}\int \tilde{T}1\,d\mu = \alpha^{-1}\langle 1, \tilde{T}^*\mu\rangle = 1$. As (1.1) holds for all $z$ we obtain $\nu \in \Omega_\vartheta$. If $I_h$ denotes the set $\{y \in A(\mathfrak{S}): y \geqq h\}$, then by theorem 1.8, the definition of $\bar{h}$ and $T^*\varphi \leqq \varphi$

$$\begin{aligned}\int \tilde{T}h\,d\mu &= \alpha\int h\,d\nu \leqq \alpha \sup\{\textstyle\int h\,d\varrho : \varrho \in \Omega_\vartheta\} \\ &= \alpha\bar{h}(\vartheta) = \alpha \inf\{y(\alpha^{-1}T^*\varphi) : y \in I_h\} \\ &\leqq \inf\{y(\varphi) : y \in I_h\} = \bar{h}(\varphi) = \textstyle\int h\,d\mu. \quad \square\end{aligned}$$

The second step in the proof of theorem 1.7 is a kind of *dominated estimate* which says that for "small" $x \in \mathscr{A}_+$ there is a "small" common bound for all $A_n x$. This vague statement is formalized for $d = 1$ by

**Theorem 1.10.** *Let* $\mathfrak{A}$, $\mathscr{S} = \{T_u\}$, $\varphi$ *be as in theorem* 1.7 *with* $d = 1$, *and* $0 \leqq \varepsilon \leqq 1$.

*For any $x \in \mathfrak{A}_+$ with $\|x\| \leqq 1$ and $\varphi(x) \leqq \varepsilon$ there exists an element $c \in \mathfrak{A}_+$ with $\|c\| \leqq 2$ and $\varphi(c) \leqq 4\varepsilon^{1/2}$ such that $A_n x \leqq c$ holds for all $n$.*

Before entering the proof recall that a subset $F$ of a convex compact space $K$ is called a *face* if it is convex and $\xi, \eta \in F$ holds for any pair $\xi, \eta \in K$ for which there exists $0 < \lambda < 1$ with $\lambda\xi + (1-\lambda)\eta \in F$. The face generated by $\xi \in K$ is the set

$$F(\xi) := \{\eta \in K : \exists \psi \in K \quad \text{and} \quad 0 < \lambda < 1 \quad \text{with } \xi = \lambda\eta + (1-\lambda)\psi\}.$$

Here $K$ will again be the state space $\mathfrak{S} \subset \mathfrak{A}^*$. For a face $F$ put

$$F^\perp = \{z \in \mathfrak{A}^{**} : z \geqq 0 \quad \text{and} \quad \langle \psi, z \rangle = 0 \ \forall \psi \in F\}.$$

We shall require the following result, contained in theorem 4.6, 5.2, and 2.4 of Effros [1963]:

**Theorem 1.11.** *If $F \subset \mathfrak{S}$ is a face, its norm closure $cl F$ is a face and $F^\perp = (cl F)^\perp$ consists of the positive elements of a left ideal in $\mathfrak{A}^{**}$ closed in the $w^*$-topology.*

*Proof of theorem 1.10.* We proceed in three steps (i)–(iii):

(i) First fix $n \geqq 1$ and assume $0 < \varepsilon < 1$. Put $\gamma = \varepsilon^{1/2}$, and let $g, h, E, \mu, \varphi$ be as above. $\varepsilon < \gamma$ and $\varphi(x) = \int x \, d\mu \leqq \varepsilon$ implies $\int (x - \gamma 1) \, d\mu < 0$. The maximal lemma then yields $\mu(E) < 1$. Set $\delta = \mu(\mathfrak{S} \setminus E)$. Let $\mu_1$ be the restriction of $\delta^{-1}\mu$ to $\mathfrak{S} \setminus E$ and let $\varphi_1$ be the barycenter of the probability measure $\mu_1$.

Our first aim is to show $(A_j x - \gamma 1)(\psi) \leqq 0$ for all $\psi \in F(\varphi_1)$ and $1 \leqq j \leqq n$. Assume this fails. Then we have $\bar{h}(\psi) > 0$. For $\psi \in F(\varphi_1)$ there exists $\xi \in \mathfrak{S}$ and $0 < \lambda < 1$ with $\varphi_1 = \lambda\psi + (1-\lambda)\xi$. As $\bar{h}$ is non negative and concave $\bar{h}(\varphi_1) > 0$. From the way $\bar{h}$ is defined there exists an $f \in A(\mathfrak{S})$ with $f \geqq \bar{h}$ and $f(\varphi) < \bar{h}(\varphi) + \delta\bar{h}(\varphi_1)$. Now a contradiction results from

$$\begin{aligned}\bar{h}(\varphi) &= \int h \, d\mu = \int_E h \, d\mu \leqq \int_E f \, d\mu = \int f \, d\mu - \int_{\mathfrak{S}-E} f \, d\mu \\ &= f(\varphi) - \delta f(\varphi_1) < \bar{h}(\varphi) + \delta\bar{h}(\varphi_1) - \delta f(\varphi_1) \leqq \bar{h}(\varphi).\end{aligned}$$

(ii) Now we want to construct a small upper bound $c_n$ for $A_j x$, $(1 \leqq j \leqq n)$: By theorem 1.11, $G := cl(F(\varphi_1))$ is a face and, by continuity, $(A_j x - \gamma 1)(\psi) \leqq 0$ holds for $\psi \in G$. $G^\perp$ consists of the positive elements of a $w^*$-closed left ideal in $\mathfrak{A}^{**}$.

Such a left ideal is of the form $\mathfrak{A}^{**}q$ for a projection $q$; see e.g. Takesaki [1979: 76]. Clearly, $G = \{\varrho \in \mathfrak{S} : \varrho(q) = 0\}$.

Put $p = 1 - q$. Assume $a \in \mathfrak{A}$ is an element with $\psi(a) \geqq 0$ for all $\psi \in G$. For any $\xi \in \mathfrak{S}$ the functional $\xi_p(\cdot) = \xi(p \cdot p)$ is a non negative multiple of an element in $G$. Hence, $\xi(pap) \geqq 0$ holds for all $\xi \in \mathfrak{S}$. This yields $pap \geqq 0$. By applying this argument to $a = \gamma 1 - A_j x$ we find $\gamma p \geqq p(A_j x)p$.

For any $t \in \mathfrak{A}_+$ we have $(p-q)\,t(p-q) \geqq 0$. Adding $t=(p+q)\,t(p+q)$ we obtain $t \leqq 2ptp + 2qtq$. Now $0 \leqq A_j x \leqq 1$ implies

$$A_j x \leqq 2p(A_j x)p + 2q(A_j x)q \leqq 2\gamma p + 2q, \quad (1 \leqq j \leqq n),$$

so that $c_n := 2\gamma p + 2q$ is an upper bound with $\|c_n\| \leqq 2$.

To estimate $\varphi(c_n)$ we use the maximal lemma:

$$\begin{aligned}\varepsilon - \varepsilon^{1/2} \geqq \varphi(x - \gamma 1) &= \int_{\mathfrak{S}\setminus E} (x-\gamma 1)\,d\mu + \int_E (x-\gamma 1)\,d\mu \\ &\geqq \delta\varphi_1(x-\gamma 1) \geqq -\delta\gamma = -\delta\varepsilon^{1/2}\end{aligned}$$

implies $\delta \geqq 1 - \varepsilon^{1/2}$ and therefore $\varphi(p) \geqq \delta\varphi_1(p) = \delta \geqq 1 - \varepsilon^{1/2}$. Hence $\varphi(c_n) \leqq 2\gamma\varphi(p) + 2(1-\varphi(p)) \leqq 4\varepsilon^{1/2}$.

(iii) The sequence $c_n \in \mathfrak{A}^{**}$ has a $w^*$-cluster point $c_\infty \in \mathfrak{A}^{**}$ with $\|c_\infty\| \leqq 2$, $A_j x \leqq c_\infty$, $(1 \leqq j < \infty)$, and $\varphi(c_\infty) \leqq 4\varepsilon^{1/2}$.

We now have a bound in $\mathfrak{A}^{**}$ and it remains to map it into $\mathfrak{A}$. Consider the canonical inclusion $\kappa$ of $\mathfrak{A}_*$ into $\mathfrak{A}^*$. Its adjoint is an order preserving isometry $\pi = \kappa^*$ of $\mathfrak{A}^{**}$ into $\mathfrak{A}$. If $z$ is an element in $\mathfrak{A}$ and $\xi_z$ its canonical embedding into $\mathfrak{A}^{**}$ (so far identified with $z$), then $\pi(\xi_z) = z$. We can take $c := \pi(c_\infty)$ as the desired bound of all $A_j x$ in $\mathfrak{A}$.

Finally observe that the assertion of the theorem is trivial for $\varepsilon = 0$ since $\varphi(x) = 0$ implies $x = 0$. Also the case $\varepsilon = 1$ reduces to the case just proved: for $\varphi(x) < 1$ we may decrease $\varepsilon$ a little and for $\varphi(x) = 1$ we have $0 \leqq x \leqq 1$ and $\varphi(1-x) = 0$ and hence $x = 1$. Then take $c = 1$. □

For general $d \geqq 1$ a dominated estimate is derived by induction:

**Theorem 1.12** (Multiparameter dominated estimate). *Let $\mathfrak{A}, \mathscr{S}, \varphi$ be as in theorem 1.7. For $x \in \mathfrak{A}_+$ there exists $c \in \mathfrak{A}_+$ with $\|c\| \leqq 2^d \|x\|$, $\varphi(c) \leqq 2^{d+2} \|x\|^{1-2^{-d}} \varphi(x)^{2^{-d}}$ and $A_v x \leqq c$ for all $v \in \mathbb{V}$.*

*Proof.* For $d = 1$ this, and even the better estimate $\varphi(c) \leqq 4\|x\|^{1/2}\varphi(x)^{1/2}$, follows from theorem 1.10, (pass to $x/\|x\|$). If the theorem has been proved for some $d \geqq 1$ and $\mathscr{S}' = \{T'_{w'}\colon w' \in \mathbb{V}'\}$ (with $\mathbb{V}' = \{0, 1, \ldots\}^{d+1}$) is a $(d+1)$-parameter semigroup we can define a $d$-parameter semigroup $\mathscr{S}$ by $T_u = T'_{(u,0)}$, $(u \in \mathbb{V})$, and a 1-parameter semigroup $\{S^0, S^1, \ldots\}$ by $S^k = T'_{(0,0,\ldots,0,k)}$, $(k \geqq 0)$. For $w' = (w_1, \ldots, w_{d+1}) = (w, w_{d+1})$ we have $A'_{w'} = A_{w_{d+1}}(S) \circ A_w$. If $c$ is the element majorizing all $A_w x$ and $c'$ the element with $A_k(S)c \leqq c'$ obtained by applying the 1-parameter case to $c$, then $A'_{w'} x \leqq c'$ holds for all $w' \in \mathbb{V}'$ and we have $\|c'\| \leqq 2\|c\| \leqq 2^{d+1}$ and

$$\begin{aligned}\varphi(c') &\leqq 4\|c\|^{1/2}\varphi(c)^{1/2}\\ &\leqq 4(2^{d/2}\|x\|^{1/2})(2^{d/2}2\|x\|^{1/2-2^{-(d+1)}}\varphi(x)^{2^{-(d+1)}})\\ &= 2^{(d+1)+2}\|x\|^{1-2^{-(d+1)}}\cdot\varphi(x)^{2^{-(d+1)}}.\end{aligned}$$

This is the assertion for $d+1$ instead of $d$. □

For the third and final step of the proof of theorem 1.7 we will need the following technical lemmas:

**Lemma 1.13.** *If $p \in \mathfrak{A}$ is a projection and $a, b \in \mathfrak{A}$ satisfy $0 \leqq a \leqq b \leqq 1$, then $\|ap\| \leqq \|bp\|^{1/2}$.*

*Proof.* $\|ap\|^2 \leqq \|a^{1/2}\|^2\|a^{1/2}p\|^2 \leqq \|pap\| \leqq \|pbp\| \leqq \|bp\|$. □

Each $z \in \mathfrak{A}$ may be uniquely written as $z = (z_1 - z_2) + i(z_3 - z_4)$ with $z_i \in \mathfrak{A}_+$ and $z_1 z_2 = 0$, $z_3 z_4 = 0$. Put $z^{++} := z_1 + z_2 + z_3 + z_4$.

**Lemma 1.14.** *If $(b_k)$ is a bounded sequence in $\mathfrak{A}$ with $s^*$-$\lim b_k = 0$, then $s^*$-$\lim b_k^{++} = 0$.*

*Proof.* If $b_k = h_k + il_k$ with self-adjoint $h_k, l_k$ then $h_k = (b_k + b_k^*)/2$ shows $s^*$-$\lim h_k = 0$. Now $|h_k|^*|h_k| = |h_k|^2 = h_k^* h_k^*$ yields $s$-$\lim |h_k| = 0$. By $h_k^+ = (|h_k| + h_k)/2$ we obtain $s$-$\lim h_k^+ = 0$. Argue similarly for $s$-$\lim h_k^- = 0$ etc. □

For the next lemma we need the following non commutative Egoroff theorem due to Saito [1967]:

**Theorem 1.15.** *If $(x_\lambda)$ is a norm-bounded net in a von -Neumann algebra $\mathfrak{A}$ with $s^*$-$\lim x_\lambda = 0$, there exists a subsequence $(x_{\lambda_j})$, $(j \geqq 1)$, such that $x_{\lambda_j}$ converges to 0 almost uniformly.*

**Lemma 1.16.** *Let $\mathfrak{A}$, $\mathcal{S}$, $\varphi$ be as in theorem 1.7. If $(x_\lambda)$ is a bounded net in $\mathfrak{A}$ with* $s^*$-$\lim x_\lambda = 0$, there exists a subsequence $(x_{\lambda_j})$ *such that $A_v x_{\lambda_j} \to 0$ almost uniformly and uniformly in $v \in \mathbb{V}$ as $j \to \infty$.*

*Proof.* First assume $x_\lambda \geqq 0$. Dividing by $2^d \sup\|x_\lambda\|$ we may also assume $\|x_\lambda\| \leqq 2^{-d}$. By the Cauchy-Schwartz inequality $\varepsilon_\lambda := \varphi(x_\lambda) = \varphi(1 x_\lambda) \leqq \varphi(1)\,\varphi(x_\lambda^* x_\lambda)$ tends to 0. By theorem 1.12 there exist elements $c_\lambda \in \mathfrak{A}_+$ with $\varphi(c_\lambda) \to 0$, $\|c_\lambda\| \leqq 1$, and $A_v c_\lambda \geqq A_v x_\lambda$ for all $v \in \mathbb{V}$. By $0 \leqq c_\lambda \leqq 1$ we obtain $0 \leqq c_\lambda^* c_\lambda \leqq c_\lambda$ and $\varphi(c_\lambda^* c_\lambda) \to 0$. As $\varphi$ is faithful and $c_\lambda$ self-adjoint, $s^*$-$\lim c_\lambda = 0$. Now theorem 1.15 asserts that there exists a subsequence $(c_{\lambda_j})$ and for each $\varepsilon > 0$ a projection $p_\varepsilon$ with $\varphi(p_\varepsilon) > 1 - \varepsilon$ such that $\|c_{\lambda_j} p_\varepsilon\| \to 0$. Lemma 1.13 then shows that $\|(A_v x_{\lambda_j}) p_\varepsilon\| \to 0$ holds uniformly for $v \in \mathbb{V}$.

In the general case we may write $x_\lambda = x_{\lambda 1} - x_{\lambda 2} + i(x_{\lambda 3} - x_{\lambda 4})$ as above and we have $0 \leqq x_{\lambda k} \leqq x_\lambda^{++}$, $(k = 1, 2, 3, 4)$, and $s^*$-lim $x_\lambda^{++} = 0$. Applying the first part of the proof to $x_\lambda^{++}$ we obtain $p_\varepsilon$ with $\|(A_v x_{\lambda_j}^{++})p_\varepsilon\| \to 0$ uniformly in $\mathbb{V}$. Lemma 1.13 now yields $\|(A_v x_{\lambda_j k})p_\varepsilon\| \leqq \|(A_v x_{\lambda_j}^{++})p_\varepsilon\|^{1/2} \to 0$. □

Finally, we can complete the proof of theorem 1.7: We apply theorem 1.5 with $\Phi = \{\varphi\}$. We may assume $Px = 0$. So there exists a norm-bounded net $(x_\lambda)$ with $s^*$-lim $x_\lambda = x$ such that each $x_\lambda$ is a finite convex combination of elements $(I - T_u)x$. By lemma 1.16 there exists for $\varepsilon > 0$ a projection $p_\varepsilon$ with $\varphi(p_\varepsilon) > 1 - \varepsilon$ such that for $\delta > 0$ there exists some $\lambda_0$ with $\|(A_v(x - x_{\lambda_0}))p_\varepsilon\| < \delta/2$ for all $v \in \mathbb{V}$.

Since $x_{\lambda_0}$ is a finite convex combination of elements $(I - T_u)x$ and $\|A_v(I - T_u)x\|$ tends to 0 for each $u$, there exists $v_0 \in \mathbb{V}$ such that $v \geqq v_0$ implies $\|A_v(I - T_u)x\| < \delta/2$ for the $u$'s in question. Then $\|A_v x_{\lambda_0}\| < \delta/2$ and hence $\|(A_v x_{\lambda_0})p_\varepsilon\| < \delta/2$. $\|(A_v x)p_\varepsilon\| < \delta$ follows. We have proved $\|(A_v x)p_\varepsilon\| \to 0$. □

**Remarks.** (1) In a similar way it can be proved that for finitely many $x^{(1)}, \ldots, x^{(K)} \in \mathfrak{A}$ there exists $p_\varepsilon$ with $\varphi(p_\varepsilon) > 1 - \varepsilon$ and $\|(A_v x^{(k)} - Px^{(k)})p_\varepsilon\| \to 0$ for $k = 1, \ldots, K$ and $v \to \infty$.

(2) Again, essentially the same arguments show that $\|(A_v x - Px)^{++}p_\varepsilon\| \to 0$. By lemma 1.13 this implies $\|(A_v s - Px)^* p_\varepsilon\| \to 0$.

**Corollary 1.17.** *Under the assumptions of theorem 1.7 we also have* $s^*$-lim $A_v x = Px$, $(v \to \infty)$.

*Proof.* We may assume $Px = 0$. Theorem 1.7 implies the existence of a projection $p_\varepsilon$ with $\varphi(p_\varepsilon) > 1 - \varepsilon$ and $\|(A_v x)p_\varepsilon\| \to 0$. Now

$$\begin{aligned}
[\alpha_\varphi(A_v x)]^2 &= \varphi(|A_v x|^2) = \varphi([p_\varepsilon + (1 - p_\varepsilon)]\,(A_v x)^*(A_v x)\,[p_\varepsilon + (1 - p_\varepsilon)]) \\
&\leqq \varphi(p_\varepsilon(A_v x)^*(A_v x)p_\varepsilon) + \varphi((1 - p_\varepsilon)|A_v x|^2) + \varphi(|A_v x|^2(1 - p_\varepsilon)) \\
&\leqq \|\,|(A_v x)p_\varepsilon|^2\,\| + 2\|A_v x\|\,(\varphi(1 - p_\varepsilon))^{1/2} \quad \text{(by Cauchy-Schwartz).}
\end{aligned}$$

The first term tends to 0, and $\varphi(1 - p_\varepsilon) < \varepsilon$. As $\varepsilon > 0$ was arbitrary $\alpha_\varphi(A_v x) \to 0$ follows. Using the second remark above the same proof yields $\alpha_\varphi^*(A_v x) \to 0$. As $\varphi$ is faithful this completes the proof. □

## Notes

To keep these notes short we refer to Takesaki [1979] for terminology from the theory of von Neumann algebras. Sinai and Anshelevich [1976] have obtained an almost uniform ergodic theorem for the case of translations in algebras of local observables, which they submitted slightly earlier than Lance [1976].

If the semigroup $\mathscr{S}$ in theorem 1.1 consists of *-homomorphisms, we can identify the projection $P$ as a conditional expectation operator: There exists a net $(A_\lambda)$ consisting of convex combinations of elements of $\mathscr{S}$ such that $\psi(A_\lambda x) \to \psi(Px)$ holds for all $\psi \in \mathfrak{A}_*$. The $A_\lambda$'s are *-homomorphisms and satisfy $A_\lambda P = PA_\lambda = P$. Hence $\psi(P((Px)y)) = \lim \psi(A_\lambda((Px)y)) = \lim \psi((Px)(A_\lambda y)) = \psi((Px)(Py))$, (since $\psi((Px)\cdot) \in \mathfrak{A}_*$). As $\psi$ was arbitrary, $P((Px)y) = (Px)(Py)$. Thus $x \to Px$ is the conditional expectation onto the subalgebra $P\mathfrak{A}$ of $\mathscr{S}$-invariant elements.

For this result actually the inequalities $T(x^* x) \geqq (Tx)^* Tx$, $(T \in \mathscr{S}, x \in \mathfrak{A})$, implied by complete positivity of $\mathscr{S}$, are sufficient, see Kümmerer-Nagel [1979]. Tischer and Wittmer [1974] asked which properties of $x$ carry over to $Px$. Frigerio and Verri [1982] considered the case where there is no faithful family of normal states.

Ergodic theorems for continuous parameter semigroups $\mathscr{S} = \{T_t: t = (t_1, \dots, t_d), t_i \geqq 0\}$ follow easily from the discrete counterparts. Conze and Dang-Ngoc [1978] have proved the almost uniform ergodic theorem for semigroups $\{T_t: t \in G\}$ indexed by elements $t$ of a connected amenable locally compact group $G$, i.e., the non commutative extension of the result of Emerson-Greenleaf. (They assume the existence of a faithful $\varphi$, for which the $T_t$ are positive $\alpha_\varphi$-contractions and satisfy $T_t 1 = 1$ and $\varphi \circ T_t = \varphi$.) They also prove local ergodic theorems in this setting, and even a local ergodic theorem, when $G$ is a Lie group.

F. J. Yeadon [1977] has extended the theorem of Lance in a different direction. He assumes the existence of a faithful normal semi-finite trace $\varrho$ on $\mathfrak{A}$ and considers a positive linear map $T$ in the space $L_1(\mathfrak{A}, \varrho)$ of (unbounded) $\varrho$-integrable operators in the sense of Segal [1952]. $L_1$ may be identified with $\mathfrak{A}_*$. If $0 \leqq Ty \leqq 1$ and $\varrho(Ty) \leqq \varrho(y)$ holds for $y \in L_1 \cap \mathfrak{A}$ with $0 \leqq y \leqq 1$, then for $x \in L_1$ the averages $A_n(T)x$ converge to some $\bar{x} \in L_1$ in the sense that for all $\varepsilon > 0$ there exists a projection $p_\varepsilon$ with $\varrho(1 - p_\varepsilon) < \varepsilon$ and $\|p_\varepsilon(A_n(T)x - \bar{x})p_\varepsilon\| \to 0$. R. Jajte [1984] has proved a result of this type for *subadditive* sequences. He considers a *-automorphism $T$ in a semifinite von Neumann algebra $\mathfrak{A}$ with a faithful semifinite normal trace $\varrho$ on $\mathfrak{A}$ and a sequence $(x_n)$ of self-adjoint elements of $L_1$ with $x_{n+k} \leqq x_n + T_*^n x_k$, $(n, k \geqq 1)$, and $\inf\{\varrho(x_n/n): n \geqq 1\} > -\infty$. He shows that $x_n/n$ converges in $L_1$ and almost uniformly (in the sense of Yeadon) to a $T_*$-invariant $\bar{x} \in L_1$.

Let $\varphi$ be a faithful normal state in a von Neumann algebra $\mathfrak{A}$ and $X$ the closure of $\mathfrak{A}_{sa}$ in $L_2(\mathfrak{A}, \varphi)$. Let $T$ be a positive linear mapping of $\mathfrak{A}$ into itself with $T1 \leqq 1$ and $\varphi \circ T \leqq \varphi$. Then $T$ can be extended to the complexification $\tilde{X}$ of the closure $X$ of $\mathfrak{A}_{sa}$ in $L_2(\mathfrak{A}, \varphi)$. For $x \in \tilde{X}$, Gol'dshtejn [1981] showed the $L_2$-convergence and the almost uniform convergence of $A_n(T)x$. He also showed convergence of $A_n(T^*)\psi$ to some $\psi_0$ in the norm of $\mathfrak{A}_*$ for $\psi \in \mathfrak{A}_*$, and that, for $\varepsilon > 0$, there is a projection $p$ with $\varphi(p) \geqq 1 - \varepsilon$ and

$$\lim_n [\sup\{|A_n(T^*)\psi(y) - \psi_0(y)| \varphi(y)^{-1}: \ y \in p\mathfrak{A}p, y > 0\}] = 0.$$

He uses the following maximal estimate: For positive $\varepsilon_n$'s and for $(x_n) \subset \mathfrak{A}_+$ with $\beta := \sum \varepsilon_n^{-1} \varphi(x_n) < 1/2$ there exists a projection $p \in \mathfrak{A}$ with $\varphi(p) \geqq 1 - 2\beta$ and $\|p(A_m(T)x_n)p\| < 2\varepsilon_n$, $(n, m \in \mathbb{N})$.

Gol'dshtejn's proof of the almost uniform ergodic theorem avoids some of the tools needed here. Jajte [1983] has generalized and simplified the argument.

Luczak [1984] considered a $C^*$-algebra $\mathfrak{A}_0$ which is ultraweakly dense in a von Neumann algebra $\mathfrak{A}$ and a *-automorphism $T$ of $\mathfrak{A}$. He studied the convergence of $N^{-1} \sum_{i=0}^{N-1} \tilde{T}_n \psi$ for "almost all" states $\psi$ of $\mathfrak{A}_0$, where $\tilde{T}_n$ are induced by $T^n$; see also Radin [1978].

Yeadon [1980] also studied mean ergodic theorems in non commutative $L_p$-spaces.

Groh and Kümmerer [1982] have proved mean ergodic theorems for semigroups of "bicontractions" in $\mathfrak{A}$. These may be regarded as non commutative generalizations of $L_1 - L_\infty$-contractions.

Corollary 1.17 was observed jointly with M. Lin.

Ayupov [1977] announced ergodic theorems in 0*-algebras.

Further references related to this section are Abdalla-Szücs [1974], Archbold [1976], Doplicher-Kastler-Størmer [1969], Saitô [1974], Lau [1976], Dang-Ngoc [1982], Radin [1973], [1976], Watanabe [1979], Szücs [1983], Ayupov [1982].

## § 9.2 Entropy and Information

We now present the theorems of Shannon-McMillan and Breiman-Ionescu-Tulcea and their multiparameter generalizations.

**1. Preliminaries and mean convergence.** $(\Omega, \mathscr{A}, \mu)$ is a probability space and $\mathscr{P}$ the family of partitions $\xi = \{A_i, i \in I\}$ of $\Omega$ into finitely many or countably many disjoint measurable sets $A_i$. We write $\xi = \{A_i\}$ for short. If $\eta = \{B_j\}$ is a second partition $\xi \vee \eta$ denotes the common *refinement*, i.e., the partition consisting of the sets $A_i \cap B_j$. $\eta$ is called finer than $\xi$ if $\xi \vee \eta = \eta$. We then write $\xi \leqq \eta$. $\xi$ is identitified with the $\sigma$-algebra generated by $\xi$. In particular $\mu(A \mid \xi) = E(1_A \mid \xi)$ is the conditional probability of $A$ given $\xi$.

The *conditional information* of $\xi$ given a sub-$\sigma$-algebra $\mathscr{B} \subset \mathscr{A}$ (or given a partition) is defined by

$$I(\xi \mid \mathscr{B}) = -\sum_{A_i \in \xi} 1_{A_i} \log \mu(A_i \mid \mathscr{B}).$$

(The choice of the logarithm is arbitrary. Let us take the natural logarithm. We put $\log 0 = -\infty$, $0 \log 0 = 0$.)

The *conditional entropy* $H(\xi \mid \mathscr{B})$ of $\xi$ given $\mathscr{B}$ is the expectation of $I(\xi \mid \mathscr{B})$. The *information* $I(\xi)$ of $\xi$ and the *entropy* $H(\xi)$ are the quantities $I(\xi \mid \mathscr{B})$, $H(\xi \mid \mathscr{B})$ obtained when $\mathscr{B}$ is the trivial $\sigma$-algebra $\{\emptyset, \Omega\}$. Clearly, $H(\xi) = -\sum \mu(A_i) \log \mu(A_i)$.

**Proposition 2.1.** *For $\xi, \eta, \zeta \in \mathscr{P}$ we have*

*(i) $I(\xi \vee \eta \mid \zeta) = I(\xi \mid \zeta) + I(\eta \mid \xi \vee \zeta)$;*

*(ii) $H(\xi \vee \eta \mid \zeta) = H(\xi \mid \zeta) + H(\eta \mid \xi \vee \zeta)$;*

*(iii) $\xi \leqq \eta$ implies $I(\xi \mid \zeta) \leqq I(\eta \mid \zeta)$ and $H(\xi \mid \zeta) \leqq H(\eta \mid \zeta)$.*

*Proof:* (i) is obtained by a direct computation, (ii) follows taking expectations, and (iii), since all terms are non negative. □

**Lemma 2.2** (Chung-Neveu). *If $\mathscr{C}_1 \subset \mathscr{C}_2 \subset \ldots$ is an increasing sequence of sub-$\sigma$-algebras of $\mathscr{A}$ and $\xi \in \mathscr{P}$, then*

$$E(\sup_n I(\xi \mid \mathscr{C}_n)) \leqq H(\xi) + 1 .$$

*Proof.* Put $I^* = \sup_n I(\xi|\mathscr{C}_n)$, and $I_i^* = \sup_n(-1_{A_i}\log\mu(A_i|\mathscr{C}_n))$. For $t>0$ consider $\mu(I_i^*>t)=\mu(\omega\in A_i\colon \inf_n \mu(A_i|\mathscr{C}_n)(\omega)<e^{-t})$.

The sets $A_{in}=\{\omega\colon \mu(A_i|\mathscr{C}_k)(\omega)\geqq e^{-t}$ for $k<n$ and $\mu(A_i|\mathscr{C}_k)(\omega)<e^{-t}\}$ are $\mathscr{C}_n$-measurable. Summing the inequalities $\int 1_{A_{in}}\mu(A_i|\mathscr{C}_n)\,d\mu \leqq e^{-t}\mu(A_{in})$ over $n$ we obtain $\mu(I_i^*>t)\leqq e^{-t}\sum_n \mu(A_{in})\leqq e^{-t}$. Hence $\mu(I_i^*>t)\leqq \operatorname{Min}(e^{-t},\mu(A_i))$. As $\{I^*>t\}$ is the disjoint union of the sets $\{I_i^*>t\}$ the desired estimate follows from

$$E(I^*) = \int_0^\infty \mu(I^*>t)\,dt$$

$$= \sum_{i=1}^\infty \int_0^\infty \mu(I_i^*>t)\,dt \leqq \sum_{i=1}^\infty \int_0^\infty \operatorname{Min}(e^{-t},\mu(A_i))\,dt$$

$$= \sum_{i=1}^\infty [-\mu(A_i)\log\mu(A_i)+\mu(A_i)] = H(\xi)+1. \quad \square$$

**Lemma 2.3.** *Let $D$ be a denumerable directed set and $\{\mathscr{B}_u, u\in D\}$ an increasing family of sub-$\sigma$-algebras of $\mathscr{A}$. If $\xi$ is a partition with $H(\xi)<\infty$ and $\mathscr{B}=\bigvee \mathscr{B}_u$ the $\sigma$-algebra generated by all $\mathscr{B}_u$, then $I(\xi|\mathscr{B}_u)$ converges to $I(\xi|\mathscr{B})$ in $L_1$-norm.*

*Proof.* Otherwise there exists an increasing sequence $u_1<u_2<u_3<\ldots$ *in* $D$ and an $\varepsilon>0$ with $\|I(\xi|\mathscr{B}_{u_k})-I(\xi|\mathscr{B})\|_1>\varepsilon$, such that for all $u\in D$ there is some $u_k\geqq u$ in the sequence. The sequence $\mathscr{C}_k:=\mathscr{B}_{u_k}$ is increasing and generates $\mathscr{B}$. Therefore the conditional probabilities $\mu(A_i|\mathscr{C}_k)$ tend in $L_1$ and stochastically to $\mu(A_i|\mathscr{B})$. Hence $\log\mu(A_i|\mathscr{C}_k)$ tends stochastically to $\log\mu(A_i|\mathscr{B})$ and Lemma 2.2. implies $\|I(\xi|\mathscr{C}_k)-I(\xi|\mathscr{B})\|_1\to 0$, a contradiction. $\square$

We now consider multiparameter groups and semigroups of measure preserving transformations in $\Omega$. We begin with the invertible case. Let $d\geqq 1$ be a fixed integer and $\{\tau_u\colon u\in\mathbb{Z}^d\}$ a group of automorphisms of $(\Omega,\mathscr{A},\mu)$. For $\xi=\{A_1,A_2,\ldots\}$, $\xi(u)$ denotes the partition $\tau_u^{-1}\xi=\{\tau_u^{-1}A_1,\tau_u^{-1}A_2,\ldots\}$, and, for $W\subset\mathbb{Z}^d$, $\xi(W)$ denotes the partition or $\sigma$-algebra $\bigvee_{w\in W}\xi(w)$. For $u=(u_1,\ldots,u_d)$, $v=(v_1,\ldots,v_d)$ we again write $u\leqq v$, when $u_i\leqq v_i$, $(i=1,\ldots,d)$, and $u<v$ when $u_i<v_i$, $(i=1,\ldots,d)$. $[u,v[$ is the "interval" $\{w\colon u\leqq w<v\}$.

$u<_\ell v$ shall mean that $u$ is smaller than $v$ in the lexicographic order. Put $0=(0,0,\ldots,0)$ and $e=(1,1,\ldots,1)$. $\pi(u)=u_1\cdot u_2\ldots u_d$ is the number of elements of $[0,u[$. $u\to\infty$ shall mean that all coordinates of $u$ tend to $\infty$. The following theorem has its roots in the fundamental work of C. Shannon [1948], which contains the proof of stochastic convergence for $d=1$ and finite $\xi$. McMillan [1953] proved $L_1$-convergence in this case and Carleson [1958] for $d=1$, $H(\xi)<\infty$. The multiparameter extension was given by Fritz [1970], and rediscovered by Thouvenot [1972], Katznelson and Weiss [1972], and Föllmer [1973].

**Theorem 2.4.** (Mean ergodic theorem for information). *Let* $\{\tau_u: u \in \mathbb{Z}^d\}$ *be a group of automorphisms of a probability space* $(\Omega, \mathscr{A}, \mu)$ *and* $H(\xi) < \infty$. *Then* $\pi(w)^{-1} I(\xi([0, w[))$ *converges in* $L_1$*-norm for* $w \to \infty$. *The limit is* $h := E(I(\xi | \xi(\{w: w <_\ell 0\})) | \mathscr{I})$ *where* $\mathscr{I}$ *is the* $\sigma$*-algebra of invariant sets. In the ergodic case the limit is* $H(\xi | \xi(\{w: w <_\ell 0\}))$.

*Proof.* For $u \in [0, w[$ put $F_{u,w} = \{v \in [0, w[: v <_\ell u\}$. Repeated application of lemma 2.1. (i) gives

$$(2.1) \qquad I(\xi([0, w[)) = \sum_{u \in [0, w[} I(\xi(u) | \xi(F_{u,w})).$$

As the transformations $\tau_u$ are measure preserving and invertible we have $\mu(\tau_u^{-1} A | B) = \mu(A | \tau_u B)$. It follows that

$$(2.2) \qquad I(\xi(u) | \xi(F_{u,w})) = I(\xi | \xi(F_{u,w} - u)) \circ \tau_u$$

where $F - u = \{v - u: v \in F\}$. The set $D$ of all pairs $(u, w)$ with $w > 0$ and $u \in [0, w[$ is a directed set with the partial order $\leqq_d$ given by $(u, w) \leqq_d (r, s)$ iff $F_{u,w} - u \subset F_{r,s} - r$, and the family of $\sigma$-algebras $\mathscr{B}_{(u,w)} := \xi(F_{u,w} - u)$ is increasing and generates $\mathscr{B} = \xi(\{v: v <_\ell 0\})$. ($(u, w)$ is large in the partial order $\leqq_d$ if $M(u, w) = \operatorname{Min}\{u_1, \ldots, u_d, w_1 - u_1, \ldots, w_d - u_d\}$ is large. If $M(u, w)$ exceeds $n$ then $F_{u,w} - u$ contains all $v <_\ell 0$ with $|v_i| \leqq n - 1$, $(i = 1, \ldots, d)$).

Lemma 2.3 implies that for any $\varepsilon > 0$ there exists an $n$ with

$$\| I(\xi | \xi(F_{u,w} - u)) - I(\xi | \mathscr{B}) \|_1 < \varepsilon \text{ for } (u, w) \in G_{n,w} := \{(u, w): M(u, w) \geqq n\}.$$

For fixed $n$, $\pi(w)^{-1} \operatorname{card}(G_{n,w}) \to 1$, $(w \to \infty)$. This, and the boundedness of the integrals $\| I(\xi | \xi(F_{u,w} - u)) \|_1 \leqq H(\xi) + 1$ (lemma 2.2), implies that

$$\| \pi(w)^{-1} \sum_{u \in [0,w[} I(\xi | \mathscr{B}) \circ \tau_u - \pi(w)^{-1} \sum_{u \in [0,w[} I(\xi | (F_{u,w} - u)) \circ \tau_u \|_1 < 2\varepsilon$$

holds for large $w$. The desired result now follows from (2.1) and (2.2) since $\pi(w)^{-1} \sum_{u \in [0,w[} I(\xi | \mathscr{B}) \circ \tau_u$ tends to $h$ in $L_1$-norm by the mean ergodic theorem. □

It is an exercise to show that the limit is $E(I(\xi | \xi([1, \infty[)) | \mathscr{I})$ for $d = 1$, and $H(\xi | \xi([1, \infty[))$ in the ergodic 1-parameter case.

*The non invertible case.* We sketch the proof that theorem 2.4 implies the analogous result for semigroups of non invertible transformations by a passage to the multiparameter analog of the "bilateral extension". The index set now is $\mathbb{V} = \{0, 1, 2, \ldots\}^d$. $\{\tau_u: u \in \mathbb{V}\}$ is a semigroup of endomorphisms of $\Omega$ with $\tau_0 =$ identity.

Let $\xi = \{A_1, A_2, \ldots\}$ with $H(\xi) < \infty$ be given. Put $X_u(\omega) = i$ if $\omega \in \tau_u^{-1} A_i$. Let $\mu'$ be the measure on $\{1, 2, \ldots\}^{\mathbb{V}} = \Omega'$ which is the distribution of

$X = (X_u: u \in \mathbb{V})$. In other words: if $V_0$ is a finite subset of $\mathbb{V}$ and $(\omega')_v$ the $v$-th coordinate of $\omega' \in \Omega'$, then

$$\mu'(\{\omega': (\omega')_v = i_v \text{ for } v \in V_0\}) = \mu(\{\omega \in \Omega: X_v(\omega) = i_v \text{ for } v \in V_0\}).$$

$\mu'$ may be extended in a unique way to a measure on $\Omega'' = \{1, 2, \ldots\}^{\mathbb{Z}^d}$ which is invariant unter the shift transformations $\theta_u$, $(u \in \mathbb{Z}^d)$. If $\xi''$ is the partition of $\Omega''$ into the sets $\{\omega'': (\omega'')_0 = i\}$, $(i = 1, 2, \ldots)$, then the joint distribution of the family $\{I(\xi([0, u[)): u \in \mathbb{V}\}$ on $\Omega$ (under $\mu$) is the same as that of the family $\{I(\xi''([0, u[)): u \in \mathbb{V}\}$ on $\Omega''$ (under $\mu''$). Applying theorem 2.4 to $\xi''$ we obtain the $L_1$-convergence of $\pi(u)^{-1} I(\xi([0, u[))$ for $u \to \infty$.

**2. Convergence a.e.** The following theorem was proved for $d = 1$ by L. Breiman [1957/60], and A. Ionescu Tulcea [1960], and for general $d$ (assuming ergodicity) by Ornstein and Weiss [1983].

**Theorem 2.5** (Pointwise ergodic theorem for information). *Let $\{\tau_u: u \in \mathbb{V}\}$ be a d-parameter semigroup of endomorphisms of $\Omega$. If $\xi$ is a finite partition $n^{-d} I(\xi([0, ne[))$ converges a.e. for $n \to \infty$.*

*Proof.* Let $\xi^n$ be the partition $\xi([0, ne[)$. If $\eta$ is a partition, then $\eta(\omega)$ shall be the element of $\eta$ which contains $\omega$. By definition $I(\xi^n)(\omega) = -\log \mu(\xi^n(\omega))$. Let $q$ be the number of elements of $\xi = \{A_1, \ldots, A_q\}$.

**Lemma 2.6.** *$h^* := \sup_{n \geqq 1} \{n^{-d} I(\xi^n)\}$ is integrable.*

*Proof.* For $t > 0$ put $B_t^n := \{\omega: -n^{-d} \log \mu(\xi^n(\omega)) > t\}$. $B_t^n$ is the union of those elements $B$ of the partition $\xi^n$ which satisfy $\mu(B) < e^{-tn^d}$. As $\xi^n$ has at most $q^{n^d}$ elements, we have $\mu(B_t^n) < (qe^{-t})^{n^d}$. Hence $\mu(h^* > t) \leqq \sum_n (qe^{-t})^{n^d} \leqq e^{-t/2}$ for $t$ large, say, for $t \geqq t_0$. The integrability follows from $\int h^* d\mu = \int_0^\infty \mu(h^* > t)\, dt \leqq t_0 + \int_{t_0}^\infty e^{-t/2} dt < \infty$. □

Now set $\bar{h}(\omega) = \limsup n^{-d} I(\xi^n)(\omega)$ and $\underline{h}(\omega) = \liminf n^{-d} I(\xi^n)(\omega)$. By lemma 2.6 both $\underline{h}$ and $\bar{h}$ are finite a.e.. Let $e^i$ be the element of $\mathbb{V}$ with $i$-th coordinate 1 and all other coordinates 0. As $\xi^{n+1}$ is finer than $\tau_{e^i}^{-1} \xi^n$ we have $I(\xi^{n+1}) \geqq I(\tau_{e^i}^{-1} \xi^n) = I(\xi^n) \circ \tau_{e^i}$. As $(n+1)^d/n^d$ tends to 1 this implies $\bar{h} \geqq \bar{h} \circ \tau_{e^i}$ and $\underline{h} \geqq \underline{h} \circ \tau_{e^i}$. Since $\underline{h}$ and $\bar{h}$ are integrable we obtain $\underline{h} = \underline{h} \circ \tau_{e^i}$ and $\bar{h} = \bar{h} \circ \tau_{e^i}$. Thus $\underline{h}$ and $\bar{h}$ are invariant under all $\tau_u$, $(u \in \mathbb{V})$. If almost sure convergence fails, there exists $\alpha > \beta$ and an invariant set $F$ of positive measure with $\bar{h} > \beta$ and $\underline{h} < \alpha$ on $F$. This means that for $\omega \in F$ there are infinitely many $n$ with $\mu(\xi^n(\omega)) > \exp(-n^d \alpha)$ and infinitely many $n$ with

(2.3) $\quad \mu(\xi^n(\omega)) < \exp(-n^d \beta)$.

Put $Q = \{1, \ldots, q\}$. For $W \subset \mathbb{V}$ the elements of $Q^W$ are called $W$-names and the $W$-name of $\omega$ is the element $\gamma = (i_u)_{u \in W}$ of $Q^W$ with $\tau_u \omega \in A_{i_u}$ for all $u \in W$. By abuse of notation $\mu(\gamma) := \mu\{\omega$: the $W$-name of $\omega$ is $\gamma\}$ is called the measure of the name $\gamma$. $W$ is called the domain of $\gamma$. If the domain is a "cube" $[w, w + ne[$ for some $w \in \mathbb{V}$ and $n \geqq 1$, we call $\gamma$ an $n$-name and $n$ its side-length. We call $\gamma$ *fat* when $\mu(\gamma) > \exp(-\alpha n^d)$. $\xi^n(\omega)$ is called fat, when the $[0, ne[$-name of $\omega$ is fat, i.e., when $\mu(\xi^n(\omega)) > \exp(-\alpha n^d)$.

Say that a name $\gamma$ is *ε-covered* by names $\{\gamma_j\}$ if the union of the domains $W_j$ of the $\gamma_j$ contains all but a fraction $< \varepsilon$ of the points of the domain $W$ of $\gamma$, and if each $\gamma_j$ agrees with $\gamma$ on $W_j \cap W$. If, moreover, the $W_j$ are contained in $W$ and disjoint we say that $\{\gamma_j\}$ *ε-tiles* $\gamma$.

**Lemma 2.7.** *For all $\varepsilon > 0$, $m \geqq 1$ and for almost all $\omega \in F$ there exists $N(\omega)$ such that for $N \geqq N(\omega)$ the $[0, Ne[$-name of $\omega$ can be ε-tiled by fat names of side-length $\geqq m$.*

*Proof.* Given any $\varepsilon', \varepsilon'' > 0$ and $K \geqq 1$ we can inductively find numbers $m = m_1 < n_1 < m_2 < n_2 < \ldots < n_K$ as follows: Put $m_1 = 1$. When $m_k$ has been determined find $n_k > m_k$ such that the set $F_k$ of points $\omega \in F$ having a fat $[0, ve[$-name for some $v$ with $m_k \leqq v \leqq n_k$ has measure $\geqq \mu(F) - \varepsilon'/K$. Then find $m_{k+1}$ with $n_k/m_{k+1} < \varepsilon''$. The intersection $F'$ of the sets $F_k$, $(k = 1, \ldots, K)$, has measure $\geqq \mu(F) - \varepsilon'$. If $\varepsilon'$ is very small most of the points of the invariant set $F$ visit $F'$ most of the time. Formally: Given $\varepsilon_1 > 0$ we assume $\varepsilon' < \varepsilon_1^2/2$. The limit $\bar{f}$ of the sequence $f_N = N^{-d} \sum_{u \in [0, Ne[} 1_{F'} \circ \tau_u$ exists a.e., (theorem 6.1.2 or 6.2.8) and satisfies $0 \leqq \bar{f} \leqq 1_F$ and $\int \bar{f} d\mu = \mu(F')$. Now $\mu(\{\omega \in F: \bar{f}(\omega) \leqq 1 - \varepsilon_1/2\}) \cdot (\varepsilon_1/2) \leqq \int (1_F - \bar{f}) d\mu < \varepsilon'$ implies that $F'' := \{\omega \in F: \bar{f}(\omega) > 1 - \varepsilon_1/2\}$ has measure $\geqq \mu(F) - \varepsilon_1$.

For $\omega \in F''$ there exists $N(\omega)$ with $N \geqq N(\omega) \Rightarrow f_N(\omega) > 1 - \varepsilon_1$. We may assume that $N(\omega)$ is much larger than $n_K$. As $F''$ was an arbitrarily large subset of $F$ it will be enough to show $N(\omega)$ has the desired property for small enough $\varepsilon_1, \varepsilon''$ and large $K$. We can then repeat the argument with $F$ replaced by $F \setminus F''$ and smaller $\varepsilon_1$. First choose $K$ with $(1 - 3^{-(d+1)})^K < \varepsilon$. We may assume $\varepsilon_1 < \varepsilon/4$. Let $R_{K+1}$ denote the set of $u \in [0, Ne[$ with $\tau_u \omega \in F'$. $R_{K+1}$ covers $1 - \varepsilon_1$ of $[0, Ne[$. Put $S_{K+1} = [0, Ne[ \setminus R_{K+1}$. Let $S'_K$ be the set of $u$'s in $[0, Ne[$ for which $[u, u + n_K e[$ is not contained in $[0, Ne[$. Choosing $N(\omega)$ large we may assume $\operatorname{card}(S'_K)/N^d < \varepsilon/8$. Put $S_K = S_{K+1} \cup S'_K$ and $R_K = [0, Ne[ \setminus S_K$. Then $s_K := N^{-d} \operatorname{card}(S_K) \leqq \varepsilon/4 + \varepsilon/8 \leqq \varepsilon/2 - \varepsilon/8$. For any $u \in R_K$ and $k = 1, \ldots, K$ there exists a cube $[u, u - ve[$ with side-length between $m_k$ and $n_k$ such that the $[u, u + ve[$-name of $\omega$ is fat. $\mathfrak{C}_k$ denotes the set of these cubes. We shall now use the Vitali covering argument to pick a disjoint subfamily of these cubes, which covers most of $[0, Ne[$. For $C = [u, u + ve[$ put $\tilde{C} = [u - ve, u + 2ve[$. $u$ is called the lower corner of $C$. Let $C_{K1}$ be one of the cubes in $\mathfrak{C}_K$ with maximal side-length, then $C_{K2}$ a largest cube in $\mathfrak{C}_K$ among those for which the lower corner is in

$R_K \setminus \tilde{C}_{K1}$, then $C_{K3}$ a largest cube in $\mathfrak{C}_K$ with lower corner in $R_K \setminus (\tilde{C}_{K1} \cup \tilde{C}_{K2})$, etc. When no such cube is available any more the chosen cubes are disjoint, and – as we shall see in a moment – they cover $3^{-(d+1)}$ of $[0, Ne[$.

Now assume $1 \leqq k < K$, and assume that the choice of disjoint cubes in $\mathfrak{C}_K, \ldots, \mathfrak{C}_{k+1}$ has been completed. At this point $[0, Ne[$ is the disjoint union of a set $R_{k+1} \subset R_K$, a set $S_{k+1} \supset S_K$ with $s_{k+1} := N^{-d} \operatorname{card}(S_{k+1}) \leqq \varepsilon/2 - \varepsilon 2^{k-K-2}$, (for $k \leqq K-2$), and of disjoint cubes selected from $\mathfrak{C}_{k+2}, \mathfrak{C}_{k+3}, \ldots, \mathfrak{C}_K$. Let $S_k'$ be the set of $u \in [0, Ne[$ for which $[u, u + n_k e[$ intersects one of the disjoint cubes chosen from $\mathfrak{C}_{k+1}, \ldots, \mathfrak{C}_K$. As their side length is $\geqq m_{k+1} \geqq (\varepsilon'')^{-1} n_k$ we see that $N^{-d} \operatorname{card}(S_K') \leqq \varepsilon 2^{-(K+3)}$ holds provided $\varepsilon''$ is small enough. (The choice of $\varepsilon''$ depends only on $K$ and $\varepsilon$). Put $S_k = S_{k+1} \cup S_k'$ and let $R_k$ be the set of points in $[0, Ne[ \setminus S_k$, which are not yet covered by one of the disjoint cubes. Then

$$R_k \subset R_{k+1}, \text{ and } s_k := N^{-d} \operatorname{card}(S_k) \leqq \frac{\varepsilon}{2} - \varepsilon^{k-K-3}.$$

Now we choose disjoint cubes $C_{k-1}, \ldots$ from $\mathfrak{C}_k$ with lower corners in $R_k$ as above. Let $U_k$ be the union of the disjoint cubes from $\mathfrak{C}_K, \ldots, \mathfrak{C}_k$, $u_k = N^{-d} \operatorname{card}(U_k)$, and $r_k = N^{-d} \operatorname{card}(R_k)$. In the case $u_{k+1} \geqq 1 - \varepsilon$ we had covered $1 - \varepsilon$ of $[0, Ne[$ even after the previous step. In the case $u_{k+1} < 1 - \varepsilon$ the fact that $[0, Ne[$ is the disjoint union of $U_{k+1}$, $S_k$ and $R_k$ implies $r_k \geqq \varepsilon/2$. Hence $R_k$ contains at least half of the points left uncovered by $U_{k+1}$. On the other hand $R_k$ is covered by $\tilde{C}_{k1} \cup \tilde{C}_{k2} \cup \ldots$. Hence $3^d(u_k - u_{k+1}) \geqq r_k \geqq (r_k + v_k)/2 \geqq (1 - u_{k+1})/3$. This implies $(1 - 3^{-(d+1)})(1 - u_{k+1}) \geqq (1 - u_k)$.

Therefore at each stage the uncovered portion is reduced by a factor $(1 - 3^{-(d+1)})$ until $1 - \varepsilon$ of $[0, Ne[$ is covered. The assertion now follows from $(1 - 3^{-(d+1)})^K < \varepsilon$. □

Now fix $m \geqq 1$ and let $\Gamma_N$ be the set of $N$-names which can be $\varepsilon$-tiled by fat names of side-length $\geqq m$.

**Lemma 2.8.** *If $m$ is sufficiently large and $\varepsilon > 0$ sufficiently small, then* $\operatorname{card}(\Gamma_N) \leqq \exp(N^d(\alpha + \beta)/2)$ *holds for all sufficiently large $N$.*

*Proof.* To specify a name in $\Gamma_N$ one can first choose disjoint cubes $W_j$ of side length $\geqq m$ covering $(1 - \varepsilon)$ of $[0, Ne[$, then specify a fat name for each domain $W_j$ and finally fill in the name arbitrarily off the chosen cubes. If there are $i$ cubes $W_j$, then $i \leqq N^d m^{-d}$ because of the disjointness. Each $W_j = [u(j), v(j)[$ is determined by its lower corner $u(j)$ and upper corner $v(j)$. There are $\binom{N^d}{i}$ possible choices of the set $\{u(1), \ldots, u(i)\}$ and $\binom{N^d}{i}$ choices of $\{v(1), \ldots, v(i)\}$. Hence the total number of ways of specifying exactly $i$ cubes $W_j$ is $\leqq \binom{N^d}{i}^2 \leqq \binom{N^d}{M(N, m)}$ with

$M(N, m) = [N^d m^{-d}] + 1$. The number of ways of choosing the cubes thus is $\leqq N^d m^{-d} \binom{N^d}{M(N, m)}$. Take the logarithm, divide by $N^d$, use Stirling's formula and let $N \to \infty$. Then the resulting expression tends to $-m^{-d} \log m^{-d} - (1 - m^{-d}) \log (1 - m^{-d})$, which is smaller than $(\beta - \alpha)/8$ for large $m$. For such an $m$ the number of choices of the $W_j$ is $\leqq \exp(N^d(\beta - \alpha)/8)$ for large $N$.

As the $W_j$ cover $(1 - \varepsilon)$ of $[0, Ne[$ there are only $q^{\varepsilon N^d} = \exp(N^d \varepsilon \log q)$ ways to fill in the names in the uncovered places. Taking $\varepsilon < (\beta - \alpha)/8 \log q$ this is $\leqq \exp(N^d(\beta - \alpha)/8)$.

The number of fat names one can put into a cube $W_j$ of side-length $n^d$ is $\leqq \exp(\alpha n^d) = \exp(\alpha \operatorname{card}(W_j))$. Hence the number of fat names one can put into a fixed choice $W_1, W_2 \ldots$ of the disjoint cubes is $\leqq \exp(N^d \alpha)$. Putting all this together the lemma is proved. □

The remainder of the proof of theorem 5.1 is easy. Let $B_N$ be the set of names in $\Gamma_N$ which have measure $< \exp(-N^d(\beta - (\beta - \alpha)/4))$. Then the measure of the set of points $\omega$ with $[0, Ne[$-name in $B_N$ is at most $\operatorname{card}(\Gamma_N) \exp(-N^d(\beta - (\beta - \alpha)/4)) \leqq \exp(-N^d(\beta - \alpha)/4)$. By Borel-Cantelli the $[0, Ne[$-name of almost all $\omega$ belongs only finitely often to $B_N$. On the other hand lemma 2.7 implies that the $[0, Ne[$-name of almost all $\omega \in F$ belongs to $\Gamma_N$ for sufficiently large $N$, and hence to $\Gamma_N \setminus B_N$. This means that eventually, for almost all $\omega \in F$, $\mu(\xi^N(\omega)) \geqq \exp(-N^d(\beta - (\beta - \alpha)/4))$, contradicting (2.3). □

**Remark.** The same argument works when $[0, Ne[$ is replaced by $[0, w[$ and $w \to \infty$ in a sector of $\mathbb{V}$. For groups $\{\tau_u, u \in \mathbb{Z}^d\}$ it also works with squares $[-Ne, Ne[$, etc. However, it is not clear if the condition that $w \to \infty$ in a sector of $\mathbb{V}$ can be dropped. The generalisation to countable partitions $\xi$ with $H(\xi) < \infty$ seems to be an open problem for $d \geqq 2$, but the case $d = 1$ has been proved for partitions with $H(\xi) < \infty$ by K.L. Chung [1961].

## Notes

The original motivation of the Shannon-McMillan theorem has been the study of the content of information provided by the observation of a random sequence of "letters" from a finite alphabet. The book of Pinsker [1960] gives an account of the work done by Shannon, McMillan, Pinsker, Dobrushin, Perez, Chintchin and others on "information stability" and the "speed of generation of information". (This is apart from the main body of information theory, dealing primarily with channels, see e.g. Csiszár-Körner [1981]).

Perez [1957] deals with more general alphabets by considering two probability measures $P, Q$ on a measurable space $(\Omega, \mathscr{A})$, and a sequence $X_1, X_2, \ldots$ which is stationary both under $P$ and $Q$. Assume that $P_n, Q_n$ are the restrictions to the $\sigma$-algebras $\mathscr{A}_n$ generated by $X_1, \ldots, X_n$ and $P_n \ll Q_n$ for all $n$, with $dP_n/dQ_n = f_n$. If $\lim n^{-1} \int \log f_n \, dP$ exists and is finite and the $X_1, X_2, \ldots$ are independent under $Q$, then $\lim n^{-1} \log f_n$ exists in the $L_1$-sense. The theorem of Shannon-McMillan is the special case where the $X_i$ have the uniform distribution on a finite set under $Q$. A. Ionescu Tulcea [1960] actually proved a.e.-

convergence in this generalized setting. S.C. Moy [1961] proved $L_1$-convergence and a.e.-convergence when $X_1, X_2, \ldots$ is Markovian with stationary transition probabilities; see also Perez [1964] and Kieffer [1973].

Let us now return to the formulation with partitions. Jacobs [1959] has proved that the theorem of Shannon-McMillan remains true if the condition of invariance of the measure $\mu$ is replaced by the condition that an invariant probability measure $\nu$ with $\mu \ll \nu$ exists. This can be achieved under assumptions of almost periodicity. Parry [1963] has an extension of the Breiman-Ionescu Tulcea result to conservative nonsingular transformations having no invariant measure. Then $n^{-1} I(\xi([0, n[))$ must be replaced by suitable ratios as in the Hurewicz case of the Chacon-Ornstein theorem.

Föllmer [1973] had already proved the multiparameter pointwise ergodic theorem for information in the special case where $\xi$ is the partition generated by the 0-coordinate of a Gibbs field with finite state space. Künsch [1981] has extended this to countable state spaces.

Kieffer [1975] has a generalization of the Shannon-McMillan theorem for the action of amenable groups. Moulin-Olagnier [1983] has a simpler proof of this result based on his "almost subadditive" mean ergodic theorem for processes $\{F_A\}$ indexed by the finite subsets $A$ of an infinite amenable group $G$ for which there exists $c > 0$ such that, for each $A$ and each decomposition $1_A = \sum_{i=1}^{m} \alpha_i 1_{A_i}$ with $\alpha_i \geqq 0$ one has $\int (F_A - \sum \alpha_i F_{A_i})^+ d\mu \leqq c \sum_{i=1}^{m} \alpha_i$. Derriennic [1983a] has a different almost subadditive ergodic theorem which yields a proof of the Breiman-Ionescu-Tulcea theorem; see § 1.5.

Pitskel' and Stepin [1971] proved a pointwise ergodic theorem for information for the group of dyadic rationals mod 1. Pitskel' [1980] has studied the asymptotic equipartition property for space homogeneous processes. (In contrast to what he writes, his results do not contradict Krengel's entropy computation since his $T_A$ is not the induced transformation). For further related work and references see Pitskel' [1971], [1978] and Nguyen-Zessin [1976], [1979], and Tempel'man [1984].

## § 9.3 Nonlinear nonexpansive mappings

Let $\mathfrak{H}$ be a real Hilbert space and $C$ a closed convex subset of $\mathfrak{H}$. A mapping $T: C \to C$ is called *nonexpansive* if $\|Tx - Ty\| \leqq \|x - y\|$ holds for all $x, y \in C$. $T$ need no longer be linear. The nonlinear ergodic theorem of Baillon [1975] asserts that $n^{-1}(T^0 + T^1 + \ldots + T^{n-1})x$ converges weakly if $T$ has a fixed point. A sufficient condition is the boundedness of $C$. Strong convergence can be obtained under supplementary conditions. We discuss this material in the multiparameter setting. Generalizations dealing with more general Banach spaces will only be reported.

### 1. Weak convergence of Cesàro averages.

**Lemma 3.1.** *Let $C \subset \mathfrak{H}$ be closed and convex and let $T$ be nonexpansive. If* $w$-$\lim x_n = p$ *and $x_n - Tx_n \to 0$ (strongly), then $p = Tp$.*

*Proof.* For any $z \in C$ we have $\|z - x_n\| \geqq \|Tz - Tx_n\|$, and hence $\|z - x_n\|^2$

$\geqq \langle Tz - Tx_n, z - x_n \rangle$. This implies $0 \leqq \langle z - Tz - (x_n - Tx_n), z - x_n \rangle$. The assumptions on $x_n$ now yield $\langle z - Tz, z - p \rangle \geqq 0$. For $0 < \lambda < 1$ put $z_\lambda = (1-\lambda)p + \lambda Tp$. Substituting $z_\lambda$ for $z$ we find $\langle z_\lambda - Tz_\lambda, p - Tp \rangle \leqq 0$. Now let $\lambda$ tend to 0. □

If $z \in C$ is a convex combination of two fixed points $x, y$ of $T$ and $Tz \neq z$, then $Tz$ cannot lie on the line connecting $x$ and $y$ because this would imply $\|Tz - Tx\| > \|x - z\|$ or $\|Tz - Ty\| > \|z - y\|$. Hence $\|x - z\| + \|z - y\| = \|x - y\| < \|x - Tz\| + \|Tz - y\| = \|Tx - Tz\| + \|Tz - Ty\| \leqq \|x - z\| + \|z - y\|$, a contradiction. Thus, the set of fixed points of a nonexpansive $T$ is a closed convex (possibly empty) subset of $C$.

After these remarks we can enter the discussion of semigroups of nonexpansive mappings. Let $d \geqq 1$ be a fixed integer, $\mathbb{V} = \{0, 1, 2, \ldots\}^d$, and let $\mathscr{S} = \{T_v : v \in \mathbb{V}\}$ be a semigroup of nonexpansive maps $T_v: C \to C$. We shall use the notation from §6.1 and write $0 = (0, \ldots, 0)$, $S_u = \sum_{v \in [0,u[} T_v$, $\pi(u) = u_1 \cdot u_2 \ldots u_d$, $A_u = \pi(u)^{-1} S_u$. $T_0$ shall be the identity $I$. The equation $T_{u+v} = T_u T_v$ describes the semigroup property. $u \to \infty$ means that all coordinates of $u$ tend to $\infty$.

If there exists one element $x_1 \in C$ for which the orbit $\mathscr{S} x_1$ is norm-bounded, then $\|T_u x_2\| - \|T_u x_1\| \leqq \|T_u x_2 - T_u x_1\| \leqq \|x_2 - x_1\|$ shows that the orbit of any bounded subset of $C$ is bounded. Then the norms of the averages $A_u x$ and also the norms of vectors like $T_w A_u x$ are bounded: there exists $M_x < \infty$ with $\|A_u x\|, \|T_u x\|, \|T_w A_u x\| \leqq M_x$ for all $u$ and $w \in \mathbb{V}$.

Let $e(i)$ denote the $i$-th unit vector in $\mathbb{V}$.

**Lemma 3.2.** *If the orbit of an element of C is norm-bounded we have* $\|A_u x - T_{e(i)} A_u x\| \to 0$, $(u \to \infty)$.

*Proof.* For any $y \in C$ and $v \in \mathbb{V}$ we have

$$\begin{aligned} 0 &\leqq \|T_v x - y\|^2 - \|T_{v+e(i)} x - T_{e(i)} y\|^2 \\ &= \|T_v x - T_{e(i)} y\|^2 - \|T_{v+e(i)} x - T_{e(i)} y\|^2 \\ &\quad + 2\langle T_v x - T_{e(i)} y, T_{e(i)} y - y \rangle + \|T_{e(i)} y - y\|^2. \end{aligned}$$

Take the sum over all $v \in [0, u[$ and multiply by $\pi(u)^{-1}$. Most of the first and second terms cancel each other. We can omit the negative terms not needed for the cancellation. Writing $V_i(u) = \{v = (v_1, \ldots, v_d) \in [0, u[ : v_i = 0\}$, we arrive at

$$0 \leqq \pi(u)^{-1} \sum_{v \in V_i(u)} \|T_v x - T_{e(i)} y\|^2 + 2\langle A_u x - T_{e(i)} y, T_{e(i)} y - y \rangle + \|T_{e(i)} y - y\|^2.$$

Put $y = A_u x$, then $\langle \ldots, \ldots \rangle = -\|T_{e(i)} y - y\|^2$ and we obtain

$$\|A_u x - T_{e(i)} A_u x\|^2 \leqq \pi(u)^{-1} \sum_{v \in V_i(u)} \|T_v x - T_{e(i)} A_u x\|^2.$$

As $\|T_v x - T_{e(i)} A_u x\|^2$ is bounded by $4M_x^2$ and $V_i(u)$ has $\pi(u)/u_i$ elements the right side is bounded by $4M_x^2 u_i^{-1}$. □

The set $F = \{x \in C\colon T_u x = x \text{ for all } u \in \mathbb{V}\}$ of fixed points of $\mathscr{S}$ is closed and convex.

**Lemma 3.3.** *If the orbit of an element of $C$ is norm-bounded $F$ is non empty. If $u(1) < u(2) < \dots$ is an increasing sequence $V'$ in $\mathbb{V}$ and $A_{u(k)}x$ converges weakly to $p$, then $p \in F$.*

*Proof.* As $\{A_u x\colon u \in \mathbb{V}\}$ is norm-bounded, it is enough to prove the second assertion. Lemma 3.2 shows that $x_n := A_{u(n)}x$ satisfies $w$-$\lim x_n = p$ and $\|x_n - T_{e(i)} x_n\| \to 0$. By Lemma 3.1, $p = T_{e(i)} p$. As $i$ was arbitrary and the $T_{e(i)}$ generate $\mathscr{S}$ we have $p \in F$. □

If $F$ is non empty $P$ shall denote the projection on $F$. This means that $Px$ is the point of $F$ closest to $x$. For $z \in C \setminus F$ the set $F$ is contained in the half space

$$\{y \in \mathfrak{H}\colon \langle y - Pz, z - Pz\rangle \leqq 0\}.$$

**Lemma 3.4.** *If $F \neq \emptyset$, then $PT_u x$ converges strongly for $u \to \infty$.*

*Proof.* For $u \leqq v$ we have

$$\|PT_v x - T_v x\| \leqq \|PT_u x - T_v x\| = \|T_{v-u} PT_u x - T_{v-u} T_u x\| \leqq \|PT_u x - T_u x\|.$$

The function $v \to \|PT_v x - T_v x\|$ is therefore nonincreasing on $\mathbb{V}$. For $\varepsilon > 0$ we can find $w(\varepsilon) \in \mathbb{V}$ such that $w(\varepsilon) \leqq u \leqq v$ implies $0 \leqq \|PT_u x - T_u x\|^2 - \|PT_v x - T_v x\|^2 < \varepsilon$.

Note that for any $y \in F$ and $z \in C$ we have $\|z - y\|^2 = \|z - Pz\|^2 + \|Pz - y\|^2 + 2\langle Pz - y, z - Pz\rangle \geqq \|z - Pz\|^2 + \|Pz - y\|^2$. If we put $y = PT_u x$ and $z = T_v x$ we obtain

$$\begin{aligned}\|PT_v x - PT_u x\|^2 &= \|Pz - y\|^2 \leqq \|z - y\|^2 - \|z - Pz\|^2 \\ &= \|T_v x - PT_u x\|^2 - \|PT_v x - T_v x\|^2 < \varepsilon.\end{aligned}$$

Hence, $PT_v x$ is a (generalized) Cauchy sequence and converges. □

We can now prove the multiparameter form of the non linear ergodic theorem of Baillon. It is a special case of a result of Brézis and Browder [1977].

**Theorem 3.5.** *If $\mathscr{S}$ is a semigroup of nonexpansive maps of a closed convex subset $C$ of a Hilbert space $\mathfrak{H}$ and $F \neq \emptyset$ (or – equivalently – the orbit of at least one element of $C$ is norm-bounded), then $A_u x$ converges weakly to some $p \in F$, $(u \to \infty)$.*

*Proof.* Let $q$ be the element of $F$ with $\|PT_u x - q\| \to 0$. It will be enough to show that if $u(1) < u(2) < \dots$ is a subsequence $V'$ of $\mathbb{V}$ for which $A_{u(n)}x$ converges weakly to some $p$, then $p = q$.

Summing the inequalities

$$\langle p - q, T_v x - PT_v x\rangle \leqq \langle PT_v x - q, T_v x - PT_v x\rangle$$
$$\leqq \|PT_v x - q\| \, \|T_v x - TP_v x\| \leqq \|PT_v x - q\| \, \|x - Px\|$$

over all $v \in [0, u[$, and dividing by $\pi(u)$ we obtain

$$\langle p - q, A_u x - \pi(u)^{-1} \sum_{v<u} PT_v x\rangle \leqq \pi(u)^{-1} \|x - Px\| \sum_{v<u} \|PT_v x - q\|.$$

The right hand side tends to 0 for $u \to \infty$. Passing to the limit in $V'$ and observing $\|\pi(u)^{-1} \sum_{v<u} PT_v x - q\| \to 0$ we find $\langle p - q, p - q\rangle \leqq 0$. Hence, $p = q$. □

**2. Generalized averages.** The observation that Cesàro averages may be replaced by more general averages in the argument above leads to some interesting related results.

We condider the family $\mathfrak{Q}$ of all matrices $Q = (q_{u,v})$ $(u, v \in \mathbb{V})$ with $q_{u,v} \geqq 0$, $\sum_v q_{u,v} = 1$ for all $u$ and

(3.1) $$\lim_{u\to\infty} \sum_v |q_{u,v} - q_{u,v+e(i)}| = 0$$

for $i = 1, \dots, d$. It is easy to see that $\lim_{u\to\infty} q_{u,v} = 0$ uniformly in $v$.

The same arguments as above yield:

**Theorem 3.5′.** *Under the assumptions of theorem 3.5 the averages $\sum_v q_{uv} T_v x$ formed with any $Q \in \mathfrak{Q}$ converge weakly to $p$ for $u \to \infty$.*

An equivalent formulation may be given with the help of Lorentz's concept of almost convergence. A generalized sequence $(x_u)$ in $\mathfrak{H}$ is called *almost convergent* if $\lim \pi(u)^{-1} \sum_{v<u} x_{v+w} = x$ uniformly in $w \in \mathbb{V}$.

**Lemma 3.6.** *If $\sum_v q_{u,v} x_v \to x$ for all $Q \in \mathfrak{Q}$, then $(x_v)$ is almost convergent to $x$. This holds for the weak topology as well.*

*Proof.* If $(x_v)$ is not almost convergent to $x$ there exists a neighborhood $U$ of $x$, a sequence $u(1) < u(2) < \dots \to \infty$ and elements $w(k) \in \mathbb{V}$ with $\pi(u(k))^{-1} \sum_{v<u(k)} x_{v+w(k)} \notin U$. For $u = u(k)$ put $q_{u,v} = \pi(u)^{-1}$ if $w(k) \leqq v < u + w(k)$ and $= 0$ elsewhere. If $u$ does not belong to the sequence put $q_{u,v} = \pi(u)^{-1}$ for $\in [0, u[$ and $= 0$ elsewhere. (It is not hard to see that the almost convergence of $(x_v)$ to $x$ conversely implies $\sum q_{uv} x_v \to x$ for all $Q$.) □

**Lemma 3.7.** *If $(x_u)$ is almost convergent to $x$ and $x_{u+e(i)} - x_u$ converges to $0$ for all $i$, then $x_u$ converges to $x$.*

*Proof.* We may assume $x = 0$. Let $U$ be a neighborhood of 0. Find a convex neighborhood $U_1$ of 0 with $U_1 + U_1 \subset U$. For some large $u \in \mathbb{V}$ the averages $y(u, w) = \pi(u)^{-1} \sum_{v<u} x_{v+w}$ belong to $U_1$ for all $w$. Let $U_2$ be a neighborhood of 0 which is so small that any sum of at most $\sigma(u) = u_1 + u_2 + \ldots + u_d$ vectors in $U_2$ belongs to $U_1$. Call $v, v' \in \mathbb{V}$ neighbors if $v' = v \pm e(i)$ for some $i$. Any two elements in $[w, w+u[$ may be connected by a string of at most $\sigma(u) = u_1 + u_2 + \ldots + u_d$ neighbors. If $w$ is so large that $x_v - x_{v'} \in U_2$ holds for any two neighbors $v, v'$ in $[w, w+u[$ and if $v \in [w, w+u[$, then $x_w - x_v$ belongs to $U_1$.

The identity $x_w = y(u, w) - \pi(u)^{-1} \sum_{v \in [w, w+u[} (x_w - x_v)$ and the convexity of $U_1$ now implies $x_w \in U$. □

Combining the lemmas with theorem 3.5′ we obtain a multiparameter extension of a result of Bruck [1978]:

**Theorem 3.8.** *Under the assumptions of theorem 3.5, $T_u x$ converges weakly iff $T_u x - T_{u+e(i)} x$ converges weakly to $0$ for $i = 1, \ldots, d$.*

**3. Strong convergence.** It follows from an example of Genel and Lindenstrauss [1975] that there exists a nonexpansive mapping in the unit ball $C$ of $\ell_2$ and an $x_0 \in C$ for which $n^{-1} \sum_{k=0}^{n-1} T^k x_0$ does not converge strongly. Thus strong convergence can only be obtained under supplementary conditions.

Again $C$ is a closed convex subset of a Hilbert space $\mathfrak{H}$. A semigroup $\mathcal{S}$ of nonexpansive mappings $T_u: C \to C$ is said to be *asymptotically isometric* on $C_0 \subset C$, if for all $x, y \in C_0$, $\lim_{u\to\infty} \|T_u x - T_{u+v} y\|$ exists uniformly in $v \in \mathbb{V}$. The proof that $T_u x$ is strongly almost convergent for $x \in C_0$ shall follow easily from the following Hilbert space theorem, which extends a result of Bruck [1978] to the multiparameter situation and requires a different proof.

**Theorem 3.9.** *If $(x_u, u \in \mathbb{V})$ is a generalized sequence in $\mathfrak{H}$ and* $\lim_{u\to\infty} \langle x_u, x_{u+v} \rangle$ *exists uniformly in $v \in \mathbb{V}$ (and is finite) then $(x_u)$ is strongly almost convergent.*

If $\eta > 0$, we call $B \subset C$ *$\eta$-covered* by sets $A_i$ if the union $A$ of the sets $A_i$ is contained in $B$ and card $(A)$/card $(B) \geqq 1 - \eta$. The proof of the following lemma is elementary and is left to the reader:

**Lemma 3.10.** *Let $\{v', v''\}$ be a two point subset of $\mathbb{V}$, then given $\eta > 0$ we can find $u^* \in \mathbb{V}$ such that any $[w, w+u[$ with $u \geqq u^*$ can be $\eta$-covered by disjoint translates $\{v'+v, v''+v\}$ of $\{v', v''\}$.*

*Proof of theorem 3.9.* The assumption implies the convergence of $\|\sum_v \alpha_v x_{v+u}\|$ to a finite limit for any $(\alpha_v)$ with $\sum_v |\alpha_v| < \infty$. Put $y(r, u) = \pi(r)^{-1} \sum_{v \in [u, u+r[} x_v$ for $r, u \in \mathbb{V}$, and $\beta = \liminf_{r \to \infty} \lim_{u \to \infty} \|y(r, u)\|$. Then we have $0 \leqq \beta \leqq \sup\{\|x_v\|: v \in \mathbb{V}\} =: M < \infty$.

We treat the case $0 < \beta < \infty$, the case $\beta = 0$ being similar and simpler. Multiplying the $x_v$ by $\beta^{-1}$ we may assume $\beta = 1$.

Let $\varepsilon > 0$ be given. By the uniform convexity of the norm there exists $\delta > 0$ so that $1 - 2\delta \leqq \|z_i\| \leqq 1 + 2\delta$, $(i = 1, 2)$, and $\|z_1 - z_2\| \geqq \varepsilon$ implies $\|(z_1 - z_2)/2\| < 1 - 2\delta$. We may assume $\delta < \varepsilon/2$.

Fix $r \in \mathbb{V}$ with $1 - \delta < \lim_u \|y(r, u)\| < 1 + \delta$ and then fix $u^\delta$ such that $1 - \delta < \|y(r, u)\| < 1 + \delta$ holds for $u \geqq u^\delta$. If $u^\delta$ is large enough the assumption in the theorem yields the following: if $(\alpha_v)$ is a vector having at most $2\pi(r)$ non zero components and satisfying $\sum|\alpha_v| \leqq 2$, then the length of any two vectors $\sum_v \alpha_v x_{v+u}$ and $\sum_v \alpha_v x_{v+u'}$ with $u, u' \geqq u^\delta$ differs by at most $\delta$.

Let us write $r \circ v = (r_1 v_1, r_2 v_2, \ldots, r_d v_d)$. Assume there exist $v', v'' \in \mathbb{V}$ with $v' \circ r, v'' \circ r \geqq u^\delta$ and $\|y(r, v' \circ r) - y(r, v'' \circ r)\| > 2\varepsilon$. Given any $\eta > 0$, the lemma implies the existence of $u^* \in \mathbb{V}$ such that any $[w, w + u[$ with $u \geqq u^*$ can be $\eta$-covered with disjoint translates of $\{v', v''\}$. If $u^*$ is large enough any $[w, w + u[$ with $u \geqq u^r := u^* \circ r$ is $2\eta$-covered by disjoint translates

$$A_v = [v' \circ r + v, v' \circ r + v + r[ \cup [v'' \circ r + v, v'' \circ r + v + r[$$

of $A_0$. From the choice of $u^\delta$ and $v' \circ r, v'' \circ r \geqq u^\delta$ we obtain the following: The vectors $y(r, v' \circ r + v)$ differ in length by less than $\delta$, and also the vectors $y(r, v' \circ r + v) - y(r, v'' \circ r + v)$ differ in length by less than $\delta$. Hence

$$1 - 2\delta < \|y(r, v' \circ r + v)\|, \|y(r, v'' \circ r + v)\| < 1 + 2\delta,$$

and $\|y(r, v' \circ r + v) - y(r, v'' \circ r + v)\| > \varepsilon$, and therefore

$$\|(y(r, v' \circ r + v) + y(r, v'' \circ \mathrm{r} + \mathrm{v}))/2\| < 1 - 2\delta.$$

Now $y(u, w)$ is the average of the vectors $x_t$ with $t \in [w, w + u[$ and the vector $(y(r, v' \circ \mathrm{r} + v) + y(r, v'' \circ r + v))/2$ is the average over the $x_t$ with $t \in A_v$. As the norms of all $x_t$ are bounded by $M < \infty$ we can achieve $\|y(u, w)\| < 1 - \delta$ for large enough $u$ and all $w$. This contradicts $\beta = 1$. We have proved the inequality $\|y(r, v' \circ r) - y(r, v'' \circ r)\| \leqq 2\varepsilon$ for $v' \circ r, v'' \circ r \geqq u^\delta$.

The remainder of the proof is easy. For large $u$ the interval $[0, u[$ may be covered, up to a small portion, by disjoint intervals $[v \circ r, v \circ r + r[$ with $v \circ r \geqq u^\delta$, and the average over each of these intervals is within $2\varepsilon$ of $y(r, v' \circ r)$. Hence $\|y(u, 0) - y(r, v' \circ r)\| < 3\varepsilon$ for large $u$ and $\|y(u', 0) - y(u, 0)\| < 6\varepsilon$ for large $u, u'$. As $\varepsilon$ was arbitrary, $y(u, 0)$ converges to some $\bar{y}$.

Clearly $\|y(r, v \circ r) - \bar{y}\| < 4\varepsilon$ for $v \circ r \geqq u^\delta$. For large enough $u$ any interval $[w, w + u[$ can be covered arbitrarily well by intervals $[v \circ r, v \circ r + r[$ with $v \circ r \geqq u^\delta$. Hence $\|y(u, w) - \bar{y}\| < 5\varepsilon$ for large $u$. As $w$ was arbitrary the convergence of $y(u, w)$ to $\bar{y}$ is uniform in $w$. □

The desired result on strong almost convergence is now an easy consequence:

**Theorem 3.11.** *If $\mathscr{S}$ as in theorem 3.5 is asymptotically isometric in $\{x\}$, then $T_u x$ is strongly almost convergent.*

*Proof.* Put $x_u = T_u x - p$ with $p \in F$. Then $\|x_u\|$ is decreasing on $\mathbb{V}$ and converges. The assumption implies that $\lim_u \|x_{u+w} - x_u\|^2$ exists uniformly in $w$. Therefore $\langle x_u, x_{u+w}\rangle$ converges uniformly in $w$. By theorem 3.9, $x_u$ is strongly almost convergent. □

*Odd T.* Let us mention at least one condition implying that $\mathscr{S}$ is asymptotically isometric in $C$. $T$ is called *odd* if $C = -C$ and $T(-x) = -Tx$ for $x \in C$. $\mathscr{S}$ is called odd if all $T_u$ are odd. In this case we find $\|T_u x + T_u y\|^2 = \|T_u x - T_u(-y)\|^2 \leqq \|x+y\|^2$. Because of $T0 = 0$, $\|T_u x\|$ is a decreasing function of $u$. The inequalities $\|T_{u+v}x \pm T_{u+v+w}y\|^2 \leqq \|T_u x \pm T_{u+w}y\|^2$ yield

$$\begin{aligned} &2|\langle T_u x, T_{u+w}y\rangle - \langle T_{u+v}x, T_{u+v+w}y\rangle| \\ &\quad \leqq (\|T_u x\|^2 - \|T_{u+v}x\|^2) + (\|T_{u+w}y\|^2 - \|T_{u+v+w}y\|^2). \end{aligned}$$

As $\|T_u x\|^2$ and $\|T_u y\|^2$ are decreasing functions of $u$ both summands are $< \varepsilon$ for large enough $u$ and all $v, w$. We obtain

$$\begin{aligned} 0 &\leqq \|T_u x - T_{u+w}y\|^2 - \|T_{u+v}x - T_{u+v+w}y\|^2 \\ &\leqq (\|T_u x\|^2 - \|T_{u+v}x\|^2) + (\|T_{u+w}y\|^2 - \|T_{u+v+w}y\|^2) + 2\varepsilon < 4\varepsilon \end{aligned}$$

for large $u$ and all $v, w$. This means that $\|T_u x - T_{u+w}y\|$ converges uniformly in $w$.

**4. Continuous parameter semigroups.** As in the linear case ergodic theorems for continuous parameter semigroups can be deduced from their discrete counterparts, but the argument is not the same. We give it only in the one parameter case although it extends easily to the case $d > 1$. Let $\mathfrak{Q}_{\mathbb{R}}$ be the family of kernels $Q = \{Q(s,t)\colon t, s \geqq 0\}$ with $Q(s,t) \geqq 0$ and $\int_0^\infty Q(s,t)\,dt = 1$, such that $Q(s,\cdot)$ has finite total variation $V(s)$ for each $s$, $\lim_{s\to\infty} V(s) = 0$, and

$$\lim_{s\to\infty} \int_0^T Q(s,t)\,dt = 0 \quad \text{for all } T < \infty.$$

A semigroup $\{T_t\colon t \geqq 0\}$ is called *strongly continuous*, if $\|T_t x - T_{t_0}x\| \to 0$, $(t \to t_0)$, holds for all $t_0 \geqq 0$ and all $x$. Then we can define the averages $A_s^Q x = \int_0^\infty Q(s,t)\,T_t x\,dt$ by approximation with Riemann sums. For $Q(s,t) = s^{-1} 1_{[0,s[}(t)$ we obtain the usual averages $s^{-1}\int_0^s T_t x\,dt$.

**Theorem 3.12.** *Let $C$ be a closed convex subset of a Hilbert space $\mathfrak{H}$, and $\{T_t, t \geqq 0\}$ a strongly continuous semigroup of nonexpansive mappings $T_t\colon C \to C$. Assume $F := \{p \in C\colon T_t p = p$ for all $t \geqq 0\} \neq \emptyset$. Then, for $Q \in \mathfrak{Q}_{\mathbb{R}}$, $x \in C$, $A_s^Q x$ converges weakly as $s \to \infty$.*

*Proof.* We follow Reich [1977]. Consider any sequence $s_n \to \infty$. For $h > 0$ put

$$q_{nk} = q_{nk}(h) = \int_{kh}^{(k+1)h} Q(s_n, t)\,dt.$$

Clearly $q_{nk} \geqq 0$ and $\sum_{k=0}^{\infty} q_{nk} = 1$. Thus the estimate

$$\sum_{k=0}^{\infty} |a_{n,k+1} - a_{n,k}| \leqq \int_0^{\infty} |Q(s_n, t+h) - Q(s_n, t)| \leqq hV(s_n)$$

shows $(q_{nk}) \in \mathfrak{Q}$. Applying theorem 3.5′ to the maps $T_k^{(h)} = T_{kh}$, $(k \geqq 0)$, we find that $A_n^{(h)} x := \sum_{k=0}^{\infty} q_{nk} T_k^{(h)} x$ converges weakly to some $q(h) \in C$.

Put $\alpha(h) = \sup\{\|x - T_t x\|\colon 0 \leqq t \leqq h\}$. Then it is simple to check $\|A_n^{(h)} x - A_{s_n}^Q x\| \leqq \alpha(h)$. This implies $\|q(h_1) - q(h_2)\| \leqq \alpha(h_1) + \alpha(h_2)$. As $\alpha(h)$ tends to 0 for $h \to 0$ we see that $q := \lim_{h \to 0} q(h)$ exists. Now $q := w\text{-}\lim A_{s_n}^Q x$ follows from $A_{s_n}^Q x - q = (A_{s_n}^Q x - A_n^{(h)} x) + (A_n^{(h)} x - q(h)) + (q(h) - q)$. As the sequence $(s_n)$ was arbitrary, the proof is complete. □

The same argument yields an extension of theorem 3.11.

## Notes

Simple proofs of theorem 3.5 and 3.5′ (for $d = 1$) have been given by Brézis and Browder [1976], Pazy [1977], and Bruck [1978]. The limit can be identified as the asymptotic center of $(T_u x)$. (The *asymptotic center* of $(x_n)$ is the element $y^*$ of $\mathfrak{H}$ minimizing $F(y) = \limsup_{n \to \infty} \|x_n - y\|$).

The weak almost convergence was observed by Bruck [1978] and Reich [1978]. Apart from their intrinsic interest, the main motivation of nonlinear ergodic theorems so far seems to be their application in the study of the asymptotic behaviour of solutions of nonlinear evolution equations, see e.g. Reich [1977]. The multiparameter extension is a special case of the results of Brézis and Browder [1977], dealing with even more general semigroups and more general Banach spaces.

The strong convergence of $A_n x = n^{-1}(T^0 + \ldots + T^{n-1})x$ for odd $T$ is due to Baillon [1976], [1976a]. Brézis-Browder [1976], [1977] extended the result, and Bruck [1978] deduced strong convergence for asymptotically isometric $T$ and gave quite a few sufficient conditions for $T$ to be asymptotically isometric. Pazy [1977] proved the strong convergence for nonexpansive $T$ with $F \neq \emptyset$ such that $(I - T)$ maps bounded closed sets into closed sets. (An omitted argument has been added in the lecture notes of Pazy [1980]). Hirano and Takahashi [1979] relaxed the condition of nonexpansiveness of $T$ and proved the weak convergence of $A_n x$ assuming $\|T^i x - T^i y\| \leqq \alpha_i \|x - y\|$ with $\alpha_i \to 1$. They also showed that the weak convergence of $A_n x$ for all $x$ implies $F \neq \emptyset$. Takahashi [1981] proved a nonlinear ergodic theorem for amenable semigroups.

If $C$ is a bounded closed convex subset of a Banach space $\mathfrak{X}$ and $T: C \to C$ nonexpansive, then $T$ need not have a fixed point. E.g. one can take $\mathfrak{X} = C([0,1])$, and $C = \{f \in \mathfrak{X}: 0 = f(0) \leqq f(\omega) \leqq f(1) = 1 \text{ for } \omega \in [0,1]\}$, and $(Tf)(\omega) = \omega \cdot f(\omega)$. Browder [1965], Kirk [1965], and Göhde [1965] proved the existence of a fixed point for uniformly convex $\mathfrak{X}$.

If $T$ has a fixed point it is not certain that $A_n x$ converges to a fixed point, even if $\mathfrak{X}$ has finite dimension. Sine gave the following example: $\mathfrak{X} = \mathbb{R}^3$ with norm $\|(a, b, c)\| = \operatorname{Max}((a^2 + b^2)^{1/2}, |c|)$, $C =$ closed unit ball and $T(a, b, c) := (-b, a, (a^2 + b^2)^{1/2})$. It can be checked that $(0, 0, 0)$ is the only fixed point and $A_n(a, b, c) \to (0, 0, a^2 + b^2)$.

A Banach space $\mathfrak{X}$ is said to have a *Fréchet differentiable norm* if $(\|x - \varepsilon y\| - \|x\|)/\varepsilon$ converges uniformly in $x$ with $\|x\| \leqq 1$ for all $y$, as $\varepsilon \to 0$.

Reich [1979] proved

**Theorem 3.13.** *Let $C$ be a closed convex subset of a uniformly convex Banach space $\mathfrak{X}$ which has a Fréchet differentiable norm. For nonexpansive $T: C \to C$ the following conditions are equivalent:*

*(a)* $F \neq \emptyset$;

*(b)* $\{T^n x, n \geqq 0\}$ *is bounded for each* $x \in \mathfrak{X}$;

*(c)* $A_n T^i x$ *converges weakly to some* $y \in C$ *uniformly in* $i = 0, 1, \ldots$

A new proof of this theorem has been given by Hirano [1980]. Baillon [1978] proved the weak convergence of $A_n x$ for $\mathfrak{X} = L_p$, $(1 < p < \infty)$. Brézis and Browder [1977] considered semigroups and general averages under supplementary conditions on $\mathfrak{X}$. Bruck [1979] has simplified Baillons proof for $\mathfrak{X} = L_p$ and extended it to uniformly rotund $\mathfrak{X}$. Reich [1983] proved a continuous time nonlinear ergodic theorem in Banach spaces assuming only strong measurability of the semigroup; see also Pazy [1979].

Rodé [1982] considers a semigroup $G$ with asymptotically invariant mean $\{\mu_\alpha\}$ and the averages $T_{\mu_\alpha}$ of a semigroup $\{T_g, g \in G\}$ of nonexpansive mappings in Hilbert space.

Hirano [1982] has a nonlinear ergodic theorem in which the condition of Fréchet differentiability of the norm is replaced by *Opial's condition*: $w\text{-}\lim x_n = x_0$ implies $\liminf \|x_n - x_0\| < \liminf \|x_n - x\|$ for all $x \neq x_0$.

Call an operator $T$ in an ordered Banach space *convex* if $T(\alpha x + (1 - \alpha) y) \leqq \alpha Tx + (1 - \alpha) Ty$ holds for all $x, y$ and $0 \leqq \alpha \leqq 1$. Rodé [1983] showed: if $T$ is nonexpansive, convex and monotonic and if $\{n^{-1} T^n 0, n \geqq 1\}$ is relatively compact, then $\lim n^{-1} T^n 0$ exists.

Djafari-Rouhani and Kakutani [1984] have recently supplied an elegant short proof of the following result containing Baillon's ergodic theorems: Let $(x_n)$ be a bounded sequence in Hilbert space and $a_n = n^{-1}(x_1 + \ldots + x_n)$. If $\|x_{i+1} - x_{j+1}\| \leqq \|x_i - x_j\|$ holds for $i, j = 0, 1, 2, \ldots$ then $(a_n)$ converges weakly. If, in addition, $\|x_{i+1} + x_{j+1}\| \leqq \|x_i + x_j\|$ holds for $i, j = 0, 1, \ldots$ then $a_n$ converges in norm.

Pazy [1979] and Baillon [1981] have written expository articles and Takahashi [1980] has a survey on fixed points. Further related references are Bruck [1981], Baillon-Bruck-Reich [1978], Kobayasi and Miyadera [1980], Miyadera [1978/79], Pazy [1978], and Gutman-Pazy [1983].

# § 9.4 Miscellanea

Here we list some further topics related to our main theme. Attempting to limit the size of the book we just provide a guide to the literature.

**1. Random sets.** Let $d \geqq 1$ be a fixed integer and let $\mathfrak{C}$ be the family of non empty compact subsets of $E = \mathbb{R}^d$. co $\mathfrak{C}$ denotes the family of non empty convex compact subsets of $E$, $\|\cdot\|$ the Euclidean norm, and $B_1$ the closed unit ball in $E$. On $\mathfrak{C}$ the *Hausdorff metric* $\varrho$ is defined by

$$\varrho(C, D) = \inf\{\varepsilon > 0: C \subset D + \varepsilon B_1 \quad \text{and} \quad D \subset C + \varepsilon B_1\}.$$

We also write $\|C\| = \sup\{\|c\|: c \in C\}$. A process $X = \{X_{i,k}: 0 \leqq i < k < \infty\}$ of $\mathfrak{C}$-valued random variables is called *subadditive* iff $i < k < l$ implies $X_{i,l}(\omega) \subset X_{i,k}(\omega) + X_{k,l}(\omega)$ everywhere. $X$ is called stationary if the joint distribution of $X$ is that of $X'$ with $X'_{i,k} = X_{i+1,k+1}$. The case of sums is the case $X_{i,k} = \sum_{j=i}^{k-1} Y_j$ where $(Y_0, Y_1, \ldots)$ is a stationary sequence of $\mathfrak{C}$-valued random variables. Schürger [1983] proved: If $X$ is a stationary and subadditive process of co $\mathfrak{C}$-valued random variables with $E(\|X_{0,1}\|) < \infty$, then $n^{-1} X_{0,n}$ converges a.e. in the Hausdorff metric.

Artstein and Vitale [1975] proved this theorem in the case of i.i.d. sums even for $\mathfrak{C}$-valued random variables. They showed that for sums the $\mathfrak{C}$-valued case can be reduced to the co $\mathfrak{C}$-valued case. Schürger showed that there exist subadditive $\mathfrak{C}$-valued processes for which $n^{-1} X_{0,n}$ fails to converge. The case of general stationary sums is due to Hess [1979], who considered also generalizations in real separable Fréchet spaces. Schürger [1983], [1984] has also "superstationary" and "almost subadditive" generalizations.

**2. Empirical distribution functions.** The Glivenko-Cantelli theorem holds, of course, also for ergodic real valued stationary processes $X_1, X_2, \ldots$. One just has to replace the strong law by the Birkhoff theorem in the proof. The assumption of ergodicity can be eliminated if one takes the conditional distribution function of $X_1$, given the $\sigma$-algebra $\mathscr{I}$ of invariant events, for the limit. Let $F_n(x, \omega) = n^{-1} \sum_{i=1}^{n} 1_{[-\infty, x]}(X_i(\omega))$ be the value of the empirical distribution function at $x$ after $n$ observations. Tucker [1959] proved: If $X_1, X_2, \ldots$ is stationary, then

$$\sup_{-\infty < x < \infty} |F_n(x, \omega) - E(1_{[-\infty, x]} \circ X_1 | \mathscr{I})(\omega)|$$

tends to 0 a.e.. Stute and Schumann [1980] extended this to stationary processes with values in $\mathbb{R}^d$, admitting also other uniformity classes than the class of intervals $[-\infty, x]$ with $x \in \mathbb{R}^d$.

**3. Martingales and ergodic theorems.** There are rather striking similarities between the proofs of pointwise ergodic theorems and martingale theorems. This has led to various attempts to unify martingale theory and ergodic theorems. For a survey see M.M. Rao [1973].

Jerison [1959] represented ergodic averages $A_n(\tau)f$ in a probability space via semi-integrable martingales in a suitable $\sigma$-finite measure space.

Rota [1961] obtained a generalization of the martingale theorem and the ergodic theorem for measure preserving flows $\{\tau_t, t \geqq 0\}$ and $f \in L_p$, $(1 < p < \infty)$, by considering *Reynolds operators* $R$. These are continuous linear operators in $L_p$ with $R(fg) = (Rf)(Rg) + R[(f - Rf)(g - Rg)]$ for $f, g \in L_p \cap L_\infty$. He proved: If $\{R_t: t > 0\}$ is a commuting family of positive contractions in $L_1$ of a probability space which are Reynolds operators and which satisfy $(sR_t - tR_s)R_s h = (s - t)R_t R_s^2 h$ for $s > t$ and $h \in L_\infty$, then $\lim_{t \to \infty} R_t f$ exists for $f \in L_p$, $(1 < p < \infty)$. He also showed that there are basically two kinds of Reynolds operators: conditional expectations and those of the form $R = (I - D)^{-1}$ where $D$ is the infinitesimal generator of a semiflow. For related work see Rota [1964], Bray [1969].

The above two approaches are restricted to measure preserving transformations. A. and C. Ionescu Tulcea [1963] have proved a maximal inequality (Theorem 4.2.6) which allows to deduce the ergodic theorem for $L_1 - L_\infty$-contractions and the martingale theorem even in the vector valued case.

Jacobs [1960] and Neveu [1965a] have given proofs of the martingale theorem using the method of proof of ergodic theorems, and Neveu showed that the decreasing martingale theorem can be considered as a limit case of an ergodic theorem of A. and C. Ionescu Tulcea.

**4. Non homogeneous Markov chains.** As the concept of an "ergodic theorem" proposed in the introduction assumes the existence of a semigroup of transformations, the theory of non homogeneous chains need not be treated here. The books of Seneta [1973] and Isaacson-Madsen [1976] treat this topic and give references. The papers of Cohn [1977], [1979] and Taylor [1978] are more recent references which came to my attention. Jacobs [1957a], [1958], [1960] has studied Markov processes with almost periodic transition probabilities.

**5. Ergodic theorems in demography.** These theorems deal with asymptotic properties of products of non negative matrices. A survey has been given by J. Cohen [1979]. In the basic model one considers a column vector $Y(t)$ with non negative components $Y_i(t)$, $(i = 1, \ldots, k)$, representing the number of females in age class $i$ at time $t \in \mathbb{Z}^+$. One assumes $Y(t+1) = x(t+1)\,Y(t)$, where $x(t)$ is a non negative matrix with $x_{i,l} = 0$ for $i \geqq 2$, $l \geqq i$. The "strong ergodic theorem" assumes $x(t)$ constant and studies the population size $\|Y(t)\| = |Y_1(t)| + \ldots + |Y_k(t)|$ and the age distribution $Y(t)/\|Y(t)\|$ via the Perron-Frobenius theory. The "weak ergodic theorem" assumes a deterministic inhomogeneous sequence $x(t)$ and shows how the age distribution of different initial populations becomes more and more similar. There are also models where the $x(t)$ are chosen at random. For the Markov chain case see Lange [1979].

Brillinger [1981] has surveyed tools and models in population mathematics.

**6. Nonlinear averages.** Convergence problems for one-dimensional infinite particle systems motivated work by Fichtner-Freudenberg [1979] on nonlinear averages:

Let $Y = (Y_i)_{i \in \mathbb{Z}}$ be a stationary ergodic sequence of non negative random variables with finite expectation and let $(f_n)$ be a sequence of functions mapping $\Omega = E^{\mathbb{Z}}$ (with $E = [0, \infty]$) into $\mathbb{R}$. One is interested in conditions on $(f_n)$ implying $f_n(Y) \to E Y_0$ in $L_1$ or at least in the weak topology. The Cesàro averages correspond to $f_n(y) = n^{-1}(y_0 + y_1 + \ldots + y_{n-1})$. But there is the need to consider nonlinear functions.

Let $\mathscr{M}$ be the set of all probability measures $\mu$ which are invariant under the shift $\theta$ and satisfy $\mu(]0, \infty[^{\mathbb{Z}}) = 1$ and $\int X_0 \, d\mu < \infty$. ($X_j(\omega) = \omega_j$ is the $j$-th coordinate of $\omega$). Let $\mathscr{F}$ be the set of measurable maps $f\colon \Omega \to E$ which are non decreasing in each coordinate and satisfy $\int f \, d\mu = \int X_0 \, d\mu$ for all $\mu \in \mathscr{M}$. E.g., the function $f^*$ with $f^*(\omega) = \limsup n^{-1}(\omega_0 + \ldots + \omega_{n-1})$ belongs to $\mathscr{F}$. If $h$ is differentiable with $0 \leqq dh(t)/dt \leqq 1$, then $f^h(\omega) = \omega_0 + h(\omega_1) - h(\omega_0)$ belongs to $\mathscr{F}$.

Call $(f_n) \subset \mathscr{F}$ weakly asymptotically $\theta$-invariant if $w\text{-}\lim(f_n - f_n \circ \theta) = 0$ holds in $L_1(\mu)$ for all $\mu \in \mathscr{M}$. Fichtner-Freudenberg show that this is equivalent to $w\text{-}\lim f_n = f^*$ in $L_1(\mu)$ for all $\mu \in \mathscr{M}$. They give sufficient conditions for the norm-convergence in $L_1(\mu)$ of $f_n$ to $f^*$.

# Supplement
# Harris Processes, Special Functions, Zero-two law

*Antoine Brunel*

This supplement deals with an important class of quasi-compact operators arising in the theory of Markov processes.

**1. Taboo potentials, integral kernels.** Let $(\Omega, \mathscr{A}, \mu)$ be a $\sigma$-finite measure space, and let $P$: $(\omega, A) \to P(\omega, A)$ be a stochastic kernel. We also denote by $P$ the operator in the space $b\mathscr{A}$ of bounded $\mathscr{A}$-measurable functions given by formula (3.1.1). $H$ denotes the space of $\mathscr{A}$-measurable functions $h$: $\Omega \to [0,1]$ with $\mu(h) := \int h\, d\mu > 0$. $I_h$ shall be the operator of multiplication by $h$.

**Definition 1.** The *taboo potential* $U_h$ associated with $h \in H$ is the operator given by

$$U_h = \sum_{n=0}^{\infty} (PI_{\mathbb{1}-h})^n P = \sum_{n=0}^{\infty} P(I_{\mathbb{1}-h}P)^n,$$

where $(I_{\mathbb{1}-h}P)^0 = I$.

Note that $U_h f$ is well defined as a measurable function with values in $[0, \infty]$ when $f$ is non negative and measurable. For $A \in \mathscr{A}$, we put $U_A = U_{1_A}$.

The Markov process associated with the stochastic kernel $P$ is called *recurrent in the sense of Harris* if the following condition $(H')$ is satisfied:

(H') $\quad U_A 1_A = \mathbb{1}$ for all $A \in \mathscr{A}$ with $\mu(A) > 0$.

As $(PI_{A^c})^n P1_A(\omega)$ is the probability that the first visit to $A$, given a start in $\omega$, happens at time $n+1$, the condition $(H')$ means that the probability of a visit in $A$ at some time $m \geqq 1$, given a start in $\omega$, is 1 for *all* $\omega \in \Omega$ and all $A$ having positive measure.

The taboo potentials $U_h$ satisfy a fundamental relation, called the *resolvent equation*:

**Proposition 2.** *For* $h, k \in H$ *with* $h \leqq k$ *the following equalities hold*

(R) $\quad U_h = U_k + U_h I_{k-h} U_k = U_k + U_k I_{k-h} U_h.$

*Proof.* Let us write $I_{1-h}P = X$, $I_{1-k}P = Y$ and $X - Y = I_{k-h}P = Z$. We denote the formal series $\sum_{n=0}^{\infty} t^n = 1 + t + t^2 + \dots$ by $G(t)$.

Then we can write

$$U_h = P \sum_{n=0}^{\infty} (Y+Z)^n = PG(Y+Z), \quad U_k = PG(Y),$$

and it suffices to verify the identity

$$G(Y+Z) = G(Y) + G(Y+Z)\, ZG(Y),$$

in which $Y$ and $Z$ are positive operators. For this it is enough to verify the identity

$$G(u+v) = G(u) + G(u+v)vG(u)$$

for formal series with non commuting variables $u$, $v$. Starting from the relations

$$G(u+v)(1-u-v) = 1 = (1-u)\, G(u)$$

it suffices to write

$$1 - u = (1-u-v) + v,$$

and to multiply this identity from the left with $G(u+v)$ and from the right with $G(u)$. □

We now derive some immediate consequences of the resolvent equation. The identity $U_h h = \mathbb{1}$ implies $U_k k = \mathbb{1}$ for all $k \in H$ with $k \geqq h$. Indeed, $U_k h \leqq U_h h = \mathbb{1}$ and (R) yield

$$\mathbb{1} = U_h h = U_k h + U_k I_{k-h} U_h h = U_k h + U_k(k-h) = U_k k.$$

For $A$ with $\mu(A) > 0$ and $0 < \theta < 1$ the resolvent equation implies

$$U_{\theta 1_A} 1_A = U_A 1_A + U_{\theta 1_A}(1-\theta) I_{1_A} U_A 1_A,$$

and if $U_A 1_A = \mathbb{1}$, it follows that

$$U_{\theta 1_A} 1_A = \mathbb{1} + (1-\theta)\, U_{\theta 1_A} 1_A.$$

This leads to $U_{\theta 1_A}(\theta 1_A) = \mathbb{1}$. Now, for $h \in H$ there exist $A \in \mathscr{A}$ with $\mu(A) > 0$ and $\theta$ with $0 < \theta < 1$ and $\theta 1_A \leqq h$. The preceding argument then proves $U_h h = \mathbb{1}$. Thus the condition

(H) $\quad U_h h = \mathbb{1} \quad$ for all $h \in H$

is equivalent to $(H')$ and we assume in the sequel that it is satisfied.

The definitions and constructions just given for a kernel extend easily to the case where $P$ is the adjoint $T^*$ of a positive contraction $T$ in $L_1(\Omega, \mathscr{A}, \mu)$. Thus $P$ acts in $L_\infty(\Omega, \mathscr{A}, \mu)$. Let us add the following remark: A kernel $P$ acts on the space of bounded measures by the formula $\nu P(A) = \int P(x, A)\, d\nu(x)$, and for $f \in b\mathscr{A}$ one has

$$\langle \nu P, f\rangle = \langle \nu, Pf\rangle = \int Pf(x)\, d\nu(x);$$

in the case $P = T^*$ we can similarly regard $P$ as an operator acting from the left on measures $\nu$ of the form $\nu = g\mu$ with $g \in L_1$ such that $\nu P = (Tg)\mu$ and

$$\langle \nu P, f\rangle = \langle Tg, f\rangle = \langle g, T^* f\rangle = \langle \nu, Pf\rangle = \int Pf(x)\, g(x)\, d\mu(x).$$

With these conventions the following material can be read either for kernels, or for operators when the bounded measures are replaced by $L_1 \cdot \mu$ and the space $b\mathscr{A}$ by $L_\infty$. It is also clear that proposition 2 applies to both kinds of operators as its proof was purely formal.

Taking $h = 0$ in (R) we obtain

$$U_0 = U_k + U_0 I_k U_k$$

and hence

$$U_0 k = \mathbb{1} + U_0 k.$$

This yields $U_0 k \equiv +\infty$ for all $k \in H$. If $P$ is the adjoint of a contraction $T$ in $L_1$, the last relation shows that $T$ is conservative and ergodic.

For $A \in \mathscr{A}$ with $\mu(A) > 0$ the operator $I_A U_A I_A$ is the *operator induced by P on A*. It is also useful to consider the *balayage operator*

$$R_h = \sum_{n=0}^{\infty} (I_{\mathbb{1}-h} P)^n$$

and to put $R_A = R_{1_A}$. For given $A$ and for $E \subset A$ with $\mu(E) > 0$ we put

$$P' = I_A U_A I_A \quad \text{and} \quad R'_E = I_A \sum_{n=0}^{\infty} (I_{A \setminus E} P')^n I_A.$$

The identity $R'_E = I_A R_E I_A$ may be verified in a formal way just like the resolvent equation. Hence, the identity $R'_E 1_E = 1_A$ yields $U'_E 1_E = P' R'_E 1_E = 1_A$. This relation expresses the Harris condition for the operator $P'$ acting on $b(\mathscr{A} \cap A)$, resp. $L_\infty(A)$. In particular it shows that the induced operator is again conservative and ergodic.

It is time now to emphasize an essential difference between the case of an abstract contraction in $L_1$ and the kernel case. In the latter case the Harris condition implies the irreducibility of the Markov chain with respect to the measure $\mu$. In other words one has, for all $\omega \in \Omega$, the absolute continuity $\mu \ll U_0(\omega, \cdot)$; this condition can also be written as $\mu \ll U_\theta(\omega, \cdot)$ for all $\omega \in \Omega$ and some $\theta \in\, ]0, 1[$. Indeed, if $U_0(\omega, A) = U_0 1_A(\omega) = 0$, then one has also $U_A 1_A(\omega) = 0$ because of $U_A \leqq U_0$.

On the other hand, we can define the notion of abstract Harris operator for a positive contraction in $L_1$ following ideas of Foguel [1979] and Foguel and Ghoussoub [1979]. To this end we recall a definition:

**Definition 3.** A positive linear operators $K: L_\infty \to L_\infty$ is called *integral kernel* (for the density $k$) if it is of the form

$$Kf(x) = \int k(x, y) f(y)\, d\mu(y),$$

where $k: \Omega \times \Omega \to \mathbb{R}^+$ is a positive $\mathscr{A} \otimes \mathscr{A}$-measurable function with $\sup_x \{\int k(x, y)\, d\mu(y)\} < \infty$. We say that $K$ is *non trivial* if $K\mathbb{1} \neq 0$.

In the special case $k(x, y) = a(x)\, b(y)$ we shall write $K = a \otimes b\mu$. Then $Kf = (\int f(y)\, b(y)\, d\mu(y))\, a$.

Using this, the abstract notion will be given by

**Definition 4.** A positive contraction $T$ in $L_1$ and its adjoint $T^* = P$ are called *Harris operators* if $(H)$ holds and there exists an $n \in \mathbb{N}$ such that $P^n$ dominates a non trivial integral kernel.

(Note, that $(H)$ holds for $P$ if and only if $T$ is conservative and ergodic).

We shall need an elementary property of integral kernels expressed by

**Lemma 5.** *Each positive linear operator in $L_\infty$ dominated by an integral kernel is itself an integral kernel.*

*Proof.* Let $Q$ and $K$ be positive linear operators in $L_\infty$, and assume that $K$ is an integral kernel dominating $Q$. We associate with $Q$ a finitely additive measure $\tilde{Q}$ on the algebra in $\Omega \times \Omega$ generated by $\mathscr{A} \times \mathscr{A}$. $\tilde{Q}$ is determined by

$$\tilde{Q}(A \times B) = \int_A Q 1_B(x)\, d\mu(x), \quad (A, B \in \mathscr{A}).$$

In the same way $K$ determines $\tilde{K}$, and clearly $\tilde{Q} \leq \tilde{K}$. But $\tilde{K}$ extends to a measure on the $\sigma$-algebra $\mathscr{A} \otimes \mathscr{A}$: For $B \in \mathscr{A} \otimes \mathscr{A}$ one has

$$\tilde{K}(B) = \int_B k(x, y)\, d\mu(x)\, d\mu(y).$$

Consequently, $\tilde{Q}$ extends to a measure on $\mathscr{A} \otimes \mathscr{A}$, too. As this measure is absolutely continuous with respect to $\tilde{K}$ it is absolutely continuous with respect to $\mu \otimes \mu$. It has a density dominated by $k$. □

From this we easily deduce

**Proposition 6.** *Assume $0 < \theta < 1$. A conservative and ergodic $P = T^*$ is a Harris operator if and only if $U_0$ dominates a non trivial integral kernel.*

*Proof.* We apply the inequality $(A+B) \wedge C \leqq A \wedge C + B \wedge C$, which holds for positive operators $A, B, C$, because $(A+B) \wedge C \leqq (A \wedge C + B) \wedge C \leqq A \wedge C + B \wedge C$.

It is sufficient to show that when $K \leqq U_\theta$ is an integral kernel with $P^n \wedge K = 0$ for all $n$, then $K = 0$. Now

$$K = U_\theta \wedge K \leqq \Big(\sum_{n=0}^{N} (1-\theta)^n P^{n+1}\Big) \wedge K + \Big(\sum_{n=N+1}^{\infty} (1-\theta)^n P^{n+1}\Big) \wedge K$$
$$\leqq \Big(\sum_{n=N+1}^{\infty} (1-\theta)^n P^{n+1}\Big) \wedge K.$$

Letting $N \to \infty$, one obtains $K = 0$. □

**2. Special functions and invariant measures.** We now introduce the class of special functions for $P$.

**Definition 7.** A non negative measurable function $f$ is called *special* if $U_h f$ belongs to $L_\infty$ for all $h \in H$. Let $\mathfrak{S}$ denote the family of special functions. A set $A \in \mathscr{A}$ is called special if $1_A \in \mathfrak{S}$.

It is expedient to begin with the case where $\Omega$ is special. Let $0 < \theta < 1$ be given. We shall establish that the assumption $\mathbb{1} \in \mathfrak{S}$ yields, for each $E \in \mathscr{A}$ with $\mu(E) > 0$, the existence of a constant $b(E) > 0$ with $U_\theta 1_E \geqq b(E)$. We can write

$$\mathbb{1} = P\mathbb{1} = P1_E + P1_{E^c} = P1_E + (PI_{E^c})\mathbb{1} = P1_E + (PI_{E^c})(P1_E + (PI_{E^c})\mathbb{1})$$
$$= \sum_{j=0}^{1} (PI_{E^c})^j P1_E + (PI_{E^c})^2 \mathbb{1},$$

and, by recurrence,

$$\mathbb{1} = \sum_{j=0}^{n} (PI_{E^c})^j P1_E + (PI_{E^c})^{n+1} \mathbb{1}.$$

Now the assumption $\mathbb{1} \in \mathfrak{S}$ implies $U_E \mathbb{1} \leqq a$ for some $a \in \mathbb{R}$. Hence

$$U_E \mathbb{1} = \sum_{j=0}^{\infty} (PI_{E^c})^j P\mathbb{1} = \sum_{j=0}^{\infty} (PI_{E^c})^j \mathbb{1} \leqq a.$$

As the sequence $(PI_{E^c})^n \mathbb{1}$ is decreasing, one obtains

$$\|(PI_{E^c})^n\| = \|(PI_{E^c})^n \mathbb{1}\| \leqq \frac{a}{n+1} \downarrow 0.$$

This yields the uniform convergence to $\mathbb{1}$ of the sequence $\sum_{j=0}^{N} (PI_{E^c})^j P1_E$. Therefore there exist $p \in \mathbb{N}$ and $b' > 0$ with

$$\sum_{j=0}^{p} P^j 1_E \geqq \sum_{j=0}^{p} (PI_{E^c})^j 1_E \geqq b'.$$

Apparently this implies the existence of $b(E) \leqq U_\theta 1_E$. We make this more precise in

**Lemma 8.** *If* $\mathbb{1} \in \mathfrak{S}$, *there exists* $0 < \varepsilon < 1$ *such that* $\mu(B) \geqq 1 - \varepsilon$ *implies* $U_\theta 1_B \geqq \varepsilon$.

*Proof.* Otherwise there exists a sequence $(A_j, j = 2, 3, \ldots)$ with $\mu(A_j) \geqq 1 - 2^{-j}$ and $U_\theta 1_{A_j} < 2^{-j}$ on a set $B_j$ with $\mu(B_j) > 0$. The set $E = \bigcap_{j \geqq 2} A_j$ then has measure $\geqq 1/2$ and $U_\theta 1_E \leqq U_\theta 1_{A_j} \leqq 2^{-j}$ on $B_j$. This contradicts $0 < b(E) \leqq U_\theta 1_E$. □

An important consequence of the preceding result is

**Theorem 9.** (Horowitz) *If* $\Omega$ *is special, then* $P$ *is a Harris operator.*

*Proof.* Let $S$ be the operator $\mathbb{1} \otimes \mu$. From § 4.1 we have

$$(U_\theta \wedge S)\mathbb{1} = \inf\{(U_\theta g + \int (\mathbb{1} - g)\, d\mu) \colon 0 \leqq g \leqq \mathbb{1}\}.$$

Fix $g$ and put $B = \{g \geqq 1/2\}$. Then

$$U_\theta g + \int (\mathbb{1} - g)\, d\mu \geqq \tfrac{1}{2} U_\theta 1_B + \int_{B^c} (\mathbb{1} - g)\, d\mu \geqq \tfrac{1}{2} U_\theta 1_B + \tfrac{1}{2}\mu(B^c).$$

If $\mu(B^c) > \varepsilon$, then $U_\theta g + \int (\mathbb{1} - g)\, d\mu \geqq \varepsilon/2$. If $\mu(B^c) \leqq \varepsilon$, then $\mu(B) \geqq 1 - \varepsilon$ and lemma 8 implies $U_\theta 1_B \geqq \varepsilon$. It follows that $U_\theta \wedge S$ is not trivial, and, hence, that $P$ is Harris. □

There is another useful property of integral kernels. Following Foguel [1979] we establish

**Lemma 10.** *Let* $Q$ *be the adjoint of a positive contraction in* $L_1$ *and* $K$ *an integral kernel. Then* $QK$ *and* $KQ$ *are integral kernels.*

*Proof.* The result is clear if $K$ is the kernel $K_0 = \mathbb{1} \otimes \mu$. Another easy case is that where the density $k(.,.)$ is bounded by some $M$, because then $K \leqq MK_0$ and the inequalities $QK \leqq MQK_0$, $KQ \leqq M(K_0 Q)$, together with lemma 5, imply the desired property. In the general case it is enough to truncate the unbounded density putting $k_n = k \wedge n$. Then the associated integral kernels $K_n$ increase and they tend to $K$, and the result follows from $QK_n \uparrow QK$ and $K_n Q \uparrow KQ$. □

We now apply this to the operator $U_\theta$. Putting $\theta' = \sqrt{\theta}$, the resolvent equation shows $U_\theta \geqq (\theta' - \theta) U_{\theta'} U_{\theta'}$. If $U_\theta$ dominates the non trivial integral kernel $K$, one can infer that

$$U_\theta \geqq (\theta' - \theta) K U_{\theta'} = (\theta' - \theta) K_1,$$

where $K_1$ is an integral kernel having a density $k_1$ which is strictly positive $\mu \otimes \mu$ almost everywhere. Indeed, if $\mu(B) > 0$, lemma 8 yields

$$K_1 1_B = K U_{\theta'} 1_B \geqq K\beta > 0.$$

Changing $k_1$ on a $\mu \otimes \mu$-nullset we can assume that $k_1$ is strictly positive everywhere.

A lemma of Harris [1956] now implies that the density $g$ of the kernel $K_1^2$ satisfies an inequality $q(x, y) \geqq a(x) b(y)$, where $a, b$ are two strictly positive measurable functions on $\Omega$. In fact, this can be verified as follows: For any integer $l > 0$ the function

$$x \to M_l(x) = \mu(\{z : k_1(x, z) \geqq 1/l\})$$

is measurable and it satisfies $\lim_{l \to \infty} M_l(x) = 1$. Let $l(x)$ be the smallest integer with $M_l(x) \geqq 3/4$. Put $a(x) = l(x)/2$. Then $a(\cdot)$ is strictly positive and the measurable set $A_x = \{k_1(x, \cdot) \geqq a(x)\}$ has measure $\mu(A_x) \geqq 3/4$. In the same way there exists a measurable function $b(\cdot) > 0$ such that $B_y = \{k_1(\cdot, y) \geqq b(y)\}$ has measure $\mu(B_y) \geqq 3/4$. Now

$$q(x, y) \geqq \int_{A_x \cap B_y} k_1(x, z) k_1(z, y)\, \mu(dz) \geqq 2a(x) b(y)\, \mu(A_x \cap A_y) \geqq a(x) b(y).$$

Now $U_\theta^2 \geqq (\theta' - \theta)^2 K_1^2$ proves that $U_{\theta^2}$ dominates an integral kernel of the form $a \otimes b\mu$. As this is true for all $\theta$ with $0 < \theta < 1$ there exist two strictly positive measurable functions $h, g$ with $U_{\theta'} \geqq h \otimes g\mu$. Consequently

$$U_\theta \geqq (\theta' - \theta)\, U_{\theta'} (h \otimes g\mu) = (\theta' - \theta)\, (U_{\theta'} h) \otimes g\mu.$$

Making use of the assumption $\mathbb{1} \in \mathfrak{S}$, we deduce $U_{\theta'} h \geqq \beta > 0$. This yields

**Proposition 11.** *If $\Omega$ is special there exists for all $\theta$ with $0 < \theta < 1$ a measure $\mu_\theta \sim \mu$ with $U_\theta \geqq \mathbb{1} \otimes \mu_\theta$.*

**Corollary 12.** *If $\Omega$ is special there exists a unique invariant probability measure $\lambda = \lambda P$ equivalent to $\mu$.*

*Proof.* The operator $Q = \theta U_\theta$ is Markovian, and the measure $v = \theta \mu_\theta$ is equivalent to $\mu$. The inequality $Q \geqq \mathbb{1} \otimes v$ implies $Q^n \geqq \mathbb{1} \otimes v$ for all $n \in \mathbb{N}$. It follows that

$$\liminf_n \int Q^n 1_A \, d\mu \geqq v(A) > 0$$

holds for all $A$ with $\mu(A) > 0$. Theorem 3.4.2 yields the existence of a probability measure $\lambda \sim \mu$ with $\lambda Q = \lambda$. Applying the resolvent equation to $U_\theta$ and to $U_{\mathbb{1}} = P$ we obtain

$$U_\theta = P + U_\theta (1 - \theta) P, \quad Q = \theta P + (1 - \theta) Q P,$$

which implies $\lambda = \theta\lambda P + (1-\theta)\lambda P = \lambda P$. The uniqueness is a consequence of the ergodicity. □

We now turn to the more general situation where there exists a special set $A$ but $\Omega$ need not be special. Let $A \in \mathscr{A}$ be a fixed set with $1_A \in \mathfrak{S}$.

Recall that the operator induced by $P$ on $A$ is $P' = I_A U_A I_A$, and that the balayage operator associated with a measurable $E \subset A$ is the operator $R'_E = I_A R_E I_A$.

**Proposition 13.** *If $A \in \mathscr{A}$ is special, the induced operator $P'$ is Harris on $A$.*

*Proof.* It follows from $U'_E = P' R'_E$ and $I_A R'_E I_A = I_A R_E I_A$, that $U'_E 1_A$ belongs to $L_\infty(A)$. The resolvent equation shows that this remains true for $U'_{\theta 1_E} 1_A$ for all $\theta$ with $0<\theta<1$. Hence $U'_h 1_A$ belongs to $L_\infty(A)$ for all $h$ with $0 \leqq h \leqq 1_A$ and $\mu(h)>0$. It remains to apply theorem 9. □

It is now natural to ask the following question: If $P$ is Harris, do there exist special sets, and does there exist an invariant measure? Indeed, we have:

**Theorem 14.** *If the operator $T$ (or $P$) is a Harris operator, there exist non zero special functions, and there exists a $\sigma$-finite invariant measure equivalent to $\mu$. This measure is unique up to a constant positive factor.*

*Proof.* 1) The construction explained just before proposition 11 shows that, for given $\theta$ with $0<\theta<1$, there exist two strictly positive measurable functions $h, g$ with $U_\theta \geqq h \otimes g\mu$. If we put $h_1 = h \wedge \frac{\theta}{2}$, then equation (R) gives

$$U_{h_1} \geqq U_{h_1}(I_{\theta-h_1}) U_\theta \geqq \frac{\theta}{2} U_{h_1} U_\theta \geqq \frac{\theta}{2} U_{h_1}(h_1 \otimes g\mu).$$

Hence $U_{h_1} \geqq \mathbb{1} \otimes \mu_1$, where $\mu_1 \sim \mu$ and $h_1 \in H$ is strictly positive. We shall now show that $h_1$ is special, and, more generally, that each non negative measurable function $f$ with $U_{h_1} f \in L_\infty$ is special. Let $f \geqq 0$ with $U_{h_1} f \leqq a < \infty$ and $h \in H$ with $h \leqq h_1$ be given. Applying (R) and using $(U_{h_1} I_{h_1-h})\mathbb{1} = U_{h_1}(h_1-h) = \mathbb{1} - U_{h_1} h \leqq \mathbb{1} - \mu_1(h)$ we obtain

$$\begin{aligned} U_h f &= \sum_{n=0}^{\infty} (U_{h_1} I_{h_1-h})^n U_{h_1} f \leqq a \sum_{n=0}^{\infty} (U_{h_1} I_{h_1-h})^n \mathbb{1} \\ &\leqq a \sum_{n=0}^{\infty} (1-\mu_1(h))^n = a/\mu_1(h) < \infty. \end{aligned}$$

Thus $U_h f$ is bounded for $h \in H$ with $h \leqq h_1$. If $h$ is arbitrary in $H$, put $h' = h \wedge h_1$. Then $h' \leqq h_1$ yields $U_h f \leqq U_{h'} f \in L_\infty$.

2) For any $\theta > 0$ the set $A = \{h_1 \geqq \theta\}$ is a special set because $1_A \leqq h_1/\theta$. By choosing $\theta$ small we can assume $\mu(A) > 0$. The second assertion now follows from proposition 13, corollary 12 and the following theorem which is of independent interest. □

**Theorem 15.** *Let $P'$ be the induced operator on a subset $A \in \mathscr{A}$. Assume there exists a finite measure $\lambda_A$ with $\lambda_A = \lambda_A P' = \lambda_A I_A$. Then the measure $\lambda = \lambda_A U_A$ satisfies $\lambda P = \lambda$. If $\sum_{n=0}^{\infty} P^n 1_A$ is strictly positive everywhere, $\lambda$ is $\sigma$-finite. If $P = T^*$ satisfies (H), $\mu(A) > 0$ and $\lambda_A \sim \mu I_A$, then $\lambda \sim \mu$, and any other P-invariant $\lambda' \sim \mu$ is a constant multiple of $\lambda$.*

*Proof.* The $P$-invariance follows from

$$\lambda = \lambda_A U_A = \lambda_A P + \lambda_A U_A I_{A^c} P = \lambda I_A P + \lambda I_{A^c} P = \lambda P.$$

To prove the $\sigma$-finiteness of $\lambda$ first check $U_A I_A \mathbb{1} \leqq \mathbb{1}$. Under the stated assumption, $\Omega$ is the union of the sets $B = B_{kn} = \{P^n 1_A > 1/k\}$, and we have

$$\begin{aligned}\lambda(B) &\leqq \langle \lambda, kP^n 1_A\rangle = k\langle \lambda P^n, 1_A\rangle = k\langle \lambda, 1_A\rangle \\ &= k\langle \lambda_A U_A, I_A \mathbb{1}\rangle \leqq k\langle \lambda_A, U_A I_A \mathbb{1}\rangle \leqq k\lambda_A(\Omega) < \infty.\end{aligned}$$

In the second assertion $\lambda \ll \mu$ follows from $P = T^*$ and $\lambda_A \ll \mu$. The ergodicity of $T$ and $\mu(A) > 0$ imply that the $T$-invariant function $f = d\lambda/d\mu$ is positive a.e.. If $f'$ is another non negative $T$-invariant (not necessarily integrable) function, then $f \wedge f' = g$ is sub-invariant. By lemma 3.3.10 and Lemma 3.3.11, $\{g > 0\}$ is $T$-absorbing. Hence $\{g > 0\} = \emptyset$ or $= \Omega$. If there exists a $\sigma$-finite invariant measure $\lambda^*$ which is not a constant multiple of $\lambda$, then the density $f'$ of a suitable multiple of $\lambda^*$ has the property $\emptyset \neq \{g > 0\} \neq \Omega$. □

**3. Zero-two laws and ergodic theorems.** We now establish a result due to Foguel [1979]; see also Foguel-Ghoussoub [1979]. It will lead us to the ergodic theorem for Harris operators. A variant will give us the celebrated zero-two law of Ornstein-Sucheston [1970].

**Lemma 16.** *Let $P, Q_1, Q_2$ be commuting positive linear operators in $L_\infty$ with $P\mathbb{1} = Q_1 \mathbb{1} = Q_2 \mathbb{1} = \mathbb{1}$. If there exist an integer $r$ and a positive operator $R$ with $R\mathbb{1} = \delta > 0$, $P^r \geqq Q_1 R$ and $P^r \geqq Q_2 R$, then* $\lim_m \|(Q_2 - Q_1) P^m\| = 0$.

*Proof.* As the sequence $\|(Q_2 - Q_1) P^m\|$ is decreasing we may assume $r = 1$. Put $Q = (Q_1 + Q_2)/2$. We can write $P = QR + S_1$, where $S_1$ is positive. Inductively, we obtain $P^n = Q^n R^n + S_n$ with $S_n \geqq 0$. Indeed, we have

$$\begin{aligned}P^{n+1} &= P(Q^n R^n + S_n) = Q^n PR^n + PS_n \\ &= Q^{n+1} R^{n+1} + (Q^n S_1 R^n + PS_n) = Q^{n+1} R^{n+1} + S_{n+1},\end{aligned}$$

where $S_{n+1} = Q^n S_1 R^n + PS_n$ is a positive operator. It follows from $P^n \mathbb{1} = \mathbb{1} = Q^n R^n \mathbb{1} + S_n \mathbb{1} = \delta^n + S_n \mathbb{1}$, that $S_n \mathbb{1} = 1 - \delta^n$.

The second step is the proof of

$$P^{nj} = Q^n T_{n,j} + S_n^j \quad \text{with } T_{nj} \geqq 0$$

by induction in $j$. If this has been verified for an integer $j \geqq 1$, we have

$$\begin{aligned} P^{n(j+1)} &= P^n(Q^n T_{n,j} + S_n^j) = Q^n(P^n T_{n,j} + R^n S_n^j) + S_n^{j+1} \\ &= Q^n T_{n,j+1} + S_n^{j+1} \end{aligned}$$

with $T_{n,j+1} = P^n T_{n,j} + R^n S_n^j \geqq 0$, $T_{n,1} = R^n$.

Put $\beta_{nj} = T_{n,j}\mathbb{1}$. By induction, $\beta_{nj}$ is a constant, because $T_{n,1}\mathbb{1} = R^n \mathbb{1} = \delta^n$. We obtain

$$P^{nj}\mathbb{1} = \mathbb{1} = \beta_{n,j} + S_n^j \mathbb{1} = \beta_{n,j} + (1 - \delta^n)^j.$$

Hence $\|T_{n,j}\| = \|T_{nj}\mathbb{1}\| = \beta_{nj} < 1$.

We now look for the limit $L$ of the decreasing sequence $\|(Q_2 - Q_1)P^n\|$. The inequality

$$\|(Q_2 - Q_1)P^{nj}\| \leqq \|(Q_2 - Q_1)Q^n\| + 2(1 - \delta^n)^j$$

shows $L \leqq \|(Q_2 - Q_1)Q^n\|$ since $\lim_j (1 - \delta^n)^j = 0$. But

$$\begin{aligned} \|(Q_2 - Q_1)Q^n\| &= \left\|(Q_2 - Q_1)\left(\frac{Q_1 + Q_2}{2}\right)^n\right\| \\ &\leqq \frac{1}{2^{n-1}} + \frac{1}{2^n}\sum_{k=0}^{n-1}\left|\binom{n}{k} - \binom{n}{k+1}\right|. \end{aligned}$$

The sequence $k \to \binom{n}{k}$ is at first increasing and then decreasing with a maximum at $\binom{n}{[n/2]}$. By Stirling's formula, $2^{-n}\binom{n}{[n/2]} = O(1/\sqrt{n})$. Hence

$$L \leqq \lim \|(Q_2 - Q_1)Q^n\| = 0. \quad \square$$

We are now ready to prove

**Theorem 17** (Ornstein-Métivier-Brunel). *For any special set $A$ for $P$ there exists a constant $M_A < \infty$ such that*

$$\left\|\sum_{n=0}^{N} P^n f\right\|_\infty \leqq M_A \|f\|_\infty$$

*holds for all $N$ and for all $f \in L_\infty(A)$ with $\lambda(f) = 0$.*

(Recall that $\lambda$ was a $\sigma$-finite invariant measure, and that $\lambda(A)$ is finite for special sets $A$).

*Proof*. For given $\theta \in ]0,1]$ let $U'_\theta$ denote the taboo potential operator in $L_\infty(A)$ associated with $P' = I_A U_A I_A$. We know from proposition 11 that there exists a measure $\lambda' \sim \lambda I_A$ on $A$ with

$$U'_\theta = \sum_{n=0}^{\infty} (1-\theta)^n (I_A U_A I_A)^{n+1} \geqq 1_A \otimes \lambda' \quad \text{on } L_\infty(A).$$

Let $\lambda_A = \lambda I_A / \lambda(A)$ be the $P'$-invariant probability measure. It is also invariant under $\theta U'_\theta$. We put

$$R = \theta 1_A \otimes \lambda', \quad E = 1_A \otimes \lambda_A.$$

Using the relations

$$R = ER = \theta U'_\theta R, \quad \theta U'_\theta E = E \theta U'_\theta = E = E^2$$

we find that

$$\theta U'_\theta \geqq R = (\theta U'_\theta) R, \quad \theta U'_\theta \geqq ER.$$

As $U'_\theta$ and $E$ commute and $R 1_A = \theta \lambda'(A) > 0$, lemma 16 applies with $P = \theta U'_\theta$, $Q_2 = \theta U'_\theta$, $r = 1$, $Q_1 = E$ and yields

$$\lim \|(\theta U'_\theta - E)(\theta U'_\theta)^n\| = \lim \|(\theta U'_\theta)^n - E\| = 0.$$

$E$ is the zero operator on the subspace $(I_A - E) L_\infty(A)$. Thus, on this subspace we have $\|(\theta U'_\theta)^n\| \to 0$. On the other hand, one has

$$P' - I_A = (P - I)(I_A + I_{A^c} U_A I_A) \quad \text{on } L_\infty(A).$$

(Apply (R) to $U_A$ and $U_0 = \mathrm{P}$).

Now take $f \in L_\infty(A)$ with $\int f d\lambda_A = 0$, then $f \in (I_A - E) L_\infty(A)$. Our considerations show the existence of an integer $n(A)$ for which $I_A - (\theta U'_\theta)^{n(A)}$ is invertible. We deduce the existence of functions $f_1, f_2, f_3 \in L_\infty(A)$ and $F \in L_\infty$ with

$$f = (I_A - (\theta U'_\theta)^{n(A)}) f_1 = (I_A - \theta U'_\theta) f_2 = (I_A - P') f_3 = (I - P) F.$$

For example $F = (I_A + I_{A^c} U_A I_A) f_3$. It is clear that $\|F\| / \|f\|$ is majorized by a constant which depends only on $A$. □

We finish our study of relations between the Harris condition and the existence of special functions with

**Theorem 18.** *If $T$ is conservative and ergodic and there exists a special set $A$ for $P = T^*$ then $T$ is a Harris operator.*

*Proof*. As $P$ is conservative and ergodic and $\mu(A) > 0$ we infer that $\lim_N \sum_{n=0}^{N} P^n 1_A = \infty$ a.e.. By Egorov's theorem there exists a set $B \subset A$ with $\mu(B) > 0$ on which the sequence $\sum_{n=0}^{N} P^n 1_A$ converges uniformly. Let $\lambda_A$ be the probability measure which is invariant under the operator $P'$ induced by $P$ on $A$. We obtain that for any $h \in H$ with $h \leqq 1_A$ there exists a smallest integer $N(h)$ with the property

$$\sum_{n=0}^{N(h)} P^n 1_A \geqq 2 M_A \lambda_A^{-1}(h) 1_B.$$

Applying theorem 17 to the function $f = 1_A - \lambda_A^{-1}(h)h$, which has norm $\|f\|_\infty \leqq \lambda_A^{-1}(h)$ we find

$$\sup_N \|\sum_{n=0}^{N} P^n(1_A - \lambda_A^{-1}(h)h)\| \leqq M_A \lambda_A^{-1}(h).$$

This yields

$$2 M_A \lambda_A^{-1}(h) 1_B \leqq \sum_{n=0}^{N(h)} P^n 1_A \leqq M_A \lambda_A^{-1}(h) + \lambda_A^{-1}(h) \sum_{n=0}^{N(h)} P^n h,$$

and, hence

$$\sum_{n=0}^{N(h)} P^n h \geqq M_A 1_B.$$

Now, for $0 < \theta < 1$, we have

$$(1) \qquad U_\theta h = \sum_{n=0}^{\infty} (1-\theta)^n P^{n+1} h \geqq \sum_{n=0}^{N(h)} (1-\theta)^n P^{n+1} h \geqq (1-\theta)^{N(h)} M_A 1_B.$$

The last inequality will allow us to prove that $U_\theta$ dominates a non trivial integral kernel. Put $K = \mathbb{1} \otimes \lambda_A$. Recall

$$(U_\theta \wedge K)\mathbb{1} = \inf\{U_\theta g + (\int (\mathbb{1} - g)\, d\lambda_A)\mathbb{1} : 0 \leqq g \leqq \mathbb{1}\}.$$

Let $C = A \cap \{g \geqq 1/2\}$. *If* $\lambda_A(C) \leqq 1/2$, then $\int (\mathbb{1} - g)\, d\lambda_A \geqq 1/2$. If $\lambda_A(C) > 1/2$, then

$$(2) \qquad \sup_N \|\sum_{n=0}^{N} P^n(1_A - \lambda_A(C)^{-1} 1_C)\| \leqq M_A \lambda_A^{-1}(C) \leqq 2 M_A.$$

Let $N_1$ be the minimal integer with $\sum_{n=0}^{N_1} P^n 1_A \geqq 4 M_A 1_B$. $N_1$ depends only on $A$ and $B$. (2) now yields $\sum_{n=0}^{N_1} P^n \lambda_A^{-1}(C) 1_C \geqq 2 M_A 1_B$. It follows that $\sum_{n=0}^{N_1} P^n g \geqq M_A 1_B$.

The minimality of $N(g)$ yields $N_1 \geqq N(g)$. Using (1) we obtain $U_\theta g \geqq (1-\theta)^{N_1} M_A 1_B$. Combining both cases we arrive at $(U_\theta \wedge K)\mathbb{1} \geqq \mathrm{Min}(\tfrac{1}{2}\mathbb{1}, (1-\theta)^{N_1} M_A 1_B) \neq 0$. $U_\theta \wedge K$ is an integral kernel because it is dominated by the integral kernel $K$. Thus $U_\theta$ dominates a non trivial integral kernel and $T$ is a Harris operator. □

The properties of uniform convergence considered above lead to a characterization of Harris operators in terms of their quasi-compactness. This notion has been studied in § 2.2. The relation to present material is made precise by

**Proposition 19.** *For a positive contraction $T$ in $L_1$ the following conditions are equivalent*

*(i) $P = T^*$ obeys a uniform ergodic theorem in the form*

$$\lim_n \left\| \frac{1}{n} \sum_{k=0}^{n-1} P^k - \mathbb{1} \otimes \lambda \right\| = 0,$$

*where $\lambda$ is a $P$-invariant probability measure equivalent to $\mu$;*

*(ii) $T$ is conservative and ergodic and $\mathbb{1}$ is special;*

*(iii) $P$ is quasi-compact, conservative and ergodic.*

*Proof.* The key part of the proof is established in § 2.2. The only new point concerns condition (ii). It is therefore sufficient to prove (ii) ⇒ (i) and (iii) ⇒ (ii). (ii) ⇒ (i) is an immediate consequence of theorem 17. (iii) ⇒ (ii): The condition of quasi-compactness yields the existence of a sequence $(Q_n)$ of compact operators of $L_\infty$ into itself with $\lim \| P^n - Q_n \|_{L_\infty} = 0$. Let $L_\infty^0$ be the subspace of $L_\infty$ consisting of the functions with $\int f \, d\lambda = 0$ and let $\pi$ be the operator $I - (\mathbb{1} \otimes \lambda)$ mapping $L_\infty$ into $L_\infty^0$. The inequalities

$$\| P^n - \pi Q_n \|_{L_\infty^0} \leqq \| \pi P^n - \pi Q_n \|_{L_\infty}$$

show $\lim \| P^n - \pi Q_n \|_{L_\infty^0} = 0$, and hence, $(I - P) L_\infty = L_\infty^0$.

As in the proof of theorem 17 the statement $\| P^n \|_{L_\infty^0} \to 0$ implies that $P$ is a Harris operator for which $\Omega$ is special. □

To complete this study it is of interest to prove, in the abstract case, a lemma established by J. Neveu [1972] for Markov chains.

**Lemma 20.** *Let $T$ be a conservative and ergodic positive contraction in $L_1$. Then the inequality*

$$U_{\theta h} 1_A \geqq (1 - \theta)^{U_A h} \quad \text{a.e.}$$

*holds for all $\theta \in \,]0, 1[$, all $A \in \mathscr{A}$ with $\mu(A) > 0$ and all $h \in H$.*

*Proof.* We first observe that $a^{Pf} \leqq P(a^f)$ holds for all $a \in \,]0, 1[$ and $0 \leqq f$. This consequence of a (generalized) Jensen inequality for $P$ can be verified as follows:

There exists a countable family $(u_n, v_n) \in \mathbb{R}^2$ with

$$a^t = \sup_n (u_n t + v_n) \quad \text{for all } t \in \mathbb{R}.$$

(It suffices to consider the tangents to the graph of the function $t \to a^t$ at the points with rational $t$.) If follows that

$$u_n f + v_n \leqq a^f$$

holds (a.e.) for all $n$. Using $P \geqq 0$ and $P\mathbb{1} = \mathbb{1}$ we obtain

$$u_n Pf + v_n \leqq P(a^f).$$

Taking the supremum over $n$ we arrive at $a^{Pf} \leqq P(a^f)$.

It is sufficient for the proof of the lemma to prove $R_{\theta h} 1_A \geqq (1-\theta)^{R_A h}$, and, morevover, it is enough to show

$$1_A \geqq (I - I_{\mathbb{1}-\theta h} P)((1-\theta)^{R_A h}),$$

since $R_{\theta h}(I - I_{\mathbb{1}-\theta h} P) = I$. To this end, it is enough to verify that $(1-\theta)^f = (\mathbb{1} - \theta h) P((1-\theta)^f)$ holds on $A^c$ for $f = R_A h$. But $f = h + PR_A h$ and the inequality $(1-\theta)^h \leqq (\mathbb{1} - \theta h)$ imply

$$(1-\theta)^h (1-\theta)^{PR_A h} \leqq (\mathbb{1} - \theta h)(1-\theta)^{PR_A h} \leqq (\mathbb{1} - \theta h) P((1-\theta)^{R_A h}). \quad \square$$

We now arrive at *a* zero-two law. We follow the proof by Foguel [1971]:

**Theorem 21.** *Let $T$ be a Markovian positive contraction in $L_1$, $P = T^*$, and let $k > 0$ be an integer. If $\| P^{n_0+k} - P^{n_0} \| < 2$ holds for some $n_0$, then*

$$\lim_n \| P^{n+k} - P^n \| = 0.$$

*Proof.* Let $W = P^{n_0+k} \wedge P^{n_0}$. We have

$$W\mathbb{1} = \inf_{0 \leqq g \leqq 1} (P^{n_0+k} g + P^{n_0}(\mathbb{1} - g)).$$

Writing $g = (\mathbb{1} + f)/2$ we find

$$\begin{aligned} W\mathbb{1} &= \inf_{-1 \leqq f \leqq 1} \left[ P^{n_0+k}\left(\frac{\mathbb{1}+f}{2}\right) + P^{n_0}\left(\frac{\mathbb{1}-f}{2}\right)\right] \\ &= \inf_{-1 \leqq f \leqq 1} [\mathbb{1} + \tfrac{1}{2}(P^{n_0+k} f - P^{n_0} f)] \\ &\geqq \mathbb{1} - \tfrac{1}{2} \| P^{n_0+k} - P^{n_0} \| = \delta > 0. \end{aligned}$$

The function $\varphi = \delta / W\mathbb{1}$ therefore belongs to $H$. If $R$ is the operator $I_\varphi W$ we obtain $W \geqq R$ and $R\mathbb{1} = \delta > 0$. Therefore we have the inequalities

$$P^{n_0+k} \geqq P^k R, \quad R\mathbb{1} = \delta > 0, \quad P^{n_0+k} \geqq IR$$

which show that the operators $P$, $Q_1 = P^k$, $Q_2 = I$, $R$ and the integer $r = n_0 - k$ satisfy the assumptions of lemma 16. This yields the desired conclusion. $\square$

The zero-two law of Ornstein-Sucheston is the deeper "pointwise" form of this result. It can be formulated as follows:

**Theorem 22.** *If, for $n = 1, 2, \ldots, P^n = T^{*n}$ are conservative and ergodic Markov operators, then we have either*

$$\text{for all } k = 1, 2, \ldots \lim_n \{ \sup_{-1 \leqq f \leqq 1} (P^{n+k}f - P^n f)\} = 0 \text{ a.e.}$$

*or*

$$\text{for all } k = 1, 2, \ldots \sup_{-1 \leqq f \leqq 1} (P^{n+k}f - P^n f) = 2 \text{ a.e.}.$$

*Proof.* Let $k > 0$ be a fixed integer and put

$$h_n := \sup_{-1 \leqq f \leqq 1} (P^{n+k}f - P^n f),$$

and $W_n = P^{n+k} \wedge P^n$. As above, we have

$$W_n \mathbb{1} = \mathbb{1} - h_n/2.$$

In particular, this shows that $0 \leqq h_n \leqq 2$. The properties $h_{n+1} \leqq h_n$ and $h_{n+1} \leqq Ph_n$ are immediate. If $g := \lim_n h_n$, we obtain $g \leqq Pg$. Hence, $g = \text{constant} = a$ with $0 \leqq a \leqq 2$. We have to establish $a \notin ]0, 2[$.

Assume $0 < a < 2$ holds. Then $W_n \mathbb{1}$ increases to $1 - a/2 > 0$. Clearly, $P^n \geqq W_n$, $P^{n+k} \geqq W_n$ and $P^{n+k} \geqq W_n P^k$. Hence $P^{n+k} \geqq W_n M$ holds for $M = (I + P^k)/2$.

If $n_1, n_2, \ldots, n_r$ is any finite sequence of integers, and $m := n_1 + n_2 + \ldots + n_r + kr$, we obtain $P^m \geqq W_{n_1} \ldots W_{n_r} M^r$. We can exploit the fact that

$$\|(I - P) M^r\| = O(1/\sqrt{r}),$$

and choose $r$ so large that $\|(I - P^k) M^r| < \varepsilon/2$, where $\varepsilon > 0$ is arbitrarily small. We can then choose $n_1, \ldots, n_r$ in such a way that, for $T_1 := W_{n_1} \cdot \ldots W_{n_r}$, one has $T_1 \mathbb{1} \neq 0$. Now $P^m = T_1 M^r + U$, where $U$ is a positive operator. By recurrence in $j$ we find that $P^{mj}$ is of the form $T_j M^r + U^j$, where $T_j$ is a positive operator. It follows from $P^{jm} \mathbb{1} = \mathbb{1} = T_j \mathbb{1} + U^j \mathbb{1}$ that $\mathbb{1} \geqq U\mathbb{1}$, and, hence, that

$$\mathbb{1} \geqq U\mathbb{1} \geqq \ldots U^j \mathbb{1} \quad \text{and} \quad 0 \neq T_1 \mathbb{1} \leqq T_j \mathbb{1}.$$

If $h := \lim_j U^j \mathbb{1}$, then $h = Uh \leqq P^m h$, which yields $h = \text{constant}$. If $h \neq 0$, then $\mathbb{1} = U\mathbb{1}$. But this contradicts to $T_j \mathbb{1} \neq 0$. Hence $U^j \mathbb{1}$ decreases to 0 a.e..

For $f \in L_\infty$ with $-\mathbb{1} \leqq f \leqq \mathbb{1}$ we can write

$$P^{jm+k}f - P^{jm}f = T_j(P^k - I) M^r f + U^j (P^k - I) f.$$

This yields

$$h_{jm} = \sup_{-1 \leqq f \leqq 1} (P^{jm+k}f - P^{jm}f) < \varepsilon/2 + 2U^j \mathbb{1}.$$

We obtain $\limsup_j h_{jm} = \lim h_n = a \leqq \varepsilon/2$. As $\varepsilon$ was arbitrarily small, this contradicts to $a > 0$. □

**4. Eigenvalues and quasi-compactness.** To complete the study of the strong ergodic theorem (Proposition 18), we shall show directly how the spectral properties of the operator $P = T^*$ are related to the quasi-compactness when $\Omega$ is a special set. This proof is based on a lemma about compactness in $L_1$ due to A. Brunel, and it has been developed by Brunel and Revuz [1974a].

**Lemma 23.** *If $(f_n)$ is a sequence which is norm-bounded in $L_1$ of a probability space $(\Omega, \mathscr{A}, \lambda)$ and which satisfies $\lim_{n,p\to\infty} \| |f_n - f_p| - \|f_n - f_p\|_1 \|_1 = 0$, then $(f_n)$ contains a Cauchy sequence.*

*Proof.* The double sequence $U_{n,p} = \int |f_n - f_p|$ is bounded. The two-dimensional Bolzano-Weierstraß theorem (see Sucheston [1959], lemma 3, p. 390) therefore yields the existence of an increasing sequence $(n_j)$ of integers such that $a = \lim_{k>j\to\infty} U_{n_j n_k}$ exists uniformly in $k > j$. Put $g_j = f_{n_j}$.

We shall prove $a = 0$. If this fails, there exists, for all $\varepsilon > 0$, an integer $M$ such that

$$\forall (k,l) \in \mathbb{N}^2,\ M < k < l \Rightarrow \int ||g_k - g_l| - a|\, d\lambda < \varepsilon^2 .$$

For such a pair $(k, l)$ we have

$$\lambda(||g_k - g_l| - a| > \varepsilon) < \varepsilon ,$$

and one can choose four integers $j_1 < j_2 < j_3 < j_4$ such that the six sets $\{|g_{j_r} - g_{j_s}| - a| \leqq \varepsilon\}$ have $\lambda$-measure $\geqq 1 - \varepsilon$. It follows that there exists at least one point $\omega \in \Omega$ for which the numbers $z_r = g_{j_r}(\omega)$ satisfy the following inequalities:

$$||z_r - z_s| - a| \leqq \varepsilon, \quad (1 \leqq r, s \leqq 4) .$$

For $a \neq 0$ and sufficiently small $\varepsilon > 0$ this yields a contradiction. Thus $(f_{n_j})$ is a Cauchy sequence. □

If $T: L_1 \to L_1$ is conservative and ergodic we have seen that the assumption "$\Omega$ is special" implies the existence of an invariant probability measure $\lambda$. Put $L_1^0 = (I - \mathbb{1} \otimes \lambda) L_1$ and $L_\infty^0 = (I - \mathbb{1} \otimes \lambda) L_\infty$. If $\mathbb{1}$ is special and we take $\theta = 1$ in the proof of theorem 17, then we have $\theta U'_\theta = P' = P$ with $A = \Omega$. The proof then shows that $I - P^n$ is an automorphism of $L_\infty^0$ for some $n$. (By an automorphism we mean a bijective linear continuous operator with continuous inverse.) Then $I - P^n = (I - P)(I + P + \ldots + P^{n-1})$ implies that $I - P$ must be an automorphism of $L_\infty^0$. By duality $I - T$ must be an automorphism of $L_1^0$.

We shall now prove:

**Proposition 24.** *If $I-T$ is an automorphism of $L_1^0$ and $z \in \mathbb{C}$ with $|z|=1$ is no eigenvalue of $T$, then any $L_1$-bounded sequence $(f_n)$ with*

$$\lim_{n,p\to\infty} \|(zI-T)(f_n-f_p)\|_1 = 0 \quad \textit{is a Cauchy sequence.}$$

*Proof.* Put $\varphi_{n,p} := |f_n - f_p| - |T(f_n - f_p)|$. We have $w_{n,p} := T|f_n - f_p| - |f_n - f_p| + \varphi_{n,p} \geqq 0$ and $\int w_{n,p}\, d\lambda = \int \varphi_{n,p}\, d\lambda$.

From $|z|=1$ we obtain $|\varphi_{n,p}| \leqq |(zI-T)(f_n-f_p)|$ and hence $\lim_{n,p\to\infty} \|\varphi_{n,p}\|_1 = 0$. This shows $w_{n,p} - \varphi_{n,p} = T(f_n - f_p) - (f_n - f_p) \to 0$, (in $L_1$). Consequently,

$$\lim_{n,p\to\infty} \|(I-T)(|f_n-f_p| - \textstyle\int |f_n-f_p|)\|_1 = 0 .$$

As $I-T$ is an automorphism in $L_1^0$, this implies $\| |f_n - f_p| - \int |f_n - f_p| \|_1 \to 0$. Now lemma 23 shows the existence of a subsequence of $(f_n)$ which converges in $L_1$. If the sequence $(f_n)$ is not Cauchy we can extract two subsequences converging to different limits $h, h'$. But then $T(h-h') = z(h-h')$ contradicts the assumption that $z$ is not an eigenvalue of $T$. □

We can now deduce

**Proposition 25.** *If $I-T$ is an automorphism of $L_1^0$, the only points on the unit circle which belong to the spectrum of $T$ are the eigenvalues.*

*Proof.* Let us first show that $(zI-T)L_1$ is a closed subspace if $z$ with $|z|=1$ is not an eigenvalue of $T$. It is enough to show that a sequence $(f_n)$ for which $(zI-T)f_n$ converges in $L_1$ to some $\varphi$ must be bounded.

If $\|f_n\|$ is not bounded there exists a subsequence with $\|f_{n_i}\| \to \infty$. Then $f_i' := f_{n_i}/\|f_{n_i}\|$ satisfies $(zI-T)f_i' \to 0$ in $L_1$ and proposition 23 shows that $f_i'$ converges in $L_1$ to some $h$. Clearly, $\|h\|=1$ and $h$ is an eigenvector with eigenvalue $z$, in contradiction to our assumption. To complete the proof let us show $(zI-T)L_1 = L_1$. If this fails there exists $0 \neq u \in L_\infty$ orthogonal to $(zI-T)L_1$. This would imply $T^* u = Pu = zu$. Thus, $z$ would be eigenvalue of $P$ and hence of $T$. □

Now everything is ready for the analysis of the structure of $T$.

**Theorem 26.** *$T$ is quasi-compact if $I-T$ is an automorphism of $L_1^0$.*

*Proof.* Let us first remark that if $I-T$ is an automorphism of $L_1^0(\lambda)$, with $\lambda$ a probability measure, then $\lambda$ is invariant. This follows since the fact that $I-P$ is an automorphism of $L_\infty^0(\lambda)$ implies $(I-P)(f - \int f d\lambda) \in L_\infty^0$ for all $f \in L_\infty^0$, and hence $\int Pf d\lambda = \int f d\lambda$. Now, if $z \neq 1$ with $|z|=1$ is an eigenvalue of $T$ and $Th = zh$, $0 \neq h \in L_1$, one has $\int h d\lambda = \int Th d\lambda = z \int h d\lambda$. Hence $\int h d\lambda = 0$ and

$h \in L_1^0$. Considering the restriction of $T$ to $L_1^0$ and writing $(zT - I) = (z-1)I + I - T$ we see that

$$(zI - T)(I - T)^{-1} = I + (z-1)(I-T)^{-1} \quad \text{on } L_1^0.$$

We can infer that the restriction of $(zI - T)$ to $L_1^0$ is invertible if $z$ is sufficiently close to 1. Thus, 1 is an isolated eigenvalue of $T$ on the unit circle. As the powers of eigenvalues on the unit circle are eigenvalues on the unit circle the preceding argument shows that $z$ is a root of unity. All eigenvalues of $T$ therefore are of the form $\exp(2n\pi i/q)$, $n = 0, 1, \ldots, q-1$. On the other hand, if $Th = zh$, $h \neq 0$, one has $T|h| \geqq |Th| = |h|$ and therefore $|h| = \text{constant} > 0$. We can assume $|h| = 1$. Passing to the reference measure $\lambda$ we can also assume $T\mathbb{1} = \mathbb{1} = P\mathbb{1}$. We can apply the results of Akcoglu-Brunel (from Section 4.1) to $\bar{z}T$, where $z = \exp(2\pi i/q)$. Clearly, the linear moduls of $\bar{z}T$ is $T$. Thus, there exists a measurable function $s$ of modulus 1 such that, for all $\varphi \in L_1$,

$$\bar{z}T\varphi = \bar{s}T(s\varphi).$$

For $\varphi = \mathbb{1}$ we obtain $Ts = \bar{z}s$ and $T\bar{s} = z\bar{s}$. One also has

$$\bar{z}Ts = \bar{s}T(s^2) = \bar{z}^2 s \quad \text{and} \quad T(\bar{s}^2) = z^2\bar{s}^2.$$

By recurrence we deduce $T(\bar{s}^n) = z^n\bar{s}^n$. For $n = q$, $T(\bar{s}^q) = \bar{s}^q$, which implies $\bar{s}^q = \mathbb{1} = s^q$. Hence $s$ is of the form

$$s = \sum_{n=0}^{q-1} 1_{A_n} \exp(2\pi i n/q)$$

where $A_0, A_1, \ldots, A_{n-1}$ is a measurable partition of the space. Writing $A_q := A_0$ we find

$$(3) \qquad Ts = \sum_{n=0}^{q-1} \exp(2\pi i n/q)\, T1_{A_n} = \bar{z}s = \sum_{n=0}^{q-1} \exp(2\pi i(n-1)/q)\, 1_{A_n}$$
$$= \sum_{n=0}^{q-1} \exp(2\pi i n/q)\, 1_{A_{n+1}}.$$

We shall now show that (3) implies $T1_{A_n} = 1_{A_{n+1}}$, $(n = 0, \ldots, q-1)$. We multiply both sides of (3) with $1_{A_{k+1}}$, $0 \leqq k \leqq q-1$, and integrate with $\lambda$. Then we find

$$\sum_{n=0}^{q-1} \exp(2\pi i n/q) \int 1_{A_{k+1}} T1_{A_n}\, d\lambda = \exp(2\pi i k/q)\, \lambda(A_{k+1}).$$

This yields

$$\int \Big(\sum_{n=0}^{q-1} \exp\left(2\pi i \frac{n-k}{q}\right) 1_{A_n}\Big) T^* 1_{A_{k+1}}\, d\lambda = \lambda(A_{k+1})$$
$$= \int T^* 1_{A_{k+1}}\, d\lambda = \int \left\{\sum_{n=0}^{q-1} \cos\left(2\pi \frac{n-k}{q}\right) 1_{A_n}\right\} T^* 1_{A_{k+1}}\, d\lambda.$$

As $\{\ldots\}$ is $\leqq 1$, it must be $= 1$ on $\{T^* 1_{A_{k+1}} > 0\}$. As $\cos(2\pi(\mathrm{n}-\mathrm{k})/q) < 1$ holds for $n \neq k$, we necessarily have $T^* 1_{A_{k+1}} \leqq 1_{A_k}$, $(k = 0, \ldots, q-1)$. But now $\mathbb{1} = T^*\mathbb{1} = \sum_{k=0}^{q-1} T^* 1_{A_{k+1}} \leqq \sum_{k=0}^{q-1} 1_{A_k} = \mathbb{1}$ shows that the desired equalities must hold.

It follows from $T 1_{A_n} = 1_{A_{n+1}}$, $(n = 0, \ldots, q-1)$, that $T^q 1_{A_0} = 1_{A_0}$. Therefore, $T$ maps $L_1(A_0)$ into $L_1(A_0)$. Let $S$ denote the restriction of $T^q$ to $L_1(A_0)$. We want to show that $I_{A_0} - S$ is an automorphism of $L_1^0(A_0)$. It is clear that $S$ is conservative and ergodic and that $I_{A_0} - S$ is injective and has a dense image in $L_1^0(A_0)$. It remains to show that $(I_{A_0} - S) L_1^0(A_0)$ is closed. Put $R = \sum_{k=0}^{q-1} T^k$. Then $R$ maps $L_1^0(A_0)$ into $L_1^0(\Omega)$ and $f = 1_{A_0} Rf$ holds for $f \in L_1^0(A_0)$. Now, if $(f_n)$ is a bounded sequence in $L_1^0(A_0)$ with $(I_{A_0} - S) f_n \to 0$ in $L_1^0(A_0)$, then the identity

$$(I - T) Rf_n = (I - T^q) f_n = (I_{A_0} - S) f_n$$

shows that $(I - T) Rf_n$ tends to $0$ in $L_1^0(\Omega)$. Hence $\|Rf_n\|_1 \to 0$, and $f_n = 1_{A_0} Rf_n \to 0$ in $L_1^0(A_0)$. We shall show that $S$ and hence $T$ is quasi-compact. We can consider $T$ and can assume that 1 is the only eigenvalue. Take $E = \mathbb{1} \otimes \lambda$. On $L_1^0$ we have $T = T - E$. The point 1 lies outside the spectrum of $T$ considered as an operator in $L_1^0$ and its spectral radius is $< 1$. It follows that

$$\lim_n \|(T - E)^n\|_{L_1^0} = 0.$$

For $f \in L_1$ we can write $f = f' + k$ with $f' \in L_1^0$ and $k \in \mathbb{C}$. Thus

$$(T - E) f = (T - E) f' = Tf' - Ef' = Tf'.$$

We obtain $(T - E)^n L_1 = (T - E)^n L_1^0$. It then follows that

$$\lim \|(T - E)^n\|_{L_1} = \lim \|(T - E)^n\|_{L_1^0} = 0.$$

As $(T - E)^n = T^n - E$, $(T\mathbb{1} = \mathbb{1})$, we arrive at $\lim \|T^n - E\|_{L_1} = 0$. In particular, $T$ is quasi-compact. □

## Notes

It has been shown in the studies of Fortet [1938] and Yosida [1939] that the validity of the uniform ergodic theorem for Markov chains is closely related to the property of quasi-compactness of the transition kernel. Doeblin [1940] and Harris [1956] have given the appropriate recurrence conditions and Harris showed that his condition (H′) implies the existence of a $\sigma$-finite invariant measure.

Orey [1962] proved a fundamental theorem, generalized by Jamison and Orey [1967] from discrete to general state spaces: If $T$ is an "aperiodic" Harris operator and $\nu$ is a signed measure with $\nu(\Omega) = 0$ then $\|\nu P^n\| \to 0$.

Ornstein [1969a] proved theorem 17 in the particular case of random walks on $\mathbb{R}$ which are recurrent in the sense of Harris, and this theorem was generalized by Métivier [1969] a bit later. The work of Orey and Ornstein motivated the theory of "bounded" sets as developed by Brunel [1971]. They were later called special sets. In particular, it was shown that the existence of special sets implies the existence of a $\sigma$-finite invariant measure, and that, for a conservative ergodic $T$, $I-T$ is an automorphism of $L_1^0$ iff $\Omega$ is special. (This theory was presented in a seminar at Minneapolis in 1969).

Horowitz [1969a], [1969b], [1979] proved the principal theorems of the abstract theory and noticed the connection with the uniform ergodic theorem.

Then, Neveu [1972] has brilliantly generalized the theory of special sets introducing the special functions and the taboo-operators and their various applications.

Brunel and Revuz [1974] have taken up these notions in their study of the connections between quasi-compactness and the Harris condition, and they used them to determine the Martin boundary of Harris recurrent random walks.

In the meantime, Ornstein and Sucheston [1970] had established the important zero-two law. A unified treatment was given by Foguel [1979] and Foguel-Ghoussoub [1979]. The present proof of theorem 18 follows Pfort [1984] who used their ideas but corrected an error in their paper.

The author of this supplement has not tried to give the shortest route to the key theorems. He has made an effort to acquaint the reader with different techniques. The exposition has been limited to the abstract point of view dealing with the operator ergodic theorems which form the main subject of this book.

There is a considerable literature on the subjects connected to this supplement, and a short guide to it shall suffice here. The books of Orey [1971] and Revuz [1975] give an exposition of the theory of Harris processes from the probabilistic point of view.

The papers of Chung [1964], Jain [1966], Jain-Jamison [1967], Winkler [1975] give contributions to Doeblin's theory of Markov chains. Other probabilistic approaches to the limit theorems have been proposed by Papangelou [1976/77], Athreya-Ney [1978], [1982], Nummelin [1978], Pitman [1974]. A special case was treated by Moy [1965], [1967].

The papers by Halmos [1947], Dowker [1951], Feldman [1962], Šidak [1967], Horowitz [1968a], Millet-Sucheston [1979] are references concerning existence of $\sigma$-finite invariant measures; see also § 3.4 and § 3.6. Millet-Sucheston considered small sets analogous to special sets.

Theorem 17 has been a useful tool for the proof of "global" or "mixed" ratio theorems, in which one considers ratios of the form

$$\sum_{i=0}^{n} \langle \nu_1, P^i h_1 \rangle / \sum_{i=0}^{n} \langle \nu_2, P^i h_2 \rangle .$$

(Note that if $h = 1_A$, then $\langle \nu, P^i h \rangle$ is the probability of a visit in $A$ at time $i$ when the initial distribution is $\nu$.)

For this topic we refer to Krengel [1966], Isaac [1967], Levitan [1970], [1971], Lin [1970a], [1972b], [1976], Foguel-Lin [1972] and Métivier [1972].

The zero-two law is closely connected to the triviality of the remote $\sigma$-algebra, established for Harris processes by Jamison-Orey [1967]. Related references are the papers of Winkler [1973], Derriennic [1976], Foguel [1976], Cohn [1979a], Lin [1982a], Revuz [1983] and Greiner-Nagel [1982].

# Bibliography

## Abbreviations

| | |
|---|---|
| AIHP | Ann. Inst. Henri Poincaré |
| AM St | Ann. Math. Stat. |
| Bull AMS | Bull. Amer. Math. Soc. |
| Can. JM | Canadian Journal Math. |
| CRAS | Comptes Rendus Acad. Sci. Paris |
| Ergebn. M. | Ergebnisse d. Math. u. Grenzgeb. |
| Isr. JM | Israel Journal Math. |
| J | Journal |
| JMS Jap. | Journal Math. Soc. Japan |
| M | Math. |
| MZ | Math. Zeitschr. |
| Pac. JM | Pacific Journal Math. |
| PAMS | Proc. Amer. Math. Soc. |
| SpLNM | Springer Lecture Notes Math. |
| TAMS | Transact. Amer. Math. Soc. |
| Tôh. MJ | Tôhoku Math. Journ. |
| ZW | Zeitschr. Wahrscheinlichkeitsth. verw. Gebiete |


Aaronson, Jon (1977): *On the ergodic theory of non-integrable functions and infinite measure spaces*. Isr. JM **27**, 163–173.

– (1979): *On the pointwise ergodic behaviour of transformations preserving infinite measures*. Isr. JM **32**, 67–82.

– (1981): *An ergodic theorem with large normalising constants*. Isr. JM **38**, 182–188.

– (1983): *The eigenvalues of non-singular transformations*. Jsr. JM **45**, 297–312.

–, and M. Keane (1982): *The visits to zero of some deterministic random walks*. Proc. London M Soc. III. Ser., **44**, 535–553.

–, M. Lin and B. Weiss (1979): *Mixing properties of Markov operators and ergodic transformations, and ergodicity of Cartesian products*. Jsr. JM **33**, 198–224.

Abdalla, S.M., and J. Szücs (1974): *On an ergodic type theorem for von Neumann algebras*. Acta Sci. M (Szeged) **36**, 167–172.

Abid, Mokhtar (1978): *Un theorème ergodique pour des processus sous-additifs et surstationnaires*. CRAS, A **287**, 149–152.

Ahmad, Ibrahim A. (1981): *On some asymptotic properties of U-statistics*. Scand. J Stat., Theory Appl. **8**, 175–182.

Akcoglu, Mustafa A. (1965): *An ergodic lemma*. PAMS **16**, 388–392.

– (1966): *Pointwise ergodic theorems*. TAMS **125**, 296–309.

– (1975): *A pointwise ergodic theorem in $L_p$-spaces*. Can. JM **27**, 1075–1082.

– (1975a): *Positive contractions of $L_1$-spaces*. MZ **143**, 5–13.

– (1979): *Pointwise ergodic theorems in $L_p$-spaces*. SpLNM **729**, 13–15.

– (1983): *Differentiation of superadditive processes*. Proc. Sherbrooke Conf. Measure Th. (1982), SpLNM **1033**, 1–11.

–, and A. Brunel (1971): *Contractions on $L_1$-spaces*. TAMS **155**, 315–325.

–, and R.V. Chacon (1965): *A convexity theorem for positive operators*. ZW **3**, 328–332.

–, – (1970): *A local ratio theorem*. Can. JM **22**, 545–552.

–, – (1970a): *Ergodic properties of operators in Lebesgue space*. Adv. in Prob. **2**, 1–47.

–, and J. Cunsolo (1970): *An ergodic theorem for semigroups*. PAMS **24**, 161–170.

–, – (1970a): *An identification of ratio ergodic limits for semi-groups*. ZW **15**, 219–229.

–, and A. del Junco (1975): *Convergence of averages of point-transformations*. PAMS **49**, 265–266.

–, – (1981): *Differentiation of n-dimensional additive processes*. Can. JM **33**, 749–768.

–, and M. Falkowitz (1983): *A general local ergodic theorem in $L_1$*. Preprint.

–, J.P. Huneke, and H. Rost (1974): *A counterexample to the Blum-Hanson theorem in general spaces*. Pac. JM **50**, 305–308.

–, and U. Krengel (1978): *A differentiation theorem for additive processes*. MZ **163**, 199–210.

–, – (1979): *A differentiation theorem in $L_p$*. MZ **169**, 31–40.

–, – (1979a): *Two examples of local ergodic divergence*. Jsr. JM **33**, 225–230.

–, – (1981): *Ergodic theorems for superadditive processes*. J Reine Angew. Math. **323**, 53–67.

–, and D.B. Miller (1976): *Dominated estimates in Hilbert spaces*. PAMS **55**, 371–375.

–, and R.W. Sharpe (1968): *Ergodic theory and boundaries.* TAMS **132**, 447–460.

–, and L. Sucheston (1972): *On operator convergence in Hilbert space and in Lebesgue space.* Periodica math. Hungar. **2**, 235–244.

–, – (1974): *On the dominated ergodic theorem in $L_p$ space.* PAMS **43**, 379–392.

–, – (1975): *Weak convergence of positive contractions implies strong convergence of averages.* ZW **32**, 139–145.

–, – (1975a): *Remarks on dilations in $L_p$-spaces.* PAMS **53**, 80–81.

–, – (1977): *Dilations of positive contractions on $L_p$-spaces.* Can. M Bull. **20**, 285–292.

–, – (1978): *A ratio ergodic theorem for superadditive processes.* ZW **44**, 269–278.

–, – (1982): *A superadditive version of Brunel's maximal ergodic lemma.* Measure theory, Oberwolfach 1981, SpLNM, **945.** 347–351.

–, – (1983): *A stochastic ergodic theorem for superadditive processes.* Erg. Th. and Dyn. Systems **3**, 335–344.

–, – (1984): *On ergodic theory and truncated limits in Banach lattices.* Measure theory, Oberwolfach 1983, SpLNM **1089**, 241–262.

–, – (1984a): *An ergodic theorem on Banach lattices.* Preprint.

–, – (1984b): *On identification of superadditive ergodic limits.* Proc. Kakutani Conf. "Contemporary Math.", 25–32.

Alaoglu, L. and G. Birkhoff (1940): *General ergodic theorems.* Ann. Math. **41**, 293–309.

Alfsen, E.M. (1971): *Compact convex sets and boundary integrals.* Springer.

Al-Hussaini, A. (1974): *A note on a mean ergodic theorem.* Ann. Polon. math. **30**, 113–117.

Ando, T. (1968): *Invariante Maße positiver Kontraktionen in $C(X)$.* Studia Math. **31**, 173–187.

Anzai, Kazuo (1977): *On the weak conditions for mean ergodic theorems.* Keio Eng. Rep. **30**, 85–92.

Archbold, R.J. (1976): *A mean ergodic theorem associated with the free group on two generators.* J. London M Soc., II. Ser. **13**, 339–345.

Aribaud, F. (1970): *Un théorème ergodique pour les espaces $L^1$.* J Funct. Anal. **5**, 395–411.

Arnold, L.K. (1968). *On σ-finite invariant measures.* ZW **9**, 85–97.

Arques, Didier and Patrick Gabriel (1977): *Sur la représentation de Rudolphe des flots filtrés ergodiques.* CRAS **284**, A 551–A 554.

Artstein, Z. and R. Vitale (1975): *A strong law of large numbers for random compact sets.* Ann. Prob. **3**, 879–882.

Assani, Idris (1984): *Quelques resultats sur les opérateurs positifs a moyennes bornées dans $L_p$.* Preprint.

– (1984a): *Sur les moyennes d'un opérateur.* Preprint.

– (1985): *On the pointwise and local ergodic theorem for non positive operators in $L_p$.* Preprint.

– (1985a): *Sur la convergence ponctuelle de quelques suites d'operateurs.* Preprint.

–, and R. Mesiar (1984): *Sur la convergence ponctuelle de $T^n f/n^\alpha$ dans $L^P$.* To appear Ann. Clermont-Ferrand **3**.

Atalla, Robert E. (1974): *On the mean convergence of Markov operators.* Proc. Edinburgh M Soc., II. Ser. **19**, 205–209.

– (1976): *On the ergodic theory of contractions.* Revista Colombiana de Math. **10**, 75–81.

– (1981): *Markov operators and quasi-Stonian spaces.* PAMS **82**, 613–618.

– (1983): *Local ergodicity of nonpositive contractions on $C(X)$.* PAMS **88**, 419–425.

– and R. Sine (1976): *Random Markov mappings.* Math. Ann. **221**, 195–199.

Atencia, E., and A. de la Torre (1982): *A dominated ergodic estimate for $L_p$-spaces with weights.* Stud. Math. **74**, 35–47.

Athreya, Krishna B., and Peter Ney (1978): *A new approach to the limit theory of recurrent Markov chains.* TAMS **245**, 493–501.

–, – (1982): *A renewal approach to the Perron-Frobenius theory of non-negative kernels on general state spaces.* MZ **179**, 507–529.

Atkinson, G. (1976): *Recurrence of co-cycles and random walks.* J London M Soc. (2), **13**, 486–488.

Axmann, D. (1980): *Struktur- und Ergodentheorie irreduzibler Operatoren auf Banachverbänden.* Diss. Univ. Tübingen.

Ayupov, Sh.A. (1977): *The ergodic theorem in $O^*$-algebras.* Dokl. Akad. Nauk USSR, 3–4 (Russian).

– (1982): *Ergodic theorems for Markov operators in Jordan algebras I, II.* Izv. Akad. Nauk. USSR, Ser. Fiz-Mat. Nauk. No. 3, 12–15 and No. 5, 7–12. (Russian).

Baillon, Jean-Bernard (1975): *Un théorème de type ergodique pour les contractions nonlinéaires dans un espace de Hilbert.* CRAS, A **280**, 1511–1514.

– (1976): *Quelques propriétés de convergence asymptotique pour les semi-groupes de contractions impaires.* CRAS, A **283**, 75–78.

– (1976a): *Quelques propriétés de convergence asymptotique pour les contractions impaires.* CRAS **283**, 587–590.

– (1978): *Comportement asymptotique des itérés de contractions non-linéaires dans les espaces* $L^p$. CRAS **286**, 157–159.

– (1981): *Nonlinear ergodic theory in Banach spaces*. Delft Progr. Rep. **6**, 87–96.

–, R.E. Bruck and S. Reich (1978): *On the asymptotic behaviour of nonexpansive mappings and semigroups in Banach spaces*. Houston J M **4**, 1–9.

Baker, R.C. (1976): *Riemann sums and Lebesgue integrals*. Quart. JM, Oxf. Ser. (2) **27**, 191–198.

Banach, Stefan (1926): *Sur la convergence presque partout des fonctionelles linéaires*. Bull. Sci. Math. (2), **50**, 27–32, 36–43.

– (1932): *Théorie des opérations linéaires*. Warsaw.

Barone, Enzo and K.P.S. Bhaskara Rao (1981): *Poincaré recurrence theorem for finitely additive measures*. Rend. Mat. Appl., VII. Sér. **1**, 521–526.

Bauer, Heinz (1981): *Probability Theory and Elements of Measure Theory*. Acad. Press.

Baum, Leonard E. and Melvin Katz (1965): *Convergence rates in the law of large numbers*. TAMS **120**, 108–123.

Baxter, Glen (1964): *An ergodic theorem with weighted averages*. J of Math. and Mech., **13**, 481–488.

– (1965): *Generalizations of the maximal ergodic theorem*. AM St **36**, 1292–1293.

Baxter, John R. and Rafael V. Chacon (1974): *A local ergodic theorem on* $L_p$. Can. JM **26**, 1206–1216.

– and J.H. Olsen (1983): *Weighted and subsequential ergodic theorems*. Can. JM **35**, 145–166.

Beboutov, N. (1942): *Markov chains with a compact state space*. Rec. Math. (Mat. Sbornik) N.S. **10** (52), 213–238.

Beck, Anatole (1963): *On the strong law of large numbers*. Ergodic Theory, Acad. Press, Ed. F.B. Wright, 21–53.

– and J.T. Schwartz (1957): *A vector valued random ergodic theorem*. PAMS **8**, 1049–1059.

Becker, M.E. (1981): *Multiparameter groups of measure-preserving transformations: a simple proof of Wiener's ergodic theorem*. Ann. Prob. **9**, 504–509.

Bellow, Alexandra (1975): *An* $L^p$*-inequality with application to ergodic theory*. Houston JM **1**, 153–159.

– (1976): *A problem in* $L^p$*-spaces*. Measure Theory, Oberwolfach, SpLNM **541**, 381–388.

– (1982): *Sur la structure des suites "mauvaises universelles" en théorie ergodique*. CRAS **294**, 55–58.

– (1984): *On "bad universal" sequences in Ergodic Theory (II)*. Proc. Sherbrooke Workshop on Meas. Th., SpLNM **1033**, 74–78.

– and V. Losert, (1984): *On sequences of density zero in ergodic theory*. Proc. Kakutani Conf. "Contemporary Mathematics".

–, – (1985): *The weighted pointwise ergodic theorem and the individual ergodic theorem along subsequences*. Preprint.

Bénozène, R. (1981): *Sommes minimales des itérés d'une contraction dans les espaces de Hilbert et de Banach*. Diss. Univ. Paris.

Berbee, Henry (1981): *Recurrence and transience for random walks with stationary increments*. ZW **56**, 531–536.

Berglund, J.F., and K.K. Hofmann, (1967): *Compact semitopological semigroups and weakly almost periodic functions*. SpLNM **42**.

–, H.D. Junghenn, and P. Milnes, (1978): *Compact right topological semigroups and generalizations of almost periodicity*. SpLNM, **663**.

Berk, K. (1968): *Ergodic theory with recurrent weights*. AM St **39**, 1107–1114.

Bewley, Truman (1971): *Extension of the Birkhoff and von Neumann ergodic theorems to semigroup actions*. AIHP, B **7**, 283–291.

Billingsley, Patrick (1965): *Ergodic Theory and Information*. Wiley.

Birkhoff, Garrett (1939): *The mean ergodic theorem*. Duke MJ **5**, 19–20.

Birkhoff, George D. (1931): *Proof of the ergodic theorem*. Proc. Nat. Acad. Sci. USA **17**, 656–660.

–, and P.A. Smith, (1928): *Structure analysis of surface transformations*. Journ. de Math. **7**, 345–379.

Bishop, Errett (1966): *An upcrossing inequality with applications*. Michigan MJ, **13**, 1–13.

– (1967): *Foundations of constructive analysis*. Mc. Graw-Hill.

– (1968): *A constructive ergodic theorem*. J of Math. and Mech. **17**, 631–639.

Blackwell, David and David Freedmann, (1964): *The tail σ-field of a Markov chain and a theorem of Orey*. AM St, **35**, 1291–1295.

Blum, Julius R., and Robert Cogburn, (1975): *On ergodic sequences of measures*. PAMS **51**, 359–365.

–, and Bennett Eisenberg, (1974): *Generalized summing sequences and the mean ergodic theorem*. PAMS **42**, 423–429

–, – (1979): *Ergodic theorems for mixing transformation groups*. Rocky Mt. JM **9**, 593–600.

–, – ,and L.-S. Hahn, (1973): *Ergodic theory and the measure of sets in the Bohr group*. Acta Sci. Math. (Szeged) **34**, 17–24.

–, and N. Friedman, (1967): *On invariant measures for classes of transformations ZW* **8**, 301–305.
–, and David L. Hanson, (1960): *On the mean ergodic theorem for subsequences.* Bull. AMS **66**, 308–311.
–, and V.J. Mizel, (1971): *On a theorem of Weyl and the ergodic theorem.* ZW **20**, 193–198.
–, and J.I, Reich, (1976): *A mean ergodic theorem for families of contractions in Hilbert space.* PAMS **61**, 183–185.
–, – (1977): *The individual ergodic theorem for p-sequences.* Isr. JM **27**, 180–184.
–, – (1977a): *On the individual ergodic theorem for K-automorphisms.* Ann. Prob. **5**, 309–314.
–, – (1982): *Pointwise ergodic theorems in L.c.a. groups.* Pac. JM **103**, 301–306.
–, and J.I. Rosenblatt, (1967): *On the moments of recurrence time.* J Math. Sci. (Delhi) **2**, 1–6.
Boclé, J. (1960): *Sur la theorie ergodique.* Ann. Inst. Fourier (Grenoble) **10**, 1–45.
Borel, Émile (1909): *Sur les probabilités dénombrables et leurs applications arithmétiques.* Rend. Circ. Mat. Palermo **26**, 247–271.
Borges, Rudolf (1966): *Zur Existenz von separablen stochastischen Prozessen.* ZW **6**, 125–128.
Bratteli, O. and D.W. Robinson, (1979): *Operator algebras and quantum statistical mechanics I. Springer.*
Bray, G. (1969): *Theoremes ergodiques de convergence en moyenne.* J Math. Anal. Appl. **25**, 471–502.
Breiman, Leo (1957/60): *The individual ergodic theorem of information theory.* AM St. **28**, 809–811; Correction **31**, 809–810.
– (1960): *The strong law of large numbers for a class of Markov chains.* AM St **31**, 801–803.
Brézis, H., and F.E. Browder, (1976): *Nonlinear ergodic theorems.* Bull. AMS **82**, 959–961.
–, – (1977): *Remarks on nonlinear ergodic theory.* Adv. Math. **25**, 165–177.
Brillinger, David. R. (1981): *Some aspects of modern population mathematics.* Can. J Stat. **9**, 173–194.
Brooks, J.K., and R.V. Chacon (1980): *Continuity and compactness of measures.* Adv. Math. **37**, 16–26.
Browder, F. (1965): *Non-expansive nonlinear operators in a Banach space.* Proc. Nat. Acad. Sci. USA **54**. 1041–1044.
Brown, James R. (1966): *Approximation theorems for Markov operators.* Pac. JM **16**, 13–23.
Bru, B., and H. Heinich, (1981): *Isometries positives et proprietes ergodiques de quelques espaces de Banach.* AIHP, B **17**, 377–405.
Bruck, Ronald E. (1978): *On the almost-convergence of iterates of a nonexpansive mapping in Hilbert space and the structure of the weak ω-limit set.* Isr. JM **29**, 1–16.
– (1979): *A simple proof of the mean ergodic theorem for nonlinear contractions in Banach spaces.* Isr. JM **32**, 107–116.
– (1981): *On the convex approximation property and the asymptotic behavior of nonlinear contractions in Banach spaces.* Isr. JM **38**, 304–314.
Bruckner, A.M. (1971): *Differentiation of integrals.* Amer. Math. Monthly **78**.
Brunel, Antoine (1963): *Sur un lemme ergodique voisin du lemme de E. Hopf, et sur une de ses applications.* CRAS **256**, 5481–5484.
– (1966): *Sur quelques problèmes de la théorie ergodique ponctuelle.* Thèse, Univ. de Paris.
– (1966a): *Sur les mesures invariantes.* ZW **5**, 300–303.
– (1970): *New conditions for existence of invariant measures in ergodic theory.* SpLNM **160**, 7–17.
– (1971): *Chaines abstraites de Markov vérifiant une condition de Orey Extension à ce cas d'un théorème ergodique de M. Métivier.* ZW **19**, 323–329.
– (1973): *Théorème ergodique ponctuel pour un semigroupe commutatif finiment engendré de contractions de $L^1$.* AIHP B, **9**, 327–343.
– (1976): *Sur les sommes d'itérés d'un opérateur positif.* SpLNM **532**, 19–34.
– and Richard Emilion, (1984): *Sur les opérateurs positifs a moyennes bornées.* CRAS I, **298**, 103–106.
– and Michael Keane, (1969): *Ergodic theorems for operator sequences.* ZW **12**, 231–240.
– and Daniel Revuz, (1974): *Un critère probabiliste de compacité des groupes.* Ann. Prob. **2**, 745–746.
–, – (1974a): *Quelques applications probabilistes de la quasi-compacité.* AIHP B, **10**, 301–337.
– and L. Sucheston, (1979): *Sur l'existence de dominants exacts pour un processus sur-additif.* CRAS A **288**, 153–155.
–, – (1984/85): *On existence of invariant elements in Banach lattices.* Preprint.
Bunimovich L.A., (1982): *Some new advances in the physical applications of ergodic theory.* Math. Research **12**, 27–33.
– and Ya.G. Sinai, (1973): *On a fundamental theorem in the theory of dispersing billards.* Math. USSR Sbornik **19**, 407–423.
Bunjakov, M.R. (1973): *Ergodic theorems for endomorphisms of abstract Boolean algebras.* V.I. Lenin. Sakharth. Polithekh. Inst. Srom. No 6 (162) Math, 124–129.

Burckel, R.B. (1970): *Weakly almost periodic functions on semigroups.* Gordon Breach.
Burke, George (1965): *A uniform ergodic theorem.* AM St **36**, 1853–1858.
Burkholder, Donald L. (1962): *Successive conditional expectations of an integrable function.* AM St **33**, 887–893.
– (1962a): *Semi-Gaussian subspaces.* TAMS **104**, 123–131.
– (1964): *Maximal inequalities as necessary conditions for almost everywhere convergence.* ZW **3**, 75–88.
– (1973): *Discussion at the end of Kingman* (1973).
– and Y.S. Chow, (1961):*Iterates of conditional expection operators.* PAMS **12**, 490–495.
Busemann, Herbert and William Feller, (1934): *Zur Differentiation der Lebesgue'schen Integrale.* Fund. Math. **22**, 226–256.
Butzer, Paul L. (1980): *The Banach-Steinhaus theorem with rates, and applications to various branches of analysis.* General Inequalities **2**, E.F. Beckenbach, Ed. ISNM **47**, Birkhäuser, 299–331.
– and U. Westphal, (1971): *The mean ergodic theorem and saturation.* Indiana Univ. MJ **20**, 1163–1174.
–, – (1972): *Ein Operatorenkalkül für das approximationstheoretische Verhalten des Ergodensatzes im Mittel.* Linear Operators and Approximation I, Butzer, Kahane, Sz-Nagy Eds., Birkhäuser, ISNM **20**, 102–113.

Calderon, Alberto P. (1953): *A general ergodic theorem.* Ann. of Math. **58**, 182–191.
– (1955): Sur les mesures invariantes. CRAS **240**, 1960–1962.
– (1968): *Ergodic theory and translation invariant operators.* Proc. Nat. Acad. Sci. USA **59**, 349–353.
Carleson, Lennart (1958): *Two remarks on the basic theorems of information theory.* Math. Scand. **6**, 175–180.
Chacon, Rafael V. (1960): *The influence of the dissipative part of a general Markov process.* PAMS **11**, 957–961.
– (1961): *On the ergodic theorem without assumption of positivity.* Bull. AMS **67**, 186–190.
– (1962): *An ergodic theorem for operators satisfying norm conditions.* J Math. Mech. **11**, 165–172.
– (1962a): *Identification of the limit of operator averages.* J Math. Mech. **11**, 961–968.
– (1962b): *Resolution of positive operators.* Bull. AMS **68**, 572–574.
– (1963): *Convergence of operator averages.* (Ergodic theory. Sympos. Tulane Univ., Ed. F.B. Wright) Academic Press, 89–120.
– (1964): *Ordinary means imply recurrent means.* Bull. AMS **70**, 796–797.
– (1964a): *A class of linear transformations.* PAMS **15**, 560–564.
– and U. Krengel, (1964): *Linear modulus of a linear operator.* PAMS **15**, 553–559.
– and J. Olsen, (1969): *Dominated estimates of positive contractions.* PAMS **20**, 266–271.
– and D.S. Ornstein, (1960): *A general ergodic theorem.* Illinois JM **4**, 153–160.
Chatard, Jaqueline (1970):*Applications des propriétés de moyenne d'un groupe localement compact à la théorie ergodique.* AIHP **6**, 307–326; erratum: AIHP **7**, 81, (1971).
– (1972): *Sur une généralization du théorème de Birkhoff.* CRAS **275**, A, 1135–1138.
– (1972a): *Applications des proprietés de moyenne d'un groupe localement compact à la théorie ergodique.* Thèse Univ. Paris VI.
Chersi, Franco (1982): *Convergence of averages and invariant measures on non-compact metrizable spaces.* Boll. Unione Mat. Ital., VI. Ser. A **1**, 269–273.
– and S. Invernizzi, (1976): *Some complements to ergodic theorems for vector-valued functions.* Boll. Unione mat. Ital., V. Ser. A **13**, 677–686.
Choksi, J.R. (1961): *Extension of a theorem of Hopf.* J London M Soc. **36**, 81–88. Corrigendum 253.
– and V.S. Prasad, (1983): *Approximation and Baire category theorems in ergodic theory.* Proc. Sherbrooke Workshop Measure Theory (1982), SpLNM **1033**, 94–113.
Choquet, Gustave and Ciprian Foias, (1975): *Solution d'un problème sur les itérés d'un opérateur positif sur $C(K)$ et propriétés de moyennes associées.* Ann. Inst. Fourier, Grenoble, **25**, 109–129.
Chung, Kai Lai (1961): *A note on the ergodic theorem of information.* AM St **32**, 612–614.
– (1964): *The general theory of Markov processes according to Doeblin.* ZW **2**, 230–250.
– (1967): *Markov chains with stationary transition probabilities.* 2nd ed., Springer.
– (1969): *A course in probability theory.* Harcourt, Brace and World.
Cohen, Joel E. (1979): *Ergodic theorems in demography.* Bull. AMS, new Ser. **1**, 275–295.
Cohen, L.W. (1940): *On the mean ergodic theorem.* Ann. Math.**41**, 505–509.
Cohn, Harry (1977): *Countable non-homogeneous Markov chains: asymptotic behaviour.* Adv. Appl. Prob. **9**, 542–552.
– (1979): *On the asymtotic events of a Markov*

*chain*. Internat. J. Math. and Math. Sci., **2**, 537–587.
– (1979a): *On the invariant events of a Markov chain*. ZW **48**, 81–96.
Coifman, Ronald, R. and Charles L. Feffermann, (1974): *Weighted norm inequalities for maximal functions and singular integrals*. Studia Math. **51**, 241–250.
Conze, Jean-Pierre (1973): *Convergence des moyennes ergodiques pour des sous-suites*. Bull. Soc. math. France **35**, 7–15.
– and N. Dang-Ngoc, (1978): *Ergodic theorems for noncommutative dynamical systems*. Invent. M **46**, 1–15.
Cornfeld, I.P., S.V. Fomin and Ya.G. Sinai, (1982): *Ergodic Theory*. Springer.
Cotlar, M. (1955): *A unified theory of Hilbert transforms and ergodic theorems*. Rev. Mat. Cuyana **1**, 105–167.
Crauel, Hans (1981): *Ergodentheorie linearer stochastischer Systeme*. Report No. **59**, Forschungsschwerpunkt Dynamische Systeme, Univ. Bremen.
Csiszár, Imre and János Körner (1981): *Information Theory: Coding Theorems for discrete memoryless channels*. Academic Press.
Cuculescu, I. and C. Foias, (1966): *An individual ergodic theorem for positive operators*. Rev. Roumaine Math. Pures Appl. **11**, 581–594.

Dang-Ngoc, Nghiem (1973): *Partie finie d'un système dynamique et deux nouvelles démonstrations du théorème de Hopf*. ZW **27**, 131–140.
– (1980): *On the integrability of the maximal ergodic theorem*. PAMS **79**, 565–570.
– (1982): *A random ergodic theorem in von Neumann algebras*. PAMS **86**, 605–608.
Dani, S.G. (1981): *Invariant measures and minimal sets of horospherical flows*. Invent. M **64**, 357–385.
– and S. Muralidharan, (1983): *On ergodic averages for affine lattice actions on tori*. Monatsh. M **96**, 17–28.
Davies, Edward Brian (1981): *One parameter semigroups*. Academic Press.
– (1982): *The harmonic functions of mean ergodic Markov semigroups*. MZ **181**, 543–552.
Davis, Burgess (1971): *Stopping rules for $S_n/n$, and the class L log L*. ZW **17**, 147–150.
– (1982): *On the integrability of the ergodic maximal function*. Studia Math. **73**, 153–167.
Davis, William J., Nassif Ghoussoub and Joram Lindenstrauss, (1981): *A lattice renorming theorem and applications to vector-valued processes*. TAMS **263**, 531–540.
Day, Mahlon M. (1942): *Ergodic theorems for Abelian semigroups*. TAMS **51**, 399–412.
– (1961): *Fixed point theorems for compact convex sets*. Illinois JM **5**, 585–590.
– (1973): *Normed linear spaces, 2nd ed. Ergebn.* M **21**, Springer.
Dean, David W. and Louis Sucheston (1966): *On invariant measures for operators*. ZW **6**, 1–9.
Dekking, F.M. (1982): *On transience and recurrence of generalized random walks*. ZW **61**, 459–465.
De Guzman, Miguel (1975): *Differentiation of integrals in $R^n$* SpLNM **481**.
De Lasnerie, Michel (1977): *Un theoreme ergodique aleatoire en moyenne*. CRAS A **285**, 285–287.
De la Torre, Alberto (1976): *A simple proof of the maximal ergodic theorem*. Can. J M **28**, 1073–1075.
– (1977): *A dominated ergodic theorem for contractions with fixed points*. Can. math. Bull. **20**, 89–91.
– (1978): *An ergodic theorem for vector valued positive contractions*. Ann. Sci. math. Que. **2**, 281–288.
– (1979): *The square function for $L_1$-$L_\infty$ contractions*. Proc. Sympos. Pure Math., **35**, Part 2, 435–438.
– (1982): *Weights in ergodic theory*. Harmonic analysis Proc. Conf., Minneapolis 1981, SpLNM **908**, 128–138.
Deleeuw, Karel und Irwing L. Glicksberg (1961): *Applications of almost periodic compactifications*. Acta math. **105**, 63–97.
Del Junco, Andrés (1977): *On the decomposition of a subadditive stochastic process*. Ann. Prob. **5**, 298–302.
– and J. Rosenblatt (1979): *Counterexamples in Ergodic Theory and Number Theory*. Math. Ann. **245**, 185–197.
– and J.M. Steele (1977): *Moving averages of ergodic processes*. Metrika, **24**, 35–43.
Dellacherie, Claude and Paul-André Meyer (1975/1980/1983): *Probabilités et potentiels I, II, III*. Hermann, Paris. Engl. Transl. North Holland Math. Studies (**29**/**72**/).
Denker, Manfred, Christian Grillenberger and Karl Sigmund (1976): *Ergodic theory on compact spaces*. SpLNM **527**.
Derriennic, Yves (1973): *On the integrability of the supremum of ergodic ratios*. Ann. Prob. **1**, 338–340.
– (1975): *Sur le théorème ergodique sous-additif*. CRAS **281**, A, 985–988.
– (1976): *Lois «zero ou deux» pour les processus de Markov-applications aux marches aléatoires*. AIHP, B, **12**, 111–129.

– (1979): *Lectures on pointwise ergodic theorems.* Dep. Math. Univ. Torún. Preprint 6.
– (1980): *Quelques applications du theoreme ergodique sous-additif.* Asterisque **74**, 183–201.
– (1983): *A remark on the time constant in first passage percolation.* Adv. Appl. Prob. **15**, 214–215.
– (1983a): *Un théorème ergodique presque sous-additif.* Ann. Prob. **11**, 669–677.
– and U. Krengel (1981): *Subadditive mean ergodic theorems.* Ergodic Theory and Dyn. Syst. **1**, 33–48.
– and Michael Lin (1973): *On invariant measures and ergodic theorems for positive operators.* J Funct. Anal. **13**, 252–267.
–, – (1984): *Sur le comportement asymptotique de puissances de convolution d'une probabilité.* AIHP, B, **20**, 127–132.

De Sam Lazaro, J. and P. A. Meyer (1975): *Questions de théorie des flots.* SpLNM **465**, 1–96.

Deshphande, M. V. and S. M. Padhye (1977): *An ergodic theorem for quasi-compact operators in locally convex spaces.* J math. phys. Sci., Madras **11**, 95–104.
–, – (1979): *Local properties of semi-groups of self-adjoint operators.* J Indian Math. Soc., New Ser. **42**, 85–94.

Dinges, Hermann (1970): *Combinatorics of partial sums.* Indiana Univ. MJ, **20**, 389–406.
– (1971): *Ein Überblick über einige neuere Ansätze zu den Gesetzen der großen Zahlen.* Proc. 4th Conf. Brasov, 37–50.
– (1974): *Stopping sequences.* Semin. Prob. VIII, SpLNM **381**, 27–36.

Direev, Yu.V. (1981): *An ergodic theorem for additive functionals.* Theory Probab. Appl. **25**, 614–617.

Dixmier, Jaques (1957): *Les algèbres d'opérateurs dans l'espaces Hilbertien.* Gauthier-Villars.

Djafari-Rouhani, Behzad and Shizuo Kakutani (1984): *Ergodic theorems for nonexpansive nonlinear operators in a Hilbert space.* Preprint.

Doeblin, Wolfgang (1938): *Sur deux problèmes de M. Kolmogoroff concernant les chaînes dénombrables.* Bull. Soc. math. France **66**, 210–220.
– (1940): *Eléments d'une theorie générale des chaînes simples constantes de Markoff.* Ann. Sci. Ecole norm. sup. III. Ser., **57**, 61–111.

Doob, Joseph L. (1938): *Stochastic processes with an integral valued parameter.* TAMS **44**, 87–150.
– (1948): *Asymptotic properties of Markov transition probabilities.* TAMS **63**, 393–421.
– (1953): *Stochastic processes.* Wiley.
– (1963): *A ratio operator limit theorem.* ZW **1**, 288–294.

Doplicher, S., D. Kastler and E. Størmer (1969): *Invariant states and asymptotic Abelianness.* J Funct. Anal. **3**, 419–434.

Dotson, W.G. jun. (1971): *Mean ergodic theorems and iterative solution of linear functional equations.* J math. Analysis Appl. **34**, 141–150.

Dowker, Yeal N. (1947): *Invariant measure and the ergodic theorems.* Duke MJ **14**, 1051–1061.
– (1950): *A new proof of the general ergodic theorem.* Acta Sci. Math. Szeged **12**, Pars B, 162–166.
– (1951): *Finite and σ-finite invariant measures.* Ann. Math. **54**, 595–608.
– (1955): *On measurable transformations in finite measure spaces.* Ann. Math., **62**, 504–516.
– (1956): *Sur les applications mesurables.* CRAS **242**, 329–331.
–, and Paul Erdös (1959): *Some examples in Ergodic Theory.* Proc. London M Soc. **9**, 227–241.

Dubins, Lester E. and Jim Pitman (1979): *A pointwise ergodic theorem for the group of rational rotations.* TAMS **251**, 299–308.

Duncan, Richard (1977): *Pointwise convergence theorems for self-adjoint and unitary contractions.* Ann. Prob. **5**, 622–626.

Dunford, Nelson (1939): *An ergodic theorem of n-parameter groups.* Proc. Nat. Acad. Sci. USA **25**, 195–196.
– (1939a): *A mean ergodic theorem.* Duke MJ **5**, 635–646.
– (1943): *Spectral theory I: Convergence to projections.* TAMS **54**, 185–217.
– (1951): *An individual ergodic theorem for non-commutative transformations.* Acta Sci. Math. (Szeged) **14**, 1–4.
– (1980): *Some ergodic theorems.* Proc. Roy. Soc. Edinburgh **85A**, 111–118.
– and J. T. Schwartz (1956): *Convergence almost everywhere of operator averages.* J. Rat. Mech. Anal. **5**, 129–178.
–, – (1958/1963): *Linear Operators I, II.* Interscience Publ.

Dvurecenskij, Anatolij and Beloslav Riecan (1980): *On the individual ergodic theorem on a logic.* Comm. Math. Univ. Carol. **21**, 385–391.

Dye, Henry (1965): *On the ergodic mixing theorem.* TAMS **18**, 123–130.

Eberlein, William F. (1949): *Abstract ergodic theorems and weak almost periodic functions.* TAMS **67**, 217–240.

– (1976): *Mean ergodic flows*. Adv. in Math. **21**, 229-232.

Effros, E. (1963): *Order ideals in a C*-algebra and its dual*. Duke MJ **30**, 391–411.

Ellis, Martin H. and Nathaniel A. Friedman (1978): *On eventually weakly wandering sequences*. Studies in probability and ergodic theory, Adv. Math., Suppl. Stud. **2**, 185–194.

–, – (1978a): *Gap sequences and eventually weakly wandering sequences*. Studies in probability and ergodic theory, Adv. Math., Suppl. Stud. **2**, 195–205.

Ellis, Robert (1957): *Locally compact transformation groups*. Duke MJ **24**, 119–126.

Emerson, William R. (1974): *The pointwise ergodic theorem for amenable groups*. Amer. JM **96**, 472–487.

– (1974a): *Large symmetric sets in amenable groups and the individual ergodic theorem*. Amer. JM **96**, 242–247.

– (1975/76): *Averaging strongly subadditive set functions in unimodular amenable groups I, II*. Pac. JM **61**, 391–400 and **64**, 353–368.

– and F.P. Greenleaf (1967): *Covering properties and Fölner conditions*. MZ **102**, 370–384.

Emilion, Richard (1982): *Convergence locale des processus sur-additifs dans* $L_p$. CRAS, I, **295**, 547–549.

– (1983): *Différentiation des processus suradditifs à plusieurs paramètres*. CRAS, I, **296**, 133–136.

– (1984): *Mean bounded operators and mean ergodic theorems*. Preprint. Results announced in CRAS I **296**, (1983), 641–643.

– (1984a): *On the pointwise ergodic theorems in* $L_p$, $(1 < p < \infty)$. Preprint.

– (1984b): *Continuity at zero of semi-groups on* $L^1$ *and differentiation of additive processes*. Preprint.

– (1984c): *Additive and superadditive local theorems*. Preprint.

– (1984d): *Semigroups in* $L_\infty$ *and local ergodic theorems*. Preprint.

– (1984e): *A general differentiation theorem for n-dimensional additive processes*. Preprint.

– and Bachar Hachem (1982): *Un theoreme ergodique local sur-additif*. CRAS, I **294**, 337–340.

–, – (1983): *Un théorème ergodique fortement suradditif à plusieurs paramètres*. CRAS **296**, I, 85–87.

England, James W. and Nathaniel F.G. Martin (1968): *On weak mixing metric automorphisms*. Bull. AMS **74**, 505–507.

Engmann, Hans (1976): *Notwendige und hinreichende Ungleichungen für die Existenz spezieller* $L^1$*-Kontraktionen*. ZW **33**, 317–329.

Van Enter, A.C.D. and J.L. Van Hemmen (1983): *The thermodynamic limit for long-range random systems*. J Statist. Physics **32**, 141–152.

Falkowitz, M. (1973): *On finite invariant measures for Markov operators*. PAMS **38**, 553–557.

– (1973a): *On weakly mixing Markov processes*. Isr. JM **14**, 221–227.

Fava, Norberto A. (1972): *Weak type inequalities for product operators*. Studia math., T, XLII., 271–288.

– and Jorge H. Nanc Lares (1978): *Norbert Wiener's ergodic theorem for convex regions*. TAMS **235**, 403–406.

Feder, Moshe (1981): *An example of local ergodic divergence*. C.R. Math. Rep. Acad. Sci. Canada **3**, 161–163.

– (1981a): *On power-bounded operators and the pointwise ergodic property*. PAMS **83**, 349–353.

Fefferman, C. and E.M. Stein (1971): *Some maximal inequalities*. Amer. JM **93**, 107–115.

Feldman, Jacob (1962): *Subinvariant measures for Markov operators*. Duke MJ **29**, 71–98.

– (1965): *Integral kernels and invariant measures for Markoff transition funcitions*. AM St **36**, 517–523.

Feller, William (1946): *A limit theorem for random variables with infinite moments*. Amer. JM **68**, 257.

Feyel, Denis (1977): *Un théorème ergodique maximal en theorie du potentiel*. CRAS A **284**, 437–439.

– (1977a): *Un théorème ergodique maximal en theorie du potentiel*. CRAS, A **284**, 753–754.

– (1977b): *Une remarque sur le lemme de Brunel*. CRAS A. **285**, 505–507.

– (1978): *Espaces complètement rèticulés de pseudo-noyaux. Applications aux résolvantes et aux semigroupes complexes. Sém. theorie potentiel, Paris, N° 3*, SpLNM **681**, 54–80.

– (1979): *Théorèmes de convergence presque-sûre, existence de semigroupes*. Adv. in Math. **34**, 145–162.

– (1982): *Convergence locale des processus sur-abéliens et sur-additifs*. CRAS I **295**, 301–303.

– (1982a): *Théorème ergodique ponctuel pour processus sur abéliens*. CRAS I, **295**, 357–358.

– (1984): *Processus abéliens associés a un semi-groupe*. MZ **187**, 305–315.

Fichtner, K.-H. and W. Freudenberg, (1979): *Weak ergodic theorems for nonlinear averaging procedures*. Math. Nachr. **90**, 39–55.

Flytzanis, Elias (1976): *On a vector-valued ergodic theorem*. Math. Balk. **6**, 64–69.

– (1978): *Vector-valued eigenfuctions of ergodic transformations*. TAMS **243**, 53–60.
– (1980): *Unitary Eigenoperators of Ergodic Transformations*. J Funct. Anal. **38**, 401–409.
Föllmer, Hans (1973): *On entropy and information gain in random fields*. ZW **26**, 207–217.
Foguel, Shaul R. (1963): *On order preserving contractions*. Isr. JM **1**, 54–59.
– (1964): *Invariant subspaces of measure preserving transformations*. Isr. JM **2**, 198–200.
– (1969): *Ergodic theory of Markov processes*. Van Nostrand Math. Studies **21**.
– (1969 a): *Ergodic decompositions of a topological space*. Isr. JM **7**, 164–167.
– (1971): *On the "zero-two" law*. Isr. JM **10**, 275–280.
– (1973): *Strong convergence of the iterates of an operator*. Isr. JM **16**, 159–161.
– (1973 a): *The ergodic theory of positive operators on continous functions*. Ann. Scuola Norm. Sup. Pisa **27**, 19–51.
– (1975): *Convergence of the iterates of a convolution*. PAMS **47**, 368–370.
– (1976):*More on the "zero-two" law*. PAMS **61**, 262–264.
– (1979): *Harris operators*. Isr. JM **33**, 281–309.
– (1980): *The Hopf decomposition of a Riesz space*. Houston JM **6**, 503–509.
– and N. A. Ghoussoub, (1979): *Ornstein-Métivier-Brunel theorem revisited*. AIHP **15**, 293–301.
– and M. Lin, (1972): *Some ratio limit theorems for Markov operators*. ZW **23**, 55–66.
– and Benjamin Weiss, (1973): *On convex power series of a conservative Markov operator*. PAMS **38**, 325–330.
Fong, Humphrey (1970): *On invariant functions for positive operators*. Collq. Math. **22**, 75–84.
– (1979): *Ratio and stochastic ergodic theorems for superadditive processes*. Can. JM **31**, 441–447.
– and M. Lin, (1976): *On the convergence of ergodic ratios for positive operators*. J math. Anal. Appl. **55**, 667–672.
– and L. Sucheston, (1971): *On the ratio ergodic theorem for semigroups*. Pac. JM **39**, 659–667.
–, – (1973): *On unaveraged convergence of positive operators in Lebesgue space*. TAMS **179**, 383–397.
–, – (1974): *On a mixing property of operators in $L_p$ spaces*. ZW **28**, 165–171.
Fortet, R. (1938): *Sur l'iteration des substitutions linéaires algebriques d'une infinité de variables et ses applications au problème des probabilités en chaîne*. Thèse de Doctorat, Paris, 1938.
– (1978): *Condition de Doeblin et quasi-compacité*. AIHP, B, **14**, 379–390.
Franken, P., D. König, U. Arndt and V. Schmidt, (1981): *Queues and point processes*. Akademie Verlag.
Freedman, David (1971): *Markov chains*. Holden Day.
Friedman, Nathaniel A. (1966): *On the Dunford-Schwartz Theorem*. ZW **5**, 226–231.
– (1970): *Introduction to Ergodic Theory*. Van Nostrand Math. Stud. **29**.
– (1979): *Eventually independent sequences*. Isr. JM **33**, 310–316.
– (1983): *Mixing on sequences*. Can. JM **35**, 339–352.
– (1984): *Higher Order partial mixing*. Proc. Kakutani Conf. "Contemporary Mathematics" AMS **26**, 111–130.
– and D. Ornstein, (1971): *On partially mixing transformations*. Indiana Univ. MJ **20**, 767–775.
–, – (1972): *On mixing and partial mixing*. Illinois JM **16**, 61–68.
Frigerio, A. and M. Verri, (1982): *Long-time asymptotic properties of dynamical semigroups on W*-algebras*. MZ **180**, 275–286.
Fritz, Jósef (1969): *Entropy of point processes*. Stud. Sci. Math. Hung. **4**, 389–399.
– (1970): *Generalization of McMillan's theorem to random set functions*. Stud. Sci. Math. Hung. **5**. 369–394.
Fukamiya, M. (1940): On the dominated ergodic theorem in $L_p$ ($p > 1$). Tôh. MJ **46**, 150–153.
Fukushima, Masatoshi (1974): *Almost polar sets and an ergodic theorem*. JMS Jap. **26**, 17–32.
– (1983): *Capacitary maximal inequalities and an ergodic theorem*. SpLNM **1021**, 130–136.
Furstenberg, Harry (1961): *Strict ergodicity and transformations of the torus*. Amer. JM **83**, 573–601.
– (1977): *Ergodic behaviour of diagonal measures and a theorem of Szemerédi on arithmetic progressions*. J d'Analyse M **31**, 204–256.
– (1981): *Recurrence in ergodic theory and combinatorial number theory*. Princeton Univ. Press.
– (1981 a): *Poincaré recurrence and number theory*. Bull. AMS, New Ser. **5**, 211–234.
– and Y. Katznelson, (1978): *An ergodic Szemerédi theorem for commuting transformations*. J d'Analyse M **34**, 275–291.
– and H. Kesten, (1960): *Products of random matrices*. AM St **31**, 457–469.
– and B. Weiss, (1977): *The finite multipliers of infinite ergodic transformations*. SpLNM **668**, Structure of Attractors, 127–132.
–, – (1978): *Topological dynamics and combinatorial number theory*. J d'Analyse M **34**, 61–85.

Gallavotti, Giovanni (1975): *Lectures on the billiard.* Lecture Notes in Physics **38**, 236–295.

– and D. Ornstein, (1974): *Billiards and Bernoulli schemes.* Comm. math. Physics **38**, 83–101.

Gantmacher, F.R. (1958): *Matrizenrechnung I.* VEB Deutscher Verlag d. Wiss.

Gaposhkin, V.F. (1981): *The local ergodic theorem for groups of unitary operators and second stationary processes.* Math. USSR, Sb. **39**, 227–242.

– (1981a): *Individual ergodic theorem for normal operators in $L_2$.* Funct. Anal. Appl. **15**, 14–18 = Funkts. Anal. Prilozh **15**, 18–22 (Russian), (1981).

Garsia, Adriano (1965): *A simple proof of E. Hopf's maximal ergodic theorem.* J Math. Mech. **14**, 381–382.

– (1967): *More about the maximal ergodic lemma of Brunel.* Proc. Nat. Acad. Sci. USA **67**, 21–24.

– (1970): *Topics in almost everywhere convergence.* Lectures in advanced math. **4**, Markham Publ. Co.

Geman, Donald and Joseph Horowitz, (1975): *Random shifts which preserve measure.* PAMS **49**, 143–150.

–, – and Joel Zinn, (1976): *Recurrence of stationary sequences.* Ann. Prob. **4**, 372–381.

Genel, A. and J. Lindenstrauss, (1975): *An example concerning fixed points.* Isr. JM **22**, 81–86.

Ghoussoub, Nassif and J.Michael Steele (1980): *Vector valued subadditive processes and applications in probability.* Ann. Prob. **8**, 83–95.

Glasner, Shmuel (1976): *Proximal flows.* SpLNM **517**.

– (1976a): *On Choquet-Deny measures.* AIHP, B, **12**, 1–10.

Godement, Roger (1948): *Les fonctions de type positif.* TAMS **63**, 1–84.

Göhde, D. (1965): *Zum Prinzip der kontraktiven Abbildung.* Math. Nachr. **30**, 251–258.

Gol'dshtejn, M.Sh. (1979): *Ein individueller Ergodensatz für substochastische Operatoren in von Neumannschen Algebren.* Dokl. Akad. Nauk USSR No. **7**, 3–5.

– (1981): *Theorems of almost everywhere convergence in von Neumann algebras.* J Oper. Theory **6**, 233–311, (Russian).

– (1981a): *An individual ergodic theorem for positive linear mappings of von Neumann algebras.* Funct. Anal. Appl. **14**, 303–304.

Goldstein, J.A., C. Radin, and R.E. Showalter, (1978): *Convergence rates for ergodic limits for semigroups and cosine functions.* Semigroup Forum **16**, 89–95.

Gologan, Radu-Nicolae (1976): *A remark on Chacon's ergodic theorem.* Rev. Roum. Math. Pur. App. **21**, 521–522.

– (1977): *A remark on the ergodic theorem with mixed conditions.* Rev. Roum. Math. Pur. Appl. **22**, 321–324.

– (1979): *On the ergodic theorem of Ornstein and Brunel for nonpositive operators.* Rev. Roum. Math. Pur. Appl. **24**, 235–239.

Graham, Virginia L. (1974): *Weakly wandering vectors for compactly generated groups of unitary operators.* JM Anal. Appl. **46**, 565–594.

Granirer, Edmond E., (1971): *On finite equivalent invariant measures for semigroups of transformations.* Duke MJ **38**, 395–408.

Gray, Robert M. and John C. Kiefer, (1980): *Asymptotically mean stationary measures.* Ann. Prob. **8**. 962–973.

Greenleaf, Frederick P. (1969): *Invariant means on topological groups and their applications*, van Nostrand Math. Studies Series **16**.

– (1973): *Ergodic theorems and the construction of summing sequences in amenable locally compact groups.* Comm. pure appl. M. **26**, 29–46.

– and William R. Emerson, (1974): *Group structure and the pointwise ergodic theorem for connected amenable groups.* Adv. Math. **14**, 153–172.

Greiner, G. and R. Nagel, (1982): *La loi "zero ou deux" et ses conséquences pour le comportement asymptotique des opérateurs positifs.* J Math. pures appliqué, **9**, Ser. 61, 261–273.

Grenander, Ulf (1952): *Stochastic processes and statistical inference. Arkiv Math.* **1**, 195–277.

Greub, Werner H. (1967): *Linear Algebra. Springer-Grundlehren* **97**, 3rd edition.

Grewe, Michael (1983): *Über konservative und dissipative Transformationen.* Diplomarbeit, Univ. Göttingen, Inst. Math. Stoch.

Grillenberger, Christian (1976): *Ensembles minimaux sans measure d'entropie maximale.* Monatsh. M **82**, 275–285.

– and U. Krengel, (1976): *On matrix summation and the pointwise ergodic theorem.* SpLNM **532**, 113–124.

–, – (1976a): *On marginal distributions and isomorphisms of stationary processes.* MZ **149**, 131–154.

–, – (1982): *On the spatial constant of superadditive set functions in $\mathbb{R}^d$.* Math. Forschung **12**, Ed. H. Michel, Akademie Verlag, 53–57.

Grimmett, Geoffrey R. (1976): *On the number of clusters in the percolation model.* J London M Soc. **13**, 346–350.

Groetsch, C.W. (1975/76): *Ergodic theory and the iterative solution of linear operator equations.* Applicable Anal. **5**, 313–321.

Groh, Ulrich and Burkhard Kümmerer, (1982): *Bibounded operators on W*-algebras*. Math. Scand. **50**, 269–285.
Guerre, S. (1978): *Procede de convergence minimale dans les espaces de Banach. Une loi des grands nombres et un théorème ergodique*. Sem. Geom. Espaces Banach, Ec. polytech., Cent. Math., 1977–1978, Exposé, 1–14.
Gundel, Horst (1979): *Weakly and strongly mixing locally compact abelian groups of measure preserving transformations with applications to abelian groups of automorphisms*. ZW **49**, 313–323.
Gundy, Richard F. (1969): *On the class L log L, martingales, and singular integrals*. Studia Math. **33**, 109–118.
Gutman, S. and A. Pazy (1983): *An ergodic theorem for semigroups of contractions*. PAMS **88**, 254–256.
Guzmán, see de Guzmán.

Haagerup, Uffe (1975): *Normal weights on W*-algebras*. J Funct. Anal. **19**, 302–317.
Hachem, Bachar (1981): *Sur le théorème ergodique surstationnaire et sous-additif dans $L_p$ $(1 \leq p < \infty)$*. CRAS, I, **292**, 837–840.
– (1982): *Quelques théorèmes ergodiques sur-additifs dans $L^p$, $(1 \leq p < \infty)$*. AIHP B **18**, 201–222.
Haïnis, Jean (1977): *A multiplicative ergodic theorem in a commutative Banach algebra*. Nonlin. Anal. **1**, 455–458.
Hajian, Arshag und Yuji Ito (1965): *Iterates of measurable transformations and Markov operators*. TAMS **117**, 371–386.
–, – (1967): *Conservative positive contractions in $L^1$*. Proc. Fifth Berkeley Symp. Math. Stat. Prob., Vol II, part II, 361–374.
–, – (1969): *Cesàro sums and measurable transformations*. J Combinatorial Theory **7**, 239–254.
–, – (1969a): *Weakly wandering sets and invariant measures for a group of transformations*. J Math. Mech. **18**, 1203–1216.
–, – (1978): *Transformations that do not accept a finite invariant measure*. Bull. AMS **84**, 417–427.
–, – and S. Kakutani (1972): *Invariant measures and orbits of dissipative transformations*. Adv. in Math. **9**, 52–65.
–, –, – (1974): *Orbits, sections and induced transformations*. Isr. JM **18**, 97–115.
– and S. Kakutani (1964): *Weakly wandering sets and invariant measures*. TAMS **110**, 136–151.
Halász, G. (1976): *Remarks on the remainder in Birkhoff's ergodic theorem*. Acta math. Acad. Sci. Hungar. **28**, 389–395.
Halmos, Paul, R. (1946): *An ergodic theorem*. Proc. Nat. Acad. Sci. USA **32**, 156–161.
– (1947): *Invariant measures*. Ann. Math. **48**, 735–754.
– (1948): *A non homogeneous ergodic theorem*. TAMS **66**, 284–288.
Hamachi, Toshihiro (1981): *On a Bernoulli shift with nonidentical factor measures*. Ergodic Th. and Dynam. Sys. **1**, 273–283.
– and Motosige Osikawa (1981): *Ergodic groups of automorphisms and Kriegers theorems*. Sem. Math. Sci. **3**, Keio Univ.
Hammersley, J. M. (1974): *Postulates for subadditive processes*. Ann. Prob. **2**, 652–680.
– and J. A. D. Welsh (1965): *First passage percolation, subadditive processes, stochastic networks, and generalized renewal theory*. Bernoulli-Bayes-Laplace Anniversary Volume. Springer.
Hanson, David L. and Gordon Pledger (1969): *On the mean ergodic theorem for weighted averages*. ZW **13**, 141–149.
– and F. T. Wright (1970): *On the existence of equivalent finite invariant measures*. ZW **14**, 200–202.
Hardy, G. H. (1949): *Divergent series*. Oxford Univ. Press.
– and J. E. Littlewood (1930): *A maximal theorem with function-theoretic applications*. Acta. Math. **54**, 81–116.
Harris, Theodore E. (1956): *The existence of stationary measures for certain Markov-processes*. Proc. 3rd. Berkeley Symp. Math. Stat. Prob. Vol. II, 113–124.
– and Herbert Robbins (1953): *Ergodic theory of Markov chains admitting an infinite invariant measure*. Proc. Nat. Acad. Sci. **39**, 860–864.
Hartman, Philip and Aurel Wintner (1941): *On the law of the iterated logarithm*. Amer. JM **63**, 169–176.
Hasegawa, S., R. Sato and S. Tsurumi (1978): *Vector valued ergodic theorems for a 1-parameter semigroup of linear operators*. Tôh. MJ II. Ser. **30**, 95–106.
Hayes, C. A. and C. Y. Pauc (1970): *Derivation and martingales*. Springer.
Heinich, Henri (1983): *Convergence des moyennes d'un opérateur positif*. CRAS **297**, 237–240.
Helmberg, Gilbert (1965): *Über die Zerlegung einer meßbaren Transformation in konservative und dissipative Bestandteile*. MZ **88**, 358–367.
– (1965a): *Über rein dissipative Transformationen*. MZ **90**, 41–53.

– (1966): *Über konservative Transformationen.* Math. Ann. **165**, 44–61.
– (1966a): *Über endliche invariante Maße auf Untermengen.* Monatsh. M **70**, 229–232.
– (1969): *Über mittlere Rückkehrzeit unter einer maßtreuen Strömung.* ZW **13**, 165–179.
– (1972): *On the converse of Hopf's ergodic theorem.* ZW **21**, 77–80.
– and F.H. Simons (1969): *On the conservative parts of Markov processes induced by a measurable transformation.* ZW **11**, 165–180.
Herglotz, Gustav (1911): *Über Potenzreihen mit positivem, reellem Teil im Einheitskreis.* Sitzungsber. Sächs. Akad. Wiss. **63**, 501–511.
Hess, Christian (1979): *Théorème ergodique et loi forte des grands nombres pour des ensembles aleatoires.* CRAS A **288**, 519–522.
Hewitt, Edwin and Karl Stromberg (1969): *Real and Abstract Analysis.* Springer, 2 nd printing.
Hiai, Fumio (1978): *Weakly mixing properties of semigroups of linear operators.* Kodai MJ **1**, 376–393.
– and R. Sato (1977): *Mean ergodic theorems for semigroups of positive linear operators.* JMS Jap. **29**, 123–134.
Hille, Einar (1945): *Remarks on ergodic theorems.* TAMS **57**, 246–269.
– and Ralph S. Phillips (1957): *Functional Analysis and Semi-groups.* Coll. Publ. AMS 2. ed. (1. ed. 1948).
Hirano, Norimichi (1980): *A proof of the mean ergodic theorem for nonexpansive mappings in Banach space.* PAMS **78**, 361–365.
– (1982): *Nonlinear ergodic theorems and weak convergence theorems.* JMS Jap. **34**, 35–46.
– and Wataru Takahashi (1979): *Nonlinear ergodic theorems for nonexpansive mappings in Hilbert spaces.* Kodai MJ **2**, 11–25.
Hofbauer, Franz and Gerhard Keller (1982): *Ergodic properties of invariant measures for piecewise monotonic transformations.* MZ **180**, 119–140.
Hoover, Th., A. Lambert and J. Quinn (1982): *The Markov process determined by a weighted composition operator.* Stud. Math. **72**, 225–235.
Hopf, Eberhard (1932): *Theory of measure and invariant integrals.* TAMS **34**, 373–393.
– (1937): *Ergodentheorie.* Ergebn. M **5**, Springer.
– (1947): *Über eine Ungleichung der Ergodentheorie.* Sitzber. Bayer. Akad. Wiss. Math. Nat. Abt. 1944, 171–176.
– (1954): *The general temporally discrete Markov process.* J Rat. Mech. Anal. **3**, 13–45.
– (1960): *On the ergodic theorem for positive linear operators.* J Reine Ang. Math. **205**, 101–106.
Horowitz, Shlomo (1968): *Some limit theorems for Markov processes.* Isr. JM **6**, 107–118.
– (1968a): *On σ-finite invariant measures for Markov processes.* Isr. JM **6**, 338–345.
– (1969): *Markov processes on a locally compact space.* Isr. JM **7**, 311–324.
– (1969a): *$L_\infty$-limit theorems for Markov processes.* Isr. JM **7**, 60–62.
– (1969b): *Strong ergodic theorems for Markov processes.* PAMS **23**, 328–334.
– (1972): *On finite invariant measures for sets of Markov operators.* PAMS **34**, 110–114.
– (1972a): *Transition probabilities and contractions of $L_\infty$.* ZW **24**, 263–274.
– (1974): *Semigroups of Markov operators.* AIHP, B, (N.S.) **10**, 155–166.
– (1979): *Pointwise convergence of the iterates of a Harris-recurrent Markov operator.* Isr. JM **33**, 177–180.
Hurewicz, Withold (1944): *Ergodic theorem without invariant measure.* Ann. Math. **45**, 192–206.

Ibragimov, I.A. and Yu.V. Linnik (1971): *Independent and stationary sequences of random variables.* Wolter-Nordhoff.
– and Y.A. Rozanov (1978): *Gaussian Random Processes.* Springer; Applic. of Math. **9**.
Ionescu Tulcea, Alexandra (1960): *Contributions to information theory for abstract alphabets.* Arkiv för Mathematik **4**, 235–247.
– (1964): *Ergodic properties of isometries in $L^p$-spaces, $1 < p < \infty$.* Bull AMS **70**, 366–371.
– (1965): *On the category of certain classes of transformations in Ergodic Theory.* TAMS **114**, 261–279.
– (1975): *An $L^p$-inequality with application to ergodic theory.* Houston JM **1**, 153–159.
– and Cassius Ionescu Tulcea (1963): *Abstract ergodic theorems.* TAMS **107**, 107–124.
– and M. Moretz (1969): *Ergodic properties of semi-Markovian operators on the $Z^1$-part.* ZW **13**, 119–122.
Ionescu Tulcea, Cassius T. (1980): *Ergodic theorems.* J. Math. Anal. Appl. **78**, 113–116.
– and G. Marinescu (1950): *Théorie ergodique pour des classes d'opérations non complètement continues.* Ann. Math. **52**, 140–147.
Irmisch, Robert (1980): *Punktweise Ergodensätze für $(C, \alpha)$-Verfahren, $0 < \alpha < 1$.* Dissertation, Fachbereich Mathematik, TH Darmstadt.
Isaac, Richard (1967): *On the ratio-limit theorem for Markov processes recurrent in the sense of Harris.* Illinois JM **11**, 608–615.
– (1973): *Theorems for conditional expectations with applications to Markov processes.* Isr. JM **16**, 362–374.

Isaacson, D.L. and R.W. Madsen (1976): *Markov chains, Theory and applications*. Wiley.

Ishitani, Hiroshi (1977): *A central limit theorem for the subadditive process and its application to products or random matrices*. Publ. R.I.M.S. Kyoto Univ. **12**, 565–575.

Istratescu, Vasile I. (1974): *On a class of operators and ergodic theory I*. Rev. Roum. Math. Pur. Appl. **19**, 411–420.

– (1977): *On a class of operators and ergodic theory II*. Proc. Fifth Conf. Prob. (Brasov 1974). Ed. Acad. R.S.R., Bucharest, 209–219.

– (1981): *Fixed point theory – an introduction*. D. Reidel, Dordrecht.

Itô, Kiyosi (1944): *On the ergodicity of a certain stationary process*. Proc. Imp. Acad. Tokyo **20**, 54–55.

– (1944a): *A kinematic theory of turbulence*. Proc. Imp. Acad. Tokyo **20**, 120–122.

– (1952): *Complex multiple Wiener integral*. Jap. JM **22**, 63–86.

Ito, Yuji (1964): *Invariant measures for Markov processes*. TAMS **110**, 152–184.

– (1965): *Uniform integrability and the pointwise ergodic theorem*. PAMS **16**, 222–227.

– (1981): *Invariant measures and the pointwise ergodic theorem*. Comment. Math. Univ. St. Pauli **30**, 193–201.

Itoh, S., H. Koerezlioglu and W. Takahashi (1973): *Ergodic theorems and averaging operators on measurable functions*. Sci. Rep. Yokohama nat. Univ., I **20**, 5–10.

Iwanik, Anzelm (1979): *Weak convergence and weighted averages for groups of operators*. Colloq. Math. **42**, 241–254.

– (1980): *Approximation theorems for stochastic operators*. Indiana Univ. MJ **29**, 415–425.

– (1981): *On pointwise convergence of Cesàro means and separation properties for Markov operators on $C(X)$*. Bull. Acad. Pol. Sci., Ser. Sci. Math. **29**, 515–520.

Jacobs, Konrad (1954): *Ein Ergodensatz für beschränkte Gruppen im Hilbertschen Raum*. Math. Ann. **128**, 340–349.

– (1955): *Periodizitätseigenschaften beschränkter Gruppen im Hilbertschen Raum*. MZ **61**, 408–428.

– (1956): *Ergodentheorie und fastperiodische Funktionen auf Halbgruppen*. MZ **64**, 298–338.

– (1957): *Fastperiodizitätseigenschaften allgemeiner Halbgruppen in Banach-Räumen*. MZ **67**, 83–92.

– (1957a): *Fastperiodische diskrete Markoffsche Prozesse von endlicher Dimension*. Abh. Math. Sem. Hamburg **21**, 194–246.

– (1958): *Fastperiodische Markoffsche Prozesse*. Math. Ann. **134**, 408–427.

– (1959): *Die Übertragung diskreter Informationen durch periodische und fastperiodische Kanäle*. Math. Ann. **137** (2), 125–135.

– (1960): *Neuere Methoden und Ergebnisse der Ergodentheorie*. Ergebn. M **29**, Springer.

– (1962–63): *Lecture Notes on Ergodic Theory*. Aarhus Univ., Mathematisk Inst.

– (1967): *On Poincaré's recurrence theorem*. Proc. Fifth Berkeley Symp. Math. Stat. Prob. Vol II, part II, 375–404.

– (1978): *Measure and integral*. Academic Press.

Jain, Naresh C. (1966): *Some limit theorems for a general Markov process*. ZW **6**, 206–223.

– and B. Jamison (1967): *Contributions to Doeblin's theory of Markov processes*. ZW **8**, 19–40.

Jajte, Ryszard (1968): *Remark on a mean ergodic theorem*. Ann. Polon. Math. **20**, 191–194.

– (1983): *Non-commutative subadditive ergodic theorems for semifinite von Neumann algebras*. Bull. Acad. Pol. Sci. Ser. Mat. Fiz. **31**, 353–360

– (1984): *A few remarks on the almost uniform ergodic theorems in von Neumann algebras*. Proc. IIIrd Conf. Prob. in Vector Spaces, Lublin 1983, SpLNM **1080**, 130–143.

Jamison, Benton (1964): *Asymptotic behavior of succesive iterates of continuous functions under a Markov operator*. J Math. Anal. Appl. **9**, 203–214.

– (1965): *Ergodic decomposition induced by certain Markov operators*. TAMS **117**, 451–468.

– (1970): *Irreducible Markov operators in $C(S)$*. PAMS **24**, 366–370.

– and Steven Orey (1967): *Tail σ-field of Markov processes recurrent in the sense of Harris*. ZW **8**, 41–48.

– and R. Sine (1969): *Irreducible almost periodic Markov operators*. J Math. Mech. **18**, 1043–1057.

–, – (1974): *Sample path convergence of stable Markov processes*. ZW **28**, 173–177.

Jerison, Meyer (1959): *Martingale formulation of ergodic theorems*. PAMS **10**, 531–539.

Jessen, Borge (1934): *On the approximation of Lebesgue integrals by Riemann sums*. Ann. Math. (2) **35**, 248–251.

–, J. Marcinkiewicz and A. Zygmund (1935): *Note on the differentiability of multiple integrals*. Fund. Math. **25**, 217–234.

Jones, Lee Kenneth (1971): *A mean ergodic theorem for weakly mixing operators*. Adv. Math. **7**, 211–216.

– (1972): *An elementary lemma on sequences of integers and its applications to functional analysis.* MZ **126**, 299–307.
– (1972a): *A short proof of Sucheston's characterization of mixing.* ZW **23**, 83–84.
– (1973): *A generalization of the mean ergodic theorem in Banach spaces.* ZW **27**, 105–107.
– and U. Krengel: *On transformations without finite invariant measure.* Adv. Math. **12**, 275–295.
– and Velimir Kuftinec (1971): *A note on the Blum-Hanson theorem.* PAMS **30**, 202–203.
– and Michael Lin (1976): *Ergodic theorems of weak mixing type.* PAMS **57**, 50–52.
–, – (1980): *Unimodular eigenvalues and weak mixing.* J Funct. Anal. **35**, 42–48.
Jones, Roger L. (1976): *The ergodic maximal function with cancellation.* Ann. Prob. **4**, 91–97.
– (1977): *Inequalities for the ergodic maximal function.* Studia Math. **60**, 111–129.
Junco, see del Junco

Kac, Marc (1947): *On the notion of recurrence in discrete stochastic processes.* AM St **53**, 1002–1010.
Kakutani, Shizuo (1938): *Iteration of linear operations in complex Banach spaces.* Proc. Imp. Acad. Tokyo **14**, 295–300.
– (1940): *Ergodic theorems and the Markoff process with a stable distribution.* Proc. Imp. Acad. Tokyo **16**, 49–54.
– (1943): *Induced measure preserving transformations.* Proc. Japan Acad. **19**, 635–641.
– (1950): *Ergodic Theory.* Proc. Int. Congr. of Math. **2**, 128–142.
– (1951): *Random ergodic theorems and Markoff processes with a stable distribution.* Proc. Sec. Berkeley Symp. Math. Stat. Prob. (1950), 247–261.
–, and Karl Petersen (1981): *The speed of convergence in the ergodic theorem.* Monatsh. M **91**, 11–18.
Kallenberg, Olav (1980): *Convergence of non-ergodic dynamical systems.* ZW **53**, 329–351.
Kamae, Teturo (1982): *A simple proof of the ergodic theorem using nonstandard analysis.* Isr. JM **42**, 284–290.
– (1983): *A characterization of weakly wandering sequences for nonsingular transformations.* Comm. Math. Univ. Sancti Pauli **32**, 55–59.
Kan, Charn-Huen (1978/1983): *Ergodic properties for Lamperti operators I, II.* Can JM, **30**, 1206-1214, and **35**, 577–585.
Kantorovič, L. (1940): *Linear operations in semi-ordered spaces.* Math. Sbornik **49**, 209–284.
Karamata, J. (1931): *Neuer Beweis und Verallgemeinerung der Tauberschen Sätze, welche die Laplacesche und Stieltjessche Transformation betreffen.* J Reine Ang. Math. **164**, 27–39.
Karlin, Samuel (1959): *Positive operators.* JM Mech. **8**, 907–937.
Kataoka, Takeshi (1981): *Note on a mean ergodic theorem.* MJ Okayama Univ. **23**, 205–206.
–, R. Sato, and H. Suzuki (1983): *Differentiation of superadditive processes.* Preprint.
Katznelson, Yitzhak and Benjamin Weiss (1972): *Commuting measure preserving transformations.* Isr. JM **12**, 161–173.
–, – (1982): *A simple proof of some ergodic theorems.* Jsr. JM **42**, 291–296.
Keane, Michael (1968): *Generalized Morse sequences.* ZW **10**, 335–353.
Keller, Gerhard (1977): *Ergodizität und K-Eigenschaft der Kollisionstransformation beim Sinai-Billard mit endlichem Horizont.* Diplomarbeit, Univ. Erlangen.
Kemeny, J.G., J.L. Snell and A.W. Knapp (1966): *Denumerable Markov chains.* van Nostrand.
Kern, Martin, Rainer Nagel and Günther Palm (1977): *Dilations of Positive Operators: Construction and Ergodic Theory.* MZ **156**, 265–277.
Kerstan, J. and K. Matthes (1965): *Gleichverteilungseigenschaften von Faltungspotenzen auf lokalkompakten Abelschen Gruppen.* Wiss. Z. Univ. Jena **14**, 457–462.
–, – and J. Mecke (1974): *Unbegrenzt teilbare Punktprozesse.* Akademie-Verlag.
Kesten, Harry (1975): *Sums of stationary processes cannot grow slower than linearly.* PAMS **49**, 205–211.
– (1982): *Percolation theory for mathematicians.* Birkhäuser, Progr. in Prob. and Stat. **2**.
Khintchine, A.I. (1933): *Zu Birkhoff's Lösung des Ergodenproblems.* Math. Ann. **107**, 485–488.
– (1934): *Eine Verschärfung des Poincaré'schen Wiederkehrsatzes.* Comp. Math. **1**, 177–179.
– (1949): *Mathematical Foundations of Statistical Mechanics.* Dover Publ.
Kieffer, John.C. (1973): *A counterexample to Perez's generalization of the Shannon-McMillan theorem.* Ann. Prob. **1**, 362–364.
– (1975): *A generalized Shannon-McMillan Theorem for the action of an amenable group on a probability space.* Ann. Prob. **3**, 1031–1037.
– (1975a): *A ratio limit theorem for strongly subadditive set functions in a locally compact amenable group.* Pac. JM **61**, 183–190.
– and J.G. Dunham (1983): *On a type of stochastic stability for a class of encoding schemes.*

IEEE Transact. Inf. Th. IT-**29**, (6), 793–797.
– and Maurice Rahe (1981): *Markov channels are asymptotically mean stationary*. Siam. JM Anal. **12**, 293–305.
–, – (1982): *The pointwise ergodic theorem for transformations whose orbits contain or are contained in the orbits of a measure preserving transformation*. Can. JM **34**, 1303–1318.
Kijima, Yoichi and Wataru Takahashi (1973): *Adjoint ergodic theorems for amenable semigroups of operators*. Science Rep. Yokohama Nat. Univ. I, **20**, 1–4.
Kim, Choo-Whan (1968): *A generalization of Ito's theorem concerning the pointwise ergodic theorem*. AM St **39**, 2145–2148.
– (1972): *Approximations of positive contractions on $L^\infty[0,1]$*. ZW **24**, 335–337.
Kin, Eijun (1972): *The General Random Ergodic Theorem I, II*. ZW **22**, 120–135, and 136–144.
Kingman, John F. C. (1968): *The ergodic theory of subadditive stochastic processes*. J Roy. Stat. Soc. B **30**, 499–510.
– (1973): *Subadditive ergodic theory*. Ann. Prob. **1**, 883–909.
– (1976): *Subadditive processes*. In: Ecole d'Eté de Probabilitès de Saint-Flour V, 1975. SpLNM **539**, 167–223.
Kipnis, Claude (1974): *Majoration des semigroupes de contractions de $L^1$ et applications*. AIHP, B **10**, 369–384.
Kirk, W. (1965): *A fixed point theorem for mappings which do not increase distances*. Amer. Math. Monthly **72**, 1000–1006.
Kirsch, W. and F. Martinelli (1982): *On the density of states of Schrödinger operators with a random potential*. J Phys. A.: Math. Gen. **15**. 2139–2156.
Kleinknecht, Ruth (1980): *Invariante Maße unter Operatorenhalbgruppen*. Diplomarbeit, Göttingen u. Clausthal.
Klimko, Eugene M. (1969): *A uniform operator ergodic theorem*. AM St **40**, 1126–1129.
– and L. Sucheston (1969): *An operator ergodic theorem for sequences of functions*. PAMS **20**, 272–276.
Kobayasi, Kazuo and Isao Miyadera (1980): *Some remarks on nonlinear ergodic theorems in Banach spaces*. Proc. Jap. Acad. A **56**, 88–92.
Koliha, Jaromir Joseph (1973): *Ergodic theory and averaging operations*. Can. JM **25**, 14–23.
– and A. P. Leung (1975): *Ergodic families of affine operators*. Math. Ann. **216**, 273–284.
Kolmogorov, Andrei Nikolaevič (1925): *Sur les fonctions harmoniques conjugées et les séries de Fourier*. Fund. Math. **7**, 23–28.
– (1930): *Sur la loi forte des grandes nombres*. CRAS **191**, 910–912.
– (1933): *Grundbegriffe der Wahrscheinlichkeitsrechnung*. Ergebn. M **2**, (Engl. transl. Foundations of Probability Theory. Chelsea, (1955)).
– (1937): *Ein vereinfachter Beweis des Birkhoff-Khintchine'schen Ergodensatzes*. Recueil M (Mat. Sborn.) N. S. **2** (44), 367–368.
Komlós, János (1967): *A generalization of a problem of Steinhaus*. Acta Math. Acad. Sci. Hungar. **18**, 217–229.
Koopman, B. O (1931): *Hamiltonian systems and linear transformations in Hilbert space*. Proc. Nat. Acad. Sci. USA **17**, 315–318.
– and J. von Neumann (1932): *Dynamical systems of continuous spectra*. Proc. Nat. Acad. Sci. USA **18**, 255–263.
Kopf, Christoph (1978): *Zur Struktur rein dissipativer Transformationen*. Archiv M **31**, 369–373.
– (1982): *Negative nonsingular transformations*. AIHP, B, **18**, 81–102.
Kopp, P. E. (1975): *Abelian ergodic theorems for vector valued functions*. Glasgow MJ **16**, 57–60.
Kovács, I. and J. Szücs (1966): *Ergodic type theorems in von Neumann algebras*. Acta Sci. Math. (Szeged) **27**, 233–246.
Kowada, Masasi (1973): *Convergence rate in the ergodic theorem for an analytic flow on the torus*. SpLNM **330**, 251–254.
Kowalsky, Hans-Joachim (1967): *Lineare Algebra*. Göschens Lehrbücher **27**, 3rd ed.
Krawczak, Michael (1985): *On upcrossing inequalities for subadditive superstationary processes*. Ergodic Theory and Dyn. Syst., to appear.
Krengel, Ulrich (1963): *Über den Absolutbetrag stetiger linearer Operatoren und seine Anwendung auf ergodische Zerlegungen*. Math. Scand. **13**, 151–187.
– (1966): *On the global limit behaviour of Markov chains and of general nonsingular Markov processes*. ZW **6**, 302–316.
– (1967): *Classification of states for operators*. Proc. Fifth Berkeley Symp. Math. Stat. Prob. Vol. II, part II, 415–429.
– (1968/69): *Darstellungssätze für Strömungen und Halbströmungen I, II*. Math. Ann. **176**, 181–190, **182**, 1–39.
– (1969 a): *A necessary and sufficient condition for the validity of the local ergodic theorem*. SpLNM **89**, 170–177.
– (1969 b): *A local ergodic theorem*. Invent. math. **6**, 329–333.
– (1970): *Transformations without finite invariant measure have finite strong generators*. SpLNM **160**, 133–157.

– (1971): *On the individual ergodic theorem for subsequences*. AM St. **42**, 1091–1095.
– (1971a): *K-flows are forward deterministic, backward completely non-deterministic stationary point processes*. J Math. Anal. Appl. **35**, 611–620.
– (1972): *Weakly wandering vectors and weakly independent partitions*. TAMS **164**, 199–226.
– (1976): *Un theoreme ergodique pour les processus sur-stationnaires*. CRAS, A **282**, 1019–1021.
– (1976a): *On Rudolph's representation of aperiodic flows*. AIHP, B, **12**, 319–338.
– (1978): *Recent progress on ergodic theorems*. Astérisque **50**, 151–192.
– (1978a): *On the speed of convergence in the ergodic theorem*. Monatsh. M **86**, 3–6.
– and M. Lin (1984): *On the range of the generator of a Markovian semigroup*. MZ **185**, 553–565.
– and Ronald Pyke (1985): *Uniform pointwise ergodic theorems for classes of averaging sets and multiparameter subadditive processes*. Preprint.
–, Rita Röttger and Ulrich Wacker (1983): *A renewal type mean ergodic theorem*. ZW **64**, 269–274.
– and L. Sucheston (1969): *Note on shift-invariant sets*. AM St **40**, 694–696.
–, – (1969a): *On mixing in infinite measure spaces*. ZW **13**, 150–164.
–, – (1978): *On semiamarts, amarts and processes with finite value*. In: Probability on Banach spaces, Adv. in Probability and Rel. Topics, editor: J. Kuelbs, **4**, 197–266.
Krieger, Wolfgang (1975): *On generators in ergodic theory*. Proc. Int. Congress of Mathematicians (Vancouver), **2**, 303–308.
Kronecker, Leopold (1884): *Näherungsweise ganzzahlige Auflösung linearer Gleichungen*. Kgl. Preuss. Akad. Wiss. Berlin, 1179–1193, 1271–1299 oder Werke III, 47–109.
Kryloff, Nicolas and Nicolas Bogoliouboff (1937): *Sur les propriétes en chaîne*. CRAS **204**, 1386–1388.
–, – (1937a): *Les propriétés ergodiques des suites des probabilités en chaîne*. CRAS **204**, 1454–1456.
–, – (1937b): *La théorie générale de la mesure dans son application de l'étude des systèmes dynamiques de la méchanique non linéaire*. Ann. Math. **38**, 65–113.
Kubo, Izumi (1969): *Quasi-flows*. Nagoya MJ **35**, 1–30.
– (1976): *Perturbed Billard Systems I: The ergodicity of the motion of a particle in a compound central field*. Nagoya MJ **61**, 1–57.
– and H. Murata (1981): *Perturbed Billard Systems II: Bernoulli properties*. Nagoya MJ **81**, 1–25.
Kubokawa, Yoshihiro (1972): *A general local ergodic theorem*. Proc. Jap. Acad. **48**, 461–465.
– (1972a): *A pointwise ergodic theorem for positive bounded operators*. Proc. Jap. Acad. **48**, 458–460.
– (1974): *A local ergodic theorem for semigroups in $L_p$*. Tôh. MJ **26**, 411–422.
– (1975): *Ergodic theorems for contraction semigroups*. JMS Jap. **27**, 184–193.
Kühne, R. (1982): *On weak mixing for semigroups*. Math. Forschung **12**, Ed. H. Michel; Akademie Verlag, 133–139.
– (1982a): *Fixpunkteigenschaften ergodischer Halbgruppen positiver Operatoren in KB-Räumen*. Math. Nachr. **108**, 252–266.
– (1982b): *Fixed points of semigroups of positive operators in KB-spaces*. Spectral Theory. Banach Center Publ. **8**, 313–319.
Kümmerer, Burkhard (1978): *An non-commutative individual ergodic theorem*. Invent. math. **46**, 139–145.
– and R. Nagel (1979): *Mean ergodic semigroups on W*-algebras*. Acta Sci. Math. (Szeged) **41**, 151–159.
Künsch, Hansrudolf (1981): *Almost sure entropy and the variational principle for random fields with unbounded state space*. ZW **58**, 69–85.
Kuipers, L. and H. Niederreiter (1974): *Uniform distribution of sequences*. Wiley.
Kurth, Rudolf (1975): *Some remarks on Poincaré's recurrence theorem*. Rend. Circ. Mat. Palermo **24**, 251–254.
Kurtz, Lynn C. and Don H. Tucker (1968): *An extended form of the mean ergodic theorem*. Pac. JM **27**, 539–545.

Lance, E. Christopher (1976): *Ergodic theorems for convex sets and operator algebras*. Invent. math. **37**, 201–214.
Landers, Dieter and Lothar Rogge (1978): *An ergodic theorem for Frechet-valued random variables*. PAMS **72**, 49–53.
Lange, Kenneth (1979): *On Cohen's stochastic generalization of the strong ergodic theorem of demography*. J Appl. Prob. **16**, 496–504.
Lasota, A. and J.A. Yorke (1973): *On the existence of invariant measures for piecewise monotonic transformations*. TAMS **186**, 481–488.
Lau, Anthony To-Ming (1976): *W*-algebras and invariant functionals*. Studia Math., T.LVI., 253–261.

Lebesgue, Henri (1910): *Sur l'integration des fonctions discontinues*. Ann. Ecole Norm. **27**, 361–450.

Ledrappier, Francois (1981): *Some relations between dimension and Lyapunov exponents*. Commun. Math. Phys. **81**, 229–238.

Leonov, V. P. (1960): *The use of the characteristic functional and semi-invariants in the ergodic theory of stationary processes*. Sov. Math. **1**, 878–881.

Lesigne, Emmanuel (1981): *Convergence de moyennes ergodiques pour des mesures diagonales*. Thèse, Univ. de Rennes.

Leviatan, D. (1974): *Saturation theorems related to the mean ergodic theorem*. Indiana Univ. MJ **24**, 87–91.

– and U. Westphal (1973): *On the mean ergodic theorem and approximation*. Mathematica, Cluj **15** (38), 83–88.

Levitan, M. L. (1970): *Some ratio limit behaviour of Markov processes*. ZW **15**, 29–50.

– (1971): *A generalized Doeblin ratio limit theorem*. AM St **42**, 904–911.

Liggett, Thomas M. (1985): *An improved subadditive ergodic theorem*. To appear Ann. Prob.

Lin, Michael (1970): *Conservative Markov processes on a topological space*. Isr. JM **8**, 165–186.

– (1970a): *Mixed ratio limit theorems for Markov processes*. Isr. JM, **8**, 357–366.

– (1971): *Mixing for Markov operators*. ZW **19**, 231–242.

– (1972): *Semi-groups of Markov operators*. Bolletino U.M.I. 4. Ser. **6**, 20–44.

– (1972a): *Mixing of Cartesian squares of positive operators*. Isr. JM **11**, 349–354.

– (1972b): *Strong ratio limit theorems for Markov processes*. AM St **43**, 569–579.

– (1974): *On the uniform ergodic theorem*. PAMS **43**, 337–340.

– (1974a): *On the uniform ergodic theorem II*. PAMS **46**, 217–225.

– (1974b): *On quasi-compact Markov operators*. Ann. Prob. **2**, 464–475.

– (1974c): *Operator repesentations of compact groups and ergodic theorems of weak-mixing type*. Unpublished; see Jones-Lin (1976/80).

– (1974d) *Convergence of the iterates of a Markov operator*. ZW **29**, 153–163.

– (1975/76): *Quasi-compactness and uniform ergodicity of Markov operators*. AIHP, B, **11**, 345–354.

– (1976): *Strong ratio limit theorems for mixing Markov operators*. AIHP, B, **12**, 181–191.

– (1977): *Ergodic properties of an operator obtained from a continuous representation*. AIHP, B, **13**, 321–331.

– (1978): *Quasi-compactness and uniform ergodicity of positive operators*. Isr. JM, **29**, 309–311.

– (1981): *On weakly mixing Markov operators and nonsingular transformations*. ZW **55**, 231–236.

– (1982): *On local ergodic convergence of semigroups and additive processes*. Isr. JM **42**, 300–308.

– (1982a): *On the "zero-two" law for conservative Markov processes*. ZW **61**, 513–525.

–, J. Montgomery and R. Sine (1977): *Change of velocity and ergodicity in flows and in Markov semigroups*. ZW **39**, 197–211.

– and R. Sine (1977): *The individual ergodic theorem for non-invariant measures*. ZW **38**, 329–331.

–, – (1979): *A spectral condition for strong convergence of Markov operators*. ZW **47**, 27–29.

–, – (1983): *Ergodic theory and the functional equation $(I\text{-}T)x = y$*. J Operator Theory **10**, 153–166.

Lloyd, Stuart P. (1963): *On certain projections in spaces of continuous functions*. Pac. JM **13**, 171–175.

– (1976): *On the mean ergodic theorem of Sine*. PAMS **56**, 121–126.

Loève, Michel (1960): *Probability Theory*. van Nostrand, 2nd ed.

Lorch, Edgar R. (1939): *Means of iterated transformations in reflexive vector spaces*. Bull. AMS **45**, 945–947.

Lorentz, G. G. (1948): *A contribution to the theory of divergent sequences*. Acta Math. **80**, 167–190.

Lotz, Heinrich P. (1968): *Über das Spektrum positiver Operatoren*. MZ **108**, 15–32.

– (1981): *Uniform ergodic theorems for Markov operators on $C(X)$*. MZ **178**, 145–156.

– and H. H. Schaefer (1968): *Über einen Satz von F. Niiro und I. Sawashima*. MZ **108**, 33–36.

Luczak, Andrzej (1974): *Ergodic theorem for a stationary process over the group of linear transformations of the line*. Bull. Soc. Sci. Lett. Lodz **24**, No. 12, 10 p.

– (1984): *Ergodic theorems for states and elements of a C*-algebra*. Preprint.

Maak, Wilhelm (1950): *Fastperiodische Funktionen*. Springer Grundlehren LXI.

– (1954): *Periodizitätseigenschaften unitärer Gruppen*. Math. Scand. **2**, 334–344.

Maharam, Dorothy (1964): *Incompressible transformations*. Fund. Math. **56**, 35–50.

– (1969): *Invariant measures and Radon-Nikodym derivatives*. TAMS **135**, 223–248.

Maker, P. (1940): *The ergodic theorem for a sequence of functions*. Duke MJ **6**, 27–30.

Marcinkiewicz, J. and A. Zygmund (1937): *Sur les fonctions independantes*. Fund. Math. **29**, 60–90.

Marcus, Brian and Karl Petersen (1979): *Balancing ergodic averages*. In: Ergodic Th., Oberwolfach 1978, SpLNM **729**, 126–143.

Margulis, G.A. (1980): *Some remarks on invariant means*. Monatsh. M **90**, 233–236.

Marstrand, John Martin (1970): *On Khintchine's conjecture about strong uniform distribution*. Proc. London M Soc. **21**, 540–556.

Maruyama, G. (1949): *The harmonic analysis of stationary stochastic processes*. Memoirs Fac. Sci. Kyushu Univ. A **4**, 45–106.

Masani, Pesi R. (1976): *Ergodic theorems for locally integrable semigroups of continuous linear operators on a Banach space*. Adv. Math. **21**, 202–228.

Mazur, S. and Orlicz, W. (1933): *Über Folgen linearer Operatoren*. Studia Math. **4**, 152–157.

McGrath, Steven A. (1976): *On the local ergodic theorems of Krengel, Kubokawa, and Terrell*. Comm. Math. Univ. Carolinae, **17**, 49–59.

– (1976a). *A pointwise abelian ergodic theorem for $L_p$-semigroups, $1 \leq p < \infty$*. J Funct. Anal. **23**, 195–198.

– (1980): *Local ergodic theorems for noncommuting semigroups*. PAMS **79**, 212–216.

– (1980a): *Some ergodic theorems f. commuting $L_1$-contractions*. Studia Math. **70**, 165–172.

McMillan, Brockway (1953): *The basic theorems of information theory*. AM St **24**, 196–216.

Menchoff, D. (1923): *Sur les séries de fonctions orthogonales I*. Fund. Math. **4**, 82–105.

Mesiar, Radko (1984): *La convergence presque sûre des T-moyennes de Cesàro*. Preprint.

Métivier, Michel (1969): *Existence of an invariant measure and an Ornstein's ergodic theorem*. AM St **40**, 79–96.

– (1972): *Théorèmes limite quotient pour chaînes de Markov récurrentes au sens de Harris*. AIHP, B, **8**, 93–105.

Meyer, Paul André (1965): *Théorie ergodique et potentiels*. Ann. Inst. Fourier, Grenoble, **15**, 89–96.

– (1965a): *Théorie ergodique et potentiels*. Identification de la limite. Ann. Inst. Fourier, Grenoble **15**, 97–102.

– (1969/70): *Travaux de H. Rost en théorie du balayage*. SpLNM **191**, 237–250.

Millet, Annie (1980): *Sur le théorème en moyenne d'Akcoglu-Sucheston*. MZ **172**, 213–237.

– and Louis Sucheston (1979): *On the existence of σ-finite invariant measures for operators*. Isr. JM **33**, 349–367.

Miyadera, I. (1978/79): *Asymptotic behavior of iterates of nonexpansive mappings in Banach spaces I, II*. Proc. Jap. Acad. A **54**, 212–214 and 318–321.

Moore, R.T. (1968): *Measurable, continuous and smooth vectors for semigroups and group representations*. Amer. Math. Soc. Memoirs **78**.

Moulin-Ollagnier, Jean (1983): *Théorème ergodique presque sous-additif et convergence en moyenne de l'information*. AIHP, B, **19**, 257–266.

Mourier, Edith (1953): *Eléments aléatoires à valeurs dans un espace de Banach*. AIHP, B, **13**, 161–244.

Moy, Shu-Teh Chen (1959): *Successive recurrence times in a stationary process*. AM St **30**, 1254–1257.

– (1960): *Equalities for stationary processes similar to an equality of Wald*. AM St **31**, 995–1000.

– (1961): *Generalization of Shannon-McMillan theorem*. Pac. JM **11**, 705–714.

– (1965): *λ-continuous Markov chains I, II*. TAMS **117**, 68–91 and **120**, 83–107.

– (1967): *Period of an irreducible positive operator*. Illinois JM **11**, 24–39.

Muckenhoupt, B. (1972): *Weighted norm inequalities for the Hardy maximal function*. TAMS **165**, 207–226.

Nagel, Rainer (1973): *Mittelergodische Halbgruppen linearer Operatoren*. Ann. Inst. Fourier, Grenoble, **23**, 75–87.

– (1974): *Ergodic and Mixing Poperties of Linear Operators*. Proc. Royal Irish Acad. **74**, 245–261.

– and Günther Palm (1982): *Lattice dilations of positive contractions on $L_p$-spaces*. Can. Math. Bull. **25**, 371–374.

–, –, and R. Derndinger (1984): *Ergodic Theory in the perspective of Functional Analysis:* 13 Lectures. To appear, SpLNM.

Natarajan, S. and K. Viswanath (1967): *On weakly stable transformations*. Sankhya Ser. A. **29**, 245–258.

Nawrotzki, Kurt (1969): *Mischungseigenschaften stationärer unbegrenzt teilbarer zufälliger Distributionen*. Wiss. Z. Fr. Schiller Univ. Jena, Math.-Nat. Reihe **18**, 397–408.

– (1982): *Ergodic theorem for random processes with embedded homogeneous flow*. Th. Prob. Appl. **26**, 388–392.

von Neumann, John (1932): *Proof of the quasi-ergodic hypothesis*. Proc. Nat. Acad. Sci. USA **18**, 70–82.

Neveu, Jacques (1961): *Sur le théorème ergodique ponctuel*. CRAS **252**, 1554–1556.

– (1964): *Potentiels markoviens discrets*. Ann. Univ. Clermont **24**, 37–89.
– (1965): *Mathematical foundations of the calculus of probability*. Holden-Day.
– (1965a): *Relations entre la théorie des martingales et la théorie ergodique*. Ann. Inst. Fourier (Grenoble) **15**, 31–42.
– (1967): *Existence of bounded invariant measures in ergodic theory*. Proc. Fifth Berkeley Symp. Math. Stat. Prob. II, (2), 461–472.
– (1968): *Processus aléatoires Gaussiens*. Les Presses de l'Univ. de Montreal.
– (1969): *Temps d'arrêt d'un système dynamique*. ZW **13**, 81–94.
– (1972): *Potentiels Markoviens récurrents des chaînes de Harris*. Ann. Inst. Fourier **22**, 85–130.
– (1972a): *Sur l'irréductibilité des chaînes de Markov*. AIHP, B, **8**, 249–254.
– (1975): *Discrete Parameter Martingales*. North Holland.
– (1976): *Processus ponctuels*. SpLNM **598**, 249–447.
– (1979): *The filling scheme and the Chacon-Ornstein theorem*. Isr. JM **33**, 368–377.
– (1983): *Courte démonstration du théorème ergodique sur-additif*. AIHP, B, **19**, 87–90.
Newton, Daniel (1966): *On Gaussian processes with simple spectrum*. ZW **5**, 207–209.
– (1968): *On a principal factor system of a normal dynamical system*. J London M Soc. **43**, 275–279.
Nguyen, Xuan Xanh (1979): *Ergodic theorems for subadditive spatial processes*. ZW **48**, 159–176.
– and Hans Zessin (1976): *Punktprozesse mit Wechselwirkung*. ZW **37**, 91–126.
–, – (1979): *Ergodic theorems for spatial processes*. ZW **48**, 133–158.
Niiro, F. and Ikuko Sawashima (1966): *On the spectral properties of positive irreducible operators in an arbitrary Banach lattice and problems of H. H. Schaefer*. Scientific Papers, College of General Education, Univ. of Tokyo, **16**, 145–183. See also Sawashima-Niiro (1973).
Nuber, J.A. (1972): *A constructive ergodic theorem*. TAMS **164**, 115–137. Erratum TAMS **216**, 393, (1976).
Nummelin, E. (1978): *A splitting technique for Harris recurrent Markov chains*. ZW **43**, 309–318.

O'Brien, George L. (1982): *The occurrence of large values in stationary sequences*. ZW **61**, 347–353.
– (1983): *Obtaining prescribed rates of convergence for the ergodic theorem*. Can. JM, to appear.
O'Brien, Robert E. jun. (1978): *Contraction semigroups, stabilization, and the mean ergodic theorem*. PAMS **71**, 89–94.
Olsen, James H. (1973): *Estimates of convex combinations of commuting isometries*. ZW **26**, 317–324.
– (1983): *A multiple sequence ergodic theorem*. Can. Math. Bull. **26**, 493–497.
Orey, Steven (1962): *An ergodic theorem for recurrent Markov chains*. ZW **1**, 174–176.
– (1971): *Limit theorems for Markov chain transition probabilities*. van Nostrand.
Ornstein, Donald S. (1960): *On invariant measures*. Bull. AMS **66**, 297–300.
– (1968): *On the pointwise behavior of iterates of a self-adjoint operator*. J Math. Mech. **18**, 473–478.
– (1969): *On a theorem of Orey*. PAMS **22**, 549–551.
– (1969a): *Random walks I, II*. TAMS **138**, 1–43, 45–60.
– (1970): *The sums of iterates of a positive operator*. Adv. in Prob. **2**, 85–115.
– (1971): *A remark on the Birkhoff ergodic theorem*. Illin. JM **15**, 77–79.
– and Louis Sucheston (1970): *An operator theorem on $L_1$ convergence to zero with applications to Markov kernels*. AM St **41**, 1631–1639.
– and B. Weiss (1983): *The Shannon-McMillan-Breiman theorem for a class of amenable groups*. Isr. JM **44**, 53–60.
Oseledec, V.I. (1965): *Markov chains, skew products and ergodic theorems for "general" dynamical systems*. Theor. Prob. Appl. **10**, 499–504.
– (1968): *A multiplicative ergodic theorem. Ljapunov characteristic numbers for dynamical systems*. Trudy Moskov. mat. Obsc. **19**, 179–210 (Russian) (1968), engl. transl. in Trans. Moscow. math. Soc. **19**, 197–231 (1969).
Osikawa, Motosige and Toshihiro Hamachi (1971): *On zero type and positive type transformations with infinite invariant measure*. Mém. Fac. Sci. Kyushu Univ., A. Math. **25**, 280–295.
Oxtoby, John C. (1952): *Ergodic sets*. Bull. AMS **58**, 116–136.

Padgett, W.L. and R.L. Taylor (1973): *Laws of large numbers for normed linear spaces and certain Fréchet spaces*. SpLNM **360**.
Pakula, Lewis and Robert Sine (1977): *On a theorem of Furstenberg and the structure of*

*topologically ergodic measures*. PAMS **65**, 52–56.

Papangelou, Fredos (1974): *On the Palm probabilities of processes of points and processes of lines*. In: Stochastic Geometry, ed. E. F. Harding and D. G. Kendall, Wiley, p. 114–147.

– (1976/77): *A martingale approach to the convergence of the iterates of a transition function*. ZW **37**, 211–226.

Parry, William (1963): *An ergodic theorem of information theory without invariant measure*. Proc. London M Soc. XIII, **52**, 605–612.

Patil, D.J. (1977): *Mean ergodic theorem in reflecive spaces*. Acta Sci. math. **39**, 135–138.

Pazy, Amnon (1977): *On the asymptotic behaviour of iterates of nonexpansive mappings in Hilbert space*. Isr. JM, **26**, 197–204.

– (1978): *The asymptotic behaviour of semigroups of nonlinear contractions having large sets of fixed points*. Proc. Royal Soc. Edinb. A **80**, 261–271.

– (1979): *Remarks on nonlinear ergodic theory in Hilbert space*. Nonlinear Anal., **3**, 863–871.

– (1980): *Some remarks on nonlinear ergodic theory in Hilbert space*. Memórias de Matemática da Universidade Federal do Rio de Janeiro 120.

Pedersen, Gert K. (1979): C*-Algebras and their automorphism groups. Academic Press.

Peller, V. V. (1978): *L'inégalité de von Neumann, la dilatation isométrique et l'approximation par isométries dans $L^p$*. CRAS, A **287**, 311–314.

Peressini, Anthony L. (1967): *Ordered topological vector spaces*. Harper and Row.

– and D. R. Sherbert (1966): *Order properties of linear mappings on sequence spaces*. Math. Ann. **165**, 318–332.

Pérez, Albert (1957): *Notions généralisées d'incertitude, d'entropie et d'information du point de vue de la théorie de martingales*. Transact. First Prague Conf. on Inf. Theory....,183–208.

– (1964): *Extensions of Shannon-McMillan's limit theorem to more general stochastic processes*. Transact. Third Prague Conf. on Inf. Theory,..., 545–574.

Pesin, Ya. B. (1977): *Characteristic Lyapunov exponents and smooth ergodic theory*. Russian Math. Surveys, **32**, 55–114.

Petersen, Karl (1979): *The converse of the dominated ergodic theorem*. J. Math. Anal. Appl. **67**, 431–436.

– (1983): *Ergodic Theory*. Cambridge studies in advanced math. **2**, Cambridge Univ. Press.

Pfaffelhuber, E. (1975): *Moving shift averages for ergodic transformations*. Metrika **22**, 97–101.

Pfort, Peter (1984): *Spezielle Funktionen und verwandte Begriffe für Markov-Operatoren*. Diplomarbeit, Univ. Göttingen, Inst. Math. Stoch.

Philipp, Walter and William F. Stout, (1975): *Almost sure invariance principles for partial sums of weakly dependent random variables*. Mem. AMS **2**, issue 2 No 161.

Phillips, Ralph S. (1943): *On weakly compact subsets of a Banach space*. Amer. JM **65**, 108–136.

Pinsker, M. S. (1960): *Information and information stability of random variables and processes*. Moscow 1960; German transl. in: Arbeiten zur Informationstheorie V, Berlin (1963).

Pitman, J. (1974): *Uniform rates of convergence for Markov chain transition probabilities*. ZW **29**, 199–227.

Pitskel, B. S. (1971): *Some remarks concerning the individual ergodic theorem of information theory*. Math. Notes **9**, 54–60, transl. of Mat. Zametki **9**, 93–103 (Russian).

– (1978): *The entropy of random fields that are homogeneous with respect to a commutative group of transformations*. Mat. Zametki **23**, 447–462; Math. Notes **23**, 242–250.

– (1980): *The Shannon-MacMillan-Breiman theorem for space-homogeneous stationary processes*. Russ. Math.Surv. **35**, 184–185.

– and A. M. Stepin (1971): *The property of entropy equidistribution of commutative groups of metric automorphisms*. Dokl. Akad. Nauk SSSR **198**, 1021–1024.

Pitt, H. R. (1942): *Some generalizations of the ergodic theorem*. Proc. Cambr. Phil. Soc. **38**, 325–343.

Poincaré, Henri (1899): *Les méthodes nouvelles de la mécanique céleste I (1892), II (1893), III (1899)*. Gauthier-Villars, Paris. Also: Dover, New York, 1957.

Pop-Stojanovic, Zoran R. (1972): *On ergodic theorem for a Banach valued random sequence*. J Austral. MSoc. **13**, 501–507.

Prizva, G. I. (1972): *An ergodic theorem for a class of Markov processes*. Select. Translat. math. Statist. Probab. **10**, 11–14.

Pulmannová, Sylvia (1982): *Individual ergodic theorem on a logic*. Math. Slovaca **32**, 413–416.

Radin, Charles (1973): *Ergodicity in von Neumann algebras*. Pac. JM **48**, 235–239.

– (1976): *A noncommutative $L_1$-mean ergodic theorem*. Adv. Math. **21**, 110–111.

– (1978): *Pointwise ergodic theory on operator algebras*. J Math. Phys. **19**, 1983–1985.

Raghunathan, M.S. (1979): *A proof of Oseledec's multiplicative ergodic theorem.* Isr. JM **32**, 356–362.

Raimi, R.A. (1964): *Minimal sets and ergodic measures in $\beta N \backslash N$.* Bull. AMS **70**, 711–712.

Rao, M.M. (1966): *Interpolation, ergodicity, and martingales.* J Math. Mech. **16**, 543–568.

– (1973): *Abstract martingales and ergodic theory.* Multivar. Anal III (Proc. Third Intern. Symp. on Mult. Anal., Wright State Univ., Dayton Ohio 1972). pp 45–60. Academic Press, N.Y.

– (1979): *Bistochastic operators.* Commentat. math., spec. Vol. II, 301–313.

Regnier, A. (1970): *Théorèmes ergodiques individuels purement topologiques.* AIHP B, **6**, 271–280.

Reich, Jacob I. (1977): *On the individual ergodic theorem for subsequences.* Ann. Prob. **5**, 1039–1046.

– (1981): *Random ergodic sequences on LCA groups.* TAMS **265**, 59–68.

Reich, Simeon (1977): *Nonlinear evolution equations and nonlinear ergodic theorems.* Nonlinear Anal., **1**, 319–330.

– (1978): *Almost convergence and nonlinear ergodic theorems.* J Approx. Th. **24**, 269–272.

– (1979): *Weak convergence theorems for nonexpansive mappings in Banach spaces.* J Math. Anal. Appl. **67**, 274–276.

– (1983): *A note on the mean ergodic theorem for nonlinear semigroups.* J Math. Anal. Appl. **91**, 547–551.

Rényi, Alfred (1963): *On stable sequences of events.* Sankhyá A, **25**, 293–302.

Révész, Pál (1960/61): *Some remarks on the random, ergodic theorem I, II.* Magyar Tud. Akad. Mat. Kutato Int. Közl. **5**, 375–381, **6**, 205–215.

Revuz, Daniel (1970): *Théorèmes limites pour les résolvantes récurrentes.* Rend. Circ. Mat. Palermo **19**, 294–300.

– (1975): *Markov Chains.* North Holland Math. Library **11**, North Holland American Elsevier. (Expanded second edition 1984).

– (1978): *Remarks on the filling scheme for recurrent Markov chains.* Duke MJ **45**, 681–689.

– (1983): *Loi du tout ou rien et comportement asymptotique pour les probaliblités de transition des processus de Markov.* AIHP, B, **19**, 9–24.

Riesz, Fréderic (1931): *Sur un théorème de maximum de MM.* Hardy et Littlewood. J London M Soc. **7**, 10–13.

– (1932): *Sur l'existence de la derivée des fonctions monotones et sur quelques problèmes qui s'y rattachent.* Acta Sci. Mat. (Szeged) **5**, 208–221.

– (1938): *Some mean ergodic theorems.* J London M Soc. **13**, 274–278.

– (1942): *Sur quelques problèmes de la théorie ergodique.* Mat. Fiz. Lapok **49**, 34–62.

– (1945): *Sur la théorie ergodique.* Comm. math. Helv. **17**, 221–239.

Robinson, D.W. and D. Ruelle, (1967): *Mean entropy of states in classical statistical mechanics.* Comm. Math. Phys. **5**, 288–300.

Rodé, Gerd (1982): *An ergodic theorem for semigroups of nonexpansive mappings in a Hilbert space.* J Math. Anal. Appl. **85**, 172–178.

– (1983): *An ergodic theorem for convex operators.* Arch. Math. **40**, 447–451.

Rolski, Tomasz (1981): *Stationary random processes associated with point processes.* Lecture Notes in Statistics **5**.

– (1981 a): *An approach to the formula $L = \lambda V$ via the theory of stationary point processes on a space of compact subsets of $R^k$.* Mathematical statistics, 1 st Pannonian Symp., Lect. Notes Stat. **8**, 220–235.

Rosenblatt, Joseph Max (1974): *Equivalent invariant measures.* Isr. JM **17**, 261–270.

– (1981): *Uniqueness of invariant means for measure preserving transformations.* TAMS **265**, 623–638.

Rosenblatt, Murray (1964): *Equicontinous Markov operators.* Theory Prob. Appl. **9**, 180–197;
Teor. Verojatnost.i. Primenen **9**, 205–222.

Rost, Hermann (1971): *Markoff-Ketten bei sich füllenden Löchern im Zustandsraum.* Ann. Inst. Fourier, Grenoble, **21**, 253–270.

– (1971 a): *Charakterisierung einer Ordnung von konischen Maßen durch positive $L^1$-Kontraktionen.* J Math. Anal. Appl. **33**, 35–42.

Rota, Gian-Carlo (1961): *Une théorie unifiée des martingales et des moyennes ergodiques.* CRAS **252**, 2064–2066.

– (1962): *An "Alternierende Verfahren" for general positive operators.* Bull. AMS **68**, 95–102.

– (1963). *On the maximal ergodic theorem for Abel-limits.* PAMS **14**, 722–723.

– (1964): *Reynolds operators.* Proc. Symp. Appl. Math. AMS **16**, 70–83.

Royden, H.L. (1968): Real Analysis. Macmillan Pub. Company.

Rozanow, Yu.A. (1967): *Stationary random processes.* Holden Day.

Rubinov, A.M (1977): *Sublinear operators and their applications.* Russian Math. Surveys **32**, (4), 115–175.

Rudin, Walter (1964): *An arithmetic property of Riemann sums.* PAMS **15**, 321–324.

Rudolph, Daniel (1976): *A two-valued step coding for ergodic flows.* MZ **150**, 201–220.

Ruelle, David (1969): *Statistical Mechanics.* Math. Phys. Monogr. Ser.; W.A. Benjamin Inc.
– (1979): *Ergodic theory of differentiable dynamical systems.* Inst. des Hautes Études Scient., Publ. Math. **50**, 275–306.
– (1982): *Characteristic exponents and invariant manifolds in Hilbert space.* Ann. Math. **155**, 243–290.
Ryll-Nardzewski, Czeslaw (1961): *Remarks on processes of calls.* Proc. 4th Berkeley Symp. Math. Stat. Prob., Vol II, 455–465.
– (1967): *On fixed points of semigroups of endomorphisms of linear spaces.* Proc. Fifth Berkeley Symp. Math. Stat. Prob. (1965–66) II (1), 55–61.
– (1975): *Topics in ergodic theory.* SpLNM **472**, 131–156.

Sachdeva, Usha (1971): *On finite invariant measures for semigroups of operators.* Can. Math. Bull., **14**, 197–206.
Saitô, K. (1967): *Non commutative extension of Lusin's theorem.* Tôh.MJ **19**, 332–340.
– (1974): *Automorphism groups of von Neumann algebras and ergodic type theorems.* Acta Sci. Math. (Szeged) **36**, 119–130.
Sakai, S. (1971): *C*-Algebras and W*-algebras.* Ergebn. M **60**, Springer.
Saks, Stanislaw (1927): *Sur les fonctionelles de M. Banach et leur application aux développement des fonctions.* Fund. Math. **10**, 186–196.
– (1934): *Remarks on the differentiability of the Lebesgue indefinite integral.* Fund. Math. **22**, 257–261.
Saleski, Alan (1980): *A note on the random mean ergodic theorem.* ZW **52**, 41–44.
Sarymsakov, T.A. (1964): *On general ergodic theory.* Dokl. Akad. Nauk SSSR **159**, (1), 25–27.
– (1966): *Proof of the general ergodic theorem.* In: Limit theorems and Statistical Inferences Izd. "Fan" Tashkent, 149–153.
Sato, Ryotaro (1973): *Ergodic properties of bounded $L_1$-operators.* PAMS **39**, 540–546.
– (1974): *Ergodic theorems for semigroups in $L_p$, $(1 < p < \infty)$.* Tôh. MJ **26**, 73–76.
– (1975): A mean ergodic theorem. Amer. math. Monthly **82**, 487–488.
– (1975a): *A note on a local ergodic theorem.* Comment. Math. Univ. Carolinae, **16**, 1–11.
– (1975b): *On mean ergodic theorems for positive operators in Lebesgue space.* JMS Jap. **27**, 207–212.
– (1976): *A mean ergodic theorem for a contraction semigroup in Lebesgue space.* Studia math. **54**, 213–219.
– (1977): *Invariant measures for ergodic semigroups of operators.* Pac. JM. **71**, 173–192.
– (1977a): *Ergodic theorems for semigroups of positive operators.* JMS Jap. **29**, 591–606.
– (1978): *On abstract mean ergodic theorems.* Tôh. MJ **30**, 575–587.
– (1978a): *Positive operators and the ergodic theorem.* Pac. JM **76**, 215–219.
– (1978b): *On local ergodic theorems for positive semigroups.* Studia math. **63**, 45–55.
– (1978c): *Contraction semigroups in Lebesgue space.* Pac. JM **78**, 251–259.
– (1979): *On abstract mean ergodic theorems.* II. MJ Okayama Univ. **21**, 141–147.
– (1979a): *The Hahn-Banach theorem implies Sine's mean ergodic theorem.* PAMS **77**, 426.
– (1979b): *Ergodic Theory and the approximate solution of simultaneous linear functional equations.* J. Math. Anal. Appl. **68**, 12–16.
– (1980): *Ratio limit theorems and applications to ergodic theory.* Studia Math., **66**, 237–245.
– (1980a): *Invariant measures and ergodic theorems for positive operators on $C(X)$ with $X$ quasi-Stonian.* MJ Okayama Univ. **22**, 77–90.
– (1980b): *Two local ergodic theorems on $L_\infty$.* JMS Jap. **32**, 415–423.
– (1981): *On a mean ergodic theorem.* PAMS **83**, 563–564.
– (1981a): *Ergodic theorems for d-parameter semigroups of Dunford-Schwartz operators.* MJ Okayama Univ. **23**, 41–57.
– (1981b): *An extrapolation theorem for contractions with fixed points.* Can. Math. Bull.**24**, 199–203.
– (1981c): *Individ. ergodic theorems for pseudo-resolvents.* MJ Okayama Univ. **23**, 59–67.
– (1982): *Maximal functions for a semiflow in an infinite measure space.* Pac. JM **100**, 437–443.
– (1983): *Individual ergodic theorems for commuting operators.* Tôh. MJ **35**, 129–135.
– (1983a): *A reverse maximal ergodic theorem.* Stud. Math. **75**, 153–160.
– (1984): *On local properties of k-parameter semiflows of nonsingular point transformations.* Acta Math. Hungar., to appear.
– (1984a): *On the ratio maximal function for an ergodic flow.* Studia Math., to appear.
Sawashima, I. and F. Niiro, (1973): *Reduction of a Sub-Markov operator to its irreducible components.* Nat. Sci. Rep. of Ochakomizu Univ. **24**, 35–59.
Sawyer, S. (1966): *Maximal inequalities of weak type.* Ann. Math., **84**, 157–174.
Schaefer, Helmut H. (1966): *Topological vector spaces.* Macmillan.
– (1974): *Banach lattices and positive operators.* Springer Grundlehren **215**.

Scheller, Hartmut (1965): *Induzierte dynamische Systeme*. Diplomarbeit, Göttingen, Inst. Math. Stoch.

Schmidt, Klaus (1982): *Two applications of a theorem of Glimm and Effros in ergodic theory*. Math. Forschung **12**, Ed. H. Michel, Akademie Verlag, 173–180.

Schmitt, Bernard (1972): *Théorème ergodique ponctuel pour les suites uniformes*. AIHP, B, **8**, 387–394.

Schürger, Klaus (1983): *Ergodic theorems for subadditive superstationary families of convex compact random sets*. ZW **62**, 125–135.

– (1984): *An almost subadditive superstationary ergodic theorem*. preprint.

Segal, I.E. (1952): *A non-commutative extension of abstract integration*. Ann. Math. **57**, 401–457.

Seneta, Eugene (1973): *Nonnegative Matrices (and Markov chains)*. George Allen and Unwin Ltd. 2nd. ed. Springer (1981).

Shannon, Claude E. (1948): *A mathematical theory of communication*. Bell System Techn. J **27**, 379–423, 623–656.

Shaw, Sen-Yen (1980): *Ergodic projections of continous and discrete semigroups*. PAMS **78**, 69–76.

– (1983): *Ergodic theorems for semigroups of operators on a Grothendieck space*. Proc. Japan Acad. Ser. A Math. Sci. **59**, 132–135.

– and Charles S.C. Lin (1978): *Ergodic theorems of semigroups and application*. Bull. Inst. Math., Acad. Sinica **6**, 181–188.

Shields, Paul C. (1967): *Invariant elements for positive contractions on a Banach lattice*. MZ **96**, 189–195.

Shreider, Yu.A. (1967): *Banach functionals and ergodic theorems*. Mat. Zametki **2**, 385–394; Math. Notes **2**, 723–728.

Siđák, Z. (1967): *Classification of Markov chains with a general state space*. Trans. 4th Prague Conf. Inf. Theory etc, 547–571.

– (1976): *Bibliography on Markov chains with a general state space. Aplikace Matematiky* **21**, 365–382.

Siegmund, David (1969): *On moments of the maximum of normed partial sums*. AM St **40**, 527–531.

Simons, Frederik H. (1965): *Ein weiterer Beweis des Zerlegungssatzes für meßbare Transformationen*. MZ **89**, 247–249.

– (1971): *Recurrence properties and periodicity for Markov processes*. Proefschrift, Tech. Hogeschool Eindhoven.

– and D.A. Overdijk (1979): *Recurrent and sweep out sets for Markov processes*. Monatsh. M **86**, 305–326.

Sinai, Yakov G. (1963): *On the foundations of the ergodic hypothesis for a dynamical system of statistical mechanics*. Soviet Math. Dokl. **4**, 1818–1822.

– (1970): *Dynamical systems with elastic reflections*. Russ. Math. Surveys **25**, 137–189.

– and V.V. Anshelevich (1976): *Some problems of noncommutative ergodic theory*. Russ. Math. Surveys **31**, 157–174.

Sine, Robert C. (1968): *Geometric theory of a single Markov operator*. Pac. JM **27**, 155–166.

– (1970): *A mean ergodic theorem*. PAMS **24**, 438–439.

– (1973): *An example in the theory of stable Markov operators*. Isr. JM **14**, 283–284.

– (1974): *Convergence theorems for weakly almost periodic Markov operators*. Isr. JM **19**, 246–255.

– (1975): *Convex combinations of uniformly mean stable Markov operators*. PAMS **51**, 123–126.

– (1975a): *On local uniform mean convergence for Markov operators*. Pac. JM **60**, 247–252.

– (1976): *A note on the ergodic properties of homeomorphisms*. PAMS **57**, 169–172.

– (1976a): *Sample path convergence of stable Markov processes II*. Indiana Univ. MJ **25**, 23–43.

Smeltzer, M.D. (1977): *Subadditive stochastic processes*. Bull. AMS **83**, 1054–1055.

Smythe, Robert T. (1973): *Strong laws of large numbers for r-dimensional arrays of random variables*. Ann. Prob. **1**, 164–170.

– (1976): *Multiparameter subadditive processes*. Ann. Prob. **4**, 772–782.

– and J.C. Wierman (1978): *First-passage percolation on the square lattice*. SpLNM **671**.

Spitzer, Frank (1964): *Principles of random walk*. van Nostrand.

Stadje, Wolfgang (1984): *Bemerkung zu einem Satz von Akcoglu und Krengel*. To appear Studia Math.

Starr, Norton (1965): *On an operator limit theorem of Rota*. AM St **36**, 1864–1866.

– (1966): *Operator limit theorems*. TAMS **121**, 90–115.

– (1971): *Majorizing operators between $L^p$-spaces and an operator extension of Lebesgue's dominated convergence theorem*. Math. Scand. **28**, 91–104.

Steele, J. Michael (1978): *Empirical discrepancies and subadditive processes*. Ann. Prob. **6**, 118–127.

– (1984): *Kingman's subadditive ergodic theorem*. ZW, to appear.

Stein, Elias M. (1961): *On limits of sequences of operators*. Ann. Math. **74**, 140–170.

– (1961a): *On the maximal ergodic theorem.* Proc. Nat. Acad. Sci. USA **47**, 1894–1897.
– (1969): *Note on the class L log L.* Studia Math. **32**, 305–310.
– (1970): *Topics in Harmonic Analysis related to the Littlewood Paley Theory.* Princeton Univ. Press.
Stout, William F. (1974): *Almost sure convergence.* Academic Press.
Stute, W. and G. Schumann, (1980): *A general Glivenko-Cantelli theorem for stationary sequences of random observations.* Scand. J Stat. **7**, 102–104.
Sucheston, Louis (1957): *A note on conservative transformations and the recurrence theorem.* Amer. JM **79**, 444–447.
– (1959): *On sequences of events of which the probalities admit a positive lower limit.* J London M Soc. **34**. 386–394.
– (1963): *On mixing and the zero-one law.* J Math. Anal. Appl. **6**, 447–456.
– (1964): *On the existence of finite invariant measures.* MZ **86**, 327–336.
– (1967): *Banach limits.* Amer. Math. Monthly, **74**, 308–311.
– (1967a): *On the ergodic theorem for positive operators I, II.* ZW **8**, 1–11 and 353–356.
– (1983): *On one-parameter proofs of almost sure convergence of multiparameter processes.* ZW **63**, 43–49.
Šujan, Štefan (1983): *Ergodic theory, entropy, and coding problems of information theory.* Kybernetica **19**, 1–67.
Sullivan, Dennis (1981): *For n > 3 there is only one finitely additive rotationally invariant measure on the n-sphere defined on all Lebesgue measurable sets.* Bull. AMS **4**, (N.S.), 121–123.
Sur, M.G. (Shur, M.G.) (1976/77): *An ergodic theorem for Markov processes I, II.* Th. Prob. Appl. **21**, 400–406 and **22**, 692–707, transl. from Teor. Verojatn. Primen **21**, 410–416, and **22**, 712–728 (Russian).
Suzuki, Hisakichi (1983): *On the two decompositions of a measure space by an operator semigroup.* MJ Okayama Univ. **25**, 87–90.
Sz.-Nagy, Béla and Ciprian Foias, (1960): *Sur les contractions de l'espace de Hilbert, IV.* Acta Sci. Mat. Szeged **21**, 251–259.
–, – (1970): *Harmonic analysis of operators on Hilbert space.* North-Holland.
Szücs, Joseph (1983): *Some weak-star ergodic theorems.* Acta Sci. Math. (Szeged) **45**, 389–394.

Takahashi, Wataru (1971): *Invariant functions for amenable semigroups of positive contractions on $L^1$.* Kodai M Sem. Rep., **23**, 131–143.
– (1971a): *Invariant ideals for amenable semigroups of Markov operators.* Kodai M Sem. Rep., **23**, 121–126.
– (1972): *Ergodic theorems for amenable semigroups of positive contractions on $L_1$.* Sci. Rep. Yokohama nat. Univ. I, **19**, 5–11.
– (1980): *Recent results in fixed point theory.* Southeast Asian Bull. M **4**, 59–85.
– (1981): *A nonlinear ergodic theorem for an amenable semigroup of nonexpansive mappings in a Hilbert space.* PAMS **81**, 253–256.
Takesaki, Masamichi (1979): *Theory of operator algebras I.* Springer.
Tanny, David (1974): *A zero-one law for stationary sequences.* ZW **30**, 139–148.
Taylor, G.C. (1978): *A topological version of some ergodic theorems.* Studia Math. **63**, 199–205.
Teicher, Henry (1967): *A dominated ergodic type theorem.* ZW **8**, 113–116.
– (1971): *Completion of a dominated ergodic theorem.* AM St **42**, 2156–2158.
Tempel'man, Arkadii A. (1962): *An ergodic theorem for random fields homogeneous in the wide sense.* Dokl. Akad. Nauk SSSR **144**, 730–733; Soviet Math. Dokl. **3**, (1962), 817–820.
– (1962a): *Ergodic theorems for homogeneous generalized stochastic fields and homogeneous stochastic fields on groups.* Litov. mat. Sbornik **1**, 195–213, (Russian).
– (1967): *Ergodic theorems for general dynamical systems.* Dokl. Akad. Nauk SSSR **176**, 790–793; Soviet Math. Dokl. **8** (1967), 1213–1216.
– (1972): *Ergodic theorems for general dynamical systems.* Trudy Moskov. Mat. Obsc. **26** = Trans. Moscow Math. Soc. **26**, 94–132.
– (1974): *Ergodic theorem for amplitude-modulated homogeneous random fields.* Lit. Mat. Sb. **14**, 221–229, Transl.: Lith. math. Trans. **14**, 698–704, (1975).
– (1984): *Specific characteristics and variational principle for homogeneous random fields.* ZW **65**, 341–365.
Terrell, Thomas R. (1970): *Local ergodic theorems for n-parameter semigroups of operators.* SpLNM **160**, 262–278.
– (1972): *The local ergodic theorem and semigroups of nonpositive operators.* J Funct. Anal. **10**, 424–429.
– (1972a): *A ratio ergodic theorem for operator semigroups.* Bollettino UMI **6**, 175–180.
Thorin, G.O. (1948): *Convexity theorems.* Comm. Sém. Math. Univ. Lund **9**.
Thouvenot, Jean-Paul (1972): *Convergence en*

*moyenne de l'information pour l'action de $Z^2$.* ZW **24**, 135–137.

Tischer, Jürgen and Hans Wittmer (1974): *Invarianz gewisser Operatorenklassen unter nichtkommutativen ergodischen Mitteln.* Manuscripta Math. **13**, 73–81.

Torre, see De la Torre

Totoki, Haruo (1964): *The mixing property of Gaussian flows.* Memoirs Fac. Sci. Kyushu Univ. A, **18**, 136–139.

– (1970): *Ergodic theory.* Matematisk Inst. Aarhus Univ. Lecture Notes Ser. **14**.

Tsurumi, Shigeru (1954): *On general ergodic theorems.* Tôh. MJ **6**, 264–273.

– (1958): *On the recurrence theorems in ergodic theory.* Proc. Japan Acad. **34**, 208–211.

– (1958a): *On the ergodic theorems concerning Markov processes.* Tôh. MJ **10**, 146–164.

– (1972): *On random ergodic theorems for a random quasi-semigroup of linear contractions.* Proc. Japan Acad. **48**, 149–152.

Tucker, Howard G. (1959): *A generalization of the Glivenko-Cantelli theorem.* AM St **30**, 828–830.

Ulam, Stanislaw M. and John v. Neumann (1945): *Random ergodic theorems.* Bull. AMS **51**, 660.

Urbanik, Kazimierz (1967): *Lectures on prediction theory.* SpLNM **44**.

Veech, William A. (1982): *Gauss measures for transformations on the space of interval exchange maps.* Ann. Math. II. Ser. **115**, 201–242.

Versik, A. M. (1962): *Spectral and metric isomorphism of some normal dynamical systems.* Sov. Math. **3**, 693–696.

Wacker, Ulrich (1982): *Maximalergodensätze und ihre Anwendung auf lokale Ergodensätze und Erneuerungssätze für Operatoren.* Diplomarbeit, Inst. Math. Stoch., Göttingen.

– (1983): *Grenzwertsätze für nichtadditive schwach abhängige Prozesse.* Dissert., Univ. Göttingen.

– (1984): *On nonadditive processes.* To appear, Erg. Theory and Dyn. Systems.

Walters, Peter (1982): *An introduction to ergodic theory.* Springer.

Watanabe, Seiji (1979): *Ergodic theorems for dynamical semi-groups on operator algebras.* Hokkaido MJ **8**, 176–190.

Weiner, H. (1968): *Invariant measures and Cesàro summability.* Pac. JM **25**, 621–629.

Weiss, Benjamin (1981): *Orbit equivalence of nonsingular actions.* Monogr. l'enseignement math. **29**, 77–107.

Von Weizsäcker, Heinrich (1974): *Sublineare Abbildungen und ein Konvergenzsatz von Banach.* Math. Ann. **212**, 165–171.

West, T. T. (1968): *Weakly compact monothetic semigroups of operators in Banach spaces.* Proc. Roy. Irish Acad. **67**, A, 27–37.

Westman, J. J. (1980): *Sums of dependent random variables.* J Math. Anal. Appl. **77**, 120–131.

Weyl, Hermann (1916): *Über die Gleichverteilung von Zahlen mod 1.* Math. Ann. **77**, 313–352.

Widder, D. (1946): *The Laplace transform.* Princeton Univ. Press.

Wiener, Norbert (1939): *The ergodic theorem.* Duke MJ **5**, 1–18.

– and Aurel Wintner (1941): *Harmonic analysis and ergodic theory.* Amer. JM **63**, 415–426.

–, – (1941a): *On the ergodic dynamics of almost periodic systems.* Amer. JM **63**, 794–824.

Winkler, William (1973): *A note on the continuous parameter zero-two law.* Ann. Prob. **1**, 341–344.

– (1975): *Doeblin's and Harris' theory of Markov – processes.* ZW **31**, 79–88.

Wolff, M. (1971): *Vektorwertige invariante Masse von rechtsamenablen Halbgruppen positiver Operatoren.* MZ **120**, 265–276.

Wolfowitz, J. (1967): *The moments of recurrence time.* PAMS **18**, 613–614.

Woś, Janusz (1982): *Random ergodic theorems for sub-Markovian operators.* Studia Math. **74**, 191–212.

Woyczynski, Wojbor A. (1978): *Geometry and martingales in Banach spaces II: Independent increments.* Adv. in Prob. **4**, Ed. J. Kuelbs.

Wright, Fred B. (1960): *The converse of the individual ergodic theorem.* PAMS **11**, 415–420.

– (1961): *The recurrence theorem.* Amer. Math. Monthly **68**, 247–248.

Yeadon, F. J. (1977): *Ergodic theorems for semifinite von Neumann algebras I.* J London M Soc., II. Ser. **16**, 326–332.

– (1980): *Ergodic theorems for semifinite von Neumann algebras II.* Math. Proc. Camb. Philos. Soc. **88**, 135–147.

Yoshimoto, Takeshi (1976): *Induced contraction semigroups and random ergodic theorems.* Dissertationes math., Warszawa **139**, 41 p.

– (1977): *On the random ergodic theorem.* Studia math. **61**, 231–237.

– (1979): *Vector valued ergodic theorems for operators satisfying norm conditions.* Pac. JM **85**, 485–499.

– (1980): *Some inequalities for ergodic power functions.* Acta Math. Hung. **36**, 19–24.

– (1982): *Pointwise ergodic theorems and function classes $M_p^\alpha$*. Studia Math. **72**, 253–271.

Yosida, Kosaku (1938): *Mean ergodic theorem in Banach spaces*. Proc. Imp. Acad. Tokyo **14**, 292–294.

– (1939): *Quasi-completely continuous linear functional operators*. Jap. JM **25**, 297–301.

– (1940): *Ergodic theorems of Birkhoff-Khintchine's type*. Jap. JM **17**, 31–36.

– (1940a): *An abstract treatment of the individual ergodic theorems*. Proc. Imp. Acad. Tokyo **16**, 280–284.

– (1974): *Functional Analysis*. Springer Grundlehren **123**, 4th edition.

– and S. Kakutani (1939): *Birkhoff's ergodic theorem and the maximal ergodic theorem*. Proc. Imp. Akad. Tokyo **15**, 165–168.

–, – (1941): *Operator-theoretical treatment of Markoff's process and mean ergodic theorem*. Ann. Math.. **42**, 188–228.

Yu, Hise (1972): *Mean ergodic theorem to hold for a harmonizable process*. J Korean M Soc. **9**, 45–48.

Zakharevich, M.I. (1978): *Characteristic exponents and a vector-valued ergodic theorem*. Vestn. Leningr. Univ. Mat. Mekh. Astron. No. **2**, 28–34, (Russian).

Zeller K. and W. Beekmann (1970): *Theorie der Limitierungsverfahren*. Ergebn. M **15**, II. Aufl., Springer.

Zimmer, Robert J. (1976): *Ergodic actions with generalized discrete spectrum*. Ill. JM **20**, 555–588.

Zygmund, Antoni (1935): *Trigonometric series I, II*. Cambridge Univ. Press, reprinted 1977.

– (1951): *An individual ergodic theorem for non-commutative transformations*. Acta Sci. Math. (Szeged) **14**, 103–110.

# Notation

Notation used only in a single section need not appear in this list

## General Notation

| | |
|---|---|
| $\exists$ | there exists |
| $\forall$ | for all |
| $:=$ | equal by definition |
| $\Rightarrow$ | logical implication |
| $\Leftrightarrow$ | logical equivalence |
| *iff* | if and only if |
| $\square$ | end of proof |
| $\mathbb{R}$ | reals |
| $\mathbb{Z}$ | integers |
| $\mathbb{N}$ | $\{1, 2, \ldots\}$ |
| $\mathbb{Q}$ | rationals |
| $\mathbb{C}$ | complex numbers |
| $\mathbb{V}$ | $\{0, 1, 2, \ldots\}^d$, 195 |
| $a \vee b$ | Max $(a, b)$ |
| $a \wedge b$ | Min $(a, b)$ |
| $a^+$ | $a \vee 0$ |
| $a^-$ | $(-a) \vee 0$ |
| $Re(\ )$ | real part |
| $A^c$ | complement of $A$ |
| cl | closure, 4, 71 |
| co | convex hull, 71 |
| $\overline{\text{co}}$ | cl co, 71 |
| lin | linear span, 71 |
| card $(A)$ | number of elements of $A$ |
| $\perp$ | orthogonality, 4 |
| $F^{\perp}$ | orthogonal complement of $F$ |
| $[u, v[$ | interval, 195 |
| $\langle f, h\rangle$ | scalar product, 3, 71 |
| $A \Delta B$ | symmetric difference, 3 |
| $I$ | identity, 3 |
| $\mathbb{1}$ | constant function 1, 3 |
| $1_A$ | indicator function of $A$, 3 |
| $\mathfrak{X}^+$ | positive cone of $\mathfrak{X}$, 3 |
| $A \oplus B$ | direct sum, 71 |
| ker | kernel, 80, 83 |
| dom | domain, 83 |
| $C(\Omega)$ | continuous functions, 2 |
| $C_b(\Omega)$ | bounded continuous functions, 111, 223 |
| w-lim | weak limit, 71 |
| w*-lim | weak*-limit, 71 |
| wo-lim, w*o-lim, so-lim | limits in the weak, weak* and strong operator topologies, 80 |
| l.c.t.v.s. | 75 |
| $\lVert T\rVert$ | norm of operator $T$, 3 |
| $\mathfrak{X}^*$ | adjoint (dual) space, 71 |
| $T^*$ | adjoint operator, 3, 71 |
| $E^J$ | $E$-valued functions on $J$, 22 |
| $\mathscr{L}(\mathfrak{X})$ | bounded linear operators in $\mathfrak{X}$, 3 |
| $e^i (= e(i))$ | $i$-th unit vector in $\mathbb{R}^d$, 206, 289 |
| $\sigma(T)$ | spectrum, 90 |
| $r(T)$ | spectral radius, 90 |
| $R_\lambda$, $R(\lambda, T)$ | resolvents, 83, 90 |

## Measure theory

| | |
|---|---|
| $\mathscr{L}_p = \mathscr{L}_p(\Omega, \mathscr{A}, \mu)$ | 2 |
| $\lVert f\rVert_p$ | 2 |
| $L_p = L_p(\Omega, \mathscr{A}, \mu)$ | 2 |
| $L_p(B)$ | $\{f \in L_p: \{f \neq 0\} \subset B\}$, 117, 137 |
| $\ell_p$ | $L_p$ with counting measure on $\mathbb{Z}^+$ |
| $L_p^{\mathfrak{X}}$ | 167 |
| $\nu \ll \mu$ | 3 |
| $\nu \sim \mu$ | 3 |
| $\nu = g\mu = g \cdot \mu$ | $d\nu/d\mu = g$, 127 |
| $\tilde{L}_1$ | $\{g\mu: g \in L_1\}$, 114 |
| $L \log^m L$ | 51,54 |
| $\varepsilon_\omega$ | measure with mass 1 in $\omega$, 177 |
| $E(f \mid \mathscr{F}) = E_\mu(f \mid \mathscr{F}) = E_{\mathscr{F}} f$ | conditional expectation, 5 |

| | |
|---|---|
| $b\mathscr{A}'$ | bounded $\mathscr{A}$-measurable functions, 113 |
| $\mathscr{M} = \mathscr{M}(\mathscr{A})$ | signed measures on $\mathscr{A}$, 113 |
| $\mathscr{M}_1^+$ | probability measures on $\mathscr{A}$ |
| $\mu \times \nu, \prod \mu_j$ | product measure, 10, 23 |
| $\mathscr{A} \otimes \mathscr{B}, \mathscr{A}^J$ | product-$\sigma$-algebra, 10, 23 |

## Basic ergodic theoretic notation

| | |
|---|---|
| $\mu \circ \tau^{-1}(A)$ | $\mu(\tau^{-1}A)$, 1 |
| $S_n f = S_n(T)f$ | $\sum_{k=0}^{n-1} T^k f$, 4 |
| $A_n f = A_n(T)f$ | $n^{-1} S_n f$, 4 |
| $A_n, S_n$ | in multiparameter case, 195 |
| $M_n^S f$ | Max $(S_1 f, \ldots, S_n f)$, 7 |
| $M_n f$ | $\mathrm{Max}_{k \leqq n} A_k f$, 7 |
| $S_{a,b} f$ | $\int_a^b T_t f\, dt$, 58, 216 |
| $A_{s,t}$ | $(t-s)^{-1} S_{s,t}$, 83, multiparameter case, 216 |
| $\pi(u)$ | $u_1 \cdot u_2 \cdot \ldots u_d$ for $u = (u_1, u_2, \ldots, u_d)$, 195 |
| $M_{0,\infty} f$ | $\sup\{A_{0,a} f: 0 < a < \infty\}$, 59 |
| $R_a(n) f$ | 58 |
| $M^{\mathscr{T}}, M_\lambda^{\mathscr{T}}$ | 63 |
| $\mathscr{I}$ | invariant $\sigma$-algebra, 5, 203 |
| $\tau_A$ | induced transformation, 20 |
| $r_A, r_A(\omega)$ | return time, 19 |
| $A_{\mathrm{ret}}, A_{\mathrm{inf}}$ | 16 |
| $C$ | conservative part, 16, 17, 116, 117 |
| $\mathscr{C}$ | absorbing sets in $C$, 126 |
| $D$ | dissipative part, 17, 116 |
| $\tilde{C}$ | maximal support of finite invariant measure, 141 |
| $\tilde{D}$ | $\Omega \setminus \tilde{C}$, 141 |

## Mean ergodic theory

| | |
|---|---|
| $F = F(T), F(\mathscr{S})$ | fixed space, 73, 77 |
| $F_*, F_*(T), F_*(\mathscr{S})$ | fixed space of adjoint operator or semigroup, 73, 77 |
| $N, N_*$ | 73, 77 |
| $\mathfrak{X}_{me}$ | mean ergodic subspace, 73 |
| $\mathfrak{X}_{fl}$ | flight vectors, 94, 105 |
| $\mathfrak{X}_{rev}$ | reversible vectors, 105 |
| $\mathfrak{X}_{uds}$ | subspace with unimodular discrete spectrum, 94, 106 |
| $\mathfrak{X}_{ap}$ | almost periodic vectors, 110 |
| $D^*(M)$ | upper density, 95 |
| $D_*(M)$ | lower density, 95 |
| $D(M)$ | density, 95 |
| $D$-lim | convergence in density, 95 |
| $\lVert \cdot \rVert$-$\overline{\mathrm{co}}$ | 87 |

## Processes

| | |
|---|---|
| $X_i(\omega)$ | $i$-th coordinate of $\omega$, 22 |
| $\Omega^+$ | $E^{\mathbb{Z}^+}$, 23 |
| $\mathscr{A}(m, n)$ | 24 |
| $\mathscr{A}_\infty, \mathscr{A}_{-\infty}$ | remote $\sigma$-algebras, 27 |
| $P_A$ | conditional probability, 28 |
| $P(\omega, A)$ | substochastic kernel, 113 |
| $F_{i,k}, F_I$ | sub-or superadditive processes, 35, 201 |
| $\bar{F}_I$ | $\lambda(I)^{-1} F_I$, 202 |

## Contractions in $L_1$

| | |
|---|---|
| $I_h, I_E$ | multiplication by $h$, by $1_E$, 124, 301 |
| $\lVert T \rVert_{1,\infty}$ | 51 |
| $T_E, T_\alpha$ | 124 |
| $\Psi_H^n, \Psi_H$ | 122 |
| $\psi_E^n, \psi_E$ | 125 |
| $H_E$ | 126 |
| $\mathbf{T}, \lvert T \rvert$ | linear modulus, 159 |
| $f \xrightarrow{n} g$ | 120 |
| $H$ | non zero measurable functions $h: \Omega \to [0,1]$. |
| $U_h$ | taboo potential operator, 301 |
| $U_A$ | $U_{1_A}$ |
| $R_h$ | 303 |
| $C_0, D_0, C^*, D^*$ | 230 |
| $\mathfrak{S}$ | special functions, 305 |
| $L_1^0$ | $\{f \in L_1: \int f\, d\mu = 0\}$. |

## Other special notation

# Index

(In general, names are listed only for detailed quotations.)